STATA DATA-MANAGEMENT
REFERENCE MANUAL
RELEASE 12

A Stata Press Publication
StataCorp LP
College Station, Texas

The suggested citation for this software is

StataCorp. 2011. *Stata: Release 12*. Statistical Software. College Station, TX: StataCorp LP.

Table of contents

Cross-referencing the documentation

When reading this manual, you will find references to other Stata manuals. For example,

[U] **26 Overview of Stata estimation commands**
[R] **regress**
[XT] **xtreg**

The first example is a reference to chapter 26, *Overview of Stata estimation commands*, in the *User's Guide*; the second is a reference to the `regress` entry in the *Base Reference Manual*; and the third is a reference to the `xtreg` entry in the *Longitudinal-Data/Panel-Data Reference Manual*.

All the manuals in the Stata Documentation have a shorthand notation:

[GSM]	*Getting Started with Stata for Mac*
[GSU]	*Getting Started with Stata for Unix*
[GSW]	*Getting Started with Stata for Windows*
[U]	*Stata User's Guide*
[R]	*Stata Base Reference Manual*
[D]	*Stata Data-Management Reference Manual*
[G]	*Stata Graphics Reference Manual*
[XT]	*Stata Longitudinal-Data/Panel-Data Reference Manual*
[MI]	*Stata Multiple-Imputation Reference Manual*
[MV]	*Stata Multivariate Statistics Reference Manual*
[P]	*Stata Programming Reference Manual*
[SEM]	*Stata Structural Equation Modeling Reference Manual*
[SVY]	*Stata Survey Data Reference Manual*
[ST]	*Stata Survival Analysis and Epidemiological Tables Reference Manual*
[TS]	*Stata Time-Series Reference Manual*
[I]	*Stata Quick Reference and Index*
[M]	*Mata Reference Manual*

Detailed information about each of these manuals may be found online at

http://www.stata-press.com/manuals/

Title

> **intro** — Introduction to data-management reference manual

Description

This entry describes this manual and what has changed since Stata 11. See the next entry, [D] **data management**, for an introduction to Stata's data-management capabilities.

Remarks

This manual documents most of Stata's data-management features and is referred to as the [D] manual. Some specialized data-management features are documented in such subject-specific reference manuals as [MI] *Stata Multiple-Imputation Reference Manual*, [TS] *Stata Time-Series Reference Manual*, [ST] *Stata Survival Analysis and Epidemiological Tables Reference Manual*, and [XT] *Stata Longitudinal-Data/Panel-Data Reference Manual*.

Following this entry, [D] **data management** provides an overview of data management in Stata and of Stata's data-management commands. The other parts of this manual are arranged alphabetically. If you are new to Stata's data-management features, we recommend that you read the following first:

> [D] **data management** — Introduction to data-management commands
>
> [U] **12 Data**
>
> [U] **13 Functions and expressions**
>
> [U] **11.5 by varlist: construct**
>
> [U] **21 Inputting and importing data**
>
> [U] **22 Combining datasets**
>
> [U] **23 Working with strings**
>
> [U] **25 Working with categorical data and factor variables**
>
> [U] **24 Working with dates and times**
>
> [U] **16 Do-files**

You can see that most of the suggested reading is in [U]. That is because [U] provides overviews of most Stata features, whereas this is a reference manual and provides details on the usage of specific commands. You will get an overview of features for combining data from [U] **22 Combining datasets**, but the details of performing a match-merge (merging the records of two files by matching the records on a common variable) will be found here, in [D] **merge**.

Stata is continually being updated, and Stata users are always writing new commands. To ensure that you have the latest features, you should install the most recent official update; see [R] **update**.

What's new

This section is intended for previous Stata users. If you are new to Stata, you may as well skip it.

1. **Automatic memory management**, which means that you no longer have to set memory and never again will you be told that there is no room because you set too little! Stata automatically adjusts its memory usage up and down according to current requirements.

1

The memory manager is tunable. We recommend the default settings. See [D] **memory** if you are interested.

Old do-files can still `set memory`. Stata merely responds, "`set memory` ignored".

2. **Excel files, importing and exporting.** And the new import preview tool lets you see the data before you import them. See [D] **import excel**.

3. **EBCDIC files, importing.** And you can convert between EBCDIC and ASCII formats; see [D] **infile (fixed format)** and [D] **filefilter**.

4. **ODBC connection strings, importing and exporting** and **ODBC support for Oracle Solaris.** See [D] **odbc**.

5. **PDF files, exporting of graphs and logs.** You can directly create PDFs from your Stata results. See [G-2] **graph export** and [R] **translate**.

6. **Business dates** allow you to define your own calendars so that they display correctly and lags and leads work as they should. You could create file `lse.stbcal` that recorded the days the London Stock Exchange is open (or closed) and then Stata would understand format `%tblse` just as it understands the usual date format `%td`. Once you define a calendar, Stata deeply understands it. You can, for instance, easily convert between `%tblse` and `%td` values. See [D] **datetime business calendars**.

7. **Improved documentation for date and time variables.** Anyone who has ever been puzzled by Stata's date and time variables, which is to say, anyone who uses them, should see [D] **datetime**, [D] **datetime translation**, and [D] **datetime display formats**.

8. **Renaming groups of variables** is now easy using `rename`'s new syntax that is 100% compatible with its old syntax. You can change names, swap names, renumber indices within variable names, and more. See [D] **rename group**.

9. **New functions**,

   a. **Tukey's Studentized range**, cumulative and inverse, `tukeyprob()` and `invtukeyprob()`.

   b. **Dunnett's multiple range**, cumulative and inverse, `dunnettprob()` and `invdunnettprob()`.

   c. **New date conversion functions** `dofb()` and `bofd()` convert between business dates and standard calendar dates. See [D] **datetime business calendars**.

   See [D] **functions**.

10. **New Stata commands getmata and putmata** make it easy to transfer your data into Mata, manipulate them, and then transfer them back to Stata. `getmata` and `putmata` are especially designed for interactive use. See [D] **putmata**.

11. **New Stata commands import sasxport, export sasxport, and import sasxport, describe** replace existing commands `fdause`, `fdasave`, and `fdadescribe`. `fdause`, `fdasave`, and `fdadescribe` are understood as synonyms. See [D] **import sasxport**.

12. **xshell** support for Mac. See [D] **shell**.

For a complete list of all the new features in Stata 12, see [U] **1.3 What's new**.

Also see

[U] **1.3 What's new**

[R] **intro** — Introduction to base reference manual

Title

data management — Introduction to data-management commands

Description

This manual, called [D], documents Stata's data-management features. See Mitchell (2010) for additional information and examples on data management in Stata.

Data management for statistical applications refers not only to classical data management—sorting, merging, appending, and the like—but also to data reorganization because the statistical routines you will use assume that the data are organized in a certain way. For example, statistical commands that analyze longitudinal data, such as `xtreg`, generally require that the data be in long rather than wide form, meaning that repeated values are recorded not as extra variables, but as extra observations.

Here are the basics everyone should know:

[D] **use**	Load Stata dataset
[D] **save**	Save Stata dataset
[D] **describe**	Describe data in memory or in file
[D] **codebook**	Describe data contents
[D] **inspect**	Display simple summary of data's attributes
[D] **count**	Count observations satisfying specified conditions
[D] **data types**	Quick reference for data types
[D] **missing values**	Quick reference for missing values
[D] **datetime**	Date and time values and variables
[D] **list**	List values of variables
[D] **edit**	Browse or edit data with Data Editor
[D] **varmanage**	Manage variable labels, formats, and other properties
[D] **rename**	Rename variable
[D] **format**	Set variables' output format
[D] **label**	Manipulate labels

You will need to create and drop variables, and here is how:

[D] **generate**	Create or change contents of variable
[D] **functions**	Functions
[D] **egen**	Extensions to generate
[D] **drop**	Eliminate variables or observations
[D] **clear**	Clear memory

3

For inputting or importing data, see

[D] **use**	Load Stata dataset
[D] **sysuse**	Use shipped dataset
[D] **webuse**	Use dataset from Stata website
[D] **input**	Enter data from keyboard
[D] **insheet**	Read text data created by a spreadsheet
[D] **import**	Overview of importing data into Stata
[D] **import excel**	Import and export Excel files
[D] **import sasxport**	Import and export datasets in SAS XPORT format
[D] **infile (fixed format)**	Read text data in fixed format with a dictionary
[D] **infile (free format)**	Read unformatted text data
[D] **infix (fixed format)**	Read text data in fixed format
[D] **odbc**	Load, write, or view data from ODBC sources
[D] **xmlsave**	Export or import dataset in XML format
[D] **hexdump**	Display hexadecimal report on file
[D] **icd9**	ICD-9-CM diagnostic and procedure codes

and for exporting data, see

[D] **save**	Save Stata dataset
[D] **export**	Overview of exporting data from Stata
[D] **outfile**	Export dataset in text format
[D] **outsheet**	Write spreadsheet-style dataset
[D] **import excel**	Import and export Excel files
[D] **import sasxport**	Import and export datasets in SAS XPORT format
[D] **odbc**	Load, write, or view data from ODBC sources

The ordering of variables and observations (sort order) can be important; see

[D] **order**	Reorder variables in dataset
[D] **sort**	Sort data
[D] **gsort**	Ascending and descending sort

To reorganize or combine data, see

[D] **merge**	Merge datasets
[D] **append**	Append datasets
[D] **reshape**	Convert data from wide to long form and vice versa
[D] **collapse**	Make dataset of summary statistics
[D] **contract**	Make dataset of frequencies and percentages
[D] **fillin**	Rectangularize dataset
[D] **expand**	Duplicate observations
[D] **expandcl**	Duplicate clustered observations
[D] **stack**	Stack data
[D] **joinby**	Form all pairwise combinations within groups
[D] **xpose**	Interchange observations and variables
[D] **cross**	Form every pairwise combination of two datasets

In the above list, we particularly want to direct your attention to [D] **reshape**, a useful command that beginners often overlook.

For random sampling, see

[D] **sample**	Draw random sample
[D] **drawnorm**	Draw sample from multivariate normal distribution

For file manipulation, see

[D] **type**	Display contents of a file
[D] **erase**	Erase a disk file
[D] **copy**	Copy file from disk or URL
[D] **cd**	Change directory
[D] **dir**	Display filenames
[D] **mkdir**	Create directory
[D] **rmdir**	Remove directory
[D] **cf**	Compare two datasets
[D] **changeeol**	Convert end-of-line characters of text file
[D] **filefilter**	Convert text or binary patterns in a file
[D] **checksum**	Calculate checksum of file
[D] **zipfile**	Compress and uncompress files and directories in zip archive format

The entries above are important. The rest are useful when you need them:

[D] **datasignature**	Determine whether data have changed
[D] **type**	Display contents of a file
[D] **notes**	Place notes in data
[D] **label language**	Labels for variables and values in multiple languages
[D] **labelbook**	Label utilities
[D] **encode**	Encode string into numeric and vice versa
[D] **recode**	Recode categorical variables
[D] **ipolate**	Linearly interpolate (extrapolate) values
[D] **destring**	Convert string variables to numeric variables and vice versa
[D] **mvencode**	Change missing values to numeric values and vice versa
[D] **pctile**	Create variable containing percentiles
[D] **range**	Generate numerical range
[D] **by**	Repeat Stata command on subsets of the data
[D] **statsby**	Collect statistics for a command across a by list
[D] **compress**	Compress data in memory
[D] **recast**	Change storage type of variable
[D] **datetime display formats**	Display formats for dates and times
[D] **datetime translation**	String to numeric date translation functions
[D] **bcal**	Business calendar file manipulation
[D] **datetime business calendars**	Business calendars
[D] **datetime business calendars creation**	Business calendars creation

[D] **assert**	Verify truth of claim
[D] **clonevar**	Clone existing variable
[D] **compare**	Compare two variables
[D] **corr2data**	Create dataset with specified correlation structure
[D] **ds**	List variables matching name patterns or other characteristics
[D] **duplicates**	Report, tag, or drop duplicate observations
[D] **isid**	Check for unique identifiers
[D] **lookfor**	Search for string in variable names and labels
[D] **memory**	Memory management
[D] **putmata**	Put Stata variables into Mata and vice versa
[D] **obs**	Increase the number of observations in a dataset
[D] **rename group**	Rename groups of variables
[D] **separate**	Create separate variables
[D] **shell**	Temporarily invoke operating system
[D] **snapshot**	Save and restore data snapshots
[D] **split**	Split string variables into parts

There are some real jewels in the above, such as [D] **notes**, [D] **compress**, and [D] **assert**, which you will find particularly useful.

Reference

Mitchell, M. N. 2010. *Data Management Using Stata: A Practical Handbook.* College Station, TX: Stata Press.

Also see

[D] **intro** — Introduction to data-management reference manual

[R] **intro** — Introduction to base reference manual

Title

> **append** — Append datasets

Syntax

> <u>app</u>end using *filename* [*filename* ...] [, *options*]

You may enclose *filename* in double quotes and must do so if *filename* contains blanks or other special characters.

options	Description
<u>generate</u>(*newvar*)	*newvar* marks source of resulting observations
<u>keep</u>(*varlist*)	keep specified variables from appending dataset(s)
<u>nol</u>abel	do not copy value-label definitions from dataset(s) on disk
<u>nonote</u>s	do not copy notes from dataset(s) on disk
force	append string to numeric or numeric to string without error

Menu

Data > Combine datasets > Append datasets

Description

append appends Stata-format datasets stored on disk to the end of the dataset in memory. If any *filename* is specified without an extension, .dta is assumed.

Stata can also join observations from two datasets into one; see [D] **merge**. See [U] **22 Combining datasets** for a comparison of append, merge, and joinby.

Options

generate(*newvar*) specifies the name of a variable to be created that will mark the source of observations. Observations from the master dataset (the data in memory before the append command) will contain 0 for this variable. Observations from the first using dataset will contain 1 for this variable; observations from the second using dataset will contain 2 for this variable; and so on.

keep(*varlist*) specifies the variables to be kept from the using dataset. If keep() is not specified, all variables are kept.

The *varlist* in keep(*varlist*) differs from standard Stata varlists in two ways: variable names in *varlist* may not be abbreviated, except by the use of wildcard characters, and you may not refer to a range of variables, such as price-weight.

nolabel prevents Stata from copying the value-label definitions from the disk dataset into the dataset in memory. Even if you do not specify this option, label definitions from the disk dataset never replace definitions already in memory.

nonotes prevents notes in the using dataset from being incorporated into the result. The default is to incorporate notes from the using dataset that do not already appear in the master data.

8

force allows string variables to be appended to numeric variables and vice versa, resulting in missing values from the using dataset. If omitted, append issues an error message; if specified, append issues a warning message.

Remarks

The disk dataset must be a Stata-format dataset; that is, it must have been created by save (see [D] **save**).

▷ Example 1

We have two datasets stored on disk that we want to combine. The first dataset, called even.dta, contains the sixth through eighth positive even numbers. The second dataset, called odd.dta, contains the first five positive odd numbers. The datasets are

```
. use even
(6th through 8th even numbers)

. list
```

	number	even
1.	6	12
2.	7	14
3.	8	16

```
. use odd
(First five odd numbers)

. list
```

	number	odd
1.	1	1
2.	2	3
3.	3	5
4.	4	7
5.	5	9

We will append the even data to the end of the odd data. Because the odd data are already in memory (we just used them above), we type append using even. The result is

```
. append using even

. list
```

	number	odd	even
1.	1	1	.
2.	2	3	.
3.	3	5	.
4.	4	7	.
5.	5	9	.
6.	6	.	12
7.	7	.	14
8.	8	.	16

Because the `number` variable is in both datasets, the variable was extended with the new data from the file `even.dta`. Because there is no variable called odd in the new data, the additional observations on odd were forward-filled with *missing* (.). Because there is no variable called `even` in the original data, the first observations on `even` were back-filled with missing.

◁

▷ Example 2

The order of variables in the two datasets is irrelevant. Stata always appends variables by name:

```
. use http://www.stata-press.com/data/r12/odd1
(First five odd numbers)

. describe

Contains data from http://www.stata-press.com/data/r12/odd1.dta
  obs:             5                          First five odd numbers
 vars:             2                          9 Jan 2011 08:41
 size:            40
```

variable name	storage type	display format	value label	variable label
odd	float	%9.0g		Odd numbers
number	float	%9.0g		

```
Sorted by:  number

. describe using http://www.stata-press.com/data/r12/even

Contains data from http://www.stata-press.com/data/r12/even
  obs:             3                          6th through 8th even numbers
 vars:             2                          9 Jan 2011 08:43
 size:            27
```

variable name	storage type	display format	value label	variable label
number	byte	%9.0g		
even	float	%9.0g		Even numbers

```
Sorted by:  number

. append using http://www.stata-press.com/data/r12/even

. list
```

	odd	number	even
1.	1	1	.
2.	3	2	.
3.	5	3	.
4.	7	4	.
5.	9	5	.
6.	.	6	12
7.	.	7	14
8.	.	8	16

The results are the same as those in the first example.

◁

When Stata appends two datasets, the definitions of the dataset in memory, called the *master* dataset, override the definitions of the dataset on disk, called the *using* dataset. This extends to value labels, variable labels, characteristics, and date–time stamps. If there are conflicts in numeric storage types, the more precise storage type will be used regardless of whether this storage type was in the *master* dataset or the *using* dataset. If a variable is stored as a string in one dataset that is longer than in the other, the longer `str#` storage type will prevail.

❑ Technical note

If a variable is a string in one dataset and numeric in the other, Stata issues an error message unless the `force` option is specified. If `force` is specified, Stata issues a warning message before appending the data. If the using dataset contains the string variable, the combined dataset will have numeric missing values for the appended data on this variable; the contents of the string variable in the using dataset are ignored. If the using dataset contains the numeric variable, the combined dataset will have empty strings for the appended data on this variable; the contents of the numeric variable in the using dataset are ignored.

❑

▷ Example 3

Because Stata has five numeric variable types—`byte`, `int`, `long`, `float`, and `double`—you may attempt to append datasets containing variables with the same name but of different numeric types; see [U] **12.2.2 Numeric storage types**.

Let's `describe` the datasets in the example above:

```
. describe using http://www.stata-press.com/data/r12/odd
Contains data from http://www.stata-press.com/data/r12/odd
  obs:            5                          First five odd numbers
  vars:           2                          9 Jan 2011 08:50
  size:          60

              storage  display    value
variable name   type   format     label    variable label

number          float  %9.0g
odd             float  %9.0g                Odd numbers

Sorted by:
. describe using http://www.stata-press.com/data/r12/even
Contains data from http://www.stata-press.com/data/r12/even
  obs:            3                          6th through 8th even numbers
  vars:           2                          9 Jan 2011 08:43
  size:          27

              storage  display    value
variable name   type   format     label    variable label

number          byte   %9.0g
even            float  %9.0g                Even numbers

Sorted by:  number
```

```
. describe using http://www.stata-press.com/data/r12/oddeven
Contains data from http://www.stata-press.com/data/r12/oddeven
  obs:             8                          First five odd numbers
  vars:            3                          9 Jan 2011 08:53
  size:          128

              storage  display   value
variable name   type   format    label    variable label

number          float  %9.0g
odd             float  %9.0g              Odd numbers
even            float  %9.0g              Even numbers

Sorted by:
```

The number variable was stored as a float in odd.dta but as a byte in even.dta. Because float is the more precise storage type, the resulting dataset, oddeven.dta, had number stored as a float. Had we instead appended odd.dta to even.dta, number would still have been stored as a float:

```
. use http://www.stata-press.com/data/r12/even, clear
(6th through 8th even numbers)
. append using http://www.stata-press.com/data/r12/odd
number was byte now float
. describe
Contains data from http://www.stata-press.com/data/r12/even.dta
  obs:             8                          6th through 8th even numbers
  vars:            3                          9 Jan 2011 08:43
  size:           96

              storage  display   value
variable name   type   format    label    variable label

number          float  %9.0g
even            float  %9.0g              Even numbers
odd             float  %9.0g              Odd numbers

Sorted by:
     Note:  dataset has changed since last saved
```

◁

▷ Example 4

Suppose that we have a dataset in memory containing the variable educ, and we have previously given a label variable educ "Education Level" command so that the variable label associated with educ is "Education Level". We now append a dataset called newdata.dta, which also contains a variable named educ, except that its variable label is "Ed. Lev". After appending the two datasets, the educ variable is still labeled "Education Level". See [U] **12.6.2 Variable labels**.

◁

▷ Example 5

Assume that the values of the educ variable are labeled with a value label named educlbl. Further assume that in newdata.dta, the values of educ are also labeled by a value label named educlbl. Thus there is one definition of educlbl in memory and another (although perhaps equivalent) definition in newdata.dta. When you append the new data, you will see the following:

```
. append using newdata
label educlbl already defined
```

If one label in memory and another on disk have the same name, `append` warns you of the problem and sticks with the definition currently in memory, ignoring the definition in the disk file.

◁

❑ Technical note

When you `append` two datasets that both contain definitions of the same value label, the codings may not be equivalent. That is why Stata warns you with a message like "label educlbl already defined". If you do not know that the two value labels are equivalent, you should convert the value-labeled variables into string variables, append the data, and then construct a new coding. `decode` and `encode` make this easy:

```
. use newdata, clear
. decode educ, gen(edstr)
. drop educ
. save newdata, replace
. use basedata
. decode educ, gen(edstr)
. drop educ
. append using newdata
. encode edstr, gen(educ)
. drop edstr
```

See [D] **encode**.

You can specify the `nolabel` option to force `append` to ignore all the value-label definitions in the incoming file, whether or not there is a conflict. In practice, you will probably never want to do this.

❑

▷ Example 6

Suppose that we have several datasets containing the populations of counties in various states. We can use `append` to combine these datasets all at once and use the `generate()` option to create a variable identifying from which dataset each observation originally came.

```
. use http://www.stata-press.com/data/r12/capop
. list
```

	county	pop
1.	Los Angeles	9878554
2.	Orange	2997033
3.	Ventura	798364

```
. append using http://www.stata-press.com/data/r12/ilpop
> http://www.stata-press.com/data/r12/txpop, generate(state)
. label define statelab 0 "CA" 1 "IL" 2 "TX"
. label values state statelab
```

. list

	county	pop	state
1.	Los Angeles	9878554	CA
2.	Orange	2997033	CA
3.	Ventura	798364	CA
4.	Cook	5285107	IL
5.	DeKalb	103729	IL
6.	Will	673586	IL
7.	Brazos	152415	TX
8.	Johnson	149797	TX
9.	Harris	4011475	TX

◁

Also see

[D] **save** — Save Stata dataset

[D] **use** — Load Stata dataset

[D] **cross** — Form every pairwise combination of two datasets

[D] **joinby** — Form all pairwise combinations within groups

[D] **merge** — Merge datasets

[U] **22 Combining datasets**

Title

> **assert** — Verify truth of claim

Syntax

<u>as</u>sert *exp* $\left[\textit{if}\right]$ $\left[\textit{in}\right]$ $\left[\,,\ \underline{r}c0\ \underline{n}ull\right]$

by is allowed; see [D] **by**.

Description

assert verifies that *exp* is true. If it is true, the command produces no output. If it is not true, assert informs you that the "assertion is false" and issues a return code of 9; see [U] **8 Error messages and return codes**.

Options

rc0 forces a return code of 0, even if the assertion is false.

null forces a return code of 8 on null assertions.

Remarks

assert is seldom used interactively because it is easier to use inspect, summarize, or tabulate to look for evidence of errors in the dataset. These commands, however, require you to review the output to spot the error. assert is useful because it tells Stata not only what to do but also what you can expect to find. Groups of assertions are often combined in a do-file to certify data. If the do-file runs all the way through without complaining, every assertion in the file is true.

```
. do myassert
. use trans, clear
(xplant data)
. assert sex=="m" | sex=="f"
. assert packs==0 if !smoker
. assert packs>0 if smoker
. sort patient date
. by patient: assert sex==sex[_n-1] if _n>1
. by patient: assert abs(bp-bp[_n-1]) < 20 if bp< . & bp[_n-1]< .
. by patient: assert died==0 if _n!=_N
. by patient: assert died==0 | died==1 if _n==_N
. by patient: assert n_xplant==0 | n_xplant==1 if _n==_N
. assert inval==int(inval)
.
.
end of do-file
```

▷ Example 1

You receive data from Bob, a coworker. He has been working on the dataset for some time, and it has now been delivered to you for analysis. Before analyzing the data, you (smartly) verify that the data are as Bob claims. In Bob's memo, he claims that 1) the dataset reflects the earnings of 522 employees, 2) the earnings are only for full-time employees, 3) the variable female is coded 1 for female and 0 otherwise, and 4) the variable exp contains the number of years, or fraction thereof, on the job. You assemble the following do-file:

```
use frombob, clear
assert _N==522
assert sal>=6000 & sal<=125000
assert female==1 | female==0
gen work=sum(female==1)
assert work[_N]>0
replace work=sum(female==0)
assert work[_N]>0
drop work
assert exp>=0 & exp<=40
```

Let's go through these assertions one by one. After using the data, you assert that _N equals 522. Remember, _N reflects the total number of observations in the dataset; see [U] **13.4 System variables (_variables)**. Bob said it was 522, so you check it. Bob's second claim was that the data are for only full-time employees. You know that everybody in your company makes a salary between $6,000 and $125,000, so you check that the salary figures are within this range. Bob's third assertion was that the female variable was coded zero or one.

You add something more. You know that your company employs both males and females, so you check that there are some of each. You create a variable called work equal to the running sum of female observations and then verify that the last observation of this variable is greater than zero. You then repeat the process for males and discard the work variable. Finally, you verify that the exp variable is never negative and is never larger than 40.

You save the above file as check.do, and here is what happens when you run it:

```
. do check
. use frombob, clear
(5/21 data)
. assert _N==522
. assert sal>6000 & sal<=125000
14 contradictions in 522 observations
assertion is false
r(9);
end of do-file
r(9);
```

Everything went fine until you checked the salary variable, when Stata told you that there were 14 contradictions to your assertion and stopped the do-file. Seeing this, you now interactively summarize the sal variable and discover that 14 people have missing salaries. You dash off a memo to Bob asking him why these data are missing.

◁

▷ Example 2

Bob responds quickly. There was a mistake in reading the salaries for the consumer relations division. He says it's fixed. You believe him but check with your do-file again. This time you type run instead of do, suppressing all the output:

```
. run check
.
```

Even though you suppressed the output, if there had been any contradictions, the messages would have printed. check.do ran fine, so all its assertions are true.

◁

❏ Technical note

assert is especially useful when you are processing large amounts of data in a do-file and wish to verify that all is going as expected. The error here may not be in the data but in the do-file itself. For instance, your do-file is rolling along, and it has just merged two datasets that it created by subsetting some other data. If everything has gone right so far, every observation should have merged. Include the line

```
assert _merge==3
```

to verify the correctness of the merge. If all the observations did not merge, the assertion will be false, and your do-file will stop.

As another example, you are combining data from many sources, and you know that after the first two datasets are combined, every individual's sex should be defined. So, you include the line

```
assert sex< .
```

in your do-file. Experienced Stata users include many assertions in their do-files when they process data.

❏

❏ Technical note

assert is smart in how it evaluates expressions. When you type something like assert _N==522 or assert work[_N]>0, assert knows that the expression needs to be evaluated only once. When you type assert female==1 | female==0, assert knows that the expression needs to be evaluated once for each observation in the dataset.

Here are some more examples demonstrating assert's intelligence.

```
by female:  assert _N==100
```

asserts that there should be 100 observations for every unique value of female. The expression is evaluated once per by-group.

```
by female:  assert work[_N]>0
```

asserts that the last observation on work in every by-group should be greater than zero. It is evaluated once per by-group.

```
    by female:  assert work>0
```

is evaluated once for each observation in the dataset and, in that sense, is formally equivalent to
assert work>0.

❏

Also see

[P] **capture** — Capture return code

[P] **confirm** — Argument verification

[U] **16 Do-files**

Title

bcal — Business calendar file manipulation

Syntax

List business calendars used by the data currently in memory

> bcal c̲heck [*varlist*] [, rc0]

List filenames and directories of available business calendars

> bcal dir [*pattern*]

Describe the specified business calendar

> bcal d̲escribe *calname*

Load the business calendar

> bcal load *calname*

where

> *varlist* is a list of variable names to be checked for whether they use business calendars. If not specified, all variables are checked.

> *pattern* is the name of a business calendar possibly containing wildcards * and ?. If *pattern* is not specified, all available business calendar names are listed.

> *calname* is the name of a business calendar either as a name or as a datetime format; for example, *calname* could be simple or %tbsimple.

Menu

Data > Other utilities > Business calendar utilities

Data > Variables Manager

Description

See [D] **datetime business calendars** for an introduction to business calendars and dates.

bcal check lists the business calendars used by the data currently in memory, if any.

bcal dir *pattern* lists filenames and directories of all available business calendars matching *pattern*, or all business calendars if *pattern* is not specified.

bcal describe *calname* presents a description of the specified business calendar.

bcal load *calname* loads the specified business calendar. Business calendars load automatically when needed, and thus use of bcal load is never required. bcal load is used by programmers writing their own business calendars. bcal load *calname* forces immediate loading of a business calendar and displays output, including any error messages due to improper calendar construction.

Option

rc0 specifies that bcal check is to exit without error (return 0) even if some calendars do not exist or have errors. Programmers can then access the results bcal check saves in r() to get even more details about the problems. If you wish to suppress bcal dir, precede the bcal check command with capture and specify the rc0 option if you wish to access the r() results.

Remarks

bcal check reports on any %tb formats used by the data currently in memory:

```
. bcal check
        %tbsimple:  defined, used by variable
                    mydate
```

bcal dir reports on business calendars available:

```
. bcal dir
  1 calendar file found:
        simple:  C:\Program Files\Stata12\ado\base\s\simple.stbcal
```

bcal describe reports on an individual calendar.

```
. bcal describe simple

  Business calendar simple (format %tbsimple):

     purpose:  Example for manual

       range:  01nov2011  30nov2011
               18932          18961    in %td units
                   0             19    in %tbsimple units

      center:  01nov2011
               18932                   in %td units
                   0                   in %tbsimple units

     omitted:      10                  days
                 121.8                 approx. days/year

    included:      20                  days
                 243.5                 approx. days/year
```

bcal load is used by programmers writing new stbcal-files. See [D] **datetime business calendars creation**.

Saved results

bcal check saves the following in r():

Macros
    r(defined)   business calendars used, stbcal-file exists, and file contains no errors
    r(undefined)   business calendars used, but no stbcal-files exist for them

Warning to programmers: Specify the rc0 option to access these returned results. By default, bcal check returns code 459 if a business calendar does not exist or if a business calendar exists but has errors; in such cases, the results are not saved.

`bcal describe` saves the following in `r()`:

Scalars
`r(min_date_td)`	calendar's minimum date in %td units
`r(max_date_td)`	calendar's maximum date in %td units
`r(ctr_date_td)`	calendar's zero date in %td units
`r(min_date_tb)`	calendar's minimum date in %tb units
`r(max_date_tb)`	calendar's maximum date in %tb units
`r(omitted)`	total number of days omitted from calendar
`r(included)`	total number of days included in calendar

Macros
`r(name)`	pure calendar name (for example, `nyse`)
`r(purpose)`	short description of calendar's purpose

`bcal load` saves the same results in `r()` as `bcal describe`, except it does not save `r(omitted)` and `r(included)`.

Methods and formulas

`bcal` is implemented as an ado-file.

Also see

[D] **datetime business calendars** — Business calendars

[D] **datetime business calendars creation** — Business calendars creation

Title

> **by** — Repeat Stata command on subsets of the data

Syntax

> by *varlist* :   *stata_cmd*
>
> <u>by</u>sort *varlist* :   *stata_cmd*

The above diagrams show by and bysort as they are typically used. The full syntax of the commands is

> by *varlist*₁ $\left[(varlist_2)\right]$ $\left[\,,\, \underline{s}\text{ort rc0}\right]$:   *stata_cmd*
>
> <u>by</u>sort *varlist*₁ $\left[(varlist_2)\right]$ $\left[\,,\, \text{rc0}\right]$:   *stata_cmd*

Description

Most Stata commands allow the by prefix, which repeats the command for each group of observations for which the values of the variables in *varlist* are the same. by without the sort option requires that the data be sorted by *varlist*; see [D] **sort**.

Stata commands that work with the by prefix indicate this immediately following their syntax diagram by reporting, for example, "by is allowed; see [D] **by**" or "bootstrap, by, etc., are allowed; see [U] **11.1.10 Prefix commands**".

by and bysort are really the same command; bysort is just by with the sort option.

The *varlist*₁ (*varlist*₂) syntax is of special use to programmers. It verifies that the data are sorted by *varlist*₁ *varlist*₂ and then performs a by as if only *varlist*₁ were specified. For instance,

> by pid (time): gen growth = (bp - bp[_n-1])/bp

performs the generate by values of pid but first verifies that the data are sorted by pid and time within pid.

Options

sort specifies that if the data are not already sorted by *varlist*, by should sort them.

rc0 specifies that even if the *stata_cmd* produces an error in one of the by-groups, then by is still to run the *stata_cmd* on the remaining by-groups. The default action is to stop when an error occurs. rc0 is especially useful when *stata_cmd* is an estimation command and some by-groups have insufficient observations.

Remarks

> ## Example 1

```
. use http://www.stata-press.com/data/r12/autornd
(1978 Automobile Data)

. keep in 1/20
(54 observations deleted)

. by mpg: egen mean_w = mean(weight)
not sorted
r(5);

. sort mpg

. by mpg: egen mean_w = mean(weight)

. list
```

	make	weight	mpg	mean_w
1.	Cad. Eldorado	4000	15	3916.667
2.	AMC Pacer	3500	15	3916.667
3.	Chev. Impala	3500	15	3916.667
4.	Buick Electra	4000	15	3916.667
5.	Cad. Deville	4500	15	3916.667
6.	Buick Riviera	4000	15	3916.667
7.	Buick LeSabre	3500	20	3350
8.	Chev. Monte Carlo	3000	20	3350
9.	Buick Skylark	3500	20	3350
10.	Buick Century	3500	20	3350
11.	AMC Spirit	2500	20	3350
12.	AMC Concord	3000	20	3350
13.	Buick Regal	3500	20	3350
14.	Chev. Malibu	3000	20	3350
15.	Chev. Nova	3500	20	3350
16.	Cad. Seville	4500	20	3350
17.	Buick Opel	2000	25	2500
18.	Chev. Monza	3000	25	2500
19.	Chev. Chevette	2000	30	2000
20.	Dodge Colt	2000	30	2000

by requires that the data be sorted. In the above example, we could have typed by mpg, sort: egen mean_w = mean(weight) or bysort mpg: egen mean_w = mean(weight) rather than the separate sort; all would yield the same results.

◁

For more examples, see [U] **11.1.2 by varlist:**, [U] **11.5 by varlist: construct**, and [U] **27.2 The by construct**. For extended introductions with detailed examples, see Cox (2002) and Mitchell (2010, chap. 7).

References

Cox, N. J. 2002. Speaking Stata: How to move step by: step. *Stata Journal* 2: 86–102.

Mitchell, M. N. 2010. *Data Management Using Stata: A Practical Handbook*. College Station, TX: Stata Press.

Also see

[D] **sort** — Sort data

[D] **statsby** — Collect statistics for a command across a by list

[P] **byable** — Make programs byable

[P] **foreach** — Loop over items

[P] **forvalues** — Loop over consecutive values

[P] **while** — Looping

[U] **11.1.2 by varlist:**

[U] **11.1.10 Prefix commands**

[U] **11.4 varlists**

[U] **11.5 by varlist: construct**

[U] **27.2 The by construct**

Title

cd — Change directory

Syntax

Stata for Windows

cd

cd ["]*directory_name*["]

cd ["]*drive*:["]

cd ["]*drive*:*directory_name*["]

pwd

Stata for Mac and Stata Unix

cd

cd ["]*directory_name*["]

pwd

If your *directory_name* contains embedded spaces, remember to enclose it in double quotes.

Description

Stata for Windows: cd changes the current working directory to the specified drive and directory. pwd is equivalent to typing cd without arguments; both display the name of the current working directory. Note: You can shell out to a DOS window; see [D] **shell**. However, typing ! cd *directory_name* does not change Stata's current directory; use the cd command to change directories.

Stata for Mac and Stata for Unix: cd (synonym chdir) changes the current working directory to *directory_name* or, if *directory_name* is not specified, the home directory. pwd displays the path of the current working directory.

Remarks

Remarks are presented under the following headings:

> *Stata for Windows*
> *Stata for Mac*
> *Stata for Unix*

Stata for Windows

When you start Stata for Windows, your current working directory is set to the *Start in* directory specified in **Properties**. If you want to change this, see [GSW] **C.1 The Windows Properties Sheet**. You can always see what your working directory is by looking at the status bar at the bottom of the Stata window.

Once you are in Stata, you can change your directory with the cd command.

```
. cd
c:\data

. cd city
c:\data\city

. cd d:
D:\

. cd kande
D:\kande

. cd "additional detail"
D:\kande\additional detail

. cd c:
C:\

. cd data\city
C:\data\city

. cd \a\b\c\d\e\f\g
C:\a\b\c\d\e\f\g

. cd ..
C:\a\b\c\d\e\f

. cd ...
C:\a\b\c\d

. cd ....
C:\a
```

When we typed cd d:, we changed to the current directory of the D drive. We navigated our way to d:\kande\additional detail with three commands: cd d:, then cd kande, and then cd "additional detail". The double quotes around "additional detail" are necessary because of the space in the directory name. We could have changed to this directory in one command: cd "d:\kande\additional detail".

Notice the last three cd commands in the example above. You are probably familiar with the cd .. syntax to move up one directory from where you are. The last two cd commands above let you move up more than one directory: cd ... is shorthand for 'cd ..\..' and cd .... is shorthand for 'cd ..\..\..'. These shorthand cd commands are not limited to Stata; they will work in your DOS windows under Windows as well.

Stata for Mac

Read [U] **11.6 Filenaming conventions** for a description of how filenames are written in a command language before reading this entry.

Invoking an application and then changing folders is an action foreign to most Mac users. If it is foreign to you, you can ignore cd and pwd. However, they can be useful. You can see the current folder (where Stata saves files and looks for files) by typing pwd. You can change the current folder by using cd or by selecting **File > Change Working Directory...**. Stata's cd understands '~' as an abbreviation for the home directory, so you can type things like cd ~/data.

```
. pwd
/Users/bill/proj
. cd "~/data/city"
/Users/bill/data/city

. _
```

If you now wanted to change to "/Users/bill/data/city/ny", you could type cd ny. If you wanted instead to change to "/Users/bill/data", you could type 'cd ..'.

Stata for Unix

cd and pwd are equivalent to Unix's cd and pwd commands. Like csh, Stata's cd understands '~' as an abbreviation for the home directory $HOME, so you can type things like cd ~/data; see [U] **11.6 Filenaming conventions**.

```
. pwd
/usr/bill/proj
. cd ~/data/city
/usr/bill/data/city

. _
```

If you now wanted to change to /usr/bill/data/city/ny, you could type cd ny. If you wanted instead to change to /usr/bill/data, you could type 'cd ..'.

Also see

[D] **copy** — Copy file from disk or URL

[D] **dir** — Display filenames

[D] **erase** — Erase a disk file

[D] **mkdir** — Create directory

[D] **rmdir** — Remove directory

[D] **shell** — Temporarily invoke operating system

[D] **type** — Display contents of a file

[U] **11.6 Filenaming conventions**

Title

cf — Compare two datasets

Syntax

cf *varlist* using *filename* [, <u>a</u>ll <u>v</u>erbose]

Menu

Data > Data utilities > Compare two datasets

Description

cf compares *varlist* of the dataset in memory (the master dataset) with the corresponding variables in *filename* (the using dataset). cf returns nothing (that is, a return code of 0) if the specified variables are identical and a return code of 9 if there are any differences. Only the variable values are compared. Variable labels, value labels, notes, characteristics, etc., are not compared.

Options

all displays the result of the comparison for each variable in *varlist*. Unless all is specified, only the results of the variables that differ are displayed.

verbose gives a detailed listing, by variable, of each observation that differs.

Remarks

cf produces messages having the following form:

> *varname*: does not exist in using
> *varname*: ___ in master but ___ in using
> *varname*: ___ mismatches
> *varname*: match

An example of the second message is "str4 in master but float in using". Unless all is specified, the fourth message does not appear—silence indicates matches.

▷ Example 1

We think the dataset in memory is identical to mydata.dta, but we are unsure. We want to understand any differences before continuing:

```
. cf _all using mydata

. _
```

28

All the variables in the master dataset are in mydata.dta, and these variables are the same in both datasets. We might see instead

```
. cf _all using mydata
            mpg:  2 mismatches
       headroom:  does not exist in using
   displacement:  does not exist in using
      gear_ratio:  does not exist in using
r(9);
```

Two changes were made to the mpg variable, and the headroom, displacement, and gear_ratio variables do not exist in mydata.dta.

To see the result of each comparison, we could append the all option to our command:

```
. cf _all using mydata, all
           make:  match
          price:  match
            mpg:  2 mismatches
          rep78:  match
       headroom:  does not exist in using
          trunk:  match
         weight:  match
         length:  match
           turn:  match
   displacement:  does not exist in using
     gear_ratio:  does not exist in using
        foreign:  match
r(9);
```

For more details on the mismatches, we can use the verbose option:

```
. cf _all using mydata, verbose
            mpg:  2 mismatches
                  obs  1. 22 in master; 33 in using
                  obs  2. 17 in master; 33 in using
       headroom:  does not exist in using
   displacement:  does not exist in using
     gear_ratio:  does not exist in using
r(9);
```

This example shows us exactly which two observations for mpg differ, as well as the value stored in each dataset.

◁

▷ Example 2

We want to compare a group of variables in the dataset in memory against the same group of variables in mydata.dta.

```
. cf mpg headroom using mydata
            mpg: 2 mismatches
       headroom: does not exist in using
r(9);
```

◁

Saved results

cf saves the following in r():

Macros
  r(Nsum)   number of differences

Methods and formulas

cf is implemented as an ado-file.

If you are using Small Stata, you may get the error "too many variables" when you stipulate _all and have many variables in your dataset. (This will not happen if you are using Stata/MP, Stata/SE, or Stata/IC.) If this happens, you will have to perform the comparison with groups of variables. See example 2 for details about how to do this.

Acknowledgment

Speed improvements in cf were based on code written by David Kantor.

Reference

Gleason, J. R. 1995. dm36: Comparing two Stata data sets. *Stata Technical Bulletin* 28: 10–13. Reprinted in *Stata Technical Bulletin Reprints*, vol. 5, pp. 39–43. College Station, TX: Stata Press.

Also see

[D] **compare** — Compare two variables

Title

> **changeeol** — Convert end-of-line characters of text file

Syntax

> changeeol *filename1 filename2*, eol(*platform*) [*options*]

filename1 and *filename2* must be filenames.

Note: Double quotes may be used to enclose the filenames, and the quotes must be used if the filename contains embedded blanks.

options	Description
*eol(windows)	convert to Windows-style end-of-line characters (\r\n)
*eol(dos)	synonym for eol(windows)
*eol(unix)	convert to Unix-style end-of-line characters (\n)
*eol(mac)	convert to Mac-style end-of-line characters (\n)
*eol(classicmac)	convert to classic Mac-style end-of-line characters (\r)
replace	overwrite *filename2*
force	force to convert *filename1* to *filename2* if *filename1* is a binary file

*eol() is required.

Description

changeeol converts text file *filename1* to text file *filename2* with the specified Windows/Unix/Mac/classic Mac-style end-of-line characters. changeeol changes the end-of-line characters from one type of file to another.

Options

eol(windows | dos | unix | mac | classicmac) specifies to which platform style *filename2* is to be converted. eol() is required.

replace specifies that *filename2* be replaced if it already exists.

force specifies that *filename1* be converted if it is a binary file.

Remarks

changeeol uses hexdump to determine whether *filename1* is ASCII or binary. If it is binary, changeeol will refuse to convert it unless the force option is specified.

Examples

Windows:

```
. changeeol orig.txt newcopy.txt, eol(windows)
```

Unix:

```
. changeeol orig.txt newcopy.txt, eol(unix)
```

Mac:

```
. changeeol orig.txt newcopy.txt, eol(mac)
```

Classic Mac:

```
. changeeol orig.txt newcopy.txt, eol(classicmac)
```

Also see

[D] **filefilter** — Convert text or binary patterns in a file

[D] **hexdump** — Display hexadecimal report on file

Title

checksum — Calculate checksum of file

Syntax

checksum *filename* [, *options*]

<u>se</u>t checksum { on | off } [, <u>perman</u>ently]

options	Description
save	save output to *filename*.sum; default is to display a report
replace	may overwrite *filename*.sum; use with save
<u>sav</u>ing(*filename2* [, replace])	save output to *filename2*; alternative to save

Description

checksum creates *filename*.sum files for later use by Stata when it reads files over a network. These optional files are used to reduce the chances of corrupted files going undetected. Whenever Stata reads file *filename*.*suffix* over a network, whether by use, net, update, etc., it also looks for *filename*.sum. If Stata finds that file, Stata reads it and uses its contents to verify that the first file was received without error. If there are errors, Stata informs the user that the file could not be read.

set checksum on tells Stata to verify that files downloaded over a network have been received without error.

set checksum off, which is the default, tells Stata to bypass the file verification.

❏ Technical note

checksum calculates a CRC checksum following the POSIX 1003.2 specification and displays the file size in bytes. checksum produces the same results as the Unix cksum command. Comparing the checksum of the original file with the received file guarantees the integrity of the received file.

When comparing Stata's checksum results with those of Unix, do not confuse Unix's sum and cksum commands. Unix's cksum and Stata's checksum use a more robust algorithm than that used by Unix's sum. In some Unix operating systems, there is no cksum command, and the more robust algorithm is obtained by specifying an option with sum.

❏

Options

save saves the output of the checksum command to the text file *filename*.sum. The default is to display a report but not create a file.

replace is for use with save; it permits Stata to overwrite an existing *filename*.sum file.

saving(*filename2* [, replace]) is an alternative to save. It saves the output in the specified filename. You must supply a file extension if you want one, because none is assumed.

permanently specifies that, in addition to making the change right now, the checksum setting be remembered and become the default setting when you invoke Stata.

Remarks

▷ Example 1

Say that you wish to put a dataset on your homepage so that colleagues can use it over the Internet by typing

```
. use http://www.myuni.edu/department/~joe/mydata
```

mydata.dta is important, and even though the chances of the file mydata.dta being corrupted by the Internet are small, you wish to guard against that. The solution is to create the checksum file named mydata.sum and place that on your homepage. Your colleagues need type nothing different, but now Stata will verify that all goes well. When they use the file, they will see either

```
. use http://www.myuni.edu/department/~joe/mydata
(important data from joe)
```

or

```
. use http://www.myuni.edu/department/~joe/mydata
file transmission error (checksums do not match)
http://www.myuni.edu/department/~joe/mydata.dta not downloaded
r(639);
```

To make the checksum file, change to the directory where the file is located and type

```
. checksum mydata.dta, save
Checksum for mydata.dta = 263508742, size = 4052
file mydata.sum saved
```

◁

▷ Example 2

Let's use checksum on the auto dataset that is shipped with Stata. We will load the dataset and save it to our current directory.

```
. use http://www.stata-press.com/data/r12/auto
(1978 Automobile Data)
. save auto
file auto.dta saved
. checksum auto.dta
Checksum for auto.dta = 2039025784, size = 5949
```

We see the report produced by checksum, but we decide to save this information to a file.

```
. checksum auto.dta, save
. type auto.sum
1 5949 2039025784
```

The first number is the version number (possibly used for future releases). The second number is the file's size in bytes, which can be used with the checksum value to ensure that the file transferred without corruption. The third number is the checksum value. Although two different files can have the same checksum value, two files with the same checksum value almost certainly could not have the same file size.

This example is admittedly artificial. Typically, you would use checksum to verify that no file transmission error occurred during a web download. If you want to verify that your own data are unchanged, using datasignature is better; see [D] **datasignature**.

◁

Saved results

checksum saves the following in r():

Scalars
r(version)	checksum version number
r(filelen)	length of file in bytes
r(checksum)	checksum value

Also see

[D] **use** — Load Stata dataset

[R] **net** — Install and manage user-written additions from the Internet

[D] **datasignature** — Determine whether data have changed

Title

> **clear** — Clear memory

Syntax

```
clear

clear [ mata | results | matrix | programs | ado ]

clear [ all | * ]
```

Description

`clear`, by itself, removes data and value labels from memory and is equivalent to typing

```
. version 12
. drop _all                 (see [D] drop)
. label drop _all           (see [D] label)
```

`clear mata` removes Mata functions and objects from memory and is equivalent to typing

```
. version 12
. mata: mata clear          (see [M-3] mata clear)
```

`clear results` eliminates saved results from memory and is equivalent to typing

```
. version 12
. return clear              (see [P] return)
. ereturn clear             (see [P] return)
. sreturn clear             (see [P] return)
. _return drop _all         (see [P] _return)
```

`clear matrix` eliminates from memory all matrices created by Stata's `matrix` command; it does not eliminate Mata matrices from memory. `clear matrix` is equivalent to typing

```
. version 12
. return clear              (see [P] return)
. ereturn clear             (see [P] return)
. sreturn clear             (see [P] return)
. _return drop _all         (see [P] _return)
. matrix drop _all          (see [P] matrix utility)
. estimates drop _all       (see [R] estimates)
```

`clear programs` eliminates all programs from memory and is equivalent to typing

```
. version 12
. program drop _all         (see [P] program)
```

`clear ado` eliminates all automatically loaded ado-file programs from memory (but not programs defined interactively or by do-files). It is equivalent to typing

```
. version 12
. program drop _allado       (see [P] program)
```

`clear all` and `clear *` are synonyms. They remove all data, value labels, matrices, scalars, constraints, clusters, saved results, sersets, and Mata functions and objects from memory. They also close all open files and postfiles, clear the class system, close any open Graph windows and dialog boxes, drop all programs from memory, and reset all timers to zero. They are equivalent to typing

```
. version 12
. drop _all                     (see [D] drop)
. label drop _all               (see [D] label)
. matrix drop _all              (see [P] matrix utility)
. scalar drop _all              (see [P] scalar)
. constraint drop _all          (see [R] constraint)
. cluster drop _all             (see [MV] cluster utility)
. file close _all               (see [P] file)
. postutil clear                (see [P] postfile)
. _return drop _all             (see [P] _return)
. discard                       (see [P] discard)
. program drop _all             (see [P] program)
. timer clear                   (see [P] timer)
. mata: mata clear              (see [M-3] mata clear)
```

Remarks

You can clear the entire dataset without affecting macros and programs by typing `clear`. You can also type `clear all`. This command has the same result as `clear` by itself but also clears matrices, scalars, constraints, clusters, saved results, sersets, Mata, the class system, business calendars, and programs; closes all open files and postfiles; closes all open Graph windows and dialog boxes; and resets all timers to zero.

▷ Example 1

We load the `bpwide` dataset to correct a mistake in the data.

```
. use http://www.stata-press.com/data/r12/bpwide
(fictional blood pressure data)
. list in 1/5
```

	patient	sex	agegrp	bp_bef~e	bp_after
1.	1	Male	30-45	143	153
2.	2	Male	30-45	163	170
3.	3	Male	30-45	153	168
4.	4	Male	30-45	153	142
5.	5	Male	30-45	146	141

```
. replace bp_after = 145 in 3
(1 real change made)
```

We made another mistake. We meant to change the value of `bp_after` in observation 4. It is easiest to begin again.

```
. clear
. use http://www.stata-press.com/data/r12/bpwide
(fictional blood pressure data)
```

◁

Methods and formulas

clear is implemented as an ado-file.

Also see

[D] **drop** — Eliminate variables or observations

[P] **discard** — Drop automatically loaded programs

[U] **11 Language syntax**

[U] **13 Functions and expressions**

Title

> **clonevar** — Clone existing variable

Syntax

clonevar *newvar* = *varname* $\lceil$ *if* $\rceil$ $\lceil$ *in* $\rceil$

Menu

Data $>$ Create or change data $>$ Other variable-creation commands $>$ Clone existing variable

Description

clonevar generates *newvar* as an exact copy of an existing variable, *varname*, with the same storage type, values, and display format as *varname*. *varname*'s variable label, value labels, notes, and characteristics will also be copied.

Remarks

clonevar has various possible uses. Programmers may desire that a temporary variable appear to the user exactly like an existing variable. Interactively, you might want a slightly modified copy of an original variable, so the natural starting point is a clone of the original.

$\triangleright$ Example 1

We have a dataset containing information on modes of travel. These data contain a variable named mode that identifies each observation as a specific mode of travel: air, train, bus, or car.

```
. use http://www.stata-press.com/data/r12/travel
. describe mode

              storage   display      value
variable name   type    format       label          variable label

mode            byte    %8.0g        travel         travel mode alternatives
. label list travel
travel:
          1 air
          2 train
          3 bus
          4 car
```

To create an identical variable identifying only observations that contain air or train, we could use clonevar with an if qualifier.

```
. clonevar airtrain = mode if mode == 1 | mode == 2
(420 missing values generated)
. describe mode airtrain

              storage   display      value
variable name   type    format       label          variable label

mode            byte    %8.0g        travel         travel mode alternatives
airtrain        byte    %8.0g        travel         travel mode alternatives
```

39

```
. list mode airtrain in 1/5
```

	mode	airtrain
1.	air	air
2.	train	train
3.	bus	.
4.	car	.
5.	air	air

The new `airtrain` variable has the same storage type, display format, value label, and variable label as `mode`. If `mode` had any characteristics or notes attached to it, they would have been applied to the new `airtrain` variable, too. The only differences in the two variables are their names and values for bus and car.

◁

❑ Technical note

The `if` qualifier used with the `clonevar` command in example 1 referred to the values of mode as 1 and 2. Had we wanted to refer to the values by their associated value labels, we could have typed

```
. clonevar airtrain = mode if mode == "air":travel | mode == "train":travel
```

For more details, see [U] **13.10 Label values**.

❑

Methods and formulas

`clonevar` is implemented as an ado-file.

Acknowledgments

`clonevar` was written by Nicholas J. Cox, Durham University, who in turn thanks Michael Blasnik, M. Blasnik & Associates, and Ken Higbee, StataCorp, for very helpful comments on a precursor of this command.

Also see

[D] **generate** — Create or change contents of variable

[D] **separate** — Create separate variables

Title

> **codebook** — Describe data contents

Syntax

codebook [*varlist*] [*if*] [*in*] [, *options*]

options	Description
Options	
<u>all</u>	print complete report without missing values
<u>h</u>eader	print dataset name and last saved date
<u>n</u>otes	print any notes attached to variables
<u>mv</u>	report pattern of missing values
<u>tab</u>ulate(*#*)	set tables/summary statistics threshold; default is tabulate(9)
<u>p</u>roblems	report potential problems in dataset
<u>d</u>etail	display detailed report on the variables; only with problems
<u>c</u>ompact	display compact report on the variables
dots	display a dot for each variable processed; only with compact
Languages	
<u>lang</u>uages [(*namelist*)]	use with multilingual datasets; see [D] **label language** for details

Menu

Data > Describe data > Describe data contents (codebook)

Description

codebook examines the variable names, labels, and data to produce a codebook describing the dataset.

Options

> Options

all is equivalent to specifying the header and notes options. It provides a complete report, which excludes only performing mv.

header adds to the top of the output a header that lists the dataset name, the date that the dataset was last saved, etc.

notes lists any notes attached to the variables; see [D] **notes**.

mv specifies that codebook search the data to determine the pattern of missing values. This is a CPU-intensive task.

tabulate(*#*) specifies the number of unique values of the variables to use to determine whether a variable is categorical or continuous. Missing values are not included in this count. The default is 9; when there are more than nine unique values, the variable is classified as continuous. Extended missing values will be included in the tabulation.

41

problems specifies that a summary report is produced describing potential problems that have been diagnosed:

- Variables that are labeled with an undefined value label

- Incompletely value-labeled variables

- Variables that are constant, including always missing

- Trailing, trimming, and embedded spaces in string variables

- Noninteger-valued date variables

See the discussion of these problems and advice on overcoming them following example 5.

detail may be specified only with the problems option. It specifies that the detailed report on the variables not be suppressed.

compact specifies that a compact report on the variables be displayed. compact may not be specified with any options other than dots.

dots specifies that a dot be displayed for every variable processed. dots may be specified only with compact.

Languages

languages[(*namelist*)] is for use with multilingual datasets; see [D] **label language**. It indicates that the codebook pertains to the languages in *namelist* or to all defined languages if no such list is specified as an argument to languages(). The output of codebook lists the data label and variable labels in these languages and which value labels are attached to variables in these languages.

Problems are diagnosed in all these languages, as well. The problem report does not provide details in which language problems occur. We advise you to rerun codebook for problematic variables; specify detail to produce the problem report again.

If you have a multilingual dataset but do not specify languages(), all output, including the problem report, is shown in the "active" language.

Remarks

codebook, without arguments, is most usefully combined with log to produce a printed listing for enclosure in a notebook documenting the data; see [U] **15 Saving and printing output—log files**. codebook is, however, also useful interactively, because you can specify one or a few variables.

⊳ Example 1

codebook examines the data in producing its results. For variables that codebook thinks are continuous, it presents the mean; the standard deviation; and the 10th, 25th, 50th, 75th, and 90th percentiles. For variables that it thinks are categorical, it presents a tabulation. In part, codebook makes this determination by counting the number of unique values of the variable. If the number is nine or fewer, codebook reports a tabulation; otherwise, it reports summary statistics.

codebook distinguishes the standard missing values (.) and the extended missing values (.a through .z, denoted by .*). If extended missing values are found, codebook reports the number of distinct missing value codes that occurred in that variable. Missing values are ignored with the tabulate option when determining whether a variable is treated as continuous or categorical.

```
. use http://www.stata-press.com/data/r12/educ3
(ccdb46, 52-54)

. codebook fips division, all
                Dataset:  http://www.stata-press.com/data/r12/educ3.dta
            Last saved:   6 Mar 2011 22:20

                  Label:  ccdb46, 52-54
    Number of variables:  42
 Number of observations:  956
                   Size:  145,312 bytes ignoring labels, etc.

_dta:
  1.  confirmed data with steve on 7/22
```

---

fips					state/place code

```
              type:  numeric (long)
             range:  [10060,560050]              units:  1
     unique values:  956                    missing .:  0/956

              mean:    256495
          std. dev:    156998

       percentiles:        10%       25%       50%       75%       90%
                         61462    120426    252848    391360    482530
```

---

division					Census Division

```
              type:  numeric (int)
             label:  division

             range:  [1,9]                        units:  1
     unique values:  9                       missing .:  4/956
   unique mv codes:  2                      missing .*:  2/956

        tabulation:  Freq.   Numeric  Label
                        69         1  N. Eng.
                        97         2  Mid Atl
                       202         3  E.N.C.
                        78         4  W.N.C.
                       115         5  S. Atl.
                        46         6  E.S.C.
                        89         7  W.S.C.
                        59         8  Mountain
                       195         9  Pacific
                         4         .
                         2        .a
```

Because division has nine unique nonmissing values, codebook reported a tabulation. If division had contained one more unique nonmissing value, codebook would have switched to reporting summary statistics, unless we had included the tabulate(#) option.

◁

▷ Example 2

The mv option is useful. It instructs codebook to search the data to determine patterns of missing values. Different kinds of missing values are not distinguished in the patterns.

```
. use http://www.stata-press.com/data/r12/citytemp
(City Temperature Data)

. codebook cooldd heatdd tempjan tempjuly, mv
```

---

cooldd   Cooling degree days

```
           type:  numeric (int)
          range:  [0,4389]                    units:  1
  unique values:  438                    missing .:  3/956

           mean:  1240.41
       std. dev:  937.668

    percentiles:        10%       25%       50%       75%       90%
                        411       615       940      1566      2761

 missing values:        heatdd==mv <-> cooldd==mv
                        tempjan==mv --> cooldd==mv
                        tempjuly==mv --> cooldd==mv
```

---

heatdd   Heating degree days

```
           type:  numeric (int)
          range:  [0,10816]                   units:  1
  unique values:  471                    missing .:  3/956

           mean:  4425.53
       std. dev:  2199.6

    percentiles:        10%       25%       50%       75%       90%
                       1510      2460      4950      6232      6919

 missing values:        cooldd==mv <-> heatdd==mv
                        tempjan==mv --> heatdd==mv
                        tempjuly==mv --> heatdd==mv
```

---

tempjan   Average January temperature

```
           type:  numeric (float)
          range:  [2.2,72.6]                  units:  .1
  unique values:  310                    missing .:  2/956

           mean:  35.749
       std. dev:  14.1881

    percentiles:        10%       25%       50%       75%       90%
                       20.2      25.1      31.3      47.8      55.1

 missing values:        tempjuly==mv <-> tempjan==mv
```

---

tempjuly   Average July temperature

```
           type:  numeric (float)
          range:  [58.099998,93.599998]       units:  0
  unique values:  196                    missing .:  0/956
 unique mv codes:  1                    missing .*:  2/956

           mean:  75.0538
       std. dev:  5.49504

    percentiles:        10%       25%       50%       75%       90%
                       68.8      71.8     74.25      78.7      82.3

 missing values:        tempjan==mv <-> tempjuly==mv
```

codebook reports that if `tempjan` is missing, `tempjuly` is also missing, and vice versa. In the output for the `cooldd` variable, codebook also reports that the pattern of missing values is the same for `cooldd` and `heatdd`. In both cases, the correspondence is indicated with "<->".

For `cooldd`, codebook also states that "`tempjan==mv --> cooldd==mv`". The one-way arrow means that a missing `tempjan` value implies a missing `cooldd` value but that a missing `cooldd` value does not necessarily imply a missing `tempjan` value.   ◁

Another feature of codebook—this one for numeric variables—is that it can determine the units of the variable. For instance, in the example above, `tempjan` and `tempjuly` both have units of 0.1, meaning that temperature is recorded to tenths of a degree. codebook handles precision considerations in making this determination (`tempjan` and `tempjuly` are floats; see [U] **13.11 Precision and problems therein**). If we had a variable in our dataset recorded in 100s (for example, 21,500 or 36,800), codebook would have reported the units as 100. If we had a variable that took on only values divisible by 5 (5, 10, 15, etc.), codebook would have reported the units as 5.

▷ Example 3

We can use the `label language` command (see [D] **label language**) and the `label` command (see [D] **label**) to create German value labels for our auto dataset. These labels are reported by codebook:

```
. use http://www.stata-press.com/data/r12/auto
(1978 Automobile Data)
. label language en, rename
(language default renamed en)
. label language de, new
(language de now current language)
. label data "1978 Automobile Daten"
. label variable foreign "Art Auto"
. label values foreign origin_de
. label define origin_de 0 "Innen" 1 "Ausländish"
. codebook foreign
```

```
────────────────────────────────────────────────────────────────────────────
foreign                                                              Art Auto
────────────────────────────────────────────────────────────────────────────

                  type:  numeric (byte)
                 label:  origin_de

                 range:  [0,1]                        units:  1
         unique values:  2                          missing .:  0/74

            tabulation:  Freq.   Numeric  Label
                            52         0  Innen
                            22         1  Ausländish
. codebook foreign, languages(en de)
────────────────────────────────────────────────────────────────────────────
foreign          in en:  Car type
                 in de:  Art Auto
────────────────────────────────────────────────────────────────────────────

                  type:  numeric (byte)
           label in en:  origin
           label in de:  origin_de

                 range:  [0,1]                        units:  1
         unique values:  2                          missing .:  0/74

            tabulation:  Freq. Numeric   origin        origin_de
                            52        0  Domestic      Innen
                            22        1  Foreign       Ausländish
```

With the `languages()` option, the value labels are shown in the specified active and available languages.

◁

▷ Example 4

`codebook, compact` summarizes the variables in your dataset, including variable labels. It is an alternative to the `summarize` command.

```
. use http://www.stata-press.com/data/r12/auto
(1978 Automobile Data)

. codebook, compact
Variable        Obs Unique     Mean   Min     Max  Label

make             74     74        .     .       .  Make and Model
price            74     74  6165.257  3291   15906  Price
mpg              74     21   21.2973    12      41  Mileage (mpg)
rep78            69      5  3.405797     1       5  Repair Record 1978
headroom         74      8  2.993243   1.5       5  Headroom (in.)
trunk            74     18  13.75676     5      23  Trunk space (cu. ft.)
weight           74     64  3019.459  1760    4840  Weight (lbs.)
length           74     47  187.9324   142     233  Length (in.)
turn             74     18  39.64865    31      51  Turn Circle (ft.)
displacement     74     31  197.2973    79     425  Displacement (cu. in.)
gear_ratio       74     36  3.014865  2.19    3.89  Gear Ratio
foreign          74      2  .2972973     0       1  Car type

. summarize
    Variable |       Obs        Mean    Std. Dev.       Min        Max

        make |         0
       price |        74    6165.257    2949.496       3291      15906
         mpg |        74     21.2973    5.785503         12         41
       rep78 |        69    3.405797    .9899323          1          5
    headroom |        74    2.993243    .8459948        1.5          5

       trunk |        74    13.75676    4.277404          5         23
      weight |        74    3019.459    777.1936       1760       4840
      length |        74    187.9324    22.26634        142        233
        turn |        74    39.64865    4.399354         31         51
displacement |        74    197.2973    91.83722         79        425

  gear_ratio |        74    3.014865    .4562871       2.19       3.89
     foreign |        74    .2972973    .4601885          0          1
```

◁

▷ Example 5

When `codebook` determines that neither a tabulation nor a listing of summary statistics is appropriate, for instance, for a string variable or for a numeric variable taking on many labeled values, it reports a few examples instead.

```
. use http://www.stata-press.com/data/r12/funnyvar

. codebook name
```

```
name                                                        (unlabeled)
```
```
                  type:  string (str5), but longest is str3
         unique values:  10                    missing "":  0/10
              examples:  "1 0"
                         "3"
                         "5"
                         "7"
               warning:  variable has embedded blanks
```

codebook is also on the lookout for common problems that might cause you to make errors when dealing with the data. For string variables, this includes leading, embedded, and trailing blanks. In the output above, codebook informed us that name includes embedded blanks. If name had leading or trailing blanks, it would have mentioned that, too.

When variables are value labeled, codebook performs two checks. First, if a value label *labname* is associated with a variable, codebook checks whether *labname* is actually defined. Second, it checks whether all values are value labeled. Partial labeling of a variable may mean that the label was defined incorrectly (for instance, the variable has values 0 and 1, but the value label maps 1 to "male" and 2 to "female") or that the variable was defined incorrectly (for example, a variable gender with three values). codebook checks whether date variables are integer valued.

If the problems option is specified, codebook does not provide detailed descriptions of each variable but reports only the potential problems in the data.

```
. codebook, problems
   Potential problems in dataset   http://www.stata-press.com/data/r12/funnyvar.dta
             potential problem     variables
```
```
   constant (or all missing) vars   human planet
       vars with nonexisting label   educ
        incompletely labeled vars   gender
   strvars that may be compressed   name address city country planet
   string vars with leading blanks   city country
  string vars with trailing blanks   planet
  string vars with embedded blanks   name address
        noninteger-valued date vars   birthdate
```
◁

In the example above, codebook, problems reported various potential problems with the dataset. These problems include

- Constant variables, including variables that are always missing

    Variables that are constant, taking the same value in all observations, or that are always missing, are often superfluous. Such variables, however, may also indicate problems. For instance, variables that are always missing may occur when importing data with an incorrect input specification. Such variables may also occur if you generate a new variable for a subset of the data, selected with an expression that is false for all observations.

    Advice: Carefully check the origin of constant variables. If you are saving a constant variable, be sure to compress the variable to use minimal storage.

- Variables with nonexisting value labels

    Stata treats value labels as separate objects that can be attached to one or more variables. A problem may arise if variables are linked to value labels that are not yet defined or if an incorrect value label name was used.

    Advice: Attach the correct value label or label define the value label; see [D] **label**.

- Incompletely labeled variables

    A variable is called "incompletely value labeled" if the variable is value labeled but no mapping is provided for some values of the variable. An example is a variable with values 0, 1, and 2 and value labels for 1, 2, and 3. This situation usually indicates an error, either in the data or in the value label.

    Advice: Change either the data or the value label.

- String variables that may be compressed

    The storage space used by a string variable is determined by its data type; see [D] **data types**. For instance, the storage type str20 indicates that 20 bytes are used per observation. If the declared storage type exceeds your requirements, memory and disk space is wasted.

    Advice: Use compress to store the data as compactly as possible.

- String variables with leading or trailing blanks

    In most applications, leading and trailing spaces do not affect the meaning of variables but are probably side effects from importing the data or from data manipulation. Spurious leading and trailing spaces force Stata to use more memory than required. In addition, manipulating strings with leading and trailing spaces is harder.

    Advice: Remove leading and trailing blanks from a string variable s by typing

        replace s = trim(s)

    See [D] **functions**.

- String variables with embedded blanks

    String variables with embedded blanks are often appropriate; however, sometimes they indicate problems importing the data.

    Advice: Verify that blanks are meaningful in the variables.

- Noninteger-valued date variables

    Stata's date formats were designed for use with integer values but will work with noninteger values.

    Advice: Carefully inspect the nature of the noninteger values. If noninteger values in a variable are the consequence of roundoff error, you may want to round the variable to the nearest integer.

        replace time = round(time)

Of course, more problems not reported by codebook are possible. These might include

- Numerical data stored as strings

    After importing data into Stata, you may discover that some string variables can actually be interpreted as numbers. Stata can do much more with numerical data than with string data. Moreover, string representation usually makes less efficient use of computer resources. destring will convert string variables to numeric.

A string variable may contain a "field" with numeric information. An example is an address variable that contains the street name followed by the house number. The Stata string functions can extract the relevant substring.

• Categorical variables stored as strings

Most statistical commands do not allow string variables. Moreover, string variables that take only a limited number of distinct values are an inefficient storage method. Use value-labeled numeric values instead. These are easily created with encode.

• Duplicate observations

See [D] **duplicates**.

• Observations that are always missing

Drop observations that are missing for all variables in *varlist* using the rownonmiss() egen function:

    egen nobs = rownonmiss(varlist)

    drop if nobs==0

Specify _all for *varlist* if only observations that are always missing should be dropped.

Saved results

codebook saves the following lists of variables with potential problems in r():

Macros
r(cons)	constant (or missing)
r(labelnotfound)	undefined value labeled
r(notlabeled)	value labeled but with unlabeled categories
r(str_type)	compressible
r(str_leading)	leading blanks
r(str_trailing)	trailing blanks
r(str_embedded)	embedded blanks
r(realdate)	noninteger dates

Methods and formulas

codebook is implemented as an ado-file.

Reference

Long, J. S. 2009. *The Workflow of Data Analysis Using Stata.* College Station, TX: Stata Press.

Also see

[D] **describe** — Describe data in memory or in file

[D] **ds** — List variables matching name patterns or other characteristics

[D] **inspect** — Display simple summary of data's attributes

[D] **labelbook** — Label utilities

[D] **notes** — Place notes in data

[D] **split** — Split string variables into parts

[U] **15 Saving and printing output—log files**

Title

> **collapse** — Make dataset of summary statistics

Syntax

> collapse *clist* $\big[$ *if* $\big]$ $\big[$ *in* $\big]$ $\big[$ *weight* $\big]$ $\big[$, *options* $\big]$

where *clist* is either

> $\big[$ *(stat)* $\big]$ *varlist* $\big[$ $\big[$ *(stat)* $\big]$ ... $\big]$
>
> $\big[$ *(stat)* $\big]$ *target_var=varname* $\big[$ *target_var=varname* ... $\big]$ $\big[$ $\big[$ *(stat)* $\big]$ ... $\big]$

or any combination of the *varlist* and *target_var* forms, and *stat* is one of

mean	means (default)	sepoisson	standard error of the mean, Poisson
median	medians		(sqrt(mean))
p1	1st percentile	sum	sums
p2	2nd percentile	rawsum	sums, ignoring optionally specified weight
...	3rd–49th percentiles	count	number of nonmissing observations
p50	50th percentile (same as median)	max	maximums
...	51st–97th percentiles	min	minimums
p98	98th percentile	iqr	interquartile range
p99	99th percentile	first	first value
sd	standard deviations	last	last value
semean	standard error of the mean	firstnm	first nonmissing value
	(sd/sqrt(n))	lastnm	last nonmissing value
sebinomial	standard error of the mean, binomial		
	(sqrt(p(1-p)/n))		

If *stat* is not specified, **mean** is assumed.

options	Description
Options	
by(*varlist*)	groups over which *stat* is to be calculated
cw	casewise deletion instead of all possible observations
fast	do not restore the original dataset should the user press *Break*; programmer's command

varlist and *varname* in *clist* may contain time-series operators; see [U] **11.4.4 Time-series varlists**.

aweights, fweights, iweights, and pweights are allowed; see [U] **11.1.6 weight**, and see *Weights* below. pweights may not be used with sd, semean, sebinomial, or sepoisson. iweights may not be used with semean, sebinomial, or sepoisson. aweights may not be used with sebinomial or sepoisson.

fast does not appear in the dialog box.

Examples:

```
. collapse age educ income, by(state)
. collapse (mean) age educ (median) income, by(state)
. collapse (mean) age educ income (median) medinc=income, by(state)
. collapse (p25) gpa [fw=number], by(year)
```

Menu

Data > Create or change data > Other variable-transformation commands > Make dataset of means, medians, etc.

Description

collapse converts the dataset in memory into a dataset of means, sums, medians, etc. *clist* must refer to numeric variables exclusively.

Note: See [D] **contract** if you want to collapse to a dataset of frequencies.

Options

> [Options]

by(*varlist*) specifies the groups over which the means, etc., are to be calculated. If this option is not specified, the resulting dataset will contain 1 observation. If it is specified, *varlist* may refer to either string or numeric variables.

cw specifies casewise deletion. If cw is not specified, all possible observations are used for each calculated statistic.

The following option is available with collapse but is not shown in the dialog box:

fast specifies that collapse not restore the original dataset should the user press *Break*. fast is intended for use by programmers.

Remarks

collapse takes the dataset in memory and creates a new dataset containing summary statistics of the original data. collapse adds meaningful variable labels to the variables in this new dataset. Because the syntax diagram for collapse makes using it appear more complicated than it is, collapse is best explained with examples.

Remarks are presented under the following headings:

> *Introductory examples*
> *Variablewise or casewise deletion*
> *Weights*
> *A final example*

Introductory examples

> Example 1

Consider the following artificial data on the grade-point average (gpa) of college students:

```
. use http://www.stata-press.com/data/r12/college
. describe
Contains data from http://www.stata-press.com/data/r12/college.dta
  obs:            12
  vars:            4                          3 Jan 2011 12:05
  size:          120
```

variable name	storage type	display format	value label	variable label
gpa	float	%9.0g		gpa for this year
hour	int	%9.0g		Total academic hours
year	int	%9.0g		1 = freshman, 2 = sophomore, 3 = junior, 4 = senior
number	int	%9.0g		number of students

```
Sorted by:  year
. list, sep(4)
```

	gpa	hour	year	number
1.	3.2	30	1	3
2.	3.5	34	1	2
3.	2.8	28	1	9
4.	2.1	30	1	4
5.	3.8	29	2	3
6.	2.5	30	2	4
7.	2.9	35	2	5
8.	3.7	30	3	4
9.	2.2	35	3	2
10.	3.3	33	3	3
11.	3.4	32	4	5
12.	2.9	31	4	2

To obtain a dataset containing the 25th percentile of gpa's for each year, we type

```
. collapse (p25) gpa [fw=number], by(year)
```

We used frequency weights.

Next we want to create a dataset containing the mean of gpa and hour for each year. We do not have to type (mean) to specify that we want the mean because the mean is reported by default.

```
. use http://www.stata-press.com/data/r12/college, clear
. collapse gpa hour [fw=number], by(year)
. list
```

	year	gpa	hour
1.	1	2.788889	29.44444
2.	2	2.991667	31.83333
3.	3	3.233333	32.11111
4.	4	3.257143	31.71428

Now we want to create a dataset containing the mean and median of gpa and hour, and we want the median of gpa and hour to be stored as variables medgpa and medhour, respectively.

```
. use http://www.stata-press.com/data/r12/college, clear
. collapse (mean) gpa hour (median) medgpa=gpa medhour=hour [fw=num], by(year)
. list
```

	year	gpa	hour	medgpa	medhour
1.	1	2.788889	29.44444	2.8	29
2.	2	2.991667	31.83333	2.9	30
3.	3	3.233333	32.11111	3.3	33
4.	4	3.257143	31.71428	3.4	32

Here we want to create a dataset containing a count of gpa and hour and the minimums of gpa and hour. The minimums of gpa and hour will be stored as variables mingpa and minhour, respectively.

```
. use http://www.stata-press.com/data/r12/college, clear
. collapse (count) gpa hour (min) mingpa=gpa minhour=hour [fw=num], by(year)
. list
```

	year	gpa	hour	mingpa	minhour
1.	1	18	18	2.1	28
2.	2	12	12	2.5	29
3.	3	9	9	2.2	30
4.	4	7	7	2.9	31

Now we replace the values of gpa in 3 of the observations with missing values.

```
. use http://www.stata-press.com/data/r12/college, clear
. replace gpa = . in 2/4
(3 real changes made, 3 to missing)
. list, sep(4)
```

	gpa	hour	year	number
1.	3.2	30	1	3
2.	.	34	1	2
3.	.	28	1	9
4.	.	30	1	4
5.	3.8	29	2	3
6.	2.5	30	2	4
7.	2.9	35	2	5
8.	3.7	30	3	4
9.	2.2	35	3	2
10.	3.3	33	3	3
11.	3.4	32	4	5
12.	2.9	31	4	2

If we now want to list the data containing the mean of gpa and hour for each year, collapse uses all observations on hour for year = 1, even though gpa is missing for observations 1–3.

```
. collapse gpa hour [fw=num], by(year)
. list
```

	year	gpa	hour
1.	1	3.2	29.44444
2.	2	2.991667	31.83333
3.	3	3.233333	32.11111
4.	4	3.257143	31.71428

If we repeat this process but specify the `cw` option, `collapse` ignores all observations that have missing values.

```
. use http://www.stata-press.com/data/r12/college, clear
. replace gpa = . in 2/4
(3 real changes made, 3 to missing)
. collapse (mean) gpa hour [fw=num], by(year) cw
. list
```

	year	gpa	hour
1.	1	3.2	30
2.	2	2.991667	31.83333
3.	3	3.233333	32.11111
4.	4	3.257143	31.71428

◁

> ## Example 2

We have individual-level data from a census in which each observation is a person. Among other variables, the dataset contains the numeric variables `age`, `educ`, and `income` and the string variable `state`. We want to create a 50-observation dataset containing the means of age, education, and income for each state.

```
. collapse age educ income, by(state)
```

The resulting dataset contains means because `collapse` assumes that we want means if we do not specify otherwise. To make this explicit, we could have typed

```
. collapse (mean) age educ income, by(state)
```

Had we wanted the mean for `age` and `educ` and the median for `income`, we could have typed

```
. collapse (mean) age educ (median) income, by(state)
```

or if we had wanted the mean for `age` and `educ` and both the mean and the median for `income`, we could have typed

```
. collapse (mean) age educ income (median) medinc=income, by(state)
```

This last dataset will contain three variables containing means—`age`, `educ`, and `income`—and one variable containing the median of income—`medinc`. Because we typed `(median) medinc=income`, Stata knew to find the median for income and to store those in a variable named `medinc`. This renaming convention is necessary in this example because a variable named `income` containing the mean is also being created.

◁

Variablewise or casewise deletion

▷ Example 3

Let's assume that in our census data, we have 25,000 persons for whom age is recorded but only 15,000 for whom income is recorded; that is, income is missing for 10,000 observations. If we want summary statistics for age and income, collapse will, by default, use all 25,000 observations when calculating the summary statistics for age. If we prefer that collapse use only the 15,000 observations for which income is not missing, we can specify the cw (casewise) option:

```
. collapse (mean) age income (median) medinc=income, by(state) cw
```

◁

Weights

collapse allows all four weight types; the default is aweights. Weight normalization affects only the sum, count, sd, semean, and sebinomial statistics.

Here are the definitions for count and sum with weights:

count:
    unweighted: _N, the number of physical observations
    aweight: _N, the number of physical observations
    fweight, iweight, pweight: $W = \sum w_j$, the sum of the user-specified weights
sum:
    unweighted: $\sum x_j$, the sum of the variable
    aweight: $\sum v_j x_j$; $v_j = (w_j$ normalized to sum to _N)
    fweight, iweight, pweight: $\sum w_j x_j$

The sd statistic with weights returns the bias-corrected standard deviation, which is based on the factor $\sqrt{N/(N-1)}$, where N is the number of observations. Statistics sd, semean, sebinomial, and sepoisson are not allowed with pweighted data. Otherwise, the statistic is changed by the weights through the computation of the count (N), as outlined above.

For instance, consider a case in which there are 25 physical observations in the dataset and a weighting variable that sums to 57. In the unweighted case, the weight is not specified, and $N = 25$. In the analytically weighted case, N is still 25; the scale of the weight is irrelevant. In the frequency-weighted case, however, $N = 57$, the sum of the weights.

The rawsum statistic with aweights ignores the weight, with one exception: observations with zero weight will not be included in the sum.

▷ Example 4

Using our same census data, suppose that instead of starting with individual-level data and aggregating to the state level, we started with state-level data and wanted to aggregate to the region level. Also assume that our dataset contains pop, the population of each state.

To obtain unweighted means and medians of age and income, by region, along with the total population, we could type

```
. collapse (mean) age income (median) medage=age medinc=income (sum) pop,
> by(region)
```

To obtain weighted means and medians of age and income, by region, along with the total population and using frequency weights, we could type

```
. collapse (mean) age income (median) medage=age medinc=income (count) pop
> [fweight=pop], by(region)
```

Note: Specifying (sum) pop would not have worked because that would have yielded the pop-weighted sum of pop. Specifying (count) age would have worked as well as (count) pop because count merely counts the number of nonmissing observations. The counts here, however, are frequency-weighted and equal the sum of pop.

To obtain the same mean and medians as above, but using analytic weights, we could type

```
. collapse (mean) age income (median) medage=age medinc=income (rawsum) pop
> [aweight=pop], by(region)
```

Note: Specifying (count) pop would not have worked because, with analytic weights, count would count numbers of physical observations. Specifying (sum) pop would not have worked because sum would calculate weighted sums (with a normalized weight). The rawsum function, however, ignores the weights and sums only the specified variable, with one exception: observations with zero weight will not be included in the sum. rawsum would have worked as the solution to all three cases.   ◁

A final example

▷ Example 5

We have census data containing information on each state's median age, marriage rate, and divorce rate. We want to form a new dataset containing various summary statistics, by region, of the variables:

```
. use http://www.stata-press.com/data/r12/census5, clear
(1980 Census data by state)
. describe
Contains data from http://www.stata-press.com/data/r12/census5.dta
  obs:          50                          1980 Census data by state
  vars:          7                          6 Apr 2011 15:43
  size:       1,700
```

| | storage | display | value | |
variable name	type	format	label	variable label
state	str14	%14s		State
state2	str2	%-2s		Two-letter state abbreviation
region	int	%8.0g	cenreg	Census region
pop	long	%10.0g		Population
median_age	float	%9.2f		Median age
marriage_rate	float	%9.0g		
divorce_rate	float	%9.0g		

```
Sorted by:  region
. collapse (median) median_age marriage divorce (mean) avgmrate=marriage
> avgdrate=divorce [aw=pop], by(region)
```

```
. list
```

	region	median~e	marria~e	divorc~e	avgmrate	avgdrate
1.	NE	31.90	.0080657	.0035295	.0081472	.0035359
2.	N Cntrl	29.90	.0093821	.0048636	.0096701	.004961
3.	South	29.60	.0112609	.0065792	.0117082	.0059439
4.	West	29.90	.0089093	.0056423	.0125199	.0063464

```
. describe

Contains data
  obs:            4                          1980 Census data by state
  vars:           6
  size:          88
```

variable name	storage type	display format	value label	variable label
region	int	%8.0g	cenreg	Census region
median_age	float	%9.2f		(p 50) median_age
marriage_rate	float	%9.0g		(p 50) marriage_rate
divorce_rate	float	%9.0g		(p 50) divorce_rate
avgmrate	float	%9.0g		(mean) marriage_rate
avgdrate	float	%9.0g		(mean) divorce_rate

```
Sorted by:  region
    Note:   dataset has changed since last saved
```

◁

Methods and formulas

collapse is implemented as an ado-file.

Acknowledgment

We thank David Roodman for writing collapse2, which inspired several features in collapse.

Also see

[D] **contract** — Make dataset of frequencies and percentages

[D] **egen** — Extensions to generate

[D] **statsby** — Collect statistics for a command across a by list

[R] **summarize** — Summary statistics

Title

> **compare** — Compare two variables

Syntax

> compare *varname*$_1$ *varname*$_2$ $\left[\,\textit{if}\,\right]$ $\left[\,\textit{in}\,\right]$

by is allowed; see [D] **by**.

Menu

Data > Data utilities > Compare two variables

Description

compare reports the differences and similarities between *varname*$_1$ and *varname*$_2$.

Remarks

▷ Example 1

One of the more useful accountings made by compare is the pattern of missing values:

```
. use http://www.stata-press.com/data/r12/fullauto
(Automobile Models)
. compare rep77 rep78
```

	count	minimum	difference average	maximum
rep77<rep78	16	-3	-1.3125	-1
rep77=rep78	43			
rep77>rep78	7	1	1	1
jointly defined	66	-3	-.2121212	1
rep77 missing only	3			
jointly missing	5			
total	74			

We see that both rep77 and rep78 are missing in 5 observations and that rep77 is also missing in 3 more observations.

◁

❑ Technical note

compare may be used with numeric variables, string variables, or both. When used with string variables, the summary of the differences (minimum, average, maximum) is not reported. When used with string and numeric variables, the breakdown by <, =, and > is also suppressed.

58

Stata does not normally attach any special meaning to the string ".", but some Stata users use the string "." to mean missing value.

❑

Methods and formulas

compare is implemented as an ado-file.

Also see

[D] **cf** — Compare two datasets

[D] **codebook** — Describe data contents

[D] **inspect** — Display simple summary of data's attributes

Title

compress — Compress data in memory

Syntax

compress [*varlist*]

Menu

Data > Data utilities > Optimize variable storage

Description

compress attempts to reduce the amount of memory used by your data.

Remarks

compress reduces the size of your dataset by considering demoting

doubles	to	longs, ints, or bytes
floats	to	ints or bytes
longs	to	ints or bytes
ints	to	bytes
strings	to	shorter strings

See [D] **data types** for an explanation of these storage types.

compress leaves your data logically unchanged but (probably) appreciably smaller. compress never makes a mistake, results in loss of precision, or hacks off strings.

▷ Example 1

If you do not specify a *varlist*, compress considers demoting all the variables in your dataset, so typing compress by itself is usually enough:

```
. use http://www.stata-press.com/data/r12/compxmpl
. compress
mpg was float now byte
price was long now int
yenprice was double now long
weight was double now int
make was str26 now str17

. _
```

If there are no compression possibilities, compress does nothing. For instance, typing compress again results in

```
. compress

. _
```

◁

Also see

[D] **data types** — Quick reference for data types

[D] **recast** — Change storage type of variable

Title

> **contract** — Make dataset of frequencies and percentages

Syntax

contract *varlist* [*if*] [*in*] [*weight*] [, *options*]

options	Description
Options	
freq(*newvar*)	name of frequency variable; default is _freq
cfreq(*newvar*)	create cumulative frequency variable
percent(*newvar*)	create percentage variable
cpercent(*newvar*)	create cumulative percentage variable
float	generate percentage variables as type float
format(*format*)	display format for new percentage variables; default is format(%8.2f)
zero	include combinations with frequency zero
nomiss	drop observations with missing values

fweights are allowed; see [U] **11.1.6 weight**.

Menu

Data > Create or change data > Other variable-transformation commands > Make dataset of frequencies

Description

contract replaces the dataset in memory with a new dataset consisting of all combinations of *varlist* that exist in the data and a new variable that contains the frequency of each combination.

Options

> Options

freq(*newvar*) specifies a name for the frequency variable. If not specified, _freq is used.

cfreq(*newvar*) specifies a name for the cumulative frequency variable. If not specified, no cumulative frequency variable is created.

percent(*newvar*) specifies a name for the percentage variable. If not specified, no percentage variable is created.

cpercent(*newvar*) specifies a name for the cumulative percentage variable. If not specified, no cumulative percentage variable is created.

float specifies that the percentage variables specified by percent() and cpercent() will be generated as variables of type float. If float is not specified, these variables will be generated as variables of type double. All generated variables are compressed to the smallest storage type possible without loss of precision; see [D] **compress**.

61

format(*format*) specifies a display format for the generated percentage variables specified by percent() and cpercent(). If format() is not specified, these variables will have the display format %8.2f.

zero specifies that combinations with frequency zero be included.

nomiss specifies that observations with missing values on any variable in *varlist* be dropped. If nomiss is not specified, all observations possible are used.

Remarks

contract takes the dataset in memory and creates a new dataset containing all combinations of *varlist* that exist in the data and a new variable that contains the frequency of each combination.

Sometimes you may want to collapse a dataset into frequency form. Several observations that have identical values on one or more variables will be replaced by one such observation, together with the frequency of the corresponding set of values. For example, in certain generalized linear models, the frequency of some combination of values is the response variable, so you would need to produce that response variable. The set of covariate values associated with each frequency is sometimes called a covariate class or covariate pattern. Such collapsing is reversible for the variables concerned, because the original dataset can be reconstituted by using expand (see [D] **expand**) with the variable containing the frequencies of each covariate class.

▷ Example 1

Suppose that we wish to collapse the auto dataset to a set of frequencies of the variables rep78, which takes values 1, 2, 3, 4, and 5, and foreign, which takes values labeled 'Domestic' and 'Foreign'.

```
. use http://www.stata-press.com/data/r12/auto
(1978 Automobile Data)
. contract rep78 foreign
. list
```

	rep78	foreign	_freq
1.	1	Domestic	2
2.	2	Domestic	8
3.	3	Domestic	27
4.	3	Foreign	3
5.	4	Domestic	9
6.	4	Foreign	9
7.	5	Domestic	2
8.	5	Foreign	9
9.	.	Domestic	4
10.	.	Foreign	1

By default, contract uses the variable name _freq for the new variable that contains the frequencies. If _freq is in use, you will be reminded to specify a new variable name via the freq() option.

Specifying the zero option requests that combinations with frequency zero also be listed.

```
. use http://www.stata-press.com/data/r12/auto, clear
(1978 Automobile Data)
. contract rep78 foreign, zero
```

```
. list
```

	rep78	foreign	_freq
1.	1	Domestic	2
2.	1	Foreign	0
3.	2	Domestic	8
4.	2	Foreign	0
5.	3	Domestic	27
6.	3	Foreign	3
7.	4	Domestic	9
8.	4	Foreign	9
9.	5	Domestic	2
10.	5	Foreign	9
11.	.	Domestic	4
12.	.	Foreign	1

◁

Methods and formulas

contract is implemented as an ado-file.

Acknowledgments

contract was written by Nicholas J. Cox of Durham University (Cox 1998). The cfreq(), percent(), cpercent(), float, and format() options were written by Roger Newson, Imperial College London.

Reference

Cox, N. J. 1998. dm59: Collapsing datasets to frequencies. *Stata Technical Bulletin* 44: 2–3. Reprinted in *Stata Technical Bulletin Reprints*, vol. 8, pp. 20–21. College Station, TX: Stata Press.

Also see

[D] **expand** — Duplicate observations

[D] **collapse** — Make dataset of summary statistics

[D] **duplicates** — Report, tag, or drop duplicate observations

Title

> **copy** — Copy file from disk or URL

Syntax

> copy *filename*$_1$ *filename*$_2$ $\left[\; , \; options \right]$

filename$_1$ may be a filename or a URL. *filename*$_2$ may be the name of a file or a directory. If *filename*$_2$ is a directory name, *filename*$_1$ will be copied to that directory. *filename*$_2$ may *not* be a URL.

Note: Double quotes may be used to enclose the filenames, and the quotes must be used if the filename contains embedded blanks.

options	Description
<u>p</u>ublic	make *filename*$_2$ readable by all
<u>t</u>ext	interpret *filename*$_1$ as text file and translate to native text format
replace	may overwrite *filename*$_2$

replace does not appear in the dialog box.

Description

copy copies *filename*$_1$ to *filename*$_2$.

Options

public specifies that *filename*$_2$ be readable by everyone; otherwise, the file will be created according to the default permissions of your operating system.

text specifies that *filename*$_1$ be interpreted as a text file and be translated to the native form of text files on your computer. Computers differ on how end-of-line is recorded: Unix systems record one line-feed character, Windows computers record a carriage-return/line-feed combination, and Mac computers record just a carriage return. text specifies that *filename*$_1$ be examined to determine how it has end-of-line recorded and that the line-end characters be switched to whatever is appropriate for your computer when the copy is made.

There is no reason to specify text when copying a file already on your computer to a different location because the file would already be in your computer's format.

Do not specify text unless you know that the file is a text file; if the file is binary and you specify text, the copy will be useless. Most word processors produce binary files, not text files. The term *text*, as it is used here, specifies a particular ASCII way of recording textual information.

When other parts of Stata read text files, they do not care how lines are terminated, so there is no reason to translate end-of-line characters on that score. You specify text because you may want to look at the file with other software.

The following option is available with copy but is not shown in the dialog box:

replace specifies that *filename*$_2$ be replaced if it already exists.

Remarks

Examples:

Windows:

```
. copy orig.dta newcopy.dta
. copy mydir\orig.dta .
. copy orig.dta ../../
. copy "my document" "copy of document"
. copy ..\mydir\doc.txt document\doc.tex
. copy http://www.stata.com/examples/simple.dta simple.dta
. copy http://www.stata.com/examples/simple.txt simple.txt, text
```

Mac and Unix:

```
. copy orig.dta newcopy.dta
. copy mydir/orig.dta .
. copy orig.dta ../../
. copy "my document" "copy of document"
. copy ../mydir/doc.txt document/doc.tex
. copy http://www.stata.com/examples/simple.dta simple.dta
. copy http://www.stata.com/examples/simple.txt simple.txt, text
```

Also see

[D] **cd** — Change directory

[D] **dir** — Display filenames

[D] **erase** — Erase a disk file

[D] **mkdir** — Create directory

[D] **rmdir** — Remove directory

[D] **shell** — Temporarily invoke operating system

[D] **type** — Display contents of a file

[U] **11.6 Filenaming conventions**

Title

> **corr2data** — Create dataset with specified correlation structure

Syntax

> corr2data *newvarlist* [, *options*]

options	Description
Main	
clear	replace the current dataset
double	generate variable type as double; default is float
n(#)	# of observations to be generated; default is current number
sds(*vector*)	standard deviations of generated variables
corr(*matrix* | *vector*)	correlation matrix
cov(*matrix* | *vector*)	covariance matrix
cstorage(full)	correlation/covariance structure is stored as a symmetric $k \times k$ matrix
cstorage(lower)	correlation/covariance structure is stored as a lower triangular matrix
cstorage(upper)	correlation/covariance structure is stored as an upper triangular matrix
forcepsd	force the covariance/correlation matrix to be positive semidefinite
means(*vector*)	means of generated variables; default is means(0)
Options	
seed(#)	seed for random-number generator

Menu

Data > Create or change data > Other variable-creation commands > Create dataset with specified correlation

Description

corr2data adds new variables with specified covariance (correlation) structure to the existing dataset or creates a new dataset with a specified covariance (correlation) structure. Singular covariance (correlation) structures are permitted. The purpose of this is to allow you to perform analyses from summary statistics (correlations/covariances and maybe the means) when these summary statistics are all you know and summary statistics are sufficient to obtain results. For example, these summary statistics are sufficient for performing analysis of t tests, variance, principal components, regression, and factor analysis. The recommended process is

. clear	(clear memory)
. corr2data ..., n(#) cov(...) ...	(create artificial data)
. regress ...	(use artificial data appropriately)

However, for factor analyses and principal components, the commands factormat and pcamat allow you to skip the step of using corr2data; see [MV] **factor** and [MV] **pca**.

The data created by corr2data are artificial; they are not the original data, and it is not a sample from an underlying population with the summary statistics specified. See [D] **drawnorm** if you want to generate a random sample. In a sample, the summary statistics will differ from the population values and will differ from one sample to the next.

The dataset corr2data creates is suitable for one purpose only: performing analyses when all that is known are summary statistics and those summary statistics are sufficient for the analysis at hand. The artificial data tricks the analysis command into producing the desired result. The analysis command, being by assumption only a function of the summary statistics, extracts from the artificial data the summary statistics, which are the same summary statistics you specified, and then makes its calculation based on those statistics.

If you doubt whether the analysis depends only on the specified summary statistics, you can generate different artificial datasets by using different seeds of the random-number generator (see the seed() option below) and compare the results, which should be the same within rounding error.

Options

⌐ Main ⌐

clear specifies that it is okay to replace the dataset in memory, even though the current dataset has not been saved on disk.

double specifies that the new variables be stored as Stata doubles, meaning 8-byte reals. If double is not specified, variables are stored as floats, meaning 4-byte reals. See [D] **data types**.

n(#) specifies the number of observations to be generated; the default is the current number of observations. If n(#) is not specified or is the same as the current number of observations, corr2data adds the new variables to the existing dataset; otherwise, corr2data replaces the dataset in memory.

sds(*vector*) specifies the standard deviations of the generated variables. sds() may not be specified with cov().

corr(*matrix* | *vector*) specifies the correlation matrix. If neither corr() nor cov() is specified, the default is orthogonal data.

cov(*matrix* | *vector*) specifies the covariance matrix. If neither corr() nor cov() is specified, the default is orthogonal data.

cstorage(full | lower | upper) specifies the storage mode for the correlation or covariance structure in corr() or cov(). The following storage modes are supported:

full specifies that the correlation or covariance structure is stored (recorded) as a symmetric $k \times k$ matrix.

lower specifies that the correlation or covariance structure is recorded as a lower triangular matrix. With k variables, the matrix should have $k(k+1)/2$ elements in the following order:

$$C_{11} \; C_{21} \; C_{22} \; C_{31} \; C_{32} \; C_{33} \; \ldots \; C_{k1} \; C_{k2} \; \ldots \; C_{kk}$$

upper specifies that the correlation or covariance structure is recorded as an upper triangular matrix. With k variables, the matrix should have $k(k+1)/2$ elements in the following order:

$$C_{11} \; C_{12} \; C_{13} \; \ldots \; C_{1k} \; C_{22} \; C_{23} \; \ldots C_{2k} \; \ldots \; C_{(k-1k-1)} \; C_{(k-1k)} \; C_{kk}$$

Specifying cstorage(full) is optional if the matrix is square. cstorage(lower) or cstorage(upper) is required for the vectorized storage methods. See *Storage modes for correlation and covariance matrices* in [D] **drawnorm** for examples.

forcepsd modifies the matrix *C* to be positive semidefinite (psd) and to thus be a proper covariance matrix. If *C* is not positive semidefinite, it will have negative eigenvalues. By setting the negative eigenvalues to 0 and reconstructing, we obtain the least-squares positive-semidefinite approximation to *C*. This approximation is a singular covariance matrix.

means(*vector*) specifies the means of the generated variables. The default is means(0).

seed(*#*) specifies the seed of the random-number generator used to generate data. *#* defaults to 0. The random numbers generated inside corr2data do not affect the seed of the standard random-number generator.

Remarks

corr2data is designed to enable analyses of correlation (covariance) matrices by commands that expect variables rather than a correlation (covariance) matrix. corr2data creates variables with exactly the correlation (covariance) that you want to analyze. Apart from means and covariances, all aspects of the data are meaningless. Only analyses that depend on the correlations (covariances) and means produce meaningful results. Thus you may perform a paired *t* test ([R] **ttest**) or an ordinary regression analysis ([R] **regress**), etc.

If you are not sure that a statistical result depends only on the specified summary statistics and not on other aspects of the data, you can generate different datasets, each having the same summary statistics but other different aspects, by specifying the seed() option. If the statistical results differ beyond what is attributable to roundoff error, then using corr2data is inappropriate.

▷ Example 1

We first run a regression using the auto dataset.

```
. use http://www.stata-press.com/data/r12/auto
(1978 Automobile Data)

. regress weight length trunk
```

Source	SS	df	MS				Number of obs =	74
							F(2, 71) =	303.95
Model	39482774.4	2	19741387.2				Prob > F =	0.0000
Residual	4611403.95	71	64949.3513				R-squared =	0.8954
							Adj R-squared =	0.8925
Total	44094178.4	73	604029.841				Root MSE =	254.85

weight	Coef.	Std. Err.	t	P>|t|	[95% Conf. Interval]	
length	33.83435	1.949751	17.35	0.000	29.94666	37.72204
trunk	-5.83515	10.14957	-0.57	0.567	-26.07282	14.40252
_cons	-3258.84	283.3547	-11.50	0.000	-3823.833	-2693.846

Suppose that, for some reason, we no longer have the auto dataset. Instead, we know the means and covariance matrices of weight, length, and trunk, and we want to do the same regression again. The matrix of means is

```
. mat list M

M[1,3]
            weight      length       trunk
_cons    3019.4595   187.93243   13.756757
```

and the covariance matrix is

```
. mat list V

symmetric V[3,3]
            weight      length       trunk
weight    604029.84
length    16370.922   495.78989
 trunk    2234.6612   69.202518   18.296187
```

To do the regression analysis in Stata, we need to create a dataset that has the specified correlation structure.

```
. corr2data x y z, n(74) cov(V) means(M)

. regress x y z

    Source |       SS       df       MS              Number of obs =      74
-----------+------------------------------           F(  2,    71) =  303.95
     Model | 39482773.3        2  19741386.6         Prob > F      =  0.0000
  Residual | 4611402.75       71  64949.3345         R-squared     =  0.8954
-----------+------------------------------           Adj R-squared =  0.8925
     Total |   44094176       73  604029.809         Root MSE      =  254.85

-----------+----------------------------------------------------------------
         x |      Coef.   Std. Err.      t    P>|t|     [95% Conf. Interval]
-----------+----------------------------------------------------------------
         y |   33.83435   1.949751    17.35   0.000     29.94666    37.72204
         z |  -5.835155   10.14957    -0.57   0.567    -26.07282    14.40251
     _cons |   -3258.84   283.3546   -11.50   0.000    -3823.833   -2693.847
----------------------------------------------------------------------------
```

The results from the regression based on the generated data are the same as those based on the real data.

◁

Methods and formulas

corr2data is implemented as an ado-file.

Two steps are involved in generating the desired dataset. The first step is to generate a zero-mean, zero-correlated dataset. The second step is to apply the desired correlation structure and the means to the zero-mean, zero-correlated dataset. In both steps, we take into account that, given any matrix $\mathbf{A}$ and any vector of variables $\mathbf{X}$, $\mathrm{Var}(\mathbf{A}'\mathbf{X}) = \mathbf{A}'\mathrm{Var}(\mathbf{X})\mathbf{A}$.

Reference

Cappellari, L., and S. P. Jenkins. 2006. Calculation of multivariate normal probabilities by simulation, with applications to maximum simulated likelihood estimation. *Stata Journal* 6: 156–189.

Also see

[D] **drawnorm** — Draw sample from multivariate normal distribution

[D] **data types** — Quick reference for data types

Title

> **count** — Count observations satisfying specified conditions

Syntax

<u>coun</u>t [*if*] [*in*]

by is allowed; see [D] **by**.

Menu

Data > Data utilities > Count observations satisfying condition

Description

count counts the number of observations that satisfy the specified conditions. If no conditions are specified, count displays the number of observations in the data.

Remarks

count may strike you as an almost useless command, but it can be one of Stata's handiest.

▷ Example 1

How many times have you obtained a statistical result and then asked yourself how it was possible? You think a moment and then mutter aloud, "Wait a minute. Is income ever *negative* in these data?" or "Is sex ever equal to *3*?" count can quickly answer those questions:

```
. use http://www.stata-press.com/data/r12/countxmpl
(1980 Census data by state)
. count
  641
. count if income<0
    0
. count if sex==3
    1
. by division: count if sex==3

-> division = New England
    0

-> division = Mountain
    0

-> division = Pacific
    1
```

We have 641 observations. income is never negative. sex, however, takes on the value 3 once. When we decompose the count by division, we see that it takes on that odd value in the Pacific division.

◁

Saved results

count saves the following in r():

Scalars
    r(N)   number of observations

Also see

[R] **tabulate oneway** — One-way tables of frequencies

Title

> **cross** — Form every pairwise combination of two datasets

Syntax

> cross using *filename*

Menu

Data > Combine datasets > Form every pairwise combination of two datasets

Description

cross forms every pairwise combination of the data in memory with the data in *filename*. If *filename* is specified without a suffix, .dta is assumed.

Remarks

This command is rarely used; also see [D] **joinby**, [D] **merge**, and [D] **append**.

Crossing refers to merging two datasets in every way possible. That is, the first observation of the data in memory is merged with every observation of *filename*, followed by the second, and so on. Thus the result will have $N_1 N_2$ observations, where N_1 and N_2 are the number of observations in memory and in *filename*, respectively.

Typically, the datasets will have no common variables. If they do, such variables will take on only the values of the data in memory.

▷ Example 1

We wish to form a dataset containing all combinations of three age categories and two sexes to serve as a stub. The three age categories are 20, 30, and 40. The two sexes are male and female:

```
. input str6 sex
            sex
  1. male
  2. female
  3. end
. save sex
file sex.dta saved
. drop _all
. input agecat
          agecat
  1. 20
  2. 30
  3. 40
  4. end
. cross using sex
```

```
. list
```

	agecat	sex
1.	20	male
2.	30	male
3.	40	male
4.	20	female
5.	30	female
6.	40	female

◁

Methods and formulas

cross is implemented as an ado-file.

References

Baum, C. F. 2009. *An Introduction to Stata Programming*. College Station, TX: Stata Press.

Franklin, C. H. 2006. Stata tip 29: For all times and all places. *Stata Journal* 6: 147–148.

Also see

[D] **save** — Save Stata dataset

[D] **append** — Append datasets

[D] **fillin** — Rectangularize dataset

[D] **joinby** — Form all pairwise combinations within groups

[D] **merge** — Merge datasets

Title

> **data types** — Quick reference for data types

Description

This entry provides a quick reference for data types allowed by Stata. See [U] **12 Data** for details.

Remarks

Storage type	Minimum	Maximum	Closest to 0 without being 0	Bytes
byte	-127	100	± 1	1
int	$-32{,}767$	32,740	± 1	2
long	$-2{,}147{,}483{,}647$	2,147,483,620	± 1	4
float	$-1.70141173319 \times 10^{38}$	$1.70141173319 \times 10^{38}$	$\pm 10^{-38}$	4
double	$-8.9884656743 \times 10^{307}$	$8.9884656743 \times 10^{307}$	$\pm 10^{-323}$	8

Precision for float is 3.795×10^{-8}.

Precision for double is 1.414×10^{-16}.

String storage type	Maximum length	Bytes
str1	1	1
str2	2	2
...	.	.
...	.	.
...	.	.
str244	244	244

Also see

[D] **compress** — Compress data in memory

[D] **destring** — Convert string variables to numeric variables and vice versa

[D] **encode** — Encode string into numeric and vice versa

[D] **format** — Set variables' output format

[D] **recast** — Change storage type of variable

[U] **12.2.2 Numeric storage types**

[U] **12.4.4 String storage types**

[U] **12.5 Formats: Controlling how data are displayed**

[U] **13.11 Precision and problems therein**

Title

datasignature — Determine whether data have changed

Syntax

```
datasignature

datasignature set [ , reset ]

datasignature confirm [ , strict ]

datasignature report

datasignature set, saving( filename[ , replace ]) [reset]

datasignature confirm using filename [ , strict ]

datasignature report using filename

datasignature clear
```

Menu

Data > Other utilities > Manage data signature

Description

These commands calculate, display, save, and verify checksums of the data, which taken together form what is called a *signature*. An example signature is 162:11(12321):2725060400:4007406597. That signature is a function of the values of the variables and their names, and thus the signature can be used later to determine whether a dataset has changed.

datasignature without arguments calculates and displays the signature of the data in memory.

datasignature set does the same, and it stores the signature as a characteristic in the dataset. You should save the dataset afterward so that the signature becomes a permanent part of the dataset.

datasignature confirm verifies that, were the signature recalculated this instant, it would match the one previously set. datasignature confirm displays an error message and returns a nonzero return code if the signatures do not match.

datasignature report displays a full report comparing the previously set signature to the current one.

In the above, the signature is stored in the dataset and accessed from it. The signature can also be stored in a separate, small file.

datasignature set, saving(*filename*) calculates and displays the signature and, in addition to storing it as a characteristic in the dataset, also saves the signature in *filename*.

datasignature confirm using *filename* verifies that the current signature matches the one stored in *filename*.

datasignature report using *filename* displays a full report comparing the current signature with the one stored in *filename*.

In all the above, if *filename* is specified without an extension, .dtasig is assumed.

datasignature clear clears the signature, if any, stored in the characteristics of the dataset in memory.

Options

reset is used with datasignature set. It specifies that even though you have previously set a signature, you want to erase the old signature and replace it with the current one.

strict is for use with datasignature confirm. It specifies that, in addition to requiring that the signatures match, you also wish to require that the variables be in the same order and that no new variables have been added to the dataset. (If any variables were dropped, the signatures would not match.)

saving(*filename* [, replace]) is used with datasignature set. It specifies that, in addition to storing the signature in the dataset, you want a copy of the signature saved in a separate file. If *filename* is specified without a suffix, .dtasig is assumed. The replace suboption allows *filename* to be replaced if it already exists.

Remarks

Remarks are presented under the following headings:

> *Using datasignature interactively*
> > *Example 1: Verification at a distance*
> > *Example 2: Protecting yourself from yourself*
> > *Example 3: Working with assistants*
> > *Example 4: Working with shared data*
> *Using datasignature in do-files*
> *Interpreting data signatures*
> *The logic of data signatures*

Using datasignature interactively

datasignature is useful in the following cases:

1. You and a coworker, separated by distance, have both received what is claimed to be the same dataset. You wish to verify that it is.

2. You work interactively and realize that you could mistakenly modify your data. You wish to guard against that.

3. You want to give your dataset to an assistant to improve the labels and the like. You wish to verify that the data returned to you are the same data.

4. You work with an important dataset served on a network drive. You wish to verify that others have not changed it.

Example 1: Verification at a distance

You load the data and type

```
. datasignature
74:12(71728):3831085005:1395876116
```

Your coworker does the same with his or her copy. You compare the two signatures.

Example 2: Protecting yourself from yourself

You load the data and type

```
. datasignature set
74:12(71728):3831085005:1395876116     (data signature set)
. save, replace
```

From then on, you periodically type

```
. datasignature confirm
(data unchanged since 19feb2011 14:24)
```

One day, however, you check and see the message:

```
. datasignature confirm
(data unchanged since 19feb2011 14:24, except 2 variables have been added)
```

You can find out more by typing

```
. datasignature report
(data signature set on Monday 19feb2011 14:24)
```

Data signature summary

```
1. Previous data signature     74:12(71728):3831085005:1395876116
2. Same data signature today   (same as 1)
3. Full data signature today   74:14(113906):1142538197:2410350265
```

Comparison of current data with previously set data signature

variables	number	notes
original # of variables	12	(values unchanged)
added variables	2	(1)
dropped variables	0	
resulting # of variables	14	

(1) Added variables are agesquared logincome.

You could now either drop the added variables or decide to incorporate them:

```
. datasignature set
data signature already set -- specify option -reset-
r(198)
. datasignature set, reset
74:14(113906):1142538197:2410350265      (data signature reset)
```

Concerning the detailed report, three data signatures are reported: 1) the stored signature, 2) the signature that would be calculated today on the basis of the same variables in their original order, and 3) the signature that would be calculated today on the basis of all the variables and in their current order.

`datasignature confirm` knew that new variables had been added because signature 1 was equal to signature 2. If some variables had been dropped, however, `datasignature confirm` would not be able to determine whether the remaining variables had changed.

Example 3: Working with assistants

You give your dataset to an assistant to have variable labels and the like added. You wish to verify that the returned data are the same data.

Saving the signature with the dataset is inadequate here. Your assistant, having your dataset, could change both your data and the signature and might even do that in a desire to be helpful. The solution is to save the signature in a separate file that you do not give to your assistant:

```
. datasignature set, saving(mycopy)
  74:12(71728):3831085005:1395876116        (data signature set)
  (file mycopy.dtasig saved)
```

You keep file `mycopy.dtasig`. When your assistant returns the dataset to you, you use it and compare the current signature to what you have stored in `mycopy.dtasig`:

```
. datasignature confirm using mycopy
  (data unchanged since 19feb2011 15:05)
```

By the way, the signature is a function of the following:

- The number of observations and number of variables in the data

- The values of the variables

- The names of the variables

- The order in which the variables occur in the dataset

- The storage types of the individual variables

The signature is not a function of variable labels, value labels, notes, and the like.

Example 4: Working with shared data

You work on a dataset served on a network drive, which means that others could change the data. You wish to know whether this occurs.

The solution here is the same as working with an assistant: you save the signature in a separate, private file on your computer,

```
. datasignature set, saving(private)
  74:12(71728):3831085005:1395876116        (data signature set)
  (file private.dtasig saved)
```

and then you periodically check the signature by typing

```
. datasignature confirm using private
  (data unchanged since 15mar2011 11:22)
```

Using datasignature in do-files

`datasignature confirm` aborts with error if the signatures do not match:

```
. datasignature confirm
data have changed since 19feb2011 15:05
r(9);
```

This means that, if you use `datasignature confirm` in a do-file, execution of the do-file will be stopped if the data have changed.

You may want to specify the `strict` option. `strict` adds two more requirements: that the variables be in the same order and that no new variables have been added. Without `strict`, these are not considered errors:

```
. datasignature confirm
  (data unchanged since 19feb2011 15:22)
. datasignature confirm, strict
  (data unchanged since 19feb2011 15:05, but order of variables has changed)
r(9);
```

and

```
. datasignature confirm
  (data unchanged since 19feb2011 15:22, except 1 variable has been added)
. datasignature confirm, strict
  (data unchanged since 19feb2011 15:22, except 1 variable has been added)
r(9);
```

If you keep logs of your analyses, issuing `datasignature` or `datasignature confirm` immediately after loading each dataset is a good idea. This way, you have a permanent record that you can use for comparison.

Interpreting data signatures

An example signature is `74:12(71728):3831085005:1395876116`. The components are

1. 74, the number of observations;

2. 12, the number of variables;

3. 71728, a checksum function of the variable names and the order in which they occur; and

4. 3831085005 and 1395876116, checksum functions of the values of the variables, calculated two different ways.

Two signatures are equal only if all their components are equal.

Two different datasets will probably not have the same signature, and it is even more unlikely that datasets containing similar values will have equal signatures. There are two data checksums, but do not read too much into that. If either data checksum changes, even just a little, the data have changed. Whether the change in the checksum is large or small—or in one, the other, or both—signifies nothing.

The logic of data signatures

The components of a data signature are known as checksums. The checksums are many-to-one mappings of the data onto the integers. Let's consider the checksums of `auto.dta` carefully.

The data portion of `auto.dta` contains 38,184 bytes. There are 256^{38184} such datasets or, equivalently, 2^{305472}. The first checksum has 2^{48} possible values, and it can be proven that those values are equally distributed over the 2^{305472} datasets. Thus there are $2^{305472}/2^{48} - 1 = 2^{305424} - 1$ datasets that have the same first checksum value as `auto.dta`. The same can be said for the second checksum. It would be difficult to prove, but we believe that the two checksums are conditionally independent, being based on different bit shifts and bit shuffles of the same data. Of the $2^{305424} - 1$ datasets that have the same first checksum as `auto.dta`, the second checksum should be equally distributed over them. Thus there are about $2^{305376} - 1$ datasets with the same first and second checksums as `auto.dta`.

Now let's consider those $2^{305376} - 1$ other datasets. Most of them look nothing like `auto.dta`. The checksum formulas guarantee that a change of one variable in 1 observation will lead to a change in the calculated result if the value changed is stored in 4 or fewer bytes, and they nearly guarantee it in other cases. When it is not guaranteed, the change cannot be subtle—"Chevrolet" will have to change to binary junk, or a double-precision 1 to $-6.476678983751e+301$, and so on. The change will be easily detected if you `summarize` your data and just glance at the minimums and maximums. If the data look at all like `auto.dta`, which is unlikely, they will look like a corrupted version.

More interesting are offsetting changes across observations. For instance, can you change one variable in 1 observation and make an offsetting change in another observation so that, taken together, they will go undetected? You can fool one of the checksums, but fooling both of them simultaneously will prove difficult. The basic rule is that the more changes you make, the easier it is to create a dataset with the same checksums as `auto.dta`, but by the time you've done that, the data will look nothing like `auto.dta`.

Saved results

`datasignature` without arguments and `datasignature set` save the following in `r()`:

Macros
    r(datasignature)   the signature

`datasignature confirm` saves the following in `r()`:

Scalars
    r(k_added)   number of variables added
Macros
    r(datasignature)   the signature

`datasignature confirm` aborts execution if the signatures do not match and so then returns nothing except a return code of 9.

`datasignature report` saves the following in `r()`:

Scalars
    r(datetime)   %tc date–time when set
    r(changed)   . if r(k_dropped) $\neq$ 0, otherwise
                               0 if data have not changed, 1 if data have changed
    r(reordered)   1 if variables reordered, 0 if not reordered,
                               . if r(k_added) $\neq$ 0 | r(k_dropped) $\neq$ 0
    r(k_original)   number of original variables
    r(k_added)   number of added variables
    r(k_dropped)   number of dropped variables
Macros
    r(origdatasignature)   original signature
    r(curdatasignature)   current signature on same variables, if it can be calculated
    r(fulldatasignature)   current full-data signature
    r(varsadded)   variable names added
    r(varsdropped)   variable names dropped

`datasignature clear` saves nothing in `r()` but does clear it.

`datasignature set` stores the signature in the following characteristics:

Characteristic
    _dta[datasignature_si]   signature
    _dta[datasignature_dt]   %tc date–time when set in %21x format
    _dta[datasignature_vl1]   part 1, original variables
    _dta[datasignature_vl2]   part 2, original variables, if necessary
    etc.

To access the original variables stored in _dta[datasignature_vl1], etc., from an ado-file, code

```
mata: ado_fromlchar("vars", _dta", "datasignature_vl")
```

Thereafter, the original variable list would be found in 'vars'.

Methods and formulas

datasignature is implemented using _datasignature; see [P] **_datasignature**.

Reference

Gould, W. W. 2006. Stata tip 35: Detecting whether data have changed. *Stata Journal* 6: 428–429.

Also see

[P] **_datasignature** — Determine whether data have changed

[P] **signestimationsample** — Determine whether the estimation sample has changed

Title

> **datetime** — Date and time values and variables

Syntax

Syntax is presented under the following headings:

Types of dates and their human readable forms (HRFs)
Stata internal form (SIF)
HRF-to-SIF conversion functions
Displaying SIFs in HRF
Building SIFs from components
SIF-to-SIF conversion
Extracting time-of-day components from SIFs
Extracting date components from SIFs
Conveniently typing SIF values
Obtaining and working with durations
Using dates and times from other software

Also see

[D] **datetime translation**	String to numeric date translation functions
[D] **datetime display formats**	Display formats for dates and times

Types of dates and their human readable forms (HRFs)

Date type	Examples of HRFs
datetime	20jan2010 09:15:22.120
date	20jan2010, 20/01/2010, ...
weekly date	2010w2
monthly date	2010m1
quarterly date	2010q1
half-yearly date	2010h1
yearly date	2010

The styles of the HRFs in the table above are merely examples. Perhaps you prefer 2010.01.20; Jan. 20, 2010; 2010-1; etc.

With the exception of yearly dates, HRFs are usually stored in string variables. If you are reading raw data, read the HRFs into strings.

HRFs are not especially useful except for reading by humans, and thus Stata provides another way of recording dates called Stata internal form (SIF). You can convert HRF dates to SIF.

Stata internal form (SIF)

The numeric values in the table below are equivalent to the string values in the table in the previous section.

SIF type	Examples in SIF	Units
datetime/c	1,479,597,200,000	milliseconds since 01jan1960 00:00:00.000, assuming 86,400 s/day
datetime/C	1,479,596,223,000	milliseconds since 01jan1960 00:00:00.000, adjusted for leap seconds*
date	18,282	days since 01jan1960 (01jan1960 = 0)
weekly date	2,601	weeks since 1960w1
monthly date	600	months since 1960m1
quarterly date	58	quarters since 1960q1
half-yearly date	100	half-years since 1960h1
yearly date	2010	years since 0000

* SIF datetime/C is equivalent to coordinated universal time (UTC). In UTC, leap seconds are periodically inserted because the length of the mean solar day is slowly increasing. See *Why there are two SIF datetime encodings* in [D] **datetime translation**.

SIF values are stored as regular Stata numeric variables.

You can convert HRFs into SIFs by using HRF-to-SIF conversion functions; see the next section, called *HRF-to-SIF conversion functions*.

You can make the numeric SIF readable by placing the appropriate %*fmt* on the numeric variable; see *Displaying SIFs in HRF*, below.

You can convert from one SIF type to another by using SIF-to-SIF conversion functions; see *SIF-to-SIF conversion*, below.

SIF dates are convenient because you can subtract them to obtain time between dates, for example,

$$\text{datetime2} - \text{datetime1} = \text{milliseconds between datetime1 and datetime2}$$
$$\text{(divide by 1,000 to obtain seconds)}$$

$$\text{date2} - \text{date1} = \text{days between date1 and date2}$$

$$\text{week2} - \text{week1} = \text{weeks between week1 and week2}$$

$$\text{month2} - \text{month1} = \text{months between month1 and month2}$$

$$\text{half2} - \text{half1} = \text{half-years between half1 and half2}$$

$$\text{year2} - \text{year1} = \text{years between year1 and year2}$$

In the remaining text, we will use the following notation:

> *tc*: a Stata double variable containing SIF datetime/c values
>
> *tC*: a Stata double variable containing SIF datetime/C values
>
> *td*: a Stata variable containing SIF date values
>
> *tw*: a Stata variable containing SIF weekly date values
> *tm*: a Stata variable containing SIF monthly date values
> *tq*: a Stata variable containing SIF quarterly date values
> *th*: a Stata variable containing SIF half-yearly date values
> *ty*: a Stata variable containing SIF yearly date values

HRF-to-SIF conversion functions

SIF type	Function to convert HRF to SIF		Note
datetime/c	*tc* =	clock(*HRFstr*, *mask*)	*tc* must be double
datetime/C	*tC* =	Clock(*HRFstr*, *mask*)	*tC* must be double
date	*td* =	date(*HRFstr*, *mask*)	*td* may be float or long
weekly date	*tw* =	weekly(*HRFstr*, *mask*)	*tw* may be float or int
monthly date	*tm* =	monthly(*HRFstr*, *mask*)	*tm* may be float or int
quarterly date	*tq* =	quarterly(*HRFstr*, *mask*)	*tq* may be float or int
half-yearly date	*th* =	halfyearly(*HRFstr*, *mask*)	*th* may be float or int
yearly date	*ty* =	yearly(*HRFstr*, *mask*)	*ty* may be float or int

Warning: To prevent loss of precision, datetime SIFs must be stored as doubles.

Examples:

1. You have datetimes stored in the string variable `mystr`, an example being "2010.07.12 14:32". To convert to SIF datetime/c, you type

    ```
    . gen double eventtime = clock(mystr, "YMDhm")
    ```

    The mask "YMDhm" specifies the order of the datetime components. In this case, they are year, month, day, hour, and minute.

2. You have datetimes stored in `mystr`, an example being "2010.07.12 14:32:12". You type

    ```
    . gen double eventtime = clock(mystr, "YMDhms")
    ```

    Mask element s specifies seconds. In example 1, there were no seconds; in this example, there are.

3. You have datetimes stored in `mystr`, an example being "2010 Jul 12 14:32". You type

    ```
    . gen double eventtime = clock(mystr, "YMDhm")
    ```

    This is the same command that you typed in example 1. In the mask, you specify the order of the components; Stata figures out the style for itself. In example 1, months were numeric. In this example, they are spelled out (and happen to be abbreviated).

4. You have datetimes stored in `mystr`, an example being "July 12, 2010 2:32 PM". You type

   ```
   . gen double eventtime = clock(mystr, "MDYhm")
   ```

   Stata automatically looks for AM and PM, in uppercase and lowercase, with and without periods.

5. You have datetimes stored in `mystr`, an example being "7-12-10 14.32". The 2-digit year is to be interpreted as being prefixed with 20. You type

   ```
   . gen double eventtime = clock(mystr, "MD20Yhm")
   ```

6. You have datetimes stored in `mystr`, an example being "14:32 on 7/12/2010". You type

   ```
   . gen double eventtime = clock(mystr, "hm#MDY")
   ```

   The # sign between m and M means, "ignore one thing between minute and month", which in this case is the word "on". Had you omitted the # from the mask, the new variable `eventtime` would have contained missing values.

7. You have a date stored in `mystr`, an example being "22/7/2010". In this case, you want to create an SIF date instead of a datetime. You type

   ```
   . gen eventdate = date(mystr, "DMY")
   ```

   Typing

   ```
   . gen double eventtime = clock(mystr, "DMY")
   ```

   would have worked, too. Variable `eventtime` would contain a different coding from that contained by `eventdate`; namely, it would contain milliseconds from 1jan1960 rather than days (1,595,376,000,000 rather than 18,465). Datetime value 1,595,376,000,000 corresponds to 22jul2010 00:00:00.000.

See [D] **datetime translation** for more information about the HRF-to-SIF conversion functions.

Displaying SIFs in HRF

SIF type	Display format to present SIF in HRF
datetime/c	%tc
datetime/C	%tC
date	%td
weekly date	%tw
monthly date	%tm
quarterly date	%tq
half-yearly date	%th
yearly date	%ty

The display formats above are the simplest forms of each of the SIFs. You can control how each type of SIF date is displayed; see [D] **datetime display formats**.

Examples:

1. You have datetimes stored in string variable `mystr`, an example being "`2010.07.12 14:32`". To convert to SIF datetime/c and make the new variable readable when displayed, you type

   ```
   . gen double eventtime = clock(mystr, "YMDhm")
   . format eventtime %tc
   ```

2. You have a date stored in `mystr`, an example being "`22/7/2010`". To convert to an SIF date and make the new variable readable when displayed, you type

   ```
   . gen eventdate = date(mystr, "DMY")
   . format eventdate %td
   ```

Building SIFs from components

SIF type	Function to build from components
datetime/c	$tc = \mathtt{mdyhms}(M,\ D,\ Y,\ h,\ m,\ s)$
	$tc = \mathtt{dhms}(td,\ h,\ m,\ s)$
	$tc = \mathtt{hms}(h,\ m,\ s)$
datetime/C	$tC = \mathtt{Cmdyhms}(M,\ D,\ Y,\ h,\ m,\ s)$
	$tC = \mathtt{Cdhms}(td,\ h,\ m,\ s)$
	$tC = \mathtt{Chms}(h,\ m,\ s)$
date	$td = \mathtt{mdy}(M,\ D,\ Y)$
weekly date	$tw = \mathtt{yw}(Y,\ W)$
monthly date	$tm = \mathtt{ym}(Y,\ M)$
quarterly date	$tq = \mathtt{yq}(Y,\ Q)$
half-yearly date	$th = \mathtt{yh}(Y,\ H)$
yearly date	$ty = \mathtt{y}(Y)$

Warning: SIFs for datetimes must be stored as `doubles`.

Examples:

1. Your dataset has three variables, `mo`, `da`, and `yr`, with each variable containing a date component in numeric form. To convert to SIF date, you type

   ```
   . gen eventdate = mdy(mo, da, yr)
   . format eventdate %td
   ```

2. Your dataset has two numeric variables, `mo` and `yr`. To convert to SIF date corresponding to the first day of the month, you type

   ```
   . gen eventdate = mdy(mo, 1, yr)
   . format eventdate %td
   ```

3. Your dataset has two numeric variables, `da` and `yr`, and one string variable, `month`, containing the spelled-out month. In this case, do not use the building-from-component functions. Instead, construct a new string variable containing the HRF and then convert the string using the HRF-to-SIF conversion functions:

   ```
   . gen str work  = month + " " + string(da) + " " + string(yr)
   . gen eventdate = date(work, "MDY")
   . format eventdate %td
   ```

SIF-to-SIF conversion

From:	To: datetime/c	datetime/C	date
datetime/c		$tC = \mathtt{Cofc}(tc)$	$td = \mathtt{dofc}(tc)$
datetime/C	$tc = \mathtt{cofC}(tC)$		$td = \mathtt{dofC}(tC)$
date	$tc = \mathtt{cofd}(td)$	$tC = \mathtt{Cofd}(td)$	
weekly			$td = \mathtt{dofw}(tw)$
monthly			$td = \mathtt{dofm}(tm)$
quarterly			$td = \mathtt{dofq}(tq)$
half-yearly			$td = \mathtt{dofh}(th)$
yearly			$td = \mathtt{dofy}(ty)$

From:	To: weekly	monthly	quarterly
date	$tw = \mathtt{wofd}(td)$	$tm = \mathtt{mofd}(td)$	$tq = \mathtt{qofd}(td)$

From:	To: half-yearly	yearly
date	$th = \mathtt{hofd}(td)$	$ty = \mathtt{yofd}(td)$

To convert between missing entries, use two functions, going through date or datetime as appropriate. For example, quarterly of monthly is $tq = \mathtt{qofd}(\mathtt{dofm}(tm))$.

Examples:

1. You have the SIF datetime/c variable `eventtime` and wish to create the new variable `eventdate` containing just the date from the datetime variable. You type

   ```
   . gen eventdate = dofc(eventtime)
   . format eventdate %td
   ```

2. You have the SIF date variable `eventdate` and wish to create the new SIF datetime/c variable `eventtime` from it. You type

   ```
   . gen double eventtime = cofd(eventdate)
   . format eventtime %tc
   ```

   The time components of the new variable will be set to the default 00:00:00.000.

3. You have the SIF quarterly variable `eventqtr` and wish to create the new SIF date variable `eventdate` from it. You type

   ```
   . gen eventdate = dofq(eventqtr)
   . format eventdate %tq
   ```

   The new variable, `eventdate`, will contain 01jan dates for quarter 1, 01apr dates for quarter 2, 01jul dates for quarter 3, and 01oct dates for quarter 4.

4. You have the SIF datetime/c variable `admittime` and wish to create the new SIF quarterly variable `admitqtr` from it. You type

   ```
   . gen admitqtr = qofd(dofc(admittime))
   . format admitqtr %tq
   ```

   Because there is no `qofc()` function, you use `qofd(dofc())`.

Extracting time-of-day components from SIFs

Desired component	Function	Example
hour of day	$hh(tc)$ or $hhC(tC)$	14
minutes of day	$mm(tc)$ or $mmC(tC)$	42
seconds of day	$ss(tc)$ or $ssC(tC)$	57.123

Notes:

$$0 \le hh(tc) \le 23, \quad 0 \le hhC(tC) \le 23$$
$$0 \le mm(tc) \le 59, \quad 0 \le mmC(tC) \le 59$$
$$0 \le ss(tc) < 60, \quad 0 \le ssC(tC) < 61 \quad \text{(sic)}$$

Example:

1. You have the SIF datetime/c variable admittime. You wish to create the new variable admithour equal to the hour and fraction of hour within the day of admission. You type

    . gen admithour = hh(admittime) + mm(admittime)/60 + ss(admittime)/3600

Extracting date components from SIFs

Desired component	Function	Example*
calendar year	$year(td)$	2011
calendar month	$month(td)$	7
calendar day	$day(td)$	5
day of week (0=Sunday)	$dow(td)$	2
Julian day of year (1=first day)	$doy(td)$	186
week within year (1=first week)	$week(td)$	27
quarter within year (1=first quarter)	$quarter(td)$	3
half within year (1=first half)	$halfyear(td)$	2

* All examples are with td=mdy(7,5,2011).

All functions require an SIF date as an argument. To extract components from other SIFs, use the appropriate SIF-to-SIF conversion function to convert to an SIF date, for example, quarter(dofq(tq)).

Examples:

1. You wish to obtain the day of week Sunday, Monday, ..., corresponding to the SIF date variable eventdate. You type

    . gen day_of_week = dow(eventdate)

The new variable, day_of_week, contains 0 for Sunday, 1 for Monday, ..., 6 for Saturday.

2. You wish to obtain the day of week Sunday, Monday, ..., corresponding to the SIF datetime/c variable eventtime. You type

        . gen day_of_week = dow(dofc(eventtime))

3. You have the SIF date variable evdate and wish to create the new SIF date variable evdate_r from it. evdate_r will contain the same date as evdate but rounded back to the first of the month. You type

        . gen evdate_r = mdy(month(evdate), 1, year(evdate))

In the above solution, we used the date-component extraction functions month() and year() and used the build-from-components function mdy().

Conveniently typing SIF values

You can type SIF values by just typing the number, such as 16,237 or 1,402,920,000,000, as in

        . gen before = cond(hiredon < 16237, 1, 0) if if !missing(hiredon)
        . drop if admittedon < 1402920000000

Easier to type is

        . gen before = cond(hiredon < td(15jun2004), 1, 0) if !missing(hiredon)
        . drop if admittedon < tc(15jun2004 12:00:00)

You can type SIF date values by typing the date inside td(), as in td(15jun2004).

You can type SIF datetime/c values by typing the datetime inside tc(), as in tc(15jun2004 12:00:00).

td() and tc() are called pseudofunctions because they translate what you type into their numerical equivalents. Pseudofunctions require only that you specify the datetime components in the expected order, so rather than 15jun2004 above, we could have specified 15 June 2004, 15-6-2004, or 15/6/2004.

The SIF pseudofunctions and their expected component order are

Desired SIF type	Pseudofunction
datetime/c	$tc([day\text{-}month\text{-}year]\ hh{:}mm[{:}ss[.sss]])$
datetime/C	$tC([day\text{-}month\text{-}year]\ hh{:}mm[{:}ss[.sss]])$
date	$td(day\text{-}month\text{-}year)$
weekly date	$tw(year\text{-}week)$
monthly date	$tm(year\text{-}month)$
quarterly date	$tq(year\text{-}quarter)$
half-yearly date	$th(year\text{-}half)$
yearly date	none necessary; just type year

The *day-month-year* in tc() and tC() are optional. If you omit them, 01jan1960 is assumed. Doing so produces time as an offset, which can be useful in, for example,

        . gen six_hrs_later = eventtime + tc(6:00)

Obtaining and working with durations

SIF values are simply durations from 1960. SIF datetime/c values record the number of milliseconds from 1jan1960 00:00:00; SIF date values record the number of days from 1jan1960, and so on.

To obtain the time between two SIF variables—the duration—subtract them:

```
. gen days_employed = curdate - hiredate
. gen double ms_inside = discharge_time - admit_time
```

To obtain a new SIF that is equal to an old SIF before or after some amount of time, just add or subtract the desired durations:

```
. gen lastdate = hiredate + days_employed
. format lastdate %td
. gen double admit_time = discharge_time - ms_inside
. format admit_time %tc
```

Remember to use the units of the SIF variables. SIF dates are in terms of days, SIF weekly dates are in terms of weeks, etc., and SIF datetimes are in terms of milliseconds. Concerning milliseconds, it is often easier to use different units and conversion functions to convert to milliseconds:

```
. gen hours_inside = hours(discharge_time - admit_time)
. gen admit_time = discharge_time - msofhours(hours_inside)
. format admit_time %tc
```

Function `hours()` converts milliseconds to hours. Function `msofhours()` converts hours to milliseconds. The millisecond conversion functions are

Function	Purpose
hours(ms)	convert milliseconds to hours returns $ms/(60 \times 60 \times 1000)$
minutes(ms)	convert milliseconds to minutes returns $ms/(60 \times 1000)$
seconds(ms)	convert milliseconds to seconds returns $ms/1000$
msofhours(h)	convert hours to milliseconds returns $h \times 60 \times 60 \times 1000$
msofminutes(m)	convert minutes to milliseconds returns $m \times 60 \times 1000$
msofseconds(s)	convert seconds to milliseconds returns $s \times 1000$

If you plan on using returned values to add to or subtract from a datetime SIF, be sure they are stored as `doubles`.

Using dates and times from other software

Most software stores dates and times numerically as durations from some sentinel date in specified units, but they differ on the sentinel date and the units. If you have imported data, it is usually possible to adjust the numeric date and datetime values to SIF.

Converting SAS dates:

SAS provides dates measured as the number of days since 01jan1960. This is the same coding as used by Stata:

```
. gen statadate = sasdate
. format statadate %td
```

SAS provides datetimes measured as the number of seconds since 01jan1960 00:00:00, assuming 86,400 seconds/day. To convert to SIF datetime/c, type

```
. gen double statatime = (sastime*1000)
. format statatime %tc
```

It is important that variables containing SAS datetimes, such as `sastime` above, be imported into Stata as `doubles`.

Converting SPSS dates:

SPSS provides dates and datetimes measured as the number of seconds since 14oct1582 00:00:00, assuming 86,400 seconds/day. To convert to SIF datetime/c, type

```
. gen double statatime = (spsstime*1000) + tc(14oct1582 00:00)
. format statatime %tc
```

To convert to SIF date, type

```
. gen statadate = dofc((spsstime*1000) + tc(14oct1582 00:00))
. format statadate %td
```

Converting R dates:

R stores dates as days since 01jan1970. To convert to SIF date, type

```
. gen statadate = rdate - td(01jan1970)
. format statadate %td
```

R stores datetimes as the number of UTC-adjusted seconds since 01jan1970 00:00:00. To convert to SIF datetime/C, type

```
. gen double statatime = rtime - tC(01jan1970 00:00)
. format statatime %tC
```

To convert to SIF datetime/c, type

```
. gen double statatime = cofC(rtime - tC(01jan1970 00:00))
. format statatime %tc
```

There are issues of which you need to be aware when working with datetime/C values; see *Why there are two SIF datetime encodings* and *Advice on using datetime/c and datetime/C*, both in [D] **datetime translation**.

Converting Excel dates:

You are unlikely to encounter Excel numerically encoded dates. If you copy and paste a spreadsheet into Stata's editor, dates and datetimes are pasted as strings in HRF. If you use a conversion package, most know how to convert the date for you.

Excel has used different date systems across operating systems. Excel for Windows used the "1900 Date System". Excel for Mac used the "1904 Date System". More recently, Excel has been standardizing on the 1900 Date System on all operating systems.

Regardless of operating system, Excel can use either encoding. See http://support.microsoft.com/kb/214330 for instructions on converting workbooks between date systems.

Converted dates will be off by four years if you choose the wrong date system.

Converting Excel 1900-Date-System dates:

For dates on or after 01mar1900, Excel stores dates as days since 30dec1899. To convert to a Stata date,

```
. gen statadate = exceldate + td(30dec1899)
. format statadate %td
```

Excel can store dates between 01jan1900 and 28feb1900, but the formula above will not handle those two months. See http://www.cpearson.com/excel/datetime.htm for more information.

For datetimes on or after 01mar1900 00:00:00, Excel stores datetimes as days plus fraction of day since 30dec1899 00:00:00. To convert with a one-second resolution to a Stata datetime,

```
. gen statatime = round((exceltime+td(30dec1899))*86400)*1000
. format statatime %tc
```

Converting Excel 1904-Date-System dates:

For dates on or after 01jan1904, Excel stores dates as days since 01jan1904. To convert to a Stata date,

```
. gen statadate = exceldate + td(01jan1904)
. format statadate %td
```

For datetimes on or after 01jan1904 00:00:00, Excel stores datetimes as days plus fraction of day since 01jan1904 00:00:00. To convert with a one-second resolution to a Stata datetime,

```
. gen statatime = round((exceltime+td(01jan1904))*86400)*1000
. format statatime %tc
```

Converting OpenOffice Dates:

OpenOffice uses the Excel 1900 Date System described above.

Description

Syntax above provides a complete overview of Stata's date and time values. Also see [D] **datetime translation** and [D] **datetime display formats** for additional information.

Remarks

The best way to learn about Stata's date and time functions is to experiment with them using the display command; see [P] **display**.

```
. display date("5-12-1998", "MDY")
14011
. display %td date("5-12-1998", "MDY")
12may1998
```

```
. display clock("5-12-1998 11:15", "MDY hm")
1.211e+12

. display %20.0gc clock("5-12-1998 11:15", "MDY hm")
1,210,590,900,000

. display %tc clock("5-12-1998 11:15", "MDY hm")
12may1998 11:15:00
```

With display, you can specify a format in front of the expression to specify how the result is to be formatted.

Reference

Gould, W. W. 2011. Using dates and times from other software. The Stata Blog: Not Elsewhere Classified. http://blog.stata.com/2011/01/05/using-dates-and-times-from-other-software/

Also see

[D] **datetime business calendars** — Business calendars

[D] **datetime display formats** — Display formats for dates and times

[D] **datetime translation** — String to numeric date translation functions

Title

datetime business calendars — Business calendars

Syntax

Apply business calendar format

    format *varlist* %tb*calname*

Apply detailed date format with business calendar format

    format *varlist* %tb*calname*[:*datetime-specifiers*]

Convert between business dates and regular dates

    { <u>gen</u>erate | replace } *bdate* = bofd("*calname*", *regulardate*)

    { <u>gen</u>erate | replace } *regulardate* = dofb(*bdate*, "*calname*")

File *calname*.stbcal contains the business calendar definition.

Details of the syntax follow:

1. Definition.
   Business calendars are regular calendars with some dates crossed out:

		November 2011				
Su	Mo	Tu	We	Th	Fr	Sa
		1	2	3	4	X
X	7	8	9	10	11	X
X	14	15	16	17	18	X
X	21	22	23	X	X	X
X	28	29	30			

   A date that appears on the business calendar is called a business date. 11nov2011 is a business date. 12nov2011 is not a business date with respect to this calendar.

   Crossed-out dates are literally omitted. That is,

   $$18\text{nov}2011 + 1 = 21\text{nov}2011$$

   $$28\text{nov}2011 - 1 = 23\text{nov}2011$$

   Stata's lead and lag operators work the same way.

2. Business calendars are named.
   Assume that the above business calendar is named simple.

3. Business calendars are defined in files named *calname*.stbcal, such as simple.stbcal. Calendars may be supplied by StataCorp and already installed, obtained from other users directly or via the SSC, or written yourself. Stbcal-files are treated in the same way as ado-files.

   You can obtain a list of all business calendars installed on your computer by typing bcal dir; see [D] **bcal**.

4. Datetime format.

The date format associated with the business calendar named `simple` is `%tbsimple`, which is to say $\% + t + b +$ *calname*.

%	it is a format
t	it is a datetime
b	it is based on a business calendar
calname	the calendar's name

5. Format variables the usual way.

You format variables to have business calendar formats just as you format any variable, using the `format` command.

```
. format mydate %tbsimple
```

specifies that existing variable `mydate` contains values according to the business calendar named `simple`. See [D] **format**.

You may format variables `%tb`*calname* regardless of whether the corresponding stbcal-file exists. If it does not exist, the underlying numeric values will be displayed in a `%g` format.

6. Detailed date formats.

You may include detailed datetime format specifiers by placing a colon and the detail specifiers after the calendar's name.

```
. format mydate %tbsimple:CCYY.NN.DD
```

would display 21nov2011 as 2011.11.21. See [D] **datetime display formats** for detailed datetime format specifiers.

7. Reading business dates.

To read files containing business dates, ignore the business date aspect and read the files as if they contained regular dates. Convert and format those dates as `%td`; see *HRF-to-SIF conversion functions* in [D] **datetime**. Then convert the regular dates to `%tb` business dates:

```
. generate mydate = bofd("simple", regulardate)
. format mydate %tbsimple
. assert mydate!=. if regulardate!=.
```

The first statement performs the conversion.

The second statement attaches the `%tbsimple` date format to the new variable `mydate` so that it will display correctly.

The third statement verifies that all dates recorded in `regulardate` fit onto the business calendar. For instance, 12nov2011 does not appear on the `simple` calendar but, of course, it does appear on the regular calendar. If the data contained 12nov2011, that would be an error. Function `bofd()` returns missing when the date does not appear on the specified calendar.

8. More on conversion.

There are only two functions specific to business dates, `bofd()` and `dofb()`. Their definitions are

$$bdate = \text{bofd}(\text{"}calname\text{"}, regulardate)$$

$$regulardate = \text{dofb}(bdate, \text{"}calname\text{"})$$

`bofd()` returns missing if *regulardate* is missing or does not appear on the specified business calendar. `dofb()` returns missing if *bdate* contains missing.

9. Obtaining day of week, etc.

You obtain day of week, etc., by converting business dates to regular dates and then using the standard functions. To obtain the day of week of *bdate* on business calendar *calname*, type

```
. generate dow = dow(dofb(bdate, "calname"))
```

See *Extracting date components from SIFs* in [D] **datetime** for the other extraction functions.

10. Stbcal-files.

The stbcal-file for `simple`, the calendar shown below,

November 2011

Su	Mo	Tu	We	Th	Fr	Sa
		1	2	3	4	X
X	7	8	9	10	11	X
X	14	15	16	17	18	X
X	21	22	23	X	X	X
X	28	29	30			

is

───────────────────────────────── begin simple.stbcal ─────────

```
*! version 1.0.0
*  simple.stbcal

version 12
purpose "Example for manual"
dateformat dmy

range 01nov2011 30nov2011
centerdate 01nov2011

omit dayofweek (Sa Su)
omit date 24nov2011
omit date 25nov2011
```

───────────────────────────────── end simple.stbcal ─────────

This calendar was so simple that we crossed out the Thanksgiving holidays by specifying the dates to be omitted. In a real calendar, we would change the last two lines,

```
omit date 24nov2011
omit date 25nov2011
```

to read

```
omit dowinmonth -1 Th of Nov and +1
```

which says to omit the last (−1) Thursday of November in every year, and omit the day after that (+1), too. See [D] **datetime business calendars creation**.

Description

Stata provides user-definable business calendars.

Remarks

See [D] **datetime** for an introduction to Stata's date and time features.

Below we work through an example from start to finish.

Remarks are presented under the following headings:

Step 1: Read the data, date as string
Step 2: Convert date variable to %td date
Step 3: Convert %td date to %tb date
Key feature: Each business calendar has its own encoding
Key feature: Omitted dates really are omitted
Key feature: Extracting components from %tb dates
Key feature: Merging on dates

Step 1: Read the data, date as string

File `bcal_simple.raw` on our website provides data, including a date variable, that is to be interpreted according to the business calendar `simple` shown under *Syntax* above.

```
. type http://www.stata-press.com/data/r12/bcal_simple.raw
11/4/11 51
11/7/11 9
11/18/11 12
11/21/11 4
11/23/11 17
11/28/11 22
```

We begin by reading the data and then listing the result. Note that we read the date as a string variable:

```
. infile str10 sdate float x using http://www.stata-press.com/data/r12/bcal_simple
(6 observations read)
. list
```

	sdate	x
1.	11/4/11	51
2.	11/7/11	9
3.	11/18/11	12
4.	11/21/11	4
5.	11/23/11	17
6.	11/28/11	22

Step 2: Convert date variable to %td date

Now we create a Stata internal form (SIF) `%td` format date from the string date:

```
. generate rdate = date(sdate, "MD20Y")
. format rdate %td
```

See *HRF-to-SIF conversion functions* in [D] **datetime**. We verify that the conversion went well and drop the string variable of the date:

```
. list
```

	sdate	x	rdate
1.	11/4/11	51	04nov2011
2.	11/7/11	9	07nov2011
3.	11/18/11	12	18nov2011
4.	11/21/11	4	21nov2011
5.	11/23/11	17	23nov2011
6.	11/28/11	22	28nov2011

```
. drop sdate
```

Step 3: Convert %td date to %tb date

We convert the %td date to a %tbsimple date following the instructions of item 7 of *Syntax* above.

```
. generate mydate = bofd("simple", rdate)
. format mydate %tbsimple
. assert mydate!=. if rdate!=.
```

Had there been any dates that could not be converted from regular dates to simple business dates, assert would have responded, "assertion is false". Nonetheless, we will list the data to show you that the conversion went well. We would usually drop the %td encoding of the date, but we want it to demonstrate a feature below.

```
. list
```

	x	rdate	mydate
1.	51	04nov2011	04nov2011
2.	9	07nov2011	07nov2011
3.	12	18nov2011	18nov2011
4.	4	21nov2011	21nov2011
5.	17	23nov2011	23nov2011
6.	22	28nov2011	28nov2011

Key feature: Each business calendar has its own encoding

In the listing above, rdate and mydate appear to be equal. They are not:

```
. format rdate mydate %9.0g          // remove date formats
. list
```

	x	rdate	mydate
1.	51	18935	3
2.	9	18938	4
3.	12	18949	13
4.	4	18952	14
5.	17	18954	16
6.	22	18959	17

%tb dates each have their own encoding, and those encodings differ from the encoding used by %td dates. It does not matter. Neither encoding is better than the other. Neither do you need to concern yourself with the encoding. If you were curious, you could learn more about the encoding used by %tbsimple by typing bcal describe simple; see [D] **bcal**.

We will drop variable rdate and put the %tbsimple format back on variable mydate:

```
. drop rdate
. format mydate %tbsimple
```

Key feature: Omitted dates really are omitted

In *Syntax*, we mentioned that for the simple business calendar

$$18\text{nov}2011 + 1 = 21\text{nov}2011$$

$$28\text{nov}2011 - 1 = 23\text{nov}2011$$

That is true:

```
. generate tomorrow = mydate + 1
. generate yesterday = mydate - 1
. format tomorrow yesterday %tbsimple
. list
```

	x	mydate	tomorrow	yesterday
1.	51	04nov2011	07nov2011	03nov2011
2.	9	07nov2011	08nov2011	04nov2011
3.	12	18nov2011	21nov2011	17nov2011
4.	4	21nov2011	22nov2011	18nov2011
5.	17	23nov2011	28nov2011	22nov2011
6.	22	28nov2011	29nov2011	23nov2011

```
. drop tomorrow yesterday
```

Stata's lag and lead operators L.*varname* and F.*varname* work similarly.

Key feature: Extracting components from %tb dates

You extract components such as day of week, month, day, and year from business dates using the same extraction functions you use with Stata's regular %td dates, namely, dow(), month(), day(), and year(), and you use function dofb() to convert business dates to regular dates. Below we add day of week to our data, list the data, and then drop the new variable:

```
. generate dow = dow(dofb(mydate, "simple"))
. list
```

	x	mydate	dow
1.	51	04nov2011	5
2.	9	07nov2011	1
3.	12	18nov2011	5
4.	4	21nov2011	1
5.	17	23nov2011	3
6.	22	28nov2011	1

```
. drop dow
```

See *Extracting date components from SIFs* in [D] **datetime**.

Key feature: Merging on dates

It may happen that you have one dataset containing business dates and a second dataset containing regular dates, say, on economic conditions, and you want to merge them. To do that, you create a regular date variable in your first dataset and merge on that:

```
. generate rdate = dofb(mydate, "simple")
. merge 1:1 rdate using econditions, keep(match)
. drop rdate
```

Also see

[D] **bcal** — Business calendar file manipulation

[D] **datetime business calendars creation** — Business calendars creation

[D] **datetime** — Date and time values and variables

Title

datetime business calendars creation — Business calendars creation

Syntax

Business calendar *calname* and corresponding display format %tb*calname* are defined by the text file *calname*.stbcal, which contains the following:

* *comments*

version *version_of_stata*

purpose *"text"*

dateformat { ymd | ydm | myd | mdy | dym | dmy }

range *date date*

centerdate *date*

[from { *date* | . } to { *date* | . }] omit ... [*if*]

. . .

. . .

where

      omit ... may be

               omit date *pdate* [and *pmlist*]

               omit dayofweek *dowlist*

               omit dowinmonth *pm# dow* [of *monthlist*] [and *pmlist*]

      [*if*] may be

               if *restriction* [& *restriction* ...]

      restriction is one of

               dow(*dowlist*)
               month(*monthlist*)
               year(*yearlist*)

      date is a date written with the *year*, *month*, and *day* in the order specified by dateformat. For instance, if dateformat is dmy, a *date* can be 12apr2011, 12-4-2011, or 12.4.2011.

      pdate is a *date* or it is a *date* with character * substituted where the year would usually appear. If dateformat is dmy, a *pdate* can be 12apr2011, 12-4-2011, or 12.4.2011; or it can be 12apr*, 12-4-*, or 12.4.*. 12apr* means the 12th of April across all years.

      dow is a day of week, in English. It may be abbreviated to as few as 2 characters, and capitalization is irrelevant. Examples: Sunday, Mo, tu, Wed, th, Friday, saturday.

      dowlist is a *dow*, or it is a space-separated list of one or more *dow*s enclosed in parentheses. Examples: Sa, (Sa), (Sa Su).

month is a month of the year, in English, or it is a month number. It may be abbreviated to the minimum possible, and capitalization is irrelevant. Examples: `January`, `2`, `Mar`, `ap`, `may`, `6`, `Jul`, `aug`, `9`, `Octob`, `nov`, `12`.

monthlist is a *month*, or it is a space-separated list of one or more *month*s enclosed in parentheses. Examples: `Nov`, `(Nov)`, `11`, `(11)`, `(Nov Dec)`, `(11 12)`.

year is a 4-digit calendar year. Examples: `1872`, `1992`, `2011`, `2050`.

yearlist is a *year*, or it is a space-separated list of one or more *year*s enclosed in parentheses. Examples: `2011`, `(2011)`, `(2011 2012)`.

pm# is a nonzero integer preceded by a plus or minus sign. Examples: `-2`, `-1`, `+1`. *pm#* appears in `omit downinmonth` *pm# dow* of *monthlist*, where *pm#* specifies which *dow* in the month. `omit downinmonth +1 Th` means the first Thursday of the month. `omit downinmonth -1 Th` means the last Thursday of the month.

pmlist is a *pm#*, or it is a space-separated list of one or more *pm#*s enclosed in parentheses. Examples: `+1`, `(+1)`, `(+1 +2)`, `(-1 +1 +2)`. *pmlist* appears in the optional `and` *pmlist* allowed at the end of `omit` *date* and `omit downinmonth`, and it specifies additional dates to be omitted. `and +1` means and the day after. `and -1` means and the day before.

Description

Stata provides user-definable business calendars. Business calendars are provided by StataCorp and by other users, and you can write your own. This entry concerns writing your own business calendars.

See [D] **datetime business calendars** for an introduction to business calendars.

Remarks

Remarks are presented under the following headings:

> *Introduction*
> *Concepts*
> *The preliminary commands*
> *The omit commands: from/to and if*
> *The omit commands: and*
> *The omit commands: omit date*
> *The omit commands: omit dayofweek*
> *The omit commands: omit downinmonth*
> *Where to place stbcal-files*
> *How to debug stbcal-files*
> *Ideas for calendars that may not occur to you*

Introduction

A business calendar is a regular calendar with some dates crossed out, such as

		November 2011				
Su	Mo	Tu	We	Th	Fr	Sa
		1	2	3	4	X
X	7	8	9	10	11	X
X	14	15	16	17	18	X
X	21	22	23	X	X	X
X	28	29	30			

The purpose of the stbcal-file is to

1. Specify the range of dates covered by the calendar.

2. Specify the particular date that will be encoded as date 0.

3. Specify the dates from the regular calendar that are to be crossed out.

The stbcal-file for the above calendar could be as simple as

```
                                                        begin example_1.stbcal

    version 12
    range 01nov2011 30nov2011
    centerdate 01nov2011
    omit date  5nov2011
    omit date  6nov2011
    omit date 12nov2011
    omit date 13nov2011
    omit date 19nov2011
    omit date 20nov2011
    omit date 24nov2011
    omit date 25nov2011
    omit date 26nov2011
    omit date 27nov2011
                                                          end example_1.stbcal
```

In fact, this calendar can be written more compactly because we can specify to omit all Saturdays and Sundays:

```
                                                        begin example_2.stbcal

    version 12
    range 01nov2011 30nov2011
    centerdate 01nov2011
    omit dayofweek (Sa Su)
    omit date 24nov2011
    omit date 25nov2011
                                                          end example_2.stbcal
```

In this particular calendar, we are omitting 24nov2011 and 25nov2011 because of the American Thanksgiving holiday. Thanksgiving is celebrated on the fourth Thursday of November, and many businesses close on the following Friday as well. It is possible to specify rules like that in stbcal-files:

```
                                                        begin example_3.stbcal

    version 12
    range 01nov2011 30nov2011
    centerdate 01nov2011
    omit dayofweek (Sa Su)
    omit dowinmonth +4 Th of Nov and +1
                                                          end example_3.stbcal
```

Understand that this calendar is an artificial example, and it is made all the more artificial because it covers so brief a period. Real stbcal-files cover at least decades, and some cover centuries.

Concepts

You are required to specify four things in an stbcal-file:

1. the version of Stata being used,

2. the range of the calendar,

3. the center date of the calendar, and

4. the dates to be omitted.

Version.

You specify the version of Stata to ensure forward compatibility with future versions of Stata. If your calendar starts with the line `version 12`, future versions of Stata will know how to interpret the file even if the definition of the stbcal-file language has greatly changed.

Range.

A calendar is defined over a specific range of dates, and you must explicitly state what that range is. When you or others use your calendar, dates outside the range will be considered invalid, which usually means that they will be treated as missing values.

Center date.

Stata stores dates as integers. In a calendar, 57 might stand for a particular date. If it did, then $57 - 1 = 56$ stands for the day before, and $57 + 1 = 58$ stands for the day after. The previous statement works just as well if we substitute $-12{,}739$ for 57, and thus the particular values do not matter except that we must agree upon what values we wish to standardize because we will be storing these values in our datasets.

The standard is called the center date, and here center does not mean the date that corresponds to the middle of your calendar. It means the date that corresponds to the center of integers, which is to say, 0. You must choose a date within the range as the standard. The particular date you choose does not matter, but most authors choose easily remembered ones. Stata's built-in `%td` calendar uses 01jan1960, but that date will probably not be available to you because the center date must be a date on the business calendars, and most businesses were closed on 01jan1960.

It will sometimes happen that you will want to expand the range of your calendar in the future. Today, you make a calendar that covers, say 1990 to 2020, which is good enough for your purposes. Later, you need to expand the range, say back to 1970 or forward to 2030, or both. When you update your calendar, do not change the center date. This way, your new calendar will be backward compatible with your previous one.

Omitted dates.

Obviously you will need to specify the dates to be omitted. You can specify the exact dates to be omitted when need be, but whenever possible, specify the rules instead of the outcome of the rules. Rules change, so learn about the `from/to` prefix that can be used in front of `omit` commands. You can code things like

```
from 01jan1960 to 31dec1968: omit ...
from 01jan1979 to .:  omit ...
```

When specifying `from/to`, . for the first date is synonymous with the opening date of the range. . for the second date is synonymous with the closing date.

The preliminary commands

Stbcal-files should begin with these commands:

```
version version_of_stata
purpose "text"
dateformat { ymd | ydm | myd | mdy | dym | dmy }
range date date
centerdate date
```

version *version_of_stata*
  At the time of this writing, you would specify `version 12`. Better still, type command `version` in Stata to discover the version of Stata you are currently using. Specify that version, and be sure to look at the online documentation so that you use the modern syntax correctly.

purpose "*text*"
  This command is optional. The purpose of `purpose` is not to make comments in your file. If you want comments, include those with a * in front. The `purpose` sets the text that `bcal describe` *calname* will display.

dateformat { ymd | ydm | myd | mdy | dym | dmy }
  This command is optional. `dateformat ymd` is assumed if not specified. This command has nothing to do with how dates will look when variables are formatted with `%tbcalname`. This command specifies how you are typing dates in this stbcal-file on the subsequent commands. Specify the format that you find convenient.

range *date date*
  The date range was discussed in *Concepts*. You must specify it.

centerdate *date*
  The centering date was discussed in *Concepts*. You must specify it.

The omit commands: from/to and if

An stbcal-file usually contains multiple `omit` commands. The `omit` commands have the syntax

$$\left[\text{from } \{ date \,|\, . \} \text{ to } \{ date \,|\, . \}: \right] \text{ omit } \ldots \left[if \right]$$

That is, an `omit` command may optionally be preceded by `from`/`to` and may optionally contain an `if` at the end.

When you do not specify `from`/`to`, results are the same as if you specified

```
from . to .:  omit ...
```

That is, the `omit` command applies to all dates from the beginning to the end of the range. In *Introduction*, we showed the command

```
omit downinmonth +4 Th of Nov and +1
```

Our sample calendar covered only the month of November, but imagine that it covered a longer period and that the business was open on Fridays following Thanksgiving up until 1998. The Thanksgiving holidays could be coded

```
from . to 31dec1997: omit downinmonth +4 Th of Nov
from 01jan1998 to .: omit downinmonth +4 Th of Nov and +1
```

The same holidays could also be coded

```
omit downinmonth +4 Th of Nov
from 01jan1998 to .: omit downinmonth +4 Th of Nov and +1
```

We like the first style better, but understand that the same dates can be omitted from the calendars multiple times and for multiple reasons, and the result is still the same as if the dates were omitted only once.

The optional `if` also determines when the `omit` statement is operational. Let's think about the Christmas holidays. Let's say a business is closed on the 24th and 25th of December. That could be coded

```
omit date 24dec*
omit date 25dec*
```

although perhaps that would be more understandable if we coded

```
from . to .: omit date 24dec*
from . to .: omit date 25dec*
```

Remember, `from . to .` is implied when not specified. In any case, we are omitting 24dec and 25dec across all years.

Now consider a more complicated rule. The business is closed on the 24th and 25th of December if the 25th is on Tuesday, Wednesday, Thursday, or Friday. If the 25th is on Saturday or Sunday, the holidays are the preceding Friday and the following Monday. If the 25th is on Monday, the holidays are Monday and Tuesday. The rule could be coded

```
omit date 25dec* and -1       if dow(Tu We Th Fr)
omit date 25dec* and (-2 -1) if dow(Sa)
omit date 25dec* and (-3 -2) if dow(Su)
omit date 25dec* and +1       if dow(Mo)
```

The `if` clause specifies that the `omit` command is only to be executed when 25dec* is one of the specified days of the week. If 25dec* is not one of those days, the `omit` statement is ignored for that year. Our focus here is on the `if` clause. We will explain about the `and` clause in the next section.

Sometimes, you have a choice between using `from/to` or `if`. In such cases, use whichever is convenient. For instance, imagine that the Christmas holiday rule for Monday changed in 2011 and 2012. You could code

```
from . to 31dec2010: omit date 25dec* and +1 if dow(Mo)
from 01jan2011 to .: omit date ... if dow(Mo)
```

or

```
omit date 25dec* and +1 if dow(Mo) & year(2007 2008 2009 2010)
omit date ... if dow(Mo) & year(2011 2012)
```

Generally, we find `from/to` more convenient to code than `if year()`.

The omit commands: and

The other common piece of syntax that shows up on `omit` commands is `and` *pmlist*. We used it above in coding the Christmas holidays,

```
omit date 25dec* and -1      if dow(Tu We Th Fr)
omit date 25dec* and (-2 -1) if dow(Sa)
omit date 25dec* and (-3 -2) if dow(Su)
omit date 25dec* and +1      if dow(Mo)
```

`and` *pmlist* specifies a list of days also to be omitted if the date being referred to is omitted. The extra days are specified as how many days they are from the date being referred to. Please excuse the inelegant "date being referred to", but sometimes the date being referred to is implied rather than stated explicitly. For this problem, however, the date being referred to is `25dec` across a number of years. The line

```
omit date 25dec* and -1      if dow(Tu We Th Fr)
```

says to omit `25dec` and the day before if `25dec` is on a Tuesday, Wednesday, etc. The line

```
omit date 25dec* and (-2 -1) if dow(Sa)
```

says to omit `25dec` and two days before and one day before if `25dec` is Saturday. The line

```
omit date 25dec* and (-3 -2) if dow(Su)
```

says to omit `25dec` and three days before and two days before if `25dec` is Sunday. The line

```
omit date 25dec* and +1      if dow(Mo)
```

says to omit `25dec` and the day after if `25dec` is Monday.

Another `omit` command for solving a different problem reads

```
omit dowinmonth -1 We of (Nov Dec) and +1 if year(2009)
```

Please focus on the `and +1`. We are going to omit the date being referred to and the date after if the year is 2009. The date being referred to here is `-1 We of (Nov Dec)`, which is to say, the last Wednesday of November and December.

The omit commands: omit date

The full syntax of `omit date` is

$$\big[\texttt{from } \{\mathit{date}\,|\,.\} \texttt{ to } \{\mathit{date}\,|\,.\}:\big] \texttt{ omit date } \mathit{pdate} \big[\texttt{and } \mathit{pmlist}\big] \big[\mathit{if}\big]$$

You may omit specific dates,

```
omit date 25dec2010
```

or you may omit the same date across years:

```
omit date 25dec*
```

The omit commands: omit dayofweek

The full syntax of omit dayofweek is

$\left[\texttt{from} \; \{ date \,|\, . \} \; \texttt{to} \; \{ date \,|\, . \} : \right] \; \texttt{omit dayofweek} \; dowlist \; \left[if \right]$

The specified days of week (Monday, Tuesday, . . .) are omitted.

The omit commands: omit dowinmonth

The full syntax of omit dowinmonth is

$\left[\texttt{from} \; \{ date \,|\, . \} \; \texttt{to} \; \{ date \,|\, . \} : \right] \; \texttt{omit} \; pm\# \; dow \; \left[\texttt{of} \; monthlist \right] \; \left[\texttt{and} \; pmlist \right] \; \left[if \right]$

dowinmonth stands for day of week in month and refers to days such as the first Monday, second Monday, . . . , next-to-last Monday, and last Monday of a month. This is written as +1 Mo, +2 Mo, . . . , -2 Mo, and -1 Mo.

Where to place stbcal-files

Stata automatically searches for stbcal-files in the same way it searches for ado-files. Stata looks for ado-files and stbcal-files in the official Stata directories, your site's directory (SITE), your current working directory (.), your personal directory (PERSONAL), and your directory for materials written by other users (PLUS). On this writer's computer, these directories happen to be

```
. sysdir
    STATA:  C:\Program Files\Stata12\
  UPDATES:  C:\Program Files\Stata12\ado\updates\
     BASE:  C:\Program Files\Stata12\ado\base\
     SITE:  C:\Program Files\Stata12\ado\site\
     PLUS:  C:\ado\plus\
 PERSONAL:  C:\ado\personal\
 OLDPLACE:  C:\ado\
```

Place calendars that you write into ., PERSONAL, or SITE. Calendars you obtain from others using net or ssc will be placed by those commands into PLUS. See [P] **sysdir**, [R] **net**, and [R] **ssc**.

How to debug stbcal-files

Stbcal-files are loaded automatically as they are needed, and because this can happen anytime, even at inopportune moments, no output is produced. If there are errors in the file, no mention is made of the problem, and thereafter Stata simply acts as if it had never found the file, which is to say, variables with %tbcalname formats are displayed in %g format.

You can tell Stata to load a calendar file right now and to show you the output, including error messages. Type

```
. bcal load calname
```

It does not matter where calname.stbcal is stored, Stata will find it. It does not matter whether Stata has already loaded calname.stbcal, either secretly or because you previously instructed the file be loaded. It will be reloaded, you will see what you wrote, and you will see any error messages.

Ideas for calendars that may not occur to you

Business calendars obviously are not restricted to businesses, and neither do they have to be restricted to days.

Say you have weekly data and want to create a calendar that contains only Mondays. You could code

```
————————————————————————————— begin mondays.stbcal ————

version 12
purpose "Mondays only"
range 04jan1960 06jan2020
centerdate 04jan1960
omitdow (Tu We Th Fr Sa Su)

————————————————————————————— end mondays.stbcal ————
```

Say you have semimonthly data and want to include the 1st and 15th of every month. You could code

```
————————————————————————————— begin smnth.stbcal ————

version 12
purpose "Semimonthly"
range 01jan1960 15dec2020
centerdate 01jan1960
omit date 2jan*
omit date 3jan*
    .

    .
omit date 14jan*
omit date 16jan*
    .

    .
omit date 31jan*
omit date  2feb*
    .

    .
————————————————————————————— end smnth.stbcal ————
```

Forgive the ellipses, but this file will be long. Even so, you have to create it only once.

As a final example, say that you just want Stata's %td dates, but you wish they were centered on 01jan1970 rather than on 01jan1960. You could code

```
————————————————————————————— begin rectr.stbcal ————

version 12
Purpose "%td centered on 01jan1970"
range 01jan1800 31dec2999
centerdate 01jan1970
————————————————————————————— end rectr.stbcal ————
```

Also see

[D] **bcal** — Business calendar file manipulation

[D] **datetime business calendars** — Business calendars

[D] **datetime** — Date and time values and variables

Title

> **datetime display formats** — Display formats for dates and times

Syntax

The formats for displaying Stata internal form (SIF) dates and times in human readable form (HRF) are

SIF type	Display format to present SIF in HRF
datetime/c	%tc [*details*]
datetime/C	%tC [*details*]
date	%td [*details*]
weekly date	%tw [*details*]
monthly date	%tm [*details*]
quarterly date	%tq [*details*]
half-yearly date	%th [*details*]
yearly date	%ty [*details*]

The optional *details* allows you to control how results appear and is composed of a sequence of the following codes:

Code	Meaning	Output
CC	century-1	01–99
cc	century-1	1–99
YY	2-digit year	00–99
yy	2-digit year	0–99
JJJ	day within year	001–366
jjj	day within year	1–366
Mon	month	Jan, Feb, ..., Dec
Month	month	January, February, ..., December
mon	month	jan, feb, ..., dec
month	month	january, february, ..., december
NN	month	01–12
nn	month	1–12
DD	day within month	01–31
dd	day within month	1–31

DAYNAME	day of week	Sunday, Monday, ... (aligned)
Dayname	day of week	Sunday, Monday, ... (unaligned)
Day	day of week	Sun, Mon, ...
Da	day of week	Su, Mo, ...
day	day of week	sun, mon, ...
da	day of week	su, mo, ...
h	half	1–2
q	quarter	1–4
WW	week	01–52
ww	week	1–52
HH	hour	00–23
Hh	hour	00–12
hH	hour	0–23
hh	hour	0–12
MM	minute	00–59
mm	minute	0–59
SS	second	00–60 (sic, due to leap seconds)
ss	second	0–60 (sic, due to leap seconds)
.s	tenths	.0–.9
.ss	hundredths	.00–.99
.sss	thousandths	.000–.999
am	show am or pm	am or pm
a.m.	show a.m. or p.m.	a.m. or p.m.
AM	show AM or PM	AM or PM
A.M.	show A.M. or P.M.	A.M. or P.M.
.	display period	.
,	display comma	,
:	display colon	:
–	display hyphen	-
_	display space	
/	display slash	/
\	display backslash	\
!_c_	display character	_c_
+	separator (see note)	

Note: + displays nothing; it may be used to separate one code from the next to make the format more readable. + is never necessary. For instance, %tchh:MM+am and %tchh:MMam have the same meaning, as does %tc+hh+:+MM+am.

When *details* is not specified, it is equivalent to specifying

Format	Implied (fully specified) format
%tC	%tCDDmonCCYY_HH:MM:SS
%tc	%tcDDmonCCYY_HH:MM:SS
%td	%tdDDmonCCYY
%tw	%twCCYY!www
%tm	%tmCCYY!mnn
%tq	%tqCCYY!qq
%th	%thCCYY!hh
%ty	%tyCCYY

That is, typing

    . format mytimevar %tc

has the same effect as typing

    . format mytimevar %tcDDmonCCYY_HH:MM:SS

Format %tcDDmonCCYY_HH:MM:SS is interpreted as

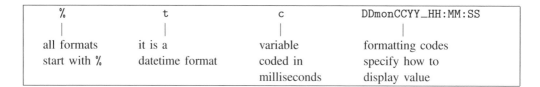

Description

Stata stores dates and times numerically in one of the eight SIFs. An SIF might be 18,282 or even 1,579,619,730,000. Place the appropriate format on it, and the 18,282 is displayed as 20jan2010 (%td). The 1,579,619,730,000 is displayed as 20jan2010 15:15:30 (%tc).

If you specify additional format characters, you can change how the result is displayed. Rather than 20jan2010, you could change it to 2010.01.20; January 20, 2010; or 1/20/10. Rather than 20jan2010 15:15:30, you could change it to 2010.01.20 15:15; January 20, 2010 3:15 pm; or Wed Jan 20 15:15:30 2010.

See [D] **datetime** for an introduction to Stata's dates and times.

Remarks

Remarks are presented under the following headings:

Specifying display formats
Times are truncated, not rounded, when displayed

Specifying display formats

Rather than using the default format 20jan2010, you could display the SIF date variable in one of these formats:

> 2010.01.20
> January 20, 2010
> 1/20/10

Likewise, rather than displaying the SIF datetime/c variable in the default format 20jan2010 15:15:30, you could display it in one of these formats:

> 2010.01.20 15:15
> January 20, 2010 3:15 pm
> Wed Jan 20 15:15:30 2010

Here is how to do it:

1. 2010.01.20
   format *mytdvar* %tdCCYY.NN.DD

2. January 20, 2010
   format *mytdvar* %tdMonth_dd,_CCYY

3. 1/20/10
   format *mytdvar* %tdnn/dd/YY

4. 2010.01.20 15:15
   format *mytcvar* %tcCCYY.NN.DD_HH:MM

5. January 20, 2010 3:15 pm
   format *mytcvar* %tcMonth_dd,_CCYY_hh:MM_am
   Code am at the end indicates that am or pm should be displayed, as appropriate.

6. Wed Jan 20 15:15:30 2010
   format *mytcvar* %tcDay_Mon_DD_HH:MM:SS_CCYY

In examples 1 to 3, the formats each begin with %td, and in examples 4 to 6, the formats begin with %tc. It is important that you specify the opening correctly—namely, as % + t + *third_character*. The third character indicates the particular SIF encoding type, which is to say, how the numeric value is to be interpreted. You specify %tc... for datetime/c variables, %tC... for datetime/C, %td... for date, and so on.

The default format for datetime/c and datetime/C variables omits the fraction of seconds; 15:15:30.000 is displayed as 15:15:30. If you wish to see the fractional seconds, specify the format

> %tcDDmonCCYY_HH:MM:SS.sss

or

> %tCDDmonCCYY_HH:MM:SS.sss

as appropriate.

Times are truncated, not rounded, when displayed

Consider the time 11:32:59.999. Other, less precise, ways of writing that time are

        11:32:59.99
        11:32:59.9
        11:32:59
        11:32

That is, when you suppress the display of more-detailed components of the time, the parts that are displayed are not rounded. Stata displays time just as a digital clock would; the time is 11:32 right up until the instant that it becomes 11:33.

Also see

[D] **datetime** — Date and time values and variables

[D] **datetime business calendars** — Business calendars

[D] **datetime translation** — String to numeric date translation functions

Title

datetime translation — String to numeric date translation functions

Syntax

The string-to-numeric date and time translation functions are

Desired SIF type	String-to-numeric translation function
datetime/c	clock(*HRFstr*, *mask* [, *topyear*])
datetime/C	Clock(*HRFstr*, *mask* [, *topyear*])
date	date(*HRFstr*, *mask* [, *topyear*])
weekly date	weekly(*HRFstr*, *mask* [, *topyear*])
monthly date	monthly(*HRFstr*, *mask* [, *topyear*])
quarterly date	quarterly(*HRFstr*, *mask* [, *topyear*])
half-yearly date	halfyearly(*HRFstr*, *mask* [, *topyear*])
yearly date	yearly(*HRFstr*, *mask* [, *topyear*])

where

HRFstr is the string value (HRF) to be translated,

topyear is described in *Working with two-digit years*, below,

and *mask* specifies the order of the date and time components and is a string composed of a sequence of these elements:

Code	Meaning
M	month
D	day within month
Y	4-digit year
19Y	2-digit year to be interpreted as 19xx
20Y	2-digit year to be interpreted as 20xx
h	hour of day
m	minutes within hour
s	seconds within minute
#	ignore one element

Blanks are also allowed in *mask*, which can make the *mask* easier to read, but they otherwise have no significance.

Examples of *masks* include

"MDY"   *HRFstr* contains month, day, and year, in that order.

"MD19Y"   means the same as "MDY" except that *HRFstr* may contain two-digit years, and when it does, they are to be treated as if they are 4-digit years beginning with 19.

"MDYhms"   *HRFstr* contains month, day, year, hour, minute, and second, in that order.

"MDY hms"   means the same as "MDYhms"; the blank has no meaning.

"MDY#hms"   means that one element between the year and the hour is to be ignored. For example, *HRFstr* contains values like "1-1-2010 at 15:23:17" or values like "1-1-2010 at 3:23:17 PM".

Description

These functions translate dates and times recorded as strings containing human readable form (HRF) to the desired Stata internal form (SIF). See [D] **datetime** for an introduction to Stata's date and time features.

Also see *Using dates and times from other software* in [D] **datetime**.

Remarks

Remarks are presented under the following headings:

> *Introduction*
> *Specifying the mask*
> *How the HRF-to-SIF functions interpret the mask*
> *Working with two-digit years*
> *Working with incomplete dates and times*
> *Translating run-together dates, such as 20060125*
> *Valid times*
> *The clock() and Clock() functions*
> *Why there are two SIF datetime encodings*
> *Advice on using datetime/c and datetime/C*
> *Determining when leap seconds occurred*
> *The date() function*
> *The other translation functions*

Introduction

The HRF-to-SIF translation functions are used to translate string HRF dates, such as "08/12/06", "12-8-2006", "12 Aug 06", "12aug2006 14:23", and "12 aug06 2:23 pm", to SIF. The HRF-to-SIF translation functions are typically used after importing or reading data. You read the date information into string variables and then the HRF-to-SIF functions translate the string into something Stata can use, namely, an SIF numeric variable.

You use generate to create the SIF variables. The translation functions are used in the expressions, such as

```
. generate double time_admitted = clock(time_admitted_str, "DMYhms")
. format time_admitted %tc
. generate date_hired = date(date_hired_str, "MDY")
. format date_hired %td
```

Every translation function—such as `clock()` and `date()` above—requires these two arguments:

1. the *HRFstr* specifying the string to be translated

2. the *mask* specifying the order in which the date and time components appear in *HRFstr*

Notes:

1. You choose the translation function `clock()`, `Clock()`, `date()`, ... according to the type of SIF value you want returned.

2. You specify the mask according to the contents of *HRFstr*.

Usually, you will want to translate an *HRFstr* containing "2006.08.13 14:23" to an SIF datetime/c or datetime/C value and translate an *HRFstr* containing "2006.08.13" to an SIF date value. If you wish, however, it can be the other way around. In that case, the detailed string would translate to an SIF date value corresponding to just the date part, 13aug2006, and the less detailed string would translate to an SIF datetime value corresponding to 13aug2006 00:00:00.000.

Specifying the mask

An argument *mask* is a string specifying the order of the date and time components in *HRFstr*. Examples of HRF date strings and the mask required to translate them include the following:

HRFstr	Corresponding mask	
01dec2006 14:22	"DMYhm"	
01-12-2006 14.22	"DMYhm"	
1dec2006 14:22	"DMYhm"	
1-12-2006 14:22	"DMYhm"	
01dec06 14:22	"DM20Yhm"	
01-12-06 14.22	"DM20Yhm"	
December 1, 2006 14:22	"MDYhm"	
2006 Dec 01 14:22	"YMDhm"	
2006-12-01 14:22	"YMDhm"	
2006-12-01 14:22:43	"YMDhms"	
2006-12-01 14:22:43.2	"YMDhms"	
2006-12-01 14:22:43.21	"YMDhms"	
2006-12-01 14:22:43.213	"YMDhms"	
2006-12-01 2:22:43.213 pm	"YMDhms"	(see note 1)
2006-12-01 2:22:43.213 pm.	"YMDhms"	
2006-12-01 2:22:43.213 p.m.	"YMDhms"	
2006-12-01 2:22:43.213 P.M.	"YMDhms"	
20061201 1422	"YMDhm"	
14:22	"hm"	(see note 2)
2006-12-01	"YMD"	
Fri Dec 01 14:22:43 CST 2006	"#MDhms#Y"	

Notes:

1. Nothing special needs to be included in *mask* to process a.m. and p.m. markers. When you include code h, the HRF-to-SIF functions automatically watch for meridian markers.

2. You specify the mask according to what is contained in *HRFstr*. If that is a subset of what the selected SIF type could record, the remaining elements are set to their defaults. clock("14:22", "hm") produces 01jan1960 14:22:00 and clock("2006-12-01", "YMD") produces 01dec2006 00:00:00. date("jan 2006", "MY") produces 01jan2006.

mask may include spaces so that it is more readable; the spaces have no meaning. Thus you can type

```
. generate double admit = clock(admitstr, "#MDhms#Y")
```

or type

```
. generate double admit = clock(admitstr, "# MD hms # Y")
```

and which one you use makes no difference.

How the HRF-to-SIF functions interpret the mask

The HRF-to-SIF functions apply the following rules when interpreting *HRFstr*:

1. For each HRF string to be translated, remove all punctuation except for the period separating seconds from tenths, hundredths, and thousandths of seconds. Replace removed punctuation with a space.

2. Insert a space in the string everywhere that a letter is next to a number, or vice versa.

3. Interpret the resulting elements according to *mask*.

For instance, consider the string

01dec2006 14:22

Under rule 1, the string becomes

01dec2006 14 22

Under rule 2, the string becomes

01 dec 2006 14 22

Finally, the HRF-to-SIF functions apply rule 3. If the mask is "DMYhm", then the functions interpret "01" as the day, "dec" as the month, and so on.

Or consider the string

Wed Dec 01 14:22:43 CST 2006

Under rule 1, the string becomes

Wed Dec 01 14 22 43 CST 2006

Applying rule 2 does not change the string. Now rule 3 is applied. If the mask is "#MDhms#Y", the translation function skips "Wed", interprets "Dec" as the month, and so on.

The # code serves a second purpose. If it appears at the end of the mask, it specifies that the rest of *string* is to be ignored. Consider translating the string

Wed Dec 01 14 22 43 CST 2006 patient 42

The mask code that previously worked when "patient 42" was not part of the string, "#MDhms#Y", will result in a missing value in this case. The functions are careful in the translation, and if the whole string is not used, they return missing. If you end the mask in #, however, the functions ignore the rest of the string. Changing the mask from "#MDhms#Y" to "#MDhms#Y#" will produce the desired result.

Working with two-digit years

Consider translating the string 01-12-06 14:22, which is to be interpreted as 01dec2006 14:22:00. The translation functions provide two ways of doing this.

The first is to specify the assumed prefix in the mask. The string 01-12-06 14:22 can be read by specifying the mask "DM20Yhm". If we instead wanted to interpret the year as 1906, we would specify the mask "DM19Yhm". We could even interpret the year as 1806 by specifying "DM18Yhm".

What if our data include 01-12-06 14:22 and include 15-06-98 11:01? We want to interpret the first year as being in 2006 and the second year as being in 1998. That is the purpose of the optional argument *topyear*:

$$clock(\textit{string}, \textit{mask} \; \big[\; , \; \textit{topyear} \big])$$

When you specify *topyear*, you are stating that when years in *string* are two digits, the full year is to be obtained by finding the largest year that does not exceed *topyear*. Thus you could code

```
. generate double timestamp = clock(timestr, "DMYhm", 2020)
```

The two-digit year 06 would be interpreted as 2006 because 2006 does not exceed 2020. The two-digit year 98 would be interpreted as 1998 because 2098 does exceed 2020.

Working with incomplete dates and times

The translation functions do not require that every component of the date and time be specified.

Translating 2006-12-01 with mask "YMD" results in 01dec2006 00:00:00.

Translating 14:22 with mask "hm" results in 01jan1960 14:22:00.

Translating 11-2006 with mask "MY" results in 01nov2006 00:00:00.

The default for a component, if not specified in the mask, is

Code	Default (if not specified)
M	01
D	01
Y	1960
h	00
m	00
s	00

Thus if you have data recording "14:22", meaning a duration of 14 hours and 22 minutes or the time 14:22 each day, you can translate it with clock(*HRFstr*, "hm"). See *Obtaining and working with durations* in [D] **datetime**.

Translating run-together dates, such as 20060125

The translation functions will translate dates and times that are run together, such as 20060125, 060125, and 20060125110215 (which is 25jan2006 11:02:15). You do not have to do anything special to translate them:

```
. display %d date("20060125", "YMD")
25jan2006
. display %td date("060125", "20YMD")
25jan2006
. display %tc clock("20060125110215", "YMDhms")
25jan2006 11:02:15
```

In a data context, you could type

```
. gen startdate = date(startdatestr, "YMD")
. gen double starttime = clock(starttimestr, "YMDhms")
```

Remember to read the original date into a string. If you mistakenly read the date as numeric, the best advice is to read the date again. Numbers such as 20060125 and 20060125110215 will be rounded unless they are stored as doubles.

If you mistakenly read the variables as numeric and have verified that rounding did not occur, you can convert the variable from numeric to string by using the string() function, which comes in one- and two-argument forms. You will need the two-argument form:

```
. gen str startdatestr = string(startdatedouble, "%10.0g")
. gen str starttimestr = string(starttimedouble, "%16.0g")
```

If you omitted the format, string() would produce 2.01e+07 for 20060125 and 2.01e+13 for 20060125110215. The format we used had a width that was 2 characters larger than the length of the integer number, although using a too-wide format does no harm.

Valid times

27:62:90 is an invalid time. If you try to convert 27:62:90 to a datetime value, you will obtain a missing value.

24:00:00 is also invalid. A correct time would be 00:00:00 of the next day.

In $hh{:}mm{:}ss$, the requirements are $0 \le hh < 24$, $0 \le mm < 60$, and $0 \le ss < 60$, although sometimes 60 is allowed. 31dec2005 23:59:60 is an invalid datetime/c but a valid datetime/C. 31dec2005 23:59:60 includes an inserted leap second.

30dec2005 23:59:60 is invalid in both datetime encodings. 30dec2005 23:59:60 did not include an inserted leap second. A correct datetime would be 31dec2005 00:00:00.

The clock() and Clock() functions

Stata provides two separate datetime encodings that we call SIF datetime/c and SIF datetime/C and that others would call "times assuming 86,400 seconds per day" and "times adjusted for leap seconds" or, equivalently, UTC times.

The syntax of the two functions is the same:

clock(*HRFstr*, *mask* [, *topyear*])

Clock(*HRFstr*, *mask* [, *topyear*])

Function `Clock()` is nearly identical to function `clock()`, except that `Clock()` returns a datetime/C value rather than a datetime/c value. For instance,

$$\text{Noon of 23nov2010} = 1{,}606{,}132{,}800{,}000 \text{ in datetime/c}$$
$$= 1{,}606{,}132{,}824{,}000 \text{ in datetime/C}$$

They differ because 24 seconds have been inserted into datetime/C between 01jan1960 and 23nov2010. Correspondingly, `Clock()` understands times in which there are leap seconds, such as 30jun1997 23:59:60. `clock()` would consider 30jun1997 23:59:60 an invalid time and so return a missing value.

Why there are two SIF datetime encodings

Stata provides two different datetime encodings, SIF datetime/c and SIF datetime/C.

SIF datetime/c assumes that there are $24 \times 60 \times 60 \times 1000$ ms per day, just as an atomic clock does. Atomic clocks count oscillations between the nucleus and the electrons of an atom and thus provide a measurement of the real passage of time.

Time of day measurements have historically been based on astronomical observation, which is a fancy way of saying that the measurements are based on looking at the sun. The sun should be at its highest point at noon, right? So however you might have kept track of time—by falling grains of sand or a wound-up spring—you would have periodically reset your clock and then gone about your business. In olden times, it was understood that the 60 seconds per minute, 60 minutes per hour, and 24 hours per day were theoretical goals that no mechanical device could reproduce accurately. These days, we have more formal definitions for measurements of time. One second is 9,192,631,770 periods of the radiation corresponding to the transition between two levels of the ground state of cesium 133. Obviously, we have better equipment than the ancients, so problem solved, right? Wrong. There are two problems: the formal definition of a second is just a little too short to use for accurately calculating the length of a day, and the Earth's rotation is slowing down.

As a result, since 1972, leap seconds have been added to atomic clocks once or twice a year to keep time measurements in synchronization with Earth's rotation. Unlike leap years, however, there is no formula for predicting when leap seconds will occur. Earth may be on average slowing down, but there is a large random component to that. As a result, leap seconds are determined by committee and announced 6 months before they are inserted. Leap seconds are added, if necessary, on the end of the day on June 30 and December 31 of the year. The exact times are designated as 23:59:60.

Unadjusted atomic clocks may accurately mark the passage of real time, but you need to understand that leap seconds are every bit as real as every other second of the year. Once a leap second is inserted, it ticks just like any other second and real things can happen during that tick.

You may have heard of terms such as GMT and UTC.

GMT is the old Greenwich Mean Time that is based on astronomical observation. GMT has been replaced by UTC.

UTC stands for coordinated universal time. It is measured by atomic clocks and is occasionally corrected for leap seconds. UTC is derived from two other times, UT1 and TAI. UT1 is the mean solar time, with which UTC is kept in sync by the occasional addition of a leap second. TAI is the atomic time on which UTC is based. TAI is a statistical combination of various atomic chronometers and even it has not ticked uniformly over its history; see http://www.ucolick.org/~sla/leapsecs/timescales.html and especially http://www.ucolick.org/~sla/leapsecs/dutc.html#TAI.

UNK is our term for the time standard most people use. UNK stands for unknown or unknowing. UNK is based on a recent time observation, probably UTC, and it just assumes that there are 86,400 seconds per day after that.

The UNK standard is adequate for many purposes, and when using it you will want to use SIF datetime/c rather than the leap second–adjusted datetime/C encoding. If you are using computer-timestamped data, however, you need to find out whether the timestamping system accounted for leap-second adjustment. Problems can arise even if you do not care about losing or gaining a second here and there.

For instance, you may import from other systems timestamp values recorded in the number of milliseconds that have passed since some agreed upon date. You may do this, but if you choose the wrong encoding scheme (choose datetime/c when you should choose datetime/C, or vice versa), more recent times will be off by 24 seconds.

To avoid such problems, you may decide to import and export data by using HRF such as "Fri Aug 18 14:05:36 CDT 2010". This method has advantages, but for datetime/C (UTC) encoding, times such as 23:59:60 are possible. Some systems will refuse to decode such times.

Stata refuses to decode 23:59:60 in the datetime/c encoding (function `clock()`) and accepts it with datetime/C (function `Clock()`). When datetime/C function `Clock()` sees a time with a 60th second, `Clock()` verifies that the time is one of the official leap seconds. Thus when translating from printable forms, try assuming datetime/c and check the result for missing values. If there are none, then you can assume your use of datetime/c was valid. If there are missing values and they are due to leap seconds and not some other error, however, you must use datetime/C `Clock()` to translate the HRF. After that, if you still want to work in datetime/c units, use function `cofC()` to translate datetime/C values into datetime/c.

If precision matters, the best way to process datetime/C data is simply to treat them that way. The inconvenience is that you cannot assume that there are 86,400 seconds per day. To obtain the duration between dates, you must subtract the two time values involved. The other difficulty has to do with dealing with dates in the future. Under the datetime/C (UTC) encoding, there is no set value for any date more than six months in the future. Below is a summary of advice.

Advice on using datetime/c and datetime/C

Stata provides two datetime encodings:

1. datetime/C, also known as UTC, which accounts for leap seconds

2. datetime/c, which ignores leap seconds (it assumes 86,400 seconds/day)

Systems vary in how they treat time variables. SAS ignores leap seconds. Oracle includes them. Stata handles either situation. Here is our advice:

- If you obtain data from a system that accounts for leap seconds, import using Stata's datetime/C encoding.

  a. If you later need to export data to a system that does not account for leap seconds, use Stata's `cofC()` function to translate time values before exporting.

  b. If you intend to `tsset` the time variable and the analysis will be at the second level or finer, just `tsset` the datetime/C variable, specifying the appropriate `delta()` if necessary—for example, `delta(1000)` for seconds.

  c. If you intend to `tsset` the time variable and the analysis will be coarser than the second level (minute, hour, etc.), create a datetime/c variable from the datetime/C variable (`generate double` *tctime* = `cofC(`*tCtime*`)`) and `tsset` that, specifying the appropriate `delta()` if necessary. You must do that because in a datetime/C variable, there are not necessarily 60 seconds in a minute; some minutes have 61 seconds.

- If you obtain data from a system that ignores leap seconds, use Stata's datetime/c encoding.

    a. If you later need to export data to a system that does account for leap seconds, use Stata's `Cofc()` function to translate time values before exporting.

    b. If you intend to `tsset` the time variable, just `tsset` it, specifying the appropriate `delta()`.

Some users prefer always to use Stata's datetime/c because `%tc` values are a little easier to work with. You can always use datetime/c if

- you do not mind having up to 1 second of error and

- you do not import or export numerical values (clock ticks) from other systems that are using leap seconds, because doing so could introduce nearly 30 seconds of error.

Remember these two things if you use datetime/C variables:

1. The number of seconds between two dates is a function of when the dates occurred. Five days from one date is not simply a matter of adding $5 \times 24 \times 60 \times 60 \times 1000$ ms. You might need to add another 1,000 ms. Three hundred sixty-five days from now might require adding 1,000 or 2,000 ms. The longer the span, the more you might have to add. The best way to add durations to datetime/C variables is to extract the components, add to them, and then reconstruct from the numerical components.

2. You cannot accurately predict datetimes more than six months into the future. We do not know what the datetime/C value of 25dec2026 00:00:00 will be because every year along the way, the International Earth Rotation Reference Systems Service (IERS) will twice announce whether a leap second will be inserted.

You can help alleviate these inconveniences. Face west and throw rocks. The benefit will be transitory only if the rocks land back on Earth, so you need to throw them really hard. We know what you are thinking, but this does not need to be a coordinated effort.

Determining when leap seconds occurred

Stata system file `leapseconds.maint` lists the dates on which leap seconds occurred. The file is updated periodically (see [R] **update**; the file is updated when you `update all`), and Stata's datetime/C functions access the file to know when leap seconds occurred.

You can access it, too. To view the file, type

```
. viewsource leapseconds.maint
```

The date() function

The syntax of the `date()` function is

`date(`*string*`, ` *mask* `[`, *topyear* `])`

The `date()` function is identical to `clock()` except that `date()` returns an SIF date value rather than a datetime value. The `date()` function is the same as `dofc(clock())`.

The other translation functions

The other translation functions are

SIF type	HRF-to-SIF translation function
weekly date	weekly(*HRFstr*, *mask* [, *topyear*])
monthly date	monthly(*HRFstr*, *mask* [, *topyear*])
quarterly date	quarterly(*HRFstr*, *mask* [, *topyear*])
half-yearly date	halfyearly(*HRFstr*, *mask* [, *topyear*])

HRFstr is the value to be translated.
mask specifies the order of the components.
topyear is described in *Working with two-digit years*, above.

These functions are rarely used because data seldom arrive in these formats.

Each of the functions translates a pair of numbers: weekly() translates a year and a week number (1–52), monthly() translates a year and a month number (1–12), quarterly() translates a year and a quarter number (1–4), and halfyearly() translates a year and a half number (1–2).

The masks allowed are far more limited than the masks for clock(), Clock(), and date():

Code	Meaning
Y	4-digit year
19Y	2-digit year to be interpreted as 19*xx*
20Y	2-digit year to be interpreted as 20*xx*
W	week number (weekly() only)
M	month number (monthly() only)
Q	quarter number (quarterly() only)
H	half-year number (halfyearly() only)

The pair of numbers to be translated must be separated by a space or punctuation. No extra characters are allowed.

Also see

[D] **datetime** — Date and time values and variables

[D] **datetime business calendars** — Business calendars

[D] **datetime display formats** — Display formats for dates and times

Title

> **describe** — Describe data in memory or in file

Syntax

Describe data in memory

    d̲escribe [*varlist*] [, *memory_options*]

Describe data in file

    d̲escribe [*varlist*] using *filename* [, *file_options*]

memory_options	Description
s̲imple	display only variable names
s̲hort	display only general information
f̲ullnames	do not abbreviate variable names
n̲umbers	display variable number along with name
replace	make dataset, not written report, of description
clear	for use with replace
va̲rlist	save r(varlist) and r(sortlist) in addition to usual saved results; programmer's option

varlist does not appear in the dialog box.

file_options	Description
s̲hort	display only general information
s̲imple	display only variable names
va̲rlist	save r(varlist) and r(sortlist) in addition to usual saved results; programmer's option

varlist does not appear in the dialog box.

Menu

describe

Data > Describe data > Describe data in memory

describe using

Data > Describe data > Describe data in file

Description

describe produces a summary of the dataset in memory or of the data stored in a Stata-format dataset.

For a compact listing of variable names, use describe, simple.

Options to describe data in memory

simple displays only the variable names in a compact format. simple may not be combined with other options.

short suppresses the specific information for each variable. Only the general information (number of observations, number of variables, size, and sort order) is displayed.

fullnames specifies that describe display the full names of the variables. The default is to present an abbreviation when the variable name is longer than 15 characters. describe using always shows the full names of the variables, so fullnames may not be specified with describe using.

numbers specifies that describe present the variable number with the variable name. If numbers is specified, variable names are abbreviated when the name is longer than eight characters. The numbers and fullnames options may not be specified together. numbers may not be specified with describe using.

replace and clear are alternatives to the options above. describe usually produces a written report, and the options above specify what the report is to contain. If you specify replace, however, no report is produced; the data in memory are instead replaced with data containing the information that the report would have presented. Each observation of the new data describes a variable in the original data; see *describe, replace* below.

clear may be specified only when replace is specified. clear specifies that the data in memory be cleared and replaced with the description information, even if the original data have not been saved to disk.

The following option is available with describe but is not shown in the dialog box:

varlist, an option for programmers, specifies that, in addition to the usual saved results, r(varlist) and r(sortlist) be saved, too. r(varlist) will contain the names of the variables in the dataset. r(sortlist) will contain the names of the variables by which the data are sorted.

Options to describe data in file

short suppresses the specific information for each variable. Only the general information (number of observations, number of variables, size, and sort order) is displayed.

simple displays only the variable names in a compact format. simple may not be combined with other options.

The following option is available with describe but is not shown in the dialog box:

varlist, an option for programmers, specifies that, in addition to the usual saved results, r(varlist) and r(sortlist) be saved, too. r(varlist) will contain the names of the variables in the dataset. r(sortlist) will contain the names of the variables by which the data are sorted.

Because Stata/MP and Stata/SE can create truly large datasets, there might be too many variables in a dataset for their names to be stored in r(varlist), given the current maximum length of macros, as determined by set maxvar. Should that occur, describe using will issue the error message "too many variables", r(103).

Remarks

Remarks are presented under the following headings:

> describe
> describe, replace

describe

If describe is typed with no operands, the contents of the dataset currently in memory are described.

The *varlist* in the describe using syntax differs from standard Stata varlists in two ways. First, you cannot abbreviate variable names; that is, you have to type displacement rather than displ. However, you can use the abbreviation character (~) to indicate abbreviations, for example, displ~. Second, you may not refer to a range of variables; specifying price-trunk is considered an error.

▷ Example 1

The basic description includes some general information on the number of variables and observations, along with a description of every variable in the dataset:

```
. use http://www.stata-press.com/data/r12/states
(State data)

. describe, numbers
Contains data from http://www.stata-press.com/data/r12/states.dta
  obs:            50                          State data
  vars:            5                          3 Jan 2011 15:17
  size:         1,100                          (_dta has notes)
```

	variable name	storage type	display format	value label	variable label
1.	state	str8	%9s		
2.	region	int	%8.0g	reg	Census Region
3.	median~e	float	%9.0g		Median Age
4.	marria~e	long	%12.0g		Marriages per 100,000
5.	divorc~e	long	%12.0g		Divorces per 100,000

```
Sorted by:  region
```

In this example, the dataset in memory comes from the file states.dta and contains 50 observations on 5 variables. The dataset is labeled "State data" and was last modified on January 3, 2011, at 15:17 (3:17 p.m.). The "_dta has notes" message indicates that a note is attached to the dataset; see [U] **12.7 Notes attached to data**.

The first variable, state, is stored as a str8 and has a display format of %9s.

The next variable, region, is stored as an int and has a display format of %8.0g. This variable has associated with it a *value label* called reg, and the variable is labeled Census Region.

The third variable, which is abbreviated median~e, is stored as a float, has a display format of %9.0g, has no value label, and has a variable label of Median Age. The variables that are abbreviated marria~e and divorc~e are both stored as longs and have display formats of %12.0g. These last two variables are labeled Marriages per 100,000 and Divorces per 100,000, respectively.

The data are sorted by region.

Because we specified the numbers option, the variables are numbered; for example, region is variable 2 in this dataset.

◁

▷ Example 2

To view the full variable names, we could omit the numbers option and specify the fullnames option.

```
. describe, fullnames
Contains data from http://www.stata-press.com/data/r12/states.dta
  obs:            50                          State data
  vars:            5                          3 Jan 2011 15:17
  size:        1,100                          (_dta has notes)
```

variable name	storage type	display format	value label	variable label
state	str8	%9s		
region	int	%8.0g	reg	Census Region
median_age	float	%9.0g		Median Age
marriage_rate	long	%12.0g		Marriages per 100,000
divorce_rate	long	%12.0g		Divorces per 100,000

```
Sorted by:  region
```

Here we did not need to specify the fullnames option to see the unabbreviated variable names because the longest variable name is 13 characters. Omitting the numbers option results in 15-character variable names being displayed.

◁

❏ Technical note

The describe listing above also shows that the size of the dataset is 1,100. If you are curious,

$$(8 + 2 + 4 + 4 + 4) \times 50 = 1100$$

The numbers 8, 2, 4, 4, and 4 are the storage requirements for a str8, int, float, long, and long, respectively; see [U] **12.2.2 Numeric storage types**. Fifty is the number of observations in the dataset.

❏

> ## Example 3

If we specify the `short` option, only general information about the data is presented:

```
. describe, short
Contains data from http://www.stata-press.com/data/r12/states.dta
    obs:            50                      State data
   vars:             5                      3 Jan 2011 15:17
   size:         1,100
Sorted by:  region
```
◁

If we specify a *varlist*, only the variables in that *varlist* are described.

> ## Example 4

Let's change datasets. The `describe` *varlist* command is particularly useful when combined with the '*' wildcard character. For instance, we can describe all the variables whose names start with pop by typing `describe pop*`:

```
. use http://www.stata-press.com/data/r12/census
(1980 Census data by state)

. describe pop*
```

variable name	storage type	display format	value label	variable label
pop	long	%12.0gc		Population
poplt5	long	%12.0gc		Pop, < 5 year
pop5_17	long	%12.0gc		Pop, 5 to 17 years
pop18p	long	%12.0gc		Pop, 18 and older
pop65p	long	%12.0gc		Pop, 65 and older
popurban	long	%12.0gc		Urban population

We can `describe` the variables state, region, and pop18p by specifying them:

```
. describe state region pop18p
```

variable name	storage type	display format	value label	variable label
state	str14	%-14s		State
region	int	%-8.0g	cenreg	Census region
pop18p	long	%12.0gc		Pop, 18 and older

◁

Typing `describe using` *filename* describes the data stored in *filename*. If an extension is not specified, `.dta` is assumed.

> Example 5

We can describe the contents of states.dta without disturbing the data that we currently have in memory by typing

```
. describe using http://www.stata-press.com/data/r12/states
Contains data                             State data
    obs:            50                    3 Jan 2011 15:17
   vars:             5
   size:         1,300
```

variable name	storage type	display format	value label	variable label
state	str8	%9s		
region	int	%8.0g	reg	Census Region
median_age	float	%9.0g		Median Age
marriage_rate	long	%12.0g		Marriages per 100,000
divorce_rate	long	%12.0g		Divorces per 100,000

```
Sorted by:  region
```

◁

describe, replace

describe with the replace option is rarely used, although you may sometimes find it convenient.

Think of describe, replace as separate from but related to describe without the replace option. Rather than producing a written report, describe, replace produces a new dataset that contains the same information a written report would. For instance, try the following:

```
. sysuse auto, clear
. describe
```
(report appears; data in memory unchanged)
```
. list
```
(visual proof that data are unchanged)
```
. describe, replace
```
(no report appears, but the data in memory are changed!)
```
. list
```
(visual proof that data are changed)

describe, replace changes the original data in memory into a dataset containing an observation for each variable in the original data. Each observation in the new data describes a variable in the original data. The new variables are

1. position, a variable containing the numeric position of the original variable (1, 2, 3, ...).

2. name, a variable containing the name of the original variable, such as "make", "price", "mpg", ...

3. type, a variable containing the storage type of the original variable, such as "str18", "int", "float", ...

4. isnumeric, a variable equal to 1 if the original variable was numeric and equal to 0 if it was string.

5. format, a variable containing the display format of the original variable, such as "%-18s", "%8.0gc", ...

6. `vallab`, a variable containing the name of the value label associated with the original variable, if any.

7. `varlab`, a variable containing the variable label of the original variable, such as `"Make and Model"`, `"Price"`, `"Mileage (mpg)"`, ...

In addition, the data contain the following characteristics:

`_dta[d_filename]`, the name of the file containing the original data.

`_dta[d_filedate]`, the date and time the file was written.

`_dta[d_N]`, the number of observations in the original data.

`_dta[d_sortedby]`, the variables on which the original data were sorted, if any.

Saved results

`describe` saves the following in `r()`:

Scalars
`r(N)`	number of observations
`r(k)`	number of variables
`r(width)`	width of dataset
`r(changed)`	flag indicating data have changed since last saved

Macros
`r(varlist)`	variables in dataset (if `varlist` specified)
`r(sortlist)`	variables by which data are sorted (if `varlist` specified)

`describe, replace` saves nothing in `r()`.

References

Cox, N. J. 1999. dm67: Numbers of missing and present values. *Stata Technical Bulletin* 49: 7–8. Reprinted in *Stata Technical Bulletin Reprints*, vol. 9, pp. 26–27. College Station, TX: Stata Press.

——. 2000. dm78: Describing variables in memory. *Stata Technical Bulletin* 56: 2–4. Reprinted in *Stata Technical Bulletin Reprints*, vol. 10, pp. 15–17. College Station, TX: Stata Press.

——. 2001a. dm67.1: Enhancements to numbers of missing and present values. *Stata Technical Bulletin* 60: 2–3. Reprinted in *Stata Technical Bulletin Reprints*, vol. 10, pp. 7–9. College Station, TX: Stata Press.

——. 2001b. dm78.1: Describing variables in memory: Update to Stata 7. *Stata Technical Bulletin* 60: 3. Reprinted in *Stata Technical Bulletin Reprints*, vol. 10, p. 17. College Station, TX: Stata Press.

Gleason, J. R. 1998. dm61: A tool for exploring Stata datasets (Windows and Macintosh only). *Stata Technical Bulletin* 45: 2–5. Reprinted in *Stata Technical Bulletin Reprints*, vol. 8, pp. 22–27. College Station, TX: Stata Press.

——. 1999. dm61.1: Update to varxplor. *Stata Technical Bulletin* 51: 2. Reprinted in *Stata Technical Bulletin Reprints*, vol. 9, p. 15. College Station, TX: Stata Press.

Also see

[D] **ds** — List variables matching name patterns or other characteristics

[D] **varmanage** — Manage variable labels, formats, and other properties

[D] **compress** — Compress data in memory

[D] **format** — Set variables' output format

[D] **label** — Manipulate labels

[D] **notes** — Place notes in data

[D] **order** — Reorder variables in dataset

[D] **rename** — Rename variable

[D] **cf** — Compare two datasets

[D] **codebook** — Describe data contents

[D] **compare** — Compare two variables

[D] **lookfor** — Search for string in variable names and labels

[SVY] **svydescribe** — Describe survey data

[U] **6 Managing memory**

Title

> **destring** — Convert string variables to numeric variables and vice versa

Syntax

Convert string variables to numeric variables

> destring [*varlist*], { generate(*newvarlist*) | replace } [*destring_options*]

Convert numeric variables to string variables

> tostring *varlist*, { generate(*newvarlist*) | replace } [*tostring_options*]

destring_options	Description
* generate(*newvarlist*)	generate $newvar_1, \ldots, newvar_k$ for each variable in *varlist*
* replace	replace string variables in *varlist* with numeric variables
ignore("*chars*")	remove specified nonnumeric characters
force	convert nonnumeric strings to missing values
float	generate numeric variables as type float
percent	convert percent variables to fractional form
dpcomma	convert variables with commas as decimals to period-decimal format

* Either generate(*newvarlist*) or replace is required.

tostring_options	Description
* generate(*newvarlist*)	generate $newvar_1, \ldots, newvar_k$ for each variable in *varlist*
* replace	replace numeric variables in *varlist* with string variables
force	force conversion ignoring information loss
format(*format*)	convert using specified format
usedisplayformat	convert using display format

* Either generate(*newvarlist*) or replace is required.

Menu

destring

Data > Create or change data > Other variable-transformation commands > Convert variables from string to numeric

tostring

Data > Create or change data > Other variable-transformation commands > Convert variables from numeric to string

133

Description

destring converts variables in *varlist* from string to numeric. If *varlist* is not specified, destring will attempt to convert all variables in the dataset from string to numeric. Characters listed in ignore() are removed. Variables in *varlist* that are already numeric will not be changed. destring treats both empty strings "" and "." as indicating sysmiss (.) and interprets the strings ".a", ".b", ..., ".z" as the extended missing values .a, .b, ..., .z; see [U] **12.2.1 Missing values**. destring also ignores any leading or trailing spaces so that, for example, " " is equivalent to "" and " . " is equivalent to ".".

tostring converts variables in *varlist* from numeric to string. The most compact string format possible is used. Variables in *varlist* that are already string will not be converted.

Options for destring

Either generate() or replace must be specified. With either option, if any string variable contains nonnumeric values not specified with ignore(), then no corresponding variable will be generated, nor will that variable be replaced (unless force is specified).

generate(*newvarlist*) specifies that a new variable be created for each variable in *varlist*. *newvarlist* must contain the same number of new variable names as there are variables in *varlist*. If *varlist* is not specified, destring attempts to generate a numeric variable for each variable in the dataset; *newvarlist* must then contain the same number of new variable names as there are variables in the dataset. Any variable labels or characteristics will be copied to the new variables created.

replace specifies that the variables in *varlist* be converted to numeric variables. If *varlist* is not specified, destring attempts to convert all variables from string to numeric. Any variable labels or characteristics will be retained.

ignore("*chars*") specifies nonnumeric characters to be removed. If any string variable contains any nonnumeric characters other than those specified with ignore(), no action will take place for that variable unless force is also specified. Note that to Stata the comma is a nonnumeric character; see also the dpcomma option below.

force specifies that any string values containing nonnumeric characters, in addition to any specified with ignore(), be treated as indicating missing numeric values.

float specifies that any new numeric variables be created initially as type float. The default is type double; see [D] **data types**. destring attempts automatically to compress each new numeric variable after creation.

percent removes any percent signs found in the values of a variable, and all values of that variable are divided by 100 to convert the values to fractional form. percent by itself implies that the percent sign, "%", is an argument to ignore(), but the converse is not true.

dpcomma specifies that variables with commas as decimal values should be converted to have periods as decimal values.

Options for tostring

Either generate() or replace must be specified. If converting any numeric variable to string would result in loss of information, no variable will be produced unless force is specified. For more details, see force below.

generate(*newvarlist*) specifies that a new variable be created for each variable in *varlist*. *newvarlist* must contain the same number of new variable names as there are variables in *varlist*. Any variable labels or characteristics will be copied to the new variables created.

replace specifies that the variables in *varlist* be converted to string variables. Any variable labels or characteristics will be retained.

force specifies that conversions be forced even if they entail loss of information. Loss of information means one of two circumstances: 1) The result of real(string(*varname*, "*format*")) is not equal to *varname*; that is, the conversion is not reversible without loss of information; 2) replace was specified, but a variable has associated value labels. In circumstance 1, it is usually best to specify usedisplayformat or format(). In circumstance 2, value labels will be ignored in a forced conversion. decode (see [D] **encode**) is the standard way to generate a string variable based on value labels.

format(*format*) specifies that a numeric format be used as an argument to the string() function, which controls the conversion of the numeric variable to string. For example, a format of %7.2f specifies that numbers are to be rounded to two decimal places before conversion to string. See *Remarks* below and [D] **functions** and [D] **format**. format() cannot be specified with usedisplayformat.

usedisplayformat specifies that the current display format be used for each variable. For example, this option could be useful when using U.S. Social Security numbers. usedisplayformat cannot be specified with format().

Remarks

Remarks are presented under the following headings:

> *destring*
> *tostring*

destring

▷ Example 1

We read in a dataset, but somehow all the variables were created as strings. The variables contain no nonnumeric characters, and we want to convert them all from string to numeric data types.

```
. use http://www.stata-press.com/data/r12/destring1
. describe
Contains data from http://www.stata-press.com/data/r12/destring1.dta
  obs:           10
  vars:           5                             3 Mar 2011 10:15
  size:          200
```

variable name	storage type	display format	value label	variable label
id	str3	%9s		
num	str3	%9s		
code	str4	%9s		
total	str5	%9s		
income	str5	%9s		

```
Sorted by:
```

```
. list
```

	id	num	code	total	income
1.	111	243	1234	543	23423
2.	111	123	2345	67854	12654
3.	111	234	3456	345	43658
4.	222	345	4567	57	23546
5.	333	456	5678	23	21432
6.	333	567	6789	23465	12987
7.	333	678	7890	65	9823
8.	444	789	8976	23	32980
9.	444	901	7654	23	18565
10.	555	890	6543	423	19234

```
. destring, replace
id has all characters numeric; replaced as int
num has all characters numeric; replaced as int
code has all characters numeric; replaced as int
total has all characters numeric; replaced as long
income has all characters numeric; replaced as long
. describe
Contains data from http://www.stata-press.com/data/r12/destring1.dta
  obs:            10
  vars:            5                          3 Mar 2011 10:15
  size:          140
```

	storage	display	value	
variable name	type	format	label	variable label
id	int	%10.0g		
num	int	%10.0g		
code	int	%10.0g		
total	long	%10.0g		
income	long	%10.0g		

```
Sorted by:
     Note:  dataset has changed since last saved
. list
```

	id	num	code	total	income
1.	111	243	1234	543	23423
2.	111	123	2345	67854	12654
3.	111	234	3456	345	43658
4.	222	345	4567	57	23546
5.	333	456	5678	23	21432
6.	333	567	6789	23465	12987
7.	333	678	7890	65	9823
8.	444	789	8976	23	32980
9.	444	901	7654	23	18565
10.	555	890	6543	423	19234

◁

> Example 2

Our dataset contains the variable `date`, which was accidentally recorded as a string because of spaces after the year and month. We want to remove the spaces. `destring` will convert it to numeric and remove the spaces.

```
. use http://www.stata-press.com/data/r12/destring2, clear
. describe date
```

variable name	storage type	display format	value label	variable label
date	str14	%10s		

```
. list date
```

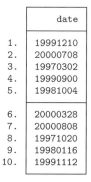

	date
1.	1999 12 10
2.	2000 07 08
3.	1997 03 02
4.	1999 09 00
5.	1998 10 04
6.	2000 03 28
7.	2000 08 08
8.	1997 10 20
9.	1998 01 16
10.	1999 11 12

```
. destring date, replace ignore(" ")
date: characters space   removed; replaced as long
. describe date
```

variable name	storage type	display format	value label	variable label
date	long	%10.0g		

```
. list date
```

	date
1.	19991210
2.	20000708
3.	19970302
4.	19990900
5.	19981004
6.	20000328
7.	20000808
8.	19971020
9.	19980116
10.	19991112

◁

> Example 3

Our dataset contains the variables `date`, `price`, and `percent`. These variables were accidentally read into Stata as string variables because they contain spaces, dollar signs, commas, and percent signs. We want to remove all these characters and create new variables for `date`, `price`, and `percent` containing numeric values. After removing the percent sign, we want to convert the `percent` variable to decimal form.

```
. use http://www.stata-press.com/data/r12/destring2, clear
. describe
Contains data from http://www.stata-press.com/data/r12/destring2.dta
  obs:            10
 vars:             3                          3 Mar 2011 22:50
 size:           280
```

variable name	storage type	display format	value label	variable label
date	str14	%10s		
price	str11	%11s		
percent	str3	%9s		

```
Sorted by:
. list
```

	date	price	percent
1.	1999 12 10	$2,343.68	34%
2.	2000 07 08	$7,233.44	86%
3.	1997 03 02	$12,442.89	12%
4.	1999 09 00	$233,325.31	6%
5.	1998 10 04	$1,549.23	76%
6.	2000 03 28	$23,517.03	35%
7.	2000 08 08	$2.43	69%
8.	1997 10 20	$9,382.47	32%
9.	1998 01 16	$289,209.32	45%
10.	1999 11 12	$8,282.49	1%

```
. destring date price percent, generate(date2 price2 percent2) ignore("$ ,%")
> percent
date: characters space removed; date2 generated as long
price: characters $ , removed; price2 generated as double
percent: characters % removed; percent2 generated as double
. describe
Contains data from http://www.stata-press.com/data/r12/destring2.dta
  obs:            10
 vars:             6                          3 Mar 2011 22:50
 size:           480
```

variable name	storage type	display format	value label	variable label
date	str14	%10s		
date2	long	%10.0g		
price	str11	%11s		
price2	double	%10.0g		
percent	str3	%9s		
percent2	double	%10.0g		

```
Sorted by:
    Note:  dataset has changed since last saved
```

```
. list
```

	date	date2	price	price2	percent	percent2
1.	1999 12 10	19991210	$2,343.68	2343.68	34%	.34
2.	2000 07 08	20000708	$7,233.44	7233.44	86%	.86
3.	1997 03 02	19970302	$12,442.89	12442.89	12%	.12
4.	1999 09 00	19990900	$233,325.31	233325.31	6%	.06
5.	1998 10 04	19981004	$1,549.23	1549.23	76%	.76
6.	2000 03 28	20000328	$23,517.03	23517.03	35%	.35
7.	2000 08 08	20000808	$2.43	2.43	69%	.69
8.	1997 10 20	19971020	$9,382.47	9382.47	32%	.32
9.	1998 01 16	19980116	$289,209.32	289209.32	45%	.45
10.	1999 11 12	19991112	$8,282.49	8282.49	1%	.01

◁

tostring

Conversion of numeric data to string equivalents can be problematic. Stata, like most software, holds numeric data to finite precision and in binary form. See the discussion in [U] **13.11 Precision and problems therein**. If no format() is specified, tostring uses the format %12.0g. This format is, in particular, sufficient to convert integers held as bytes, ints, or longs to string equivalent without loss of precision.

However, users will often need to specify a format themselves, especially when the numeric data have fractional parts and for some reason a conversion to string is required.

▷ Example 4

Our dataset contains a string month variable and numeric year and day variables. We want to convert the three variables to a %td date.

```
. use http://www.stata-press.com/data/r12/tostring, clear
. list
```

	id	month	day	year
1.	123456789	jan	10	2001
2.	123456710	mar	20	2001
3.	123456711	may	30	2001
4.	123456712	jun	9	2001
5.	123456713	oct	17	2001
6.	123456714	nov	15	2001
7.	123456715	dec	28	2001
8.	123456716	apr	29	2001
9.	123456717	mar	11	2001
10.	123456718	jul	3	2001

```
. tostring year day, replace
year was float now str4
day was float now str2
. generate date = month + "/" + day + "/" + year
. generate edate = date(date, "MDY")
. format edate %td
```

```
. list
```

	id	month	day	year	date	edate
1.	123456789	jan	10	2001	jan/10/2001	10jan2001
2.	123456710	mar	20	2001	mar/20/2001	20mar2001
3.	123456711	may	30	2001	may/30/2001	30may2001
4.	123456712	jun	9	2001	jun/9/2001	09jun2001
5.	123456713	oct	17	2001	oct/17/2001	17oct2001
6.	123456714	nov	15	2001	nov/15/2001	15nov2001
7.	123456715	dec	28	2001	dec/28/2001	28dec2001
8.	123456716	apr	29	2001	apr/29/2001	29apr2001
9.	123456717	mar	11	2001	mar/11/2001	11mar2001
10.	123456718	jul	3	2001	jul/3/2001	03jul2001

◁

Saved characteristics

Each time the destring or tostring commands are issued, an entry is made in the characteristics list of each converted variable. You can type char list to view these characteristics.

After example 3, we could use char list to find out what characters were removed by the destring command.

```
. char list
date2[destring]:        Characters removed were:  space
price2[destring]:       Characters removed were:  $ ,
percent2[destring]:     Characters removed were:  %
```

Methods and formulas

destring and tostring are implemented as ado-files.

Acknowledgment

destring and tostring were originally written by Nicholas J. Cox of Durham University.

References

Cox, N. J. 1999a. dm45.1: Changing string variables to numeric: Update. *Stata Technical Bulletin* 49: 2. Reprinted in *Stata Technical Bulletin Reprints*, vol. 9, p. 14. College Station, TX: Stata Press.

——. 1999b. dm45.2: Changing string variables to numeric: Correction. *Stata Technical Bulletin* 52: 2. Reprinted in *Stata Technical Bulletin Reprints*, vol. 9, p. 14. College Station, TX: Stata Press.

——. 2011. Speaking Stata: MMXI and all that: Handling Roman numerals within Stata. *Stata Journal* 11: 126–142.

Cox, N. J., and W. W. Gould. 1997. dm45: Changing string variables to numeric. *Stata Technical Bulletin* 37: 4–6. Reprinted in *Stata Technical Bulletin Reprints*, vol. 7, pp. 34–37. College Station, TX: Stata Press.

Cox, N. J., and J. B. Wernow. 2000a. dm80: Changing numeric variables to string. *Stata Technical Bulletin* 56: 8–12. Reprinted in *Stata Technical Bulletin Reprints*, vol. 10, pp. 24–28. College Station, TX: Stata Press.

——. 2000b. dm80.1: Update to changing numeric variables to string. *Stata Technical Bulletin* 57: 2. Reprinted in *Stata Technical Bulletin Reprints*, vol. 10, pp. 28–29. College Station, TX: Stata Press.

Also see

[D] **generate** — Create or change contents of variable

[D] **split** — Split string variables into parts

[D] **egen** — Extensions to generate

[D] **encode** — Encode string into numeric and vice versa

[D] **functions** — Functions

Title

> **dir** — Display filenames

Syntax

$\{\texttt{dir} \,|\, \texttt{ls}\}$ $\left[\,"\,\right]\left[\,\textit{filespec}\,\right]\left[\,"\,\right]$ $\left[\,, \underline{\text{w}}\text{ide}\,\right]$

Note: Double quotes must be used to enclose *filespec* if the name contains spaces.

Description

dir and ls—they work the same way—list the names of files in the specified directory; the names of the commands come from names popular on Unix and Windows computers. *filespec* may be any valid Mac, Unix, or Windows file path or file specification (see [U] **11.6 Filenaming conventions**) and may include '*' to indicate any string of characters.

Option

wide under Mac and Windows produces an effect similar to specifying /W with the DOS dir command—it compresses the resulting listing by placing more than one filename on a line. Under Unix, it produces the same effect as typing ls -F -C. Without the wide option, ls is equivalent to typing ls -F -1.

Remarks

Mac and Unix: The only difference between the Stata and Unix ls commands is that piping through the more(1) or pg(1) filter is unnecessary—Stata always pauses when the screen is full.

Windows: Other than minor differences in presentation format, there is only one difference between the Stata and DOS dir commands: the DOS /P option is unnecessary, because Stata always pauses when the screen is full.

▷ Example 1

The only real difference between the Stata dir and DOS and Unix equivalent commands is that output never scrolls off the screen; Stata always pauses when the screen is full.

If you use Stata for Windows and wish to obtain a list of all your Stata-format data files, type

```
. dir *.dta
    3.9k    7/07/00 13:51   auto.dta
    0.6k    8/04/00 10:40   cancer.dta
    3.5k    7/06/98 17:06   census.dta
    3.4k    1/25/98  9:20   hsng.dta
    0.3k    1/26/98 16:54   kva.dta
    0.7k    4/27/00 11:39   sysage.dta
    0.5k    5/09/97  2:56   systolic.dta
   10.3k    7/13/98  8:37   Household Survey.dta
```

You could also include the `wide` option:

```
. dir *.dta, wide
    3.9k auto.dta              0.6k cancer.dta           3.5k census.dta
    3.4k hsng.dta              0.3k kva.dta              0.7k sysage.dta
    0.5k systolic.dta         10.3k Household Survey.dta
```

Unix users will find it more natural to type

```
. ls *.dta
-rw-r-----  1 roger      2868 Mar  4 15:34 highway.dta
-rw-r-----  1 roger       941 Apr  5 09:43 hoyle.dta
-rw-r-----  1 roger     19312 May 14 10:36 p1.dta
-rw-r-----  1 roger     11838 Apr 11 13:26 p2.dta
```

but they could type `dir` if they preferred. Mac users may also type either command.

```
. dir *.dta
-rw-r-----  1 roger      2868 Mar  4 15:34 highway.dta
-rw-r-----  1 roger       941 Apr  5 09:43 hoyle.dta
-rw-r-----  1 roger     19312 May 14 10:36 p1.dta
-rw-r-----  1 roger     11838 Apr 11 13:26 p2.dta
```

◁

❑ Technical note

There is an extended macro function named `dir` which allows you to obtain a list of files in a macro for later processing. See *Macro extended functions for filenames and file paths* in [P] **macro**.

❑

Also see

[D] **cd** — Change directory

[D] **copy** — Copy file from disk or URL

[D] **erase** — Erase a disk file

[D] **mkdir** — Create directory

[D] **rmdir** — Remove directory

[D] **shell** — Temporarily invoke operating system

[D] **type** — Display contents of a file

[U] **11.6 Filenaming conventions**

Title

> **drawnorm** — Draw sample from multivariate normal distribution

Syntax

> drawnorm *newvarlist* [, *options*]

options	Description
Main	
clear	replace the current dataset
double	generate variable type as double; default is float
n(#)	# of observations to be generated; default is current number
sds(*vector*)	standard deviations of generated variables
corr(*matrix* | *vector*)	correlation matrix
cov(*matrix* | *vector*)	covariance matrix
cstorage(full)	correlation/covariance structure is stored as a symmetric $k \times k$ matrix
cstorage(lower)	correlation/covariance structure is stored as a lower triangular matrix
cstorage(upper)	correlation/covariance structure is stored as an upper triangular matrix
forcepsd	force the covariance/correlation matrix to be positive semidefinite
means(*vector*)	means of generated variables; default is means(0)
Options	
seed(#)	seed for random-number generator

Menu

Data > Create or change data > Other variable-creation commands > Draw sample from normal distribution

Description

drawnorm draws a sample from a multivariate normal distribution with desired means and covariance matrix. The default is orthogonal data with mean 0 and variance 1. The covariance matrix may be singular. The values generated are a function of the current random-number seed or the number specified with set seed(); see [R] **set seed**.

Options

> Main

clear specifies that the dataset in memory be replaced, even though the current dataset has not been saved on disk.

double specifies that the new variables be stored as Stata doubles, meaning 8-byte reals. If double is not specified, variables are stored as floats, meaning 4-byte reals. See [D] **data types**.

n(#) specifies the number of observations to be generated. The default is the current number of observations. If n(#) is not specified or is the same as the current number of observations, drawnorm adds the new variables to the existing dataset; otherwise, drawnorm replaces the data in memory.

sds(*vector*) specifies the standard deviations of the generated variables. sds() may not be specified with cov().

corr(*matrix | vector*) specifies the correlation matrix. If neither corr() nor cov() is specified, the default is orthogonal data.

cov(*matrix | vector*) specifies the covariance matrix. If neither cov() nor corr() is specified, the default is orthogonal data.

cstorage(full | lower | upper) specifies the storage mode for the correlation or covariance structure in corr() or cov(). The following storage modes are supported:

full specifies that the correlation or covariance structure is stored (recorded) as a symmetric $k \times k$ matrix.

lower specifies that the correlation or covariance structure is recorded as a lower triangular matrix. With k variables, the matrix should have $k(k + 1)/2$ elements in the following order:

$$C_{11} \ C_{21} \ C_{22} \ C_{31} \ C_{32} \ C_{33} \ \dots \ C_{k1} \ C_{k2} \ \dots \ C_{kk}$$

upper specifies that the correlation or covariance structure is recorded as an upper triangular matrix. With k variables, the matrix should have $k(k + 1)/2$ elements in the following order:

$$C_{11} \ C_{12} \ C_{13} \ \dots \ C_{1k} \ C_{22} \ C_{23} \ \dots C_{2k} \ \dots \ C_{(k-1k-1)} \ C_{(k-1k)} \ C_{kk}$$

Specifying cstorage(full) is optional if the matrix is square. cstorage(lower) or cstorage(upper) is required for the vectorized storage methods. See *Example 2: Storage modes for correlation and covariance matrices*.

forcepsd modifies the matrix C to be positive semidefinite (psd), and so be a proper covariance matrix. If C is not positive semidefinite, it will have negative eigenvalues. By setting negative eigenvalues to 0 and reconstructing, we obtain the least-squares positive-semidefinite approximation to C. This approximation is a singular covariance matrix.

means(*vector*) specifies the means of the generated variables. The default is means(0).

⌐ Options ⌐

seed(#) specifies the initial value of the random-number seed used by the runiform() function. The default is the current random-number seed. Specifying seed(#) is the same as typing set seed # before issuing the drawnorm command.

Remarks

▷ Example 1

Suppose that we want to draw a sample of 1,000 observations from a normal distribution $N(\mathbf{M}, \mathbf{V})$, where **M** is the mean matrix and **V** is the covariance matrix:

```
. matrix M = 5, -6, 0.5
```

```
. matrix V = (9, 5, 2 \ 5, 4, 1 \ 2, 1, 1)

. matrix list M

M[1,3]
      c1   c2   c3
r1     5   -6   .5

. matrix list V

symmetric V[3,3]
      c1   c2   c3
r1     9
r2     5    4
r3     2    1    1

. drawnorm x y z, n(1000) cov(V) means(M)
(obs 1000)

. summarize
```

Variable	Obs	Mean	Std. Dev.	Min	Max
x	1000	5.001715	3.00608	-4.572042	13.66046
y	1000	-5.980279	2.004755	-12.08166	-.0963039
z	1000	.5271135	1.011095	-2.636946	4.102734

```
. correlate, cov
(obs=1000)
```

	x	y	z
x	9.03652		
y	5.04462	4.01904	
z	2.10142	1.08773	1.02231

◁

❑ Technical note

The values generated by drawnorm are a function of the current random-number seed. To reproduce the same dataset each time drawnorm is run with the same setup, specify the same seed number in the seed() option.

❑

▷ Example 2: Storage modes for correlation and covariance matrices

The three storage modes for specifying the correlation or covariance matrix in corr2data and drawnorm can be illustrated with a correlation structure, C, of 4 variables. In full storage mode, this structure can be entered as a 4×4 Stata matrix:

```
. matrix C = ( 1.0000,  0.3232,  0.1112,  0.0066 \ ///
               0.3232,  1.0000,  0.6608, -0.1572 \ ///
               0.1112,  0.6608,  1.0000, -0.1480 \ ///
               0.0066, -0.1572, -0.1480,  1.0000 )
```

Elements within a row are separated by commas, and rows are separated by a backslash, \. We use the input continuation operator /// for convenient multiline input; see [P] **comments**. In this storage mode, we probably want to set the row and column names to the variable names:

```
.   matrix rownames C = price trunk headroom rep78
.   matrix colnames C = price trunk headroom rep78
```

This correlation structure can be entered more conveniently in one of the two vectorized storage modes. In these modes, we enter the lower triangle or the upper triangle of C in rowwise order; these two storage modes differ only in the order in which the $k(k + 1)/2$ matrix elements are recorded. The lower storage mode for C comprises a vector with $4(4 + 1)/2 = 10$ elements, that is, a 1×10 or 10×1 Stata matrix, with one row or column,

```
. matrix C = ( 1.0000,  ///
          0.3232,  1.0000,  ///
          0.1112,  0.6608,  1.0000,  ///
          0.0066, -0.1572, -0.1480,  1.0000)
```

or more compactly as

```
. matrix C = ( 1, 0.3232, 1, 0.1112, 0.6608, 1, 0.0066, -0.1572, -0.1480, 1 )
```

C may also be entered in upper storage mode as a vector with $4(4 + 1)/2 = 10$ elements, that is, a 1×10 or 10×1 Stata matrix,

```
. matrix C = ( 1.0000,  0.3232,  0.1112,  0.0066,  ///
                  1.0000,  0.6608, -0.1572,  ///
                           1.0000, -0.1480,  ///
                                    1.0000 )
```

or more compactly as

```
. matrix C = ( 1, 0.3232, 0.1112, 0.0066, 1, 0.6608, -0.1572, 1, -0.1480, 1 )
```

◁

Methods and formulas

drawnorm is implemented as an ado-file.

Results are asymptotic. The more observations generated, the closer the correlation matrix of the dataset is to the desired correlation structure.

Let $\mathbf{V} = \mathbf{A}'\mathbf{A}$ be the desired covariance matrix and $\mathbf{M}$ be the desired mean matrix. We first generate $\mathbf{X}$, such that $\mathbf{X} \sim N(\mathbf{0}, \mathbf{I})$. Let $\mathbf{Y} = \mathbf{A}'\mathbf{X} + \mathbf{M}$, then $\mathbf{Y} \sim N(\mathbf{M}, \mathbf{V})$.

Also see

[D] **corr2data** — Create dataset with specified correlation structure

[R] **set seed** — Specify initial value of random-number seed

Title

drop — Eliminate variables or observations

Syntax

Drop variables

    drop *varlist*

Drop observations

    drop if *exp*

Drop a range of observations

    drop in *range* [if *exp*]

Keep variables

    keep *varlist*

Keep observations that satisfy specified condition

    keep if *exp*

Keep a range of observations

    keep in *range* [if *exp*]

by is allowed with the second syntax of drop and the second syntax of keep; see [D] **by**.

Menu

Keep or drop variables

Data > Variables Manager

Keep or drop observations

Data > Create or change data > Keep or drop observations

Description

drop eliminates variables or observations from the data in memory.

keep works the same way as drop, except that you specify the variables or observations to be kept rather than the variables or observations to be deleted.

Warning: `drop` and `keep` are not reversible. Once you have eliminated observations, you cannot read them back in again. You would need to go back to the original dataset and read it in again. Instead of applying `drop` or `keep` for a subset analysis, consider using `if` or `in` to select subsets temporarily. This is usually the best strategy. Alternatively, applying `preserve` followed in due course by `restore` may be a good approach.

Remarks

You can clear the entire dataset by typing `drop _all` without affecting value labels, macros, and programs. (Also see [U] **12.6 Dataset, variable, and value labels**, [U] **18.3 Macros**, and [P] **program**.)

▷ Example 1

We will systematically eliminate data until, at the end, no data are left in memory. We begin by describing the data:

```
. use http://www.stata-press.com/data/r12/census11
(1980 Census data by state)

. describe
Contains data from http://www.stata-press.com/data/r12/census11.dta
  obs:            50                          1980 Census data by state
  vars:           15                          6 Apr 2011 15:43
  size:        3,300
```

variable name	storage type	display format	value label	variable label
state	str14	%-14s		State
state2	str2	%-2s		Two-letter state abbreviation
region	int	%-8.0g	cenreg	Census region
pop	long	%12.0gc		Population
poplt5	long	%12.0gc		Pop, < 5 year
pop5_17	long	%12.0gc		Pop, 5 to 17 years
pop18p	long	%12.0gc		Pop, 18 and older
pop65p	long	%12.0gc		Pop, 65 and older
popurban	long	%12.0gc		Urban population
medage	float	%9.2f		Median age
death	long	%12.0gc		Number of deaths
marriage	long	%12.0gc		Number of marriages
divorce	long	%12.0gc		Number of divorces
mrgrate	float	%9.0g		
dvcrate	float	%9.0g		

```
Sorted by:  region
```

We can eliminate all the variables with names that begin with pop by typing `drop pop*`:

```
. drop pop*
. describe
Contains data from http://www.stata-press.com/data/r12/census11.dta
  obs:            50                          1980 Census data by state
  vars:            9                          6 Apr 2011 15:43
  size:         2,100
```

variable name	storage type	display format	value label	variable label
state	str14	%-14s		State
state2	str2	%-2s		Two-letter state abbreviation
region	int	%-8.0g	cenreg	Census region
medage	float	%9.2f		Median age
death	long	%12.0gc		Number of deaths
marriage	long	%12.0gc		Number of marriages
divorce	long	%12.0gc		Number of divorces
mrgrate	float	%9.0g		
dvcrate	float	%9.0g		

```
Sorted by:  region
     Note:  dataset has changed since last saved
```

Let's eliminate more variables and then eliminate observations:

```
. drop marriage divorce mrgrate dvcrate
. describe
Contains data from http://www.stata-press.com/data/r12/census11.dta
  obs:            50                          1980 Census data by state
  vars:            5                          6 Apr 2011 15:43
  size:         1,300
```

variable name	storage type	display format	value label	variable label
state	str14	%-14s		State
state2	str2	%-2s		Two-letter state abbreviation
region	int	%-8.0g	cenreg	Census region
medage	float	%9.2f		Median age
death	long	%12.0gc		Number of deaths

```
Sorted by:  region
     Note:  dataset has changed since last saved
```

Next we will drop any observation for which medage is greater than 32.

```
. drop if medage>32
(3 observations deleted)
```

Let's drop the first observation in each region:

```
. by region: drop if _n==1
(4 observations deleted)
```

Now we drop all but the last observation in each region:

```
. by region: drop if _n !=_N
(39 observations deleted)
```

Let's now drop the first 2 observations in our dataset:

```
. drop in 1/2
(2 observations deleted)
```

Finally, let's get rid of everything:

```
. drop _all
. describe
Contains data
  obs:          0
  vars:         0
  size:         0
Sorted by:
```
◁

Typing `keep in 10/1` is the same as typing `drop in 1/9`.

Typing `keep if x==3` is the same as typing `drop if x !=3`.

`keep` is especially useful for keeping a few variables from a large dataset. Typing `keep myvar1 myvar2` is the same as typing `drop` followed by all the variables in the dataset *except* `myvar1` and `myvar2`.

❏ Technical note

In addition to dropping variables and observations, `drop _all` removes any business calendars; see [D] **datetime business calendars**.

❏

Reference

Cox, N. J. 2001. dm89: Dropping variables or observations with missing values. *Stata Technical Bulletin* 60: 7–8. Reprinted in *Stata Technical Bulletin Reprints*, vol. 10, pp. 44–46. College Station, TX: Stata Press.

Also see

[D] **clear** — Clear memory

[D] **varmanage** — Manage variable labels, formats, and other properties

[U] **11 Language syntax**

[U] **13 Functions and expressions**

Title

> **ds** — List variables matching name patterns or other characteristics

Syntax

Simple syntax

> ds [, <u>al</u>pha]

Advanced syntax

> ds [*varlist*] [, *options*]

options	Description
Main	
not	list variables not specified in *varlist*
<u>al</u>pha	list variables in alphabetical order
<u>det</u>ail	display additional details
<u>var</u>width(*#*)	display width for variable names; default is varwidth(12)
skip(*#*)	gap between variables; default is skip(2)
Advanced	
has(*spec*)	describe subset that matches *spec*
not(*spec*)	describe subset that does not match *spec*
<u>inse</u>nsitive	perform case-insensitive pattern matching
indent(*#*)	indent output; seldom used

insensitive and indent(*#*) are not shown in the dialog box.

spec	Description
<u>t</u>ype *typelist*	specified types
<u>f</u>ormat *patternlist*	display format matching *patternlist*
<u>var</u>label [*patternlist*]	variable label or variable label matching *patternlist*
<u>char</u> [*patternlist*]	characteristic or characteristic matching *patternlist*
<u>val</u>label [*patternlist*]	value label or value label matching *patternlist*

typelist used in has(type *typelist*) and not(type *typelist*) is a list of one or more types, each of which may be numeric, string, byte, int, long, float, or double, or may be a *numlist* such as 1/8 to mean "str1 str2 ... str8". Examples include

has(type int)	is of type int
has(type byte int long)	is of integer type
not(type int)	is not of type int
not(type byte int long)	is not of the integer types
has(type numeric)	is a numeric variable

152

`not(type string)`	is not a string variable (same as above)
`has(type 1/40)`	is `str1`, `str2`, ..., `str40`
`has(type numeric 1/2)`	is numeric or `str1` or `str2`

patternlist used in, for instance, `has(format `*patternlist*`)`, is a list of one or more *patterns*. A pattern is the expected text with the addition of the characters * and ?. * indicates 0 or more characters go here, and ? indicates exactly 1 character goes here. Examples include

`has(format *f)`	format is `%#.#f`
`has(format %t*)`	has time or date format
`has(format %-*s)`	is a left-justified string
`has(varl *weight*)`	variable label includes word `weight`
`has(varl *weight* *Weight*)`	variable label has `weight` or `Weight`

To match a phrase, enclose the phrase in quotes.

`has(varl "*some phrase*")`	variable label has `some phrase`

If instead you used `has(varl *some phrase*)`, then only variables having labels ending in `some` or starting with `phrase` would be listed.

Menu

Data > Describe data > Compactly list variable names

Description

ds lists variable names of the dataset currently in memory in a compact or detailed format, and lets you specify subsets of variables to be listed, either by name or by properties (for example, the variables are numeric). In addition, ds leaves behind in `r(varlist)` the names of variables selected so that you can use them in a subsequent command.

ds, typed without arguments, lists all variable names of the dataset currently in memory in a compact form.

Options

    ⌐ Main ⌐

not specifies that the variables in *varlist* not be listed. For instance, `ds pop*, not` specifies that all variables not starting with the letters pop be listed. The default is to list all the variables in the dataset or, if *varlist* is specified, the variables specified.

alpha specifies that the variables be listed in alphabetical order.

detail specifies that detailed output identical to that of describe be produced. If detail is specified, `varwidth()`, `skip()`, and `indent()` are ignored.

`varwidth(#)` specifies the display width of the variable names; the default is `varwidth(12)`.

`skip(#)` specifies the number of spaces between variable names, where all variable names are assumed to be the length of the longest variable name; the default is `skip(2)`.

⌐ Advanced ⌐

has(*spec*) and not(*spec*) select from the dataset (or from *varlist*) the subset of variables that meet or fail the specification *spec*. Selection may be made on the basis of storage type, variable label, value label, display format, or characteristics. Only one not, has(), or not() option may be specified.

has(type string) selects all string variables. Typing ds, has(type string) would list all string variables in the dataset, and typing ds pop*, has(type string) would list all string variables whose names begin with the letters pop.

has(varlabel) selects variables with defined variable labels. has(varlabel *weight*) selects variables with variable labels including the word "weight". not(varlabel) would select all variables with no variable labels.

has(vallabel) selects variables with defined value labels. has(vallabel yesno) selects variables whose value label is yesno. has(vallabel *no) selects variables whose value label ends in the letters no.

has(format *patternlist*) specifies variables whose format matches any of the patterns in *patternlist*. has(format *f) would select all variables with formats ending in f, which presumably would be all %#.#f, %0#.#f, and %-#.#f formats. has(format *f *fc) would select all ending in f or fc. not(format %t* %-t*) would select all variables except those with date or time-series formats.

has(char) selects all variables with defined characteristics. has(char problem) selects all variables with a characteristic named problem.

The following options are available with ds but are not shown in the dialog box:

insensitive specifies that the matching of the *pattern* in has() and not() be case insensitive.

indent(*#*) specifies the amount the lines are indented.

Remarks

If ds is typed without any operands, then a compact list of the variable names for the data currently in memory is displayed.

▷ Example 1

ds can be especially useful if you have a dataset with over 1,000 variables, but you may find it convenient even if you have considerably fewer variables.

```
. use http://www.stata-press.com/data/r12/educ3
(ccdb46, 52-54)

. ds
fips       popcol     medhhinc   tlf        emp        clfbls     z
crimes     perhspls   medfinc    clf        empmanuf   clfuebls   adjinc
pcrimes    perclpls   state      clffem     emptrade   famnw      perman
crimrate   prcolhs    division   clfue      empserv    fam2w      pertrade
pop25pls   medage     region     empgovt    osigind    famwsamp   perserv
pophspls   perwhite   dc         empself    osigindp   pop18pls   perother
```

◁

> Example 2

You might wonder why you would ever specify a *varlist* with this command. Remember that a *varlist* understands the '*' abbreviation character and the '–' dash notation; see [U] **11.4 varlists**.

```
. ds p*
pcrimes    pophspls   perhspls   prcolhs    pop18pls   pertrade   perother
pop25pls   popcol     perclpls   perwhite   perman     perserv
. ds popcol-clfue
popcol     perclpls   medage     medhhinc   state      region     tlf        clffem
perhspls   prcolhs    perwhite   medfinc    division   dc         clf        clfue
```
◁

> Example 3

Because the primary use of ds is to inspect the names of variables, it is sometimes useful to let ds display the variable names in alphabetical order.

```
. ds, alpha
adjinc     crimes     empmanuf   famwsamp   osigindp   perserv    pophspls
clf        crimrate   empself    fips       pcrimes    pertrade   prcolhs
clfbls     dc         empserv    medage     perclpls   perwhite   region
clffem     division   emptrade   medfinc    perhspls   pop18pls   state
clfue      emp        fam2w      medhhinc   perman     pop25pls   tlf
clfuebls   empgovt    famnw      osigind    perother   popcol     z
```
◁

Saved results

ds saves the following in r():

Macros
    r(varlist)   the varlist in the order displayed

Methods and formulas

ds is implemented as an ado-file.

Reference

Cox, N. J. 2010. Speaking Stata: Finding variables. *Stata Journal* 10: 281–296.

Also see

[D] **compress** — Compress data in memory

[D] **cf** — Compare two datasets

[D] **codebook** — Describe data contents

[D] **compare** — Compare two variables

[D] **describe** — Describe data in memory or in file

[D] **format** — Set variables' output format

[D] **label** — Manipulate labels

[D] **lookfor** — Search for string in variable names and labels

[D] **notes** — Place notes in data

[D] **order** — Reorder variables in dataset

[D] **rename** — Rename variable

Title

duplicates — Report, tag, or drop duplicate observations

Syntax

Report duplicates

duplicates report [*varlist*] [*if*] [*in*]

List one example for each group of duplicates

duplicates examples [*varlist*] [*if*] [*in*] [, *options*]

List all duplicates

duplicates list [*varlist*] [*if*] [*in*] [, *options*]

Tag duplicates

duplicates tag [*varlist*] [*if*] [*in*] , generate(*newvar*)

Drop duplicates

duplicates drop [*if*] [*in*]

duplicates drop *varlist* [*if*] [*in*] , force

options	Description
Main	
<u>compress</u>	compress width of columns in both table and display formats
<u>noc</u>ompress	use display format of each variable
fast	synonym for nocompress; no delay in output of large datasets
<u>ab</u>breviate(#)	abbreviate variable names to # characters; default is ab(8)
<u>string</u>(#)	truncate string variables to # characters; default is string(10)
Options	
<u>table</u>	force table format
<u>display</u>	force display format
<u>header</u>	display variable header once; default is table mode
<u>noheader</u>	suppress variable header
<u>header</u>(#)	display variable header every # lines
<u>clean</u>	force table format with no divider or separator lines
<u>divi</u>der	draw divider lines between columns
<u>separator</u>(#)	draw a separator line every # lines; default is separator(5)
<u>sepby</u>(*varlist*)	draw a separator line whenever *varlist* values change
<u>nol</u>abel	display numeric codes rather than label values
Summary	
mean$\big[$(*varlist*)$\big]$	add line reporting the mean for each of the (specified) variables
sum$\big[$(*varlist*)$\big]$	add line reporting the sum for each of the (specified) variables
N$\big[$(*varlist*)$\big]$	add line reporting the number of nonmissing values for each of the (specified) variables
<u>labv</u>ar(*varname*)	substitute Mean, Sum, or N for value of *varname* in last row of table
Advanced	
<u>constant</u>$\big[$(*varlist*)$\big]$	separate and list variables that are constant only once
<u>notr</u>im	suppress string trimming
<u>absolute</u>	display overall observation numbers when using by *varlist*:
nodotz	display numerical values equal to .z as field of blanks
<u>subvarname</u>	substitute characteristic for variable name in header
<u>lines</u>ize(#)	columns per line; default is linesize(79)

Menu

Data > Data utilities > Manage duplicate observations

Description

    duplicates reports, displays, lists, tags, or drops duplicate observations, depending on the subcommand specified. Duplicates are observations with identical values either on all variables if no *varlist* is specified or on a specified *varlist*.

    duplicates report produces a table showing observations that occur as one or more copies and indicating how many observations are "surplus" in the sense that they are the second (third, . . .) copy of the first of each group of duplicates.

duplicates examples lists one example for each group of duplicated observations. Each example represents the first occurrence of each group in the dataset.

duplicates list lists all duplicated observations.

duplicates tag generates a variable representing the number of duplicates for each observation. This will be 0 for all unique observations.

duplicates drop drops all but the first occurrence of each group of duplicated observations. The word drop may not be abbreviated.

Any observations that do not satisfy specified if and/or in conditions are ignored when you use report, examples, list, or drop. The variable created by tag will have missing values for such observations.

Options for duplicates examples and duplicates list

⌐ Main ¬

compress, nocompress, fast, abbreviate(#), string(#); see [D] **list**.

⌐ Options ¬

table, display, header, noheader, header(#), clean, divider, separator(#), sepby(*varlist*), nolabel; see [D] **list**.

⌐ Summary ¬

mean$\big[$(*varlist*)$\big]$, sum$\big[$(*varlist*)$\big]$, N$\big[$(*varlist*)$\big]$, labvar(*varname*); see [D] **list**.

⌐ Advanced ¬

constant$\big[$(*varlist*)$\big]$, notrim, absolute, nodotz, subvarname, linesize(#); see [D] **list**.

Option for duplicates tag

generate(*newvar*) is required and specifies the name of a new variable that will tag duplicates.

Option for duplicates drop

force specifies that observations duplicated with respect to a named *varlist* be dropped. The force option is required when such a *varlist* is given as a reminder that information may be lost by dropping observations, given that those observations may differ on any variable not included in *varlist*.

Remarks

Current data management and analysis may hinge on detecting (and sometimes dropping) duplicate observations. In Stata terms, *duplicates* are observations with identical values, either on all variables if no *varlist* is specified, or on a specified *varlist*; that is, 2 or more observations that are identical on all specified variables form a group of duplicates. When the specified variables are a set of explanatory variables, such a group is often called a *covariate pattern* or a *covariate class*.

Linguistic purists will point out that duplicate observations are strictly only those that occur in pairs, and they might prefer a more literal term, although the most obvious replacement, "replicates", already has another statistical meaning. However, the looser term appears in practice to be much more frequently used for this purpose and to be as easy to understand.

Observations may occur as duplicates through some error; for example, the same observations might have been entered more than once into your dataset. For example, some researchers deliberately enter a dataset twice. Each entry is a check on the other, and all observations should occur as identical pairs, assuming that one or more variables identify unique records. If there is just one copy, or more than two copies, there has been an error in data entry.

Or duplicate observations may also arise simply because some observations just happen to be identical, which is especially likely with categorical variables or large datasets. In this second situation, consider whether `contract`, which automatically produces a count of each distinct set of observations, is more appropriate for your problem. See [D] **contract**.

Observations unique on all variables in *varlist* occur as single copies. Thus there are no surplus observations in the sense that no observation may be dropped without losing information about the contents of observations. (Information will inevitably be lost on the frequency of such observations. Again, if recording frequency is important to you, `contract` is the better command to use.) Observations that are duplicated twice or more occur as copies, and in each case, all but one copy may be considered surplus.

This command helps you produce a dataset, usually smaller than the original, in which each observation is *unique* (literally, each occurs only once) and *distinct* (each differs from all the others). If you are familiar with Unix systems, or with sets of Unix utilities ported to other platforms, you will know the `uniq` command, which removes duplicate adjacent lines from a file, usually as part of a pipe.

▷ Example 1

Suppose that we are given a dataset in which some observations are unique (no other observation is identical on all variables) and other observations are duplicates (in each case, at least 1 other observation exists that is identical). Imagine dropping all but 1 observation from each group of duplicates, that is, dropping the surplus observations. Now all the observations are unique. This example helps clarify the difference between 1) identifying unique observations before dropping surplus copies and 2) identifying unique observations after dropping surplus copies (whether in truth or merely in imagination). `codebook` (see [D] **codebook**) reports the number of unique values for each variable in this second sense.

Suppose that we have typed in a dataset for 200 individuals. However, a simple `describe` or `count` shows that we have 202 observations in our dataset. We guess that we may have typed in 2 observations twice. `duplicates report` gives a quick report of the occurrence of duplicates:

```
. use http://www.stata-press.com/data/r12/dupxmpl
. duplicates report
Duplicates in terms of all variables
```

copies	observations	surplus
1	198	0
2	4	2

Our hypothesis is supported: 198 observations are unique (just 1 copy of each), whereas 4 occur as duplicates (2 copies of each; in each case, 1 may be dubbed surplus). We now wish to see which observations are duplicates, so the next step is to ask for a `duplicates list`.

```
. duplicates list
Duplicates in terms of all variables
```

group:	obs:	id	x	y
1	42	42	0	2
1	43	42	0	2
2	145	144	4	4
2	146	144	4	4

The records for `id` 42 and `id` 144 were evidently entered twice. Satisfied, we now issue `duplicates drop`.

```
. duplicates drop
Duplicates in terms of all variables
(2 observations deleted)
```

◁

The `report`, `list`, and `drop` subcommands of `duplicates` are perhaps the most useful, especially for a relatively small dataset. For a larger dataset with many duplicates, a full listing may be too long to be manageable, especially as you see repetitions of the same data. `duplicates examples` gives you a more compact listing in which each group of duplicates is represented by just 1 observation, the first to occur.

A subcommand that is occasionally useful is `duplicates tag`, which generates a new variable containing the number of duplicates for each observation. Thus unique observations are tagged with value 0, and all duplicate observations are tagged with values greater than 0. For checking double data entry, in which you expect just one surplus copy for each individual record, you can generate a tag variable and then look at observations with tag not equal to 1 because both unique observations and groups with two or more surplus copies need inspection.

```
. duplicates tag, gen(tag)
Duplicates in terms of all variables
```

As of Stata 11, the `browse` subcommand is no longer available. To open duplicates in the Data Browser, use the following commands:

```
. duplicates tag, generate(newvar)
. browse if newvar > 1
```

See [D] **edit** for details on the `browse` command.

Methods and formulas

`duplicates` is implemented as an ado-file.

Acknowledgments

duplicates was written by Nicholas J. Cox, Durham University, who in turn thanks Thomas Steichen, RJRT, for ideas contributed to an earlier jointly written program (Steichen and Cox 1998).

References

Jacobs, M. 1991. dm4: A duplicated value identification program. *Stata Technical Bulletin* 4: 5. Reprinted in *Stata Technical Bulletin Reprints*, vol. 1, p. 30. College Station, TX: Stata Press.

Steichen, T. J., and N. J. Cox. 1998. dm53: Detection and deletion of duplicate observations. *Stata Technical Bulletin* 41: 2–4. Reprinted in *Stata Technical Bulletin Reprints*, vol. 7, pp. 52–55. College Station, TX: Stata Press.

Wang, D. 2000. dm77: Removing duplicate observations in a dataset. *Stata Technical Bulletin* 54: 16–17. Reprinted in *Stata Technical Bulletin Reprints*, vol. 9, pp. 87–88. College Station, TX: Stata Press.

Also see

[D] **edit** — Browse or edit data with Data Editor

[D] **list** — List values of variables

[D] **codebook** — Describe data contents

[D] **contract** — Make dataset of frequencies and percentages

[D] **isid** — Check for unique identifiers

Title

> **edit** — Browse or edit data with Data Editor

Syntax

Edit using Data Editor

> <u>ed</u>it [*varlist*] [*if*] [*in*] [, <u>nol</u>abel]

Browse using Data Editor

> <u>br</u>owse [*varlist*] [*if*] [*in*] [, <u>nol</u>abel]

Menu

edit

Data > Data Editor > Data Editor (Edit)

browse

Data > Data Editor > Data Editor (Browse)

Description

edit brings up a spreadsheet-style data editor for entering new data and editing existing data. edit is a better alternative to input; see [D] **input**.

browse is similar to edit, except that modifications to the data by editing in the grid are not permitted. browse is a convenient alternative to list; see [D] **list**.

See [GS] **6 Using the Data Editor** (GSM, GSU, or GSW) for a tutorial discussion of the Data Editor. This entry provides the technical details.

Option

nolabel causes the underlying numeric values, rather than the label values (equivalent strings), to be displayed for variables with value labels; see [D] **label**.

Remarks

Remarks are presented under the following headings:

> *Modes*
> *The current observation and current variable*
> *Assigning value labels to variables*
> *Changing values of existing cells*
> *Adding new variables*
> *Adding new observations*
> *Copying and pasting*
> *Logging changes*
> *Advice*

Clicking on Stata's **Data Editor (Edit)** button is equivalent to typing `edit` by itself. Clicking on Stata's **Data Editor (Browse)** button is equivalent to typing `browse` by itself.

`edit`, typed by itself, opens the Data Editor with all observations on all variables displayed. If you specify a *varlist*, only the specified variables are displayed in the Editor. If you specify one or both of `in` *range* and `if` *exp*, only the observations specified are displayed.

Modes

We will refer to the Data Editor in the singular with `edit` and `browse` referring to two of its three modes.

Full-edit mode. This is the Editor's mode that you enter when you type `edit` or type `edit` followed by a list of variables. All features of the Editor are turned on.

Filtered mode. This is the Editor's mode that you enter when you use `edit` with or without a list of variables but include `in` *range*, `if` *exp*, or both, or if you filter the data from within the Editor. A few of the Editor's features are turned off, most notably, the ability to sort data and the ability to paste data into the Editor.

Browse mode. This is the Editor's mode that you enter when you use `browse` or when you change the Editor's mode to **Browse** after you start the Editor. The ability to type in the Editor, thereby changing data, is turned off, ensuring that the data cannot accidentally be changed. One feature that is left on may surprise you: the ability to sort data. Sorting, in Stata's mind, is not really a change to the dataset. On the other hand, if you enter using `browse` and specify `in` *range* or `if` *exp*, sorting is not allowed. You can think of this as restricted-browse mode.

Actually, the Editor does not set its mode to filtered just because you specify an `in` *range* or `if` *exp*. It sets its mode to filtered if you specify `in` or `if` and if this restriction is effective, that is, if the `in` or `if` would actually cause some data to be omitted. For instance, typing `edit if x>0` would result in unrestricted full-edit mode if x were greater than zero for all observations.

The current observation and current variable

The Data Editor looks much like a spreadsheet, with rows and columns corresponding to observations and variables, respectively. At all times, one of the cells is highlighted. This is called the current cell. The observation (row) of the current cell is called the current observation. The variable (column) of the current cell is called the current variable.

You change the current cell by clicking with the mouse on another cell or by using the arrow keys.

To help distinguish between the different types of variables in the Editor, string values are displayed in red, value labels are displayed in blue, and all other values are displayed in black. You can change the colors for strings and value labels by right-clicking on the Data Editor window and selecting **Preferences...**.

Assigning value labels to variables

You can assign a value label to a nonstring variable by right-clicking any cell on the variable column, choosing the **Value Labels** menu, and selecting a value label from the **Attach Value Label to Variable '***varname***'** menu. You can define a value label by right-clicking on the Data Editor window and selecting **Value Labels > Manage Value Labels...**. You can also accomplish these tasks by using the Properties pane; see [GS] **6 Using the Data Editor** (GSM, GSU, or GSW) for details.

Changing values of existing cells

Make the cell you wish to change the current cell. Type the new value, and press *Enter*. When updating string variables, do not type double quotes around the string. For variables that have a value label, you can right-click on the cell to display a list of values for the value label. You can assign a new value to the cell by selecting a value from the list.

❏ Technical note

Stata experts will wonder about storage types. Say that variable mpg is stored as an int and you want to change the fourth observation to contain 22.5. The Data Editor will change the storage type of the variable. Similarly, if the variable is a str4 and you type alpha, it will be changed to str5.

The Editor will not, however, change numeric variable types to strings (unless the numeric variable contains only missing values). This is intentional, as such a change could result in a loss of data and is probably the result of a mistake.

❏

Adding new variables

Go to the first empty column, and begin entering your data. The first entry that you make will create the variable and determine whether that variable is numeric or string. The variable will be given a name like var1, but you can rename it by using the Properties pane.

❏ Technical note

Stata experts: The storage type will be determined automatically. If you type a number, the created variable will be numeric; if you type a string, it will be a string. Thus if you want a string variable, be sure that your first entry cannot be interpreted as a number. A way to achieve this is to use surrounding quotes so that "123" will be taken as the string "123", not the number 123. If you want a numeric variable, do not worry about whether it is byte, int, float, etc. If a byte will hold your first number but you need a float to hold your second number, the Editor will recast the variable later.

❏

❏ Technical note

If you do not type in the first empty column but instead type in one to the right of it, the Editor will create variables for all the intervening columns.

❏

Adding new observations

Go to the first empty row, and begin entering your data. As soon as you add one cell below the last row of the dataset, an observation will be created.

❑ Technical note

If you do not enter data in the first empty row but, instead, enter data in a row below it, the Data Editor will create observations for all the intervening rows.

❑

Copying and pasting

You can copy and paste data between Stata's Data Editor and other applications.

First, select the data you wish to copy. In Stata, click on a cell and drag the mouse across other cells to select a range of cells. If you want to select an entire column, click once on the variable name at the top of that column. If you want to select an entire row, click once on the observation number at the left of that row. You can hold down the mouse button after clicking and drag to select multiple columns or rows.

Once you have selected the data, copy the data to the Clipboard. In Stata, right-click on the selected data, and select **Copy**.

You can copy data to the Clipboard from Stata with or without the variable names at the top of each column by right-clicking on the Data Editor window, selecting **Preferences...**, and checking or unchecking *Include variable names on copy to Clipboard*.

You can choose to copy either the value labels or the underlying numeric values associated with the selected data by right-clicking on the Data Editor window, selecting **Preferences...**, and checking or unchecking *Copy value labels instead of numbers*. For more information about value labels, see [U] **12.6.3 Value labels** and [D] **label**.

After you have copied data to the Clipboard from Stata's Data Editor or another spreadsheet, you can paste the data into Stata's Data Editor. First, select the top-left cell of the area into which you wish to paste the data by clicking on it once. Then right-click on the cell and select **Paste**. Stata will paste the data from the Clipboard into the Editor, overwriting any data below and to the right of the cell you selected as the top left of the paste area. If the Data Editor is in filtered mode or in browse mode, **Paste** will be disabled, meaning that you cannot paste into the Data Editor. You can have more control over how data is pasted by selecting **Paste Special...**.

❑ Technical note

If you attempt to paste one or more string values into numeric variables, the original numeric values will be left unchanged for those cells. Stata will display a message box to let you know that this has happened: "You attempted to paste one or more string values into numeric variables. The contents of these cells, if any, are unchanged."

If you see this message, you should look carefully at the data that you pasted into Stata's Data Editor to make sure that you pasted into the area that you intended. We recommend that you take a snapshot of your data before pasting into Stata's Data Editor so that you can restore the data from the snapshot if you make a mistake. See [GS] **6 Using the Data Editor** (GSM, GSU, or GSW) to read about snapshots.

❑

Logging changes

When you use edit to enter new data or change existing data, you will find output in the Stata Results window documenting the changes that you made. For example, a line of this output might be

```
. replace mpg = 22.5 in 5
```

The Editor submits a command to Stata for everything you do in it except pasting. If you are logging your results, you will have a permanent record of what you did in the Editor.

Advice

- People who care about data integrity know that editors are dangerous—it is too easy to make changes accidentally. Never use edit when you want to browse.

- Protect yourself when you edit existing data by limiting exposure. If you need to change mpg and need to see model to know which value of mpg to change, do not click on the **Data Editor** button. Instead, type edit model mpg. It is now impossible for you to change (damage) variables other than model and mpg. Furthermore, if you know that you need to change mpg only if it is missing, you can reduce your exposure even more by typing 'edit model mpg if mpg>=.'.

- Stata's Data Editor is safer than most because it logs changes to the Results window. Use this feature—look at the log afterward, and verify that the changes you made are the changes you wanted to make.

References

Brady, T. 1998. dm63: Dialog box window for browsing, editing, and entering observations. *Stata Technical Bulletin* 46: 2–6. Reprinted in *Stata Technical Bulletin Reprints*, vol. 8, pp. 28–34. College Station, TX: Stata Press.

——. 2000. dm63.1: A new version of winshow for Stata 6. *Stata Technical Bulletin* 53: 3–5. Reprinted in *Stata Technical Bulletin Reprints*, vol. 9, pp. 15–19. College Station, TX: Stata Press.

Also see

[D] **import** — Overview of importing data into Stata

[D] **input** — Enter data from keyboard

[D] **list** — List values of variables

[D] **save** — Save Stata dataset

[GSM] **6 Using the Data Editor**

[GSW] **6 Using the Data Editor**

[GSU] **6 Using the Data Editor**

Title

egen — Extensions to generate

Syntax

egen [*type*] *newvar* = *fcn* (*arguments*) [*if*] [*in*] [, *options*]

by is allowed with some of the **egen** functions, as noted below.

where depending on the *fcn*, *arguments* refers to an expression, *varlist*, or *numlist*, and the *options* are also *fcn* dependent, and where *fcn* is

anycount (*varlist*), <u>v</u>alues (*integer numlist*)
    may not be combined with **by**. It returns the number of variables in *varlist* for which values are equal to any integer value in a supplied *numlist*. Values for any observations excluded by either **if** or **in** are set to 0 (not missing). Also see anyvalue (*varname*) and anymatch (*varlist*).

anymatch (*varlist*), <u>v</u>alues (*integer numlist*)
    may not be combined with **by**. It is 1 if any variable in *varlist* is equal to any integer value in a supplied *numlist* and 0 otherwise. Values for any observations excluded by either **if** or **in** are set to 0 (not missing). Also see anyvalue (*varname*) and anycount (*varlist*).

anyvalue (*varname*) , <u>v</u>alues (*integer numlist*)
    may not be combined with **by**. It takes the value of *varname* if *varname* is equal to any integer value in a supplied *numlist* and is missing otherwise. Also see anymatch (*varlist*) and anycount (*varlist*).

concat (*varlist*) [, <u>f</u>ormat (%*fmt*) <u>d</u>ecode <u>maxl</u>ength (#) <u>p</u>unct (*pchars*)]
    may not be combined with **by**. It concatenates *varlist* to produce a string variable. Values of string variables are unchanged. Values of numeric variables are converted to string, as is, or are converted using a numeric format under the format (%*fmt*) option or decoded under the decode option, in which case maxlength () may also be used to control the maximum label length used. By default, variables are added end to end: punct (*pchars*) may be used to specify punctuation, such as a space, punct (" "), or a comma, punct (,).

count (*exp*)   (allows **by** *varlist*:)
    creates a constant (within *varlist*) containing the number of nonmissing observations of *exp*. Also see rownonmiss () and rowmiss ().

cut (*varname*), { at (#,#,...,#) | group (#) } [<u>i</u>codes <u>lab</u>el]
    may not be combined with **by**. It creates a new categorical variable coded with the left-hand ends of the grouping intervals specified in the at () option, which expects an ascending numlist.

    at (#,#,...,#) supplies the breaks for the groups, in ascending order. The list of breakpoints may be simply a list of numbers separated by commas but can also include the syntax a(b)c, meaning from a to c in steps of size b. If no breaks are specified, the command expects the group () option.

    group (#) specifies the number of equal frequency grouping intervals to be used in the absence of breaks. Specifying this option automatically invokes icodes.

    icodes requests that the codes 0, 1, 2, etc., be used in place of the left-hand ends of the intervals.

167

label requests that the integer-coded values of the grouped variable be labeled with the left-hand ends of the grouping intervals. Specifying this option automatically invokes icodes.

diff(*varlist*)

may not be combined with by. It creates an indicator variable equal to 1 if the variables in *varlist* are not equal and 0 otherwise.

ends(*strvar*) [, <u>punct</u>(*pchars*) <u>trim</u> [<u>head</u>|<u>last</u>|<u>tail</u>]]

may not be combined with by. It gives the first "word" or head (with the head option), the last "word" (with the last option), or the remainder or tail (with the tail option) from string variable *strvar*.

head, last, and tail are determined by the occurrence of *pchars*, which is by default one space (" ").

The head is whatever precedes the first occurrence of *pchars*, or the whole of the string if it does not occur. For example, the head of "frog toad" is "frog" and that of "frog" is "frog". With punct(,), the head of "frog,toad" is "frog".

The last word is whatever follows the last occurrence of *pchars* or is the whole of the string if a space does not occur. The last word of "frog toad newt" is "newt" and that of "frog" is "frog". With punct(,), the last word of "frog,toad" is "toad".

The remainder or tail is whatever follows the first occurrence of *pchars*, which will be the empty string "" if *pchars* does not occur. The tail of "frog toad newt" is "toad newt" and that of "frog" is "". With punct(,), the tail of "frog,toad" is "toad".

The trim option trims any leading or trailing spaces.

fill(*numlist*)

may not be combined with by. It creates a variable of ascending or descending numbers or complex repeating patterns. *numlist* must contain at least two numbers and may be specified using standard *numlist* notation; see [U] **11.1.8 numlist**. if and in are not allowed with fill().

group(*varlist*) [, <u>missing</u> <u>label</u> lname(*name*) <u>truncate</u>(*num*)]

may not be combined with by. It creates one variable taking on values 1, 2, ... for the groups formed by *varlist*. *varlist* may contain numeric variables, string variables, or a combination of the two. The order of the groups is that of the sort order of *varlist*. missing indicates that missing values in *varlist* (either . or "") are to be treated like any other value when assigning groups, instead of as missing values being assigned to the group missing. The label option returns integers from 1 up according to the distinct groups of *varlist* in sorted order. The integers are labeled with the values of *varlist* or the value labels, if they exist. lname() specifies the name to be given to the value label created to hold the labels; lname() implies label. The truncate() option truncates the values contributed to the label from each variable in *varlist* to the length specified by the integer argument *num*. The truncate option cannot be used without specifying the label option. The truncate option does not change the groups that are formed; it changes only their labels.

iqr(*exp*)   (allows by *varlist*:)

creates a constant (within *varlist*) containing the interquartile range of *exp*. Also see pctile().

kurt(*varname*)   (allows by *varlist*:)

returns the kurtosis (within *varlist*) of *varname*.

mad(*exp*)   (allows by *varlist*:)

returns the median absolute deviation from the median (within *varlist*) of *exp*.

max(*exp*)   (allows by *varlist*:)

creates a constant (within *varlist*) containing the maximum value of *exp*.

mdev(*exp*)   (allows by *varlist*:)
   returns the mean absolute deviation from the mean (within *varlist*) of *exp*.

mean(*exp*)   (allows by *varlist*:)
   creates a constant (within *varlist*) containing the mean of *exp*.

median(*exp*)   (allows by *varlist*:)
   creates a constant (within *varlist*) containing the median of *exp*. Also see pctile().

min(*exp*)   (allows by *varlist*:)
   creates a constant (within *varlist*) containing the minimum value of *exp*.

mode(*varname*) [, minmode maxmode nummode(*integer*) missing]   (allows by *varlist*:)
   produces the mode (within *varlist*) for *varname*, which may be numeric or string. The mode
   is the value occurring most frequently. If two or more modes exist or if *varname* contains
   all missing values, the mode produced will be a missing value. To avoid this, the minmode,
   maxmode, or nummode() option may be used to specify choices for selecting among the multiple
   modes, and the missing option will treat missing values as categories. minmode returns the
   lowest value, and maxmode returns the highest value. nummode(#) will return the #th mode,
   counting from the lowest up. Missing values are excluded from determination of the mode
   unless missing is specified. Even so, the value of the mode is recorded for observations for
   which the values of *varname* are missing unless they are explicitly excluded, that is, by if
   varname < . or if *varname* != "".

mtr(*year income*)
   may not be combined with by. It returns the U.S. marginal income tax rate for a married couple
   with taxable income *income* in year *year*, where $1930 \leq year \leq 2011$. *year* and *income* may
   be specified as variable names or constants; for example, mtr(1993 faminc), mtr(surveyyr
   28000), or mtr(surveyyr faminc). A blank or comma may be used to separate *income* from
   year.

pc(*exp*) [, prop]   (allows by *varlist*:)
   returns *exp* (within *varlist*) scaled to be a percentage of the total, between 0 and 100. The prop
   option returns *exp* scaled to be a proportion of the total, between 0 and 1.

pctile(*exp*) [, p(#)]   (allows by *varlist*:)
   creates a constant (within *varlist*) containing the #th percentile of *exp*. If p(#) is not specified,
   50 is assumed, meaning medians. Also see median().

rank(*exp*) [, field | track | unique]   (allows by *varlist*:)
   creates ranks (within *varlist*) of *exp*; by default, equal observations are assigned the average
   rank. The field option calculates the field rank of *exp*: the highest value is ranked 1, and there
   is no correction for ties. That is, the field rank is 1 + the number of values that are higher.
   The track option calculates the track rank of *exp*: the lowest value is ranked 1, and there is
   no correction for ties. That is, the track rank is 1 + the number of values that are lower. The
   unique option calculates the unique rank of *exp*: values are ranked 1, ..., #, and values and
   ties are broken arbitrarily. Two values that are tied for second are ranked 2 and 3.

rowfirst(*varlist*)
   may not be combined with by. It gives the first nonmissing value in *varlist* for each observation
   (row). If all values in *varlist* are missing for an observation, *newvar* is set to missing.

rowlast(*varlist*)
   may not be combined with by. It gives the last nonmissing value in *varlist* for each observation
   (row). If all values in *varlist* are missing for an observation, *newvar* is set to missing.

rowmax(*varlist*)
> may not be combined with by. It gives the maximum value (ignoring missing values) in *varlist* for each observation (row). If all values in *varlist* are missing for an observation, *newvar* is set to missing.

rowmean(*varlist*)
> may not be combined with by. It creates the (row) means of the variables in *varlist*, ignoring missing values; for example, if three variables are specified and, in some observations, one of the variables is missing, in those observations *newvar* will contain the mean of the two variables that do exist. Other observations will contain the mean of all three variables. Where none of the variables exist, *newvar* is set to missing.

rowmedian(*varlist*)
> may not be combined with by. It gives the (row) median of the variables in *varlist*, ignoring missing values. If all variables in *varlist* are missing for an observation, *newvar* is set to missing in that observation. Also see rowpctile().

rowmin(*varlist*)
> may not be combined with by. It gives the minimum value in *varlist* for each observation (row). If all values in *varlist* are missing for an observation, *newvar* is set to missing.

rowmiss(*varlist*)
> may not be combined with by. It gives the number of missing values in *varlist* for each observation (row).

rownonmiss(*varlist*) [, <u>s</u>trok]
> may not be combined with by. It gives the number of nonmissing values in *varlist* for each observation (row)—this is the value used by rowmean() for the denominator in the mean calculation.
>
> > String variables may not be specified unless the strok option is also specified. If strok is specified, string variables will be counted as containing missing values when they contain "". Numeric variables will be counted as containing missing when their value is "$\geq .$".

rowpctile(*varlist*) [, p(*#*)]
> may not be combined with by. It gives the *#*th percentile of the variables in *varlist*, ignoring missing values. If all variables in *varlist* are missing for an observation, *newvar* is set to missing in that observation. If p() is not specified, p(50) is assumed, meaning medians. Also see rowmedian().

rowsd(*varlist*)
> may not be combined with by. It creates the (row) standard deviations of the variables in *varlist*, ignoring missing values.

rowtotal(*varlist*) [, <u>m</u>issing]
> may not be combined with by. It creates the (row) sum of the variables in *varlist*, treating missing values as 0. If missing is specified and all values in *varlist* are missing for an observation, *newvar* is set to missing.

sd(*exp*)   (allows by *varlist*:)
> creates a constant (within *varlist*) containing the standard deviation of *exp*. Also see mean().

seq() [, <u>f</u>rom(*#*) <u>t</u>o(*#*) <u>b</u>lock(*#*)]   (allows by *varlist*:)
> returns integer sequences. Values start from from() (default 1) and increase to to() (the default is the maximum number of values) in blocks (default size 1). If to() is less than the maximum number, sequences restart at from(). Numbering may also be separate within groups defined by *varlist* or decreasing if to() is less than from(). Sequences depend on the sort order of observations, following three rules: 1) observations excluded by if or in are not

counted; 2) observations are sorted by *varlist*, if specified; and 3) otherwise, the order is that when called. No *arguments* are specified.

skew(*varname*) (allows **by** *varlist*:)

returns the skewness (within *varlist*) of *varname*.

std(*exp*) [, <u>m</u>ean(*#*) <u>s</u>td(*#*)]

may not be combined with **by**. It creates the standardized values of *exp*. The options specify the desired mean and standard deviation. The default is **mean(0)** and **std(1)**, producing a variable with mean 0 and standard deviation 1.

tag(*varlist*) [, <u>m</u>issing]

may not be combined with **by**. It tags just 1 observation in each distinct group defined by *varlist*. When all observations in a group have the same value for a summary variable calculated for the group, it will be sufficient to use just one value for many purposes. The result will be 1 if the observation is tagged and never missing, and 0 otherwise. Values for any observations excluded by either **if** or **in** are set to 0 (not missing). Hence, if **tag** is the variable produced by **egen tag = tag(***varlist***)**, the idiom **if tag** is always safe. **missing** specifies that missing values of *varlist* may be included.

total(*exp*) [, <u>m</u>issing] (allows **by** *varlist*:)

creates a constant (within *varlist*) containing the sum of *exp* treating missing as 0. If **missing** is specified and all values in *exp* are missing, *newvar* is set to missing. Also see **mean()**.

Menu

Data > Create or change data > Create new variable (extended)

Description

egen creates *newvar* of the optionally specified storage type equal to *fcn*(*arguments*). Here *fcn*() is a function specifically written for **egen**, as documented below or as written by users. Only **egen** functions may be used with **egen**, and conversely, only **egen** may be used to run **egen** functions.

Depending on *fcn*(), *arguments*, if present, refers to an expression, *varlist*, or a *numlist*, and the *options* are similarly *fcn* dependent. Explicit subscripting (using _N and _n), which is commonly used with **generate**, should not be used with **egen**; see [U] **13.7 Explicit subscripting**.

Remarks

Remarks are presented under the following headings:

> *Summary statistics*
> *Generating patterns*
> *Marking differences among variables*
> *Ranks*
> *Standardized variables*
> *Row functions*
> *Categorical and integer variables*
> *String variables*
> *U.S. marginal income tax rate*

See Mitchell (2010) for numerous examples using **egen**.

Summary statistics

The functions count(), iqr(), kurt(), mad(), max(), mdev(), mean(), median(), min(), mode(), pc(), pctile(), sd(), skew(), and total() create variables containing summary statistics. These functions take a by ... : prefix and, if specified, calculate the summary statistics within each by-group.

▷ Example 1: Without the by prefix

Without the by prefix, the result produced by these functions is a constant for every observation in the data. For instance, we have data on cholesterol levels (chol) and wish to have a variable that, for each patient, records the deviation from the average across all patients:

```
. use http://www.stata-press.com/data/r12/egenxmpl
. egen avg = mean(chol)
. generate deviation = chol - avg
```
◁

▷ Example 2: With the by prefix

These functions are most useful when the by prefix is specified. For instance, assume that our dataset includes dcode, a hospital–patient diagnostic code, and los, the number of days that the patient remained in the hospital. We wish to obtain the deviation in length of stay from the median for all patients having the same diagnostic code:

```
. use http://www.stata-press.com/data/r12/egenxmpl2, clear
. by dcode, sort: egen medstay = median(los)
. generate deltalos = los - medstay
```
◁

❏ Technical note

Distinguish carefully between Stata's sum() function and egen's total() function. Stata's sum() function creates the running sum, whereas egen's total() function creates a constant equal to the overall sum; for example,

```
. clear
. set obs 5
obs was 0, now 5
. generate a = _n
. generate sum1=sum(a)
. egen sum2=total(a)
. list
```

	a	sum1	sum2
1.	1	1	15
2.	2	3	15
3.	3	6	15
4.	4	10	15
5.	5	15	15

❏

❑ Technical note

The definitions and formulas used by these functions are the same as those used by summarize; see [R] **summarize**. For comparison with summarize, mean() and sd() correspond to the mean and standard deviation. total() is the numerator of the mean, and count() is its denominator. min() and max() correspond to the minimum and maximum. median()—or, equally well, pctile() with p(50)—is the median. pctile() with p(5) refers to the fifth percentile, and so on. iqr() is the difference between the 75th and 25th percentiles.

❑

The mode is the most common value of a dataset, whether it contains numeric or string variables. It is perhaps most useful for categorical variables (whether defined by integers or strings) or for other integer-valued values, but mode() can be applied to variables of any type. Nevertheless, the modes of continuous (or nearly continuous) variables are perhaps better estimated either from inspection of a graph of a frequency distribution or from the results of some density estimation (see [R] **kdensity**).

Missing values need special attention. It is possible that missing is the most common value in a variable (whether missing is defined by the period [.] or extended missing values [.a, .b, ..., .z] for numeric variables or the empty string [""] for string variables). However, missing values are by default excluded from determination of modes. If you wish to include them, use the missing option.

In contrast, egen mode = mode(*varname*) allows the generation of nonmissing modes for observations for which *varname* is missing. This allows use of the mode as one simple means of imputing categorical variables. If you want the mode to be missing whenever *varname* is missing, you can specify if *varname* < . or if *varname* != "" or, most generally, if !missing(*varname*).

mad() and mdev() produce alternative measures of spread. The median absolute deviation from the median and even the mean deviation will both be more resistant than the standard deviation to heavy tails or outliers, in particular from distributions with heavier tails than the normal or Gaussian. The first measure was named the MAD by Andrews et al. (1972) but was already known to K. F. Gauss in 1816, according to Hampel et al. (1986). For more historical and statistical details, see David (1998) and Wilcox (2003, 72–73).

Generating patterns

To create a sequence of numbers, simply "show" the fill() function how the sequence should look. It must be a linear progression to produce the expected results. Stata does not understand geometric progressions. To produce repeating patterns, you present fill() with the pattern twice in the *numlist*.

▷ Example 3: Sequences produced by fill()

Here are some examples of ascending and descending sequences produced by fill():

```
. clear
. set obs 12
obs was 0, now 12
. egen i=fill(1 2)
. egen w=fill(100 99)
. egen x=fill(22 17)
. egen y=fill(1 1 2 2)
. egen z=fill(8 8 8 7 7 7)
```

```
. list, sep(4)
```

	i	w	x	y	z
1.	1	100	22	1	8
2.	2	99	17	1	8
3.	3	98	12	2	8
4.	4	97	7	2	7
5.	5	96	2	3	7
6.	6	95	-3	3	7
7.	7	94	-8	4	6
8.	8	93	-13	4	6
9.	9	92	-18	5	6
10.	10	91	-23	5	5
11.	11	90	-28	6	5
12.	12	89	-33	6	5

◁

▷ Example 4: Patterns produced by fill( )

Here are examples of patterns produced by fill():

```
. clear
. set obs 12
obs was 0, now 12
. egen a=fill(0 0 1 0 0 1)
. egen b=fill(1 3 8 1 3 8)
. egen c=fill(-3(3)6 -3(3)6)
. egen d=fill(10 20 to 50    10 20 to 50)
. list, sep(4)
```

	a	b	c	d
1.	0	1	-3	10
2.	0	3	0	20
3.	1	8	3	30
4.	0	1	6	40
5.	0	3	-3	50
6.	1	8	0	10
7.	0	1	3	20
8.	0	3	6	30
9.	1	8	-3	40
10.	0	1	0	50
11.	0	3	3	10
12.	1	8	6	20

◁

▷ Example 5: seq( )

seq() creates a new variable containing one or more sequences of integers. It is useful mainly for quickly creating observation identifiers or automatically numbering levels of factors or categorical variables.

```
. clear
. set obs 12
```

In the simplest case,

```
. egen a = seq()
```

is just equivalent to the common idiom

```
. generate a = _n
```

a may also be obtained from

```
. range a 1 _N
```

(the actual value of _N may also be used).

In more complicated cases, seq() with option calls is equivalent to calls to the versatile functions int and mod.

```
. egen b = seq(), b(2)
```

produces integers in blocks of 2, whereas

```
. egen c = seq(), t(6)
```

restarts the sequence after 6 is reached.

```
. egen d = seq(), f(10) t(12)
```

shows that sequences may start with integers other than 1, and

```
. egen e = seq(), f(3) t(1)
```

shows that they may decrease.

The results of these commands are shown by

```
. list, sep(4)
```

	a	b	c	d	e
1.	1	1	1	10	3
2.	2	1	2	11	2
3.	3	2	3	12	1
4.	4	2	4	10	3
5.	5	3	5	11	2
6.	6	3	6	12	1
7.	7	4	1	10	3
8.	8	4	2	11	2
9.	9	5	3	12	1
10.	10	5	4	10	3
11.	11	6	5	11	2
12.	12	6	6	12	1

All these sequences could have been generated in one line with `generate` and with the use of the `int` and `mod` functions. The variables b through e are obtained with

```
. gen b = 1 + int((_n - 1)/2)
. gen c = 1 + mod(_n - 1, 6)
. gen d = 10 + mod(_n - 1, 3)
. gen e = 3 - mod(_n - 1, 3)
```

Nevertheless, `seq()` may save users from puzzling out such solutions or from typing in the needed values.

In general, the sequences produced depend on the sort order of observations, following three rules:

1. observations excluded by `if` or `in` are not counted;

2. observations are sorted by *varlist*, if specified; and

3. otherwise, the order is that specified when `seq()` is called.

The result of applying `seq` was not guaranteed to be identical from application to application whenever sorting was required, even with identical data, because of the indeterminacy of sorting. That is, if we sort, say, integer values, it is sufficient that all the 1s are together and are followed by all the 2s. But there is no guarantee that the order of the 1s, as defined by any other variables, will be identical from sort to sort.

◁

The `fill()` and `seq()` functions are alternatives. In essence, `fill()` requires a minimal example that indicates the kind of sequence required, whereas `seq()` requires that the rule be specified through options. There are sequences that `fill()` can produce that `seq()` cannot, and vice versa. `fill()` cannot be combined with `if` or `in`, in contrast to `seq()`, which can.

Marking differences among variables

> Example 6: diff( )

We have three measures of respondents' income obtained from different sources. We wish to create the variable `differ` equal to 1 for disagreements:

```
. use http://www.stata-press.com/data/r12/egenxmpl3, clear
. egen byte differ = diff(inc*)
. list if differ==1
```

	inc1	inc2	inc3	id	differ
10.	42,491	41,491	41,491	110	1
11.	26,075	25,075	25,075	111	1
12.	26,283	25,283	25,283	112	1
78.	41,780	41,780	41,880	178	1
100.	25,687	26,687	25,687	200	1
101.	25,359	26,359	25,359	201	1
102.	25,969	26,969	25,969	202	1
103.	25,339	26,339	25,339	203	1
104.	25,296	26,296	25,296	204	1
105.	41,800	41,000	41,000	205	1
134.	26,233	26,233	26,133	234	1

Rather than typing diff(inc*), we could have typed diff(inc1 inc2 inc3).

◁

Ranks

▷ Example 7: rank()

Most applications of rank() will be to one variable, but the argument *exp* can be more general, namely, an expression. In particular, rank(-*varname*) reverses ranks from those obtained by rank(*varname*).

The default ranking and those obtained by using one of the track, field, and unique options differ principally in their treatment of ties. The default is to assign the same rank to tied values such that the sum of the ranks is preserved. The track option assigns the same rank but resembles the convention in track events; thus, if one person had the lowest time and three persons tied for second-lowest time, their ranks would be 1, 2, 2, and 2, and the next person(s) would have rank 5. The field option acts similarly except that the highest is assigned rank 1, as in field events in which the greatest distance or height wins. The unique option breaks ties arbitrarily: its most obvious use is assigning ranks for a graph of ordered values. See also group() for another kind of "ranking".

```
. use http://www.stata-press.com/data/r12/auto, clear
(1978 Automobile Data)
. keep in 1/10
(64 observations deleted)
. egen rank = rank(mpg)
. egen rank_r = rank(-mpg)
. egen rank_f = rank(mpg), field
. egen rank_t = rank(mpg), track
. egen rank_u = rank(mpg), unique
. egen rank_ur = rank(-mpg), unique
. sort rank_u
. list mpg rank*
```

	mpg	rank	rank_r	rank_f	rank_t	rank_u	rank_ur
1.	15	1	10	10	1	1	10
2.	16	2	9	9	2	2	9
3.	17	3	8	8	3	3	8
4.	18	4	7	7	4	4	7
5.	19	5	6	6	5	5	6
6.	20	6.5	4.5	4	6	6	5
7.	20	6.5	4.5	4	6	7	4
8.	22	8.5	2.5	2	8	8	3
9.	22	8.5	2.5	2	8	9	2
10.	26	10	1	1	10	10	1

◁

Standardized variables

▷ Example 8: std( )

We have a variable called age recording the median age in the 50 states. We wish to create the standardized value of age and verify the calculation:

```
. use http://www.stata-press.com/data/r12/states1, clear
(State data)
. egen stdage = std(age)
. summarize age stdage
```

Variable	Obs	Mean	Std. Dev.	Min	Max
age	50	29.54	1.693445	24.2	34.7
stdage	50	6.41e-09	1	-3.153336	3.047044

```
. correlate age stdage
(obs=50)
```

	age	stdage
age	1.0000	
stdage	1.0000	1.0000

summarize shows that the new variable has a mean of approximately zero; 10^{-9} is the precision of a float and is close enough to zero for all practical purposes. If we wanted, we could have typed egen double stdage = std(age), making stdage a double-precision variable, and the mean would have been 10^{-16}. In any case, summarize also shows that the standard deviation is 1. correlate shows that the new variable and the original variable are perfectly correlated.

We may optionally specify the mean and standard deviation for the new variable. For instance,

```
. egen newage1 = std(age), std(2)
. egen newage2 = std(age), mean(2) std(4)
. egen newage3 = std(age), mean(2)
. summarize age newage1-newage3
```

Variable	Obs	Mean	Std. Dev.	Min	Max
age	50	29.54	1.693445	24.2	34.7
newage1	50	1.28e-08	2	-6.306671	6.094089
newage2	50	2	4	-10.61334	14.18818
newage3	50	2	1	-1.153336	5.047044

```
. correlate age newage1-newage3
(obs=50)
```

	age	newage1	newage2	newage3
age	1.0000			
newage1	1.0000	1.0000		
newage2	1.0000	1.0000	1.0000	
newage3	1.0000	1.0000	1.0000	1.0000

◁

Row functions

> Example 9: rowtotal( )

generate's sum() function creates the vertical, running sum of its argument, whereas egen's total() function creates a constant equal to the overall sum. egen's rowtotal() function, however, creates the horizontal sum of its arguments. They all treat missing as zero. However, if the missing option is specified with total() or rowtotal(), then *newvar* will contain missing values if all values of *exp* or *varlist* are missing.

```
. use http://www.stata-press.com/data/r12/egenxmpl4, clear
. egen hsum = rowtotal(a b c)
. generate vsum = sum(hsum)
. egen sum = total(hsum)
. list
```

	a	b	c	hsum	vsum	sum
1.	.	2	3	5	5	63
2.	4	.	6	10	15	63
3.	7	8	.	15	30	63
4.	10	11	12	33	63	63

◁

> Example 10: rowmean( ), rowmedian( ), rowpctile( ), rowsd( ), and rownonmiss( )

summarize displays the mean and standard deviation of a variable across observations; program writers can access the mean in r(mean) and the standard deviation in r(sd) (see [R] **summarize**). egen's rowmean() function creates the means of observations across variables. rowmedian() creates the medians of observations across variables. rowpctile() returns the #th percentile of the variables specified in *varlist*. rowsd() creates the standard deviations of observations across variables. rownonmiss() creates a count of the number of nonmissing observations, the denominator of the rowmean() calculation:

```
. use http://www.stata-press.com/data/r12/egenxmpl4, clear
. egen avg = rowmean(a b c)
. egen median = rowmedian(a b c)
. egen pct25 = rowpctile(a b c), p(25)
. egen std = rowsd(a b c)
. egen n = rownonmiss(a b c)
. list
```

	a	b	c	avg	median	pct25	std	n
1.	.	2	3	2.5	2.5	2	.7071068	2
2.	4	.	6	5	5	4	1.414214	2
3.	7	8	.	7.5	7.5	7	.7071068	2
4.	10	11	12	11	11	10	1	3

◁

▷ Example 11: rowmiss( )

rowmiss() returns $k - $ rownonmiss(), where k is the number of variables specified. rowmiss() can be especially useful for finding casewise-deleted observations caused by missing values.

```
. use http://www.stata-press.com/data/r12/auto3, clear
(1978 Automobile Data)
. correlate price weight mpg
(obs=70)
```

	price	weight	mpg
price	1.0000		
weight	0.5309	1.0000	
mpg	-0.4478	-0.7985	1.0000

```
. egen excluded = rowmiss(price weight mpg)
. list make price weight mpg if excluded !=0
```

	make	price	weight	mpg
5.	Buick Electra	.	4,080	15
12.	Cad. Eldorado	14,500	3,900	.
40.	Olds Starfire	4,195	.	24
51.	Pont. Phoenix	.	3,420	.

◁

▷ Example 12: rowmin( ), rowmax( ), rowfirst( ), and rowlast( )

rowmin(), rowmax(), rowfirst(), and rowlast() return the minimum, maximum, first, or last nonmissing value, respectively, for the specified variables within an observation (row).

```
. use http://www.stata-press.com/data/r12/egenxmpl5, clear
. egen min = rowmin(x y z)
(1 missing value generated)
. egen max = rowmax(x y z)
(1 missing value generated)
. egen first = rowfirst(x y z)
(1 missing value generated)
. egen last = rowlast(x y z)
(1 missing value generated)
. list, sep(4)
```

	x	y	z	min	max	first	last
1.	-1	2	3	-1	3	-1	3
2.	.	-6	.	-6	-6	-6	-6
3.	7	.	-5	-5	7	7	-5
4.	.	.	.	.	.	.	.
5.	4	.	.	4	4	4	4
6.	.	.	8	8	8	8	8
7.	.	3	7	3	7	3	7
8.	5	-1	6	-1	6	5	6

◁

Categorical and integer variables

▷ Example 13: anyvalue( ), anymatch( ), and anycount( )

anyvalue(), anymatch(), and anycount() are for categorical or other variables taking integer values. If we define a subset of values specified by an integer *numlist* (see [U] **11.1.8 numlist**), anyvalue() extracts the subset, leaving every other value missing; anymatch() defines an indicator variable (1 if in subset, 0 otherwise); and anycount() counts occurrences of the subset across a set of variables. Therefore, with just one variable, anymatch(*varname*) and anycount(*varname*) are equivalent.

With the auto dataset, we can generate a variable containing the high values of rep78 and a variable indicating whether rep78 has a high value:

```
. use http://www.stata-press.com/data/r12/auto, clear
(1978 Automobile Data)
. egen hirep = anyvalue(rep78), v(3/5)
(15 missing values generated)
. egen ishirep = anymatch(rep78), v(3/5)
```

Here it is easy to produce the same results with official Stata commands:

```
. generate hirep = rep78 if inlist(rep78,3,4,5)
. generate byte ishirep = inlist(rep78,3,4,5)
```

However, as the specification becomes more complicated or involves several variables, the egen functions may be more convenient.

◁

▷ Example 14: group( )

group() maps the distinct groups of a varlist to a categorical variable that takes on integer values from 1 to the total number of groups. order of the groups is that of the sort order of *varlist*. The *varlist* may be of numeric variables, string variables, or a mixture of the two. The resulting variable can be useful for many purposes, including stepping through the distinct groups easily and systematically and cleaning up an untidy ordering. Suppose that the actual (and arbitrary) codes present in the data are 1, 2, 4, and 7, but we desire equally spaced numbers, as when the codes will be values on one axis of a graph. group() maps these to 1, 2, 3, and 4.

We have a variable agegrp that takes on the values 24, 40, 50, and 65, corresponding to age groups 18–24, 25–40, 41–50, and 51 and above. Perhaps we created this coding using the recode() function (see [U] **13.3 Functions** and [U] **25 Working with categorical data and factor variables**) from another age-in-years variable:

```
. generate agegrp=recode(age,24,40,50,65)
```

We now want to change the codes to 1, 2, 3, and 4:

```
. egen agegrp2 = group(agegrp)
```

◁

▷ Example 15: group( ) with missing values

We have two categorical variables, race and sex, which may be string or numeric. We want to use ir (see [ST] **epitab**) to create a Mantel–Haenszel weighted estimate of the incidence rate. ir, however, allows only one variable to be specified in its by( ) option. We type

```
. use http://www.stata-press.com/data/r12/egenxmpl6, clear
. egen racesex = group(race sex)
(2 missing values generated)
. ir deaths smokes pyears, by(racesex)
  (output omitted )
```

The new numeric variable, racesex, will be missing wherever race or sex is missing (meaning . for numeric variables and "" for string variables), so missing values will be handled correctly. When we list some of the data, we see

```
. list race sex racesex in 1/7, sep(0)
```

	race	sex	racesex
1.	White	Female	1
2.	White	Male	2
3.	Black	Female	3
4.	Black	Male	4
5.	Black	Male	4
6.	.	Female	.
7.	Black	.	.

group( ) began by putting the data in the order of the grouping variables and then assigned the numeric codes. Observations 6 and 7 were assigned to racesex==. because, in one case, race was not known, and in the other, sex was not known. (These observations were not used by ir.)

If we wanted the unknown groups to be treated just as any other category, we could have typed

```
. egen rs2=group(race sex), missing
. list race sex rs2 in 1/7, sep(0)
```

	race	sex	rs2
1.	White	Female	1
2.	White	Male	2
3.	Black	Female	3
4.	Black	Male	4
5.	Black	Male	4
6.	.	Female	6
7.	Black	.	5

◁

The resulting variable from group( ) does not have value labels. Therefore, the values carry no indication of meaning. Interpretation requires comparison with the original *varlist*.

The label option produces a categorical variable with value labels. These value labels are either the actual values of *varname* or any value labels of *varname*, if they exist. The values of *varname* could be as long as those of one str244 variable, but value labels may be no longer than 80 characters.

String variables

Concatenation of string variables is provided in Stata. In context, Stata understands the addition symbol + as specifying concatenation or adding strings end to end. "soft" + "ware" produces "software", and given string variables s1 and s2, s1 + s2 indicates their concatenation.

The complications that may arise in practice include wanting 1) to concatenate the string versions of numeric variables and 2) to concatenate variables, together with some separator such as a space or a comma. Given numeric variables n1 and n2,

```
. generate newstr = s1 + string(n1) + string(n2) + s2
```

shows how numeric values may be converted to their string equivalents before concatenation, and

```
. generate newstr = s1 + " " + s2 + " " + s3
```

shows how spaces may be added between variables. Stata will automatically assign the most appropriate data type for the new string variables.

▷ Example 16: concat( )

concat() allows us to do everything in one line concisely.

```
. egen newstr = concat(s1 n1 n2 s2)
```

carries with it an implicit instruction to convert numeric values to their string equivalents, and the appropriate string data type is worked out within concat() by Stata's automatic promotion. Moreover,

```
. egen newstr = concat(s1 s2 s3), p(" ")
```

specifies that spaces be used as separators. (The default is to have no separation of concatenated strings.)

As an example of punctuation other than a space, consider

```
. egen fullname = concat(surname forename), p(", ")
```

Noninteger numerical values can cause difficulties, but

```
. egen newstr = concat(n1 n2), format(%9.3f) p(" ")
```

specifies the use of format %9.3f. This is equivalent to

```
. generate str1 newstr = ""
. replace newstr = string(n1,"%9.3f") + " " + string(n2,"%9.3f")
```

See [D] **functions** for more about string().

◁

As a final flourish, the decode option instructs concat() to use value labels. With that option, the maxlength() option may also be used. For more details about decode, see [D] **encode**. Unlike the decode command, however, concat() uses string(*varname*), not "", whenever values of *varname* are not associated with value labels, and the format() option, whenever specified, applies to this use of string().

▷ Example 17: ends( )

The ends(*strvar*) function is used for subdividing strings. The approach is to find specified separators by using the strpos() string function and then to extract what is desired, which either precedes or follows the separators, using the substr() string function.

By default, substrings are considered to be separated by individual spaces, so we will give definitions in those terms and then generalize.

The head of the string is whatever precedes the first space or is the whole of the string if no space occurs. This could also be called the first "word". The tail of the string is whatever follows the first space. This could be nothing or one or more words. The last word in the string is whatever follows the last space or is the whole of the string if no space occurs.

To clarify, let's look at some examples. The quotation marks here just mark the limits of each string and are not part of the strings.

	head	tail	last
"frog"	"frog"	""	"frog"
"frog toad"	"frog"	"toad"	"toad"
"frog toad newt"	"frog"	"toad newt"	"newt"
"frog toad newt"	"frog"	" toad newt"	"newt"
"frog toad newt"	"frog"	"toad newt"	"newt"

The main subtlety is that these functions are literal, so the tail of "frog toad newt", in which two spaces follow "frog", includes the second of those spaces, and is thus " toad newt". Therefore, you may prefer to use the trim option to trim the result of any leading or trailing spaces, producing "toad newt" in this instance.

The punct(*pchars*) option may be used to specify separators other than spaces. The general definitions of the head, tail, and last options are therefore interpreted in terms of whatever separator has been specified; that is, they are relative to the first or last occurrence of the separator in the string value. Thus, with punct(,) and the string "Darwin, Charles Robert", the head is "Darwin", and the tail and the last are both " Charles Robert". Note again the leading space in this example, which may be trimmed with trim. The punctuation (here the comma, ",") is discarded, just as it is with one space.

pchars, the argument of punct(), will usually, but not always, be one character. If two or more characters are specified, these must occur together; for example, punct(:;) would mean that words are separated by a colon followed by a semicolon (that is, :;). It is not implied, in particular, that the colon and semicolon are alternatives. To do that, you would have to modify the programs presented here or resort to first principles by using split; see [D] **split**.

With personal names, the head or last option might be applied to extract surnames if strings were similar to "Darwin, Charles Robert" or "Charles Robert Darwin", with the surname coming first or last. What then happens with surnames like "von Neumann" or "de la Mare"? "von Neumann, John" is no problem, if the comma is specified as a separator, but the last option is not intelligent enough to handle "Walter de la Mare" properly. For that, the best advice is to use programs specially written for person-name extraction, such as extrname (Gould 1993).

◁

U.S. marginal income tax rate

mtr(*year income*) (Schmidt 1993, 1994) returns the U.S. marginal income tax rate for a married couple with taxable income *income* in year *year*, where $1930 \leq year \leq 2011$.

▷ Example 18: mtr( )

Schmidt (1993) examines the change in the progressivity of the U.S. tax schedule over the period from 1930 to 1990. As a measure of progressivity, he calculates the difference in the marginal tax rates at the 75th and 25th percentiles of income, using a dataset of percentiles of taxable income developed by Hakkio, Rush, and Schmidt (1996). (Certain aspects of the income distribution are imputed in these data.) A subset of the data contains the following:

```
. describe
Contains data from income1.dta
  obs:            61
  vars:            4                          12 Feb 2011 03:33
  size:         1,020

              storage  display    value
variable name   type   format     label      variable label

year            float  %9.0g                 Year
inc25           float  %9.0g                 25th percentile
inc50           float  %9.0g                 50th percentile
inc75           float  %9.0g                 75th percentile

Sorted by:
. summarize
    Variable |      Obs        Mean   Std. Dev.       Min        Max

        year |       61        1960   17.75293       1930       1990
       inc25 |       61    6948.272   6891.921      819.4   27227.35
       inc50 |       61    11645.15   11550.71    1373.29   45632.43
       inc75 |       61    18166.43    18019.1    2142.33   71186.58
```

Given the series for income and the four-digit year, we can generate the marginal tax rates corresponding to the 25th and 75th percentiles of income:

```
. egen mtr25 = mtr(year inc25)
. egen mtr75 = mtr(year inc75)
. summarize mtr25 mtr75
    Variable |      Obs        Mean   Std. Dev.       Min        Max

       mtr25 |       61    .1664898   .0677949     .01125        .23
       mtr75 |       61    .2442053   .1148427     .01125    .424625
```

◁

Methods and formulas

egen is implemented as an ado-file.

Stata users have written many extra functions for egen. Type net search egen to locate Internet sources of programs.

Acknowledgments

The mtr() function of egen was written by Timothy J. Schmidt of the Federal Reserve Bank of Kansas City.

The cut function was written by David Clayton, Cambridge Institute for Medical Research, and Michael Hills, London School of Hygiene and Tropical Medicine (retired) (1999a, 1999b, 1999c).

Many of the other egen functions were written by Nicholas J. Cox, Durham University, UK.

References

Andrews, D. F., P. J. Bickel, F. R. Hampel, P. J. Huber, W. H. Rogers, and J. W. Tukey. 1972. *Robust Estimates of Location: Survey and Advances*. Princeton: Princeton University Press.

Cappellari, L., and S. P. Jenkins. 2006. Calculation of multivariate normal probabilities by simulation, with applications to maximum simulated likelihood estimation. *Stata Journal* 6: 156–189.

Clayton, D. G., and M. Hills. 1999a. dm66: Recoding variables using grouped values. *Stata Technical Bulletin* 49: 6–7. Reprinted in *Stata Technical Bulletin Reprints*, vol. 9, pp. 23–25. College Station, TX: Stata Press.

——. 1999b. dm66.1: Stata 6 version of recoding variables using grouped values. *Stata Technical Bulletin* 50: 3. Reprinted in *Stata Technical Bulletin Reprints*, vol. 9, p. 25. College Station, TX: Stata Press.

——. 1999c. dm66.2: Update of cut to Stata 6. *Stata Technical Bulletin* 51: 2–3. Reprinted in *Stata Technical Bulletin Reprints*, vol. 9, pp. 25–26. College Station, TX: Stata Press.

Cox, N. J. 1999. dm70: Extensions to generate, extended. *Stata Technical Bulletin* 50: 9–17. Reprinted in *Stata Technical Bulletin Reprints*, vol. 9, pp. 34–45. College Station, TX: Stata Press.

——. 2000. dm70.1: Extensions to generate, extended: Corrections. *Stata Technical Bulletin* 57: 2. Reprinted in *Stata Technical Bulletin Reprints*, vol. 10, p. 9. College Station, TX: Stata Press.

——. 2009. Speaking Stata: Rowwise. *Stata Journal* 9: 137–157.

Cox, N. J., and R. Goldstein. 1999a. dm72: Alternative ranking procedures. *Stata Technical Bulletin* 51: 5–7. Reprinted in *Stata Technical Bulletin Reprints*, vol. 9, pp. 48–51. College Station, TX: Stata Press.

——. 1999b. dm72.1: Alternative ranking procedures: Update. *Stata Technical Bulletin* 52: 2. Reprinted in *Stata Technical Bulletin Reprints*, vol. 9, p. 51. College Station, TX: Stata Press.

David, H. A. 1998. Early sample measures of variability. *Statistical Science* 13: 368–377.

Esman, R. M. 1998. dm55: Generating sequences and patterns of numeric data: An extension to egen. *Stata Technical Bulletin* 43: 2–3. Reprinted in *Stata Technical Bulletin Reprints*, vol. 8, pp. 4–5. College Station, TX: Stata Press.

Gould, W. W. 1993. dm13: Person name extraction. *Stata Technical Bulletin* 13: 6–11. Reprinted in *Stata Technical Bulletin Reprints*, vol. 3, pp. 25–31. College Station, TX: Stata Press.

Hakkio, C. S., M. Rush, and T. J. Schmidt. 1996. The marginal income tax rate schedule from 1930 to 1990. *Journal of Monetary Economics* 38: 117–138.

Hampel, F. R., E. M. Ronchetti, P. J. Rousseeuw, and W. A. Stahel. 1986. *Robust Statistics: The Approach Based on Influence Functions*. New York: Wiley.

Mitchell, M. N. 2010. *Data Management Using Stata: A Practical Handbook*. College Station, TX: Stata Press.

Ryan, P. 1999. dm71: Calculating the product of observations. *Stata Technical Bulletin* 51: 3–4. Reprinted in *Stata Technical Bulletin Reprints*, vol. 9, pp. 45–48. College Station, TX: Stata Press.

——. 2001. dm87: Calculating the row product of observations. *Stata Technical Bulletin* 60: 3–4. Reprinted in *Stata Technical Bulletin Reprints*, vol. 10, pp. 39–41. College Station, TX: Stata Press.

Schmidt, T. J. 1993. sss1: Calculating U.S. marginal income tax rates. *Stata Technical Bulletin* 15: 17–19. Reprinted in *Stata Technical Bulletin Reprints*, vol. 3, pp. 197–200. College Station, TX: Stata Press.

——. 1994. sss1.1: Updated U.S. marginal income tax rate function. *Stata Technical Bulletin* 22: 29. Reprinted in *Stata Technical Bulletin Reprints*, vol. 4, p. 224. College Station, TX: Stata Press.

Wilcox, R. R. 2003. *Applying Contemporary Statistical Techniques*. San Diego, CA: Academic Press.

Also see

[D] **collapse** — Make dataset of summary statistics

[D] **generate** — Create or change contents of variable

[U] **13.3 Functions**

Title

> **encode** — Encode string into numeric and vice versa

Syntax

String variable to numeric variable

> <u>en</u>code *varname* $\big[$ *if* $\big]$ $\big[$ *in* $\big]$, <u>g</u>enerate(*newvar*) $\big[$ <u>l</u>abel(*name*) <u>noe</u>xtend $\big]$

Numeric variable to string variable

> <u>de</u>code *varname* $\big[$ *if* $\big]$ $\big[$ *in* $\big]$, <u>g</u>enerate(*newvar*) $\big[$ <u>maxl</u>ength(*#*) $\big]$

Menu

encode

Data > Create or change data > Other variable-transformation commands > Encode value labels from string variable

decode

Data > Create or change data > Other variable-transformation commands > Decode strings from labeled numeric variable

Description

encode creates a new variable named *newvar* based on the string variable *varname*, creating, adding to, or just using (as necessary) the value label *newvar* or, if specified, *name*. Do not use encode if *varname* contains numbers that merely happen to be stored as strings; instead, use generate *newvar* = real(*varname*) or destring; see [U] **23.2 Categorical string variables**, *String functions* in [D] **functions**, and [D] **destring**.

decode creates a new string variable named *newvar* based on the "encoded" numeric variable *varname* and its value label.

Options for encode

generate(*newvar*) is required and specifies the name of the variable to be created.

label(*name*) specifies the name of the value label to be created or used and added to if the named value label already exists. If label() is not specified, encode uses the same name for the label as it does for the new variable.

noextend specifies that *varname* not be encoded if there are values contained in *varname* that are not present in label(*name*). By default, any values not present in label(*name*) will be added to that label.

188

Options for decode

generate(*newvar*) is required and specifies the name of the variable to be created.

maxlength(*#*) specifies how many characters of the value label to retain; *#* must be between 1 and 244. The default is maxlength(244).

Remarks

Remarks are presented under the following headings:

> *encode*
> *decode*

encode

encode is most useful in making string variables accessible to Stata's statistical routines, most of which can work only with numeric variables. encode is also useful in reducing the size of a dataset. If you are not familiar with value labels, read [U] **12.6.3 Value labels**.

The maximum number of associations within each value label is 65,536 (1,000 for Small Stata). Each association in a value label maps a string of up to 244 characters to a number. If your string has entries longer than that, only the first 244 characters are retained and are significant.

▷ Example 1

We have a dataset on high blood pressure, and among the variables is sex, a string variable containing either "male" or "female". We wish to run a regression of high blood pressure on race, sex, and age group. We type regress hbp race sex age_grp and get the message "no observations".

```
. use http://www.stata-press.com/data/r12/hbp2
. regress hbp sex race age_grp
no observations
r(2000);
```

Stata's statistical procedures cannot directly deal with string variables; as far as they are concerned, all observations on sex are missing. encode provides the solution:

```
. encode sex, gen(gender)
. regress hbp gender race age_grp
```

Source	SS	df	MS		Number of obs =	1121
					F(3, 1117) =	15.15
Model	2.01013476	3	.67004492		Prob > F =	0.0000
Residual	49.3886164	1117	.044215413		R-squared =	0.0391
					Adj R-squared =	0.0365
Total	51.3987511	1120	.045891742		Root MSE =	.21027

hbp	Coef.	Std. Err.	t	P>|t|	[95% Conf. Interval]	
gender	.0394747	.0130022	3.04	0.002	.0139633	.0649861
race	-.0409453	.0113721	-3.60	0.000	-.0632584	-.0186322
age_grp	.0241484	.00624	3.87	0.000	.0119049	.0363919
_cons	-.016815	.0389167	-0.43	0.666	-.093173	.059543

encode looks at a string variable and makes an internal table of all the values it takes on, here "male" and "female". It then alphabetizes that list and assigns numeric codes to each entry. Thus 1 becomes "female" and 2 becomes "male". It creates a new int variable (gender) and substitutes a 1 where sex is "female", a 2 where sex is "male", and a *missing* (.) where sex is *null* (""). It creates a value label (also named gender) that records the mapping 1 ↔ female and 2 ↔ male. Finally, encode labels the values of the new variable with the value label.

◁

▷ Example 2

It is difficult to distinguish the result of encode from the original string variable. For instance, in our last two examples, we typed encode sex, gen(gender). Let's compare the two variables:

```
. list sex gender in 1/4
```

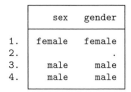

They look almost identical, although you should notice the missing value for gender in the second observation.

The difference does show, however, if we tell list to ignore the value labels and show how the data really appear:

```
. list sex gender in 1/4, nolabel
```

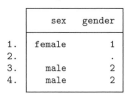

We could also ask to see the underlying value label:

```
. label list gender
gender:
            1 female
            2 male
```

gender really is a numeric variable, but because *all* Stata commands understand value labels, the variable displays as "male" and "female", just as the underlying string variable sex would.

◁

▷ Example 3

We can drastically reduce the size of our dataset by encoding strings and then discarding the underlying string variable. We have a string variable, sex, that records each person's sex as "male" and "female". Because female has six characters, the variable is stored as a str6.

We can encode the sex variable and use compress to store the variable as a byte, which takes only 1 byte. Because our dataset contains 1,130 people, the string variable takes 6,780 bytes, but the encoded variable will take only 1,130 bytes.

```
. use http://www.stata-press.com/data/r12/hbp2, clear
. describe
Contains data from http://www.stata-press.com/data/r12/hbp2.dta
  obs:          1,130
 vars:              7                          3 Mar 2011 06:47
 size:         24,860
```

variable name	storage type	display format	value label	variable label
id	str10	%10s		Record identification number
city	byte	%8.0g		
year	int	%8.0g		
age_grp	byte	%8.0g	agefmt	
race	byte	%8.0g	racefmt	
hbp	byte	%8.0g	yn	high blood pressure
sex	str6	%9s		

```
Sorted by:
. encode sex, generate(gender)
. list sex gender in 1/5
```

	sex	gender
1.	female	female
2.		.
3.	male	male
4.	male	male
5.	female	female

```
. drop sex
. rename gender sex
. compress
sex was long now byte
. describe
Contains data from http://www.stata-press.com/data/r12/hbp2.dta
  obs:          1,130
 vars:              7                          3 Mar 2011 06:47
 size:         19,210
```

variable name	storage type	display format	value label	variable label
id	str10	%10s		Record identification number
city	byte	%8.0g		
year	int	%8.0g		
age_grp	byte	%8.0g	agefmt	
race	byte	%8.0g	racefmt	
hbp	byte	%8.0g	yn	high blood pressure
sex	byte	%8.0g	gender	

```
Sorted by:
     Note:  dataset has changed since last saved
```

The size of our dataset has fallen from 24,860 bytes to 19,210 bytes.

◁

❏ Technical note

In the examples above, the value label did not exist before encode created it, because that is not required. If the value label does exist, encode uses your encoding as far as it can and adds new mappings for anything not found in your value label. For instance, if you wanted "female" to be encoded as 0 rather than 1 (possibly for use in linear regression), you could type

```
. label define gender 0 "female"
. encode sex, gen(gender)
```

You can also specify the name of the value label. If you do not, the value label is assumed to have the same name as the newly created variable. For instance,

```
. label define sexlbl 0 "female"
. encode sex, gen(gender) label(sexlbl)
```

❏

decode

decode is used to convert numeric variables with associated value labels into true string variables.

▷ Example 4

We have a numeric variable named female that records the values 0 and 1. female is associated with a value label named sexlbl that says that 0 means male and 1 means female:

```
. use http://www.stata-press.com/data/r12/hbp3, clear
. describe female
```

variable name	storage type	display format	value label	variable label
female	byte	%8.0g	sexlbl	

```
. label list sexlbl
sexlbl:
          0 male
          1 female
```

We see that female is stored as a byte. It is a numeric variable. Nevertheless, it has an associated value label describing what the numeric codes mean, so if we tabulate the variable, for instance, it appears to contain the strings "male" and "female":

```
. tabulate female
```

female	Freq.	Percent	Cum.
male	695	61.61	61.61
female	433	38.39	100.00
Total	1,128	100.00	

We can create a real string variable from this numerically encoded variable by using decode:

```
. decode female, gen(sex)
. describe sex
```

variable name	storage type	display format	value label	variable label
sex	str6	%9s		

We have a new variable called sex. It is a string, and Stata automatically created the shortest possible string. The word "female" has six characters, so our new variable is a str6. female and sex appear indistinguishable:

```
. list female sex in 1/4
```

```
      | female        sex  |
      +---------------------+
  1.  | female     female   |
  2.  |      .               |
  3.  |   male       male   |
  4.  |   male       male   |
```

But when we add `nolabel`, the difference is apparent:

```
. list female sex in 1/4, nolabel
```

```
      | female        sex  |
      +---------------------+
  1.  |      1     female   |
  2.  |      .               |
  3.  |      0       male   |
  4.  |      0       male   |
```

◁

▷ Example 5

`decode` is most useful in instances when we wish to match-merge two datasets on a variable that has been encoded inconsistently.

For instance, we have two datasets on individual states in which one of the variables (`state`) takes on values such as "CA" and "NY". The state variable was originally a string, but along the way the variable was encoded into an integer with a corresponding value label in one or both datasets.

We wish to merge these two datasets, but either 1) one of the datasets has a string variable for state and the other an encoded variable or 2) although both are numeric, we are not certain that the codings are consistent. Perhaps "CA" has been coded 5 in one dataset and 6 in another.

Because `decode` will take an encoded variable and turn it back into a string, `decode` provides the solution:

```
use first               (load the first dataset)
decode state, gen(st)   (make a string state variable)
drop state              (discard the encoded variable)
sort st                 (sort on string)
save first, replace     (save the dataset)
use second              (load the second dataset)
decode state, gen(st)   (make a string variable)
drop state              (discard the encoded variable)
sort st                 (sort on string)
merge 1:1 st using first  (merge the data)
```

◁

Also see

[D] **compress** — Compress data in memory

[D] **destring** — Convert string variables to numeric variables and vice versa

[D] **generate** — Create or change contents of variable

[U] **12.6.3 Value labels**

[U] **23.2 Categorical string variables**

Title

erase — Erase a disk file

Syntax

$\{$ erase $|$ rm $\}$ $\left[\text{"}\right]$ *filename* $\left[\text{"}\right]$

Note: Double quotes must be used to enclose *filename* if the name contains spaces.

Description

The erase command erases files stored on disk. rm is a synonym for erase for the convenience of Unix users.

Stata for Mac users: erase is permanent; the file is not moved to the Trash but is immediately removed from the disk.

Stata for Windows users: erase is permanent; the file is not moved to the Recycle Bin but is immediately removed from the disk.

Remarks

The only difference between Stata's erase (rm) command and the DOS DEL or Unix rm(1) command is that we may not specify groups of files. Stata requires that we erase files one at a time.

Mac users may prefer to discard files by dragging them to the Trash.

Windows users may prefer to discard files by dragging them to the Recycle Bin.

▷ Example 1

Stata provides seven operating system equivalent commands: cd, copy, dir, erase, mkdir, rmdir, and type, or, from the Unix perspective, cd, copy, ls, rm, mkdir, rmdir, and cat. These commands are provided for Mac users, too. Stata users can also issue any operating system command by using Stata's shell command, so you should never have to exit Stata to perform some housekeeping detail.

Suppose that we have the file mydata.dta stored on disk and we wish to permanently eliminate it:

```
. erase mydata
file mydata not found
r(601);
. erase mydata.dta

.
```

Our first attempt, erase mydata, was unsuccessful. Although Stata ordinarily supplies the file extension for you, it does not do so when you type erase. You must be explicit. Our second attempt eliminated the file. Unix users could have typed rm mydata.dta if they preferred.

◁

Also see

[D] **cd** — Change directory

[D] **copy** — Copy file from disk or URL

[D] **dir** — Display filenames

[D] **mkdir** — Create directory

[D] **rmdir** — Remove directory

[D] **shell** — Temporarily invoke operating system

[D] **type** — Display contents of a file

[U] **11.6 Filenaming conventions**

Title

> **expand** — Duplicate observations

Syntax

expand $\left[=\right] exp \left[if\right] \left[in\right] \left[, \underline{\text{gen}}\text{erate}(newvar) \right]$

Menu

Data > Create or change data > Other variable-transformation commands > Duplicate observations

Description

expand replaces each observation in the dataset with *n* copies of the observation, where *n* is equal to the required expression rounded to the nearest integer. If the expression is less than 1 or equal to *missing*, it is interpreted as if it were 1, and the observation is retained but not duplicated.

Option

generate(*newvar*) creates new variable *newvar* containing 0 if the observation originally appeared in the dataset and 1 if the observation is a duplicate. For instance, after an expand, you could revert to the original observations by typing keep if *newvar*==0.

Remarks

> Example 1

expand is, admittedly, a strange command. It can, however, be useful in tricky programs or for reformatting data for survival analysis (see examples in [ST] **epitab**). Here is a silly use of expand:

```
. use http://www.stata-press.com/data/r12/expandxmpl
. list
```

	n	x
1.	-1	1
2.	0	2
3.	1	3
4.	2	4
5.	3	5

```
. expand n
(1 negative count ignored; observation not deleted)
(1 zero count ignored; observation not deleted)
(3 observations created)
```

```
. list
```

	n	x
1.	-1	1
2.	0	2
3.	1	3
4.	2	4
5.	3	5
6.	2	4
7.	3	5
8.	3	5

The new observations are added to the end of the dataset. expand informed us that it created 3 observations. The first 3 observations were not replicated because n was less than or equal to 1. n is 2 in the fourth observation, so expand created one replication of this observation, bringing the total number of observations of this type to 2. expand created two replications of observation 5 because n is 3.

Because there were 5 observations in the original dataset and because expand adds new observations onto the end of the dataset, we could now undo the expansion by typing drop in 6/1.

◁

Also see

[D] **contract** — Make dataset of frequencies and percentages

[D] **expandcl** — Duplicate clustered observations

[D] **fillin** — Rectangularize dataset

Title

expandcl — Duplicate clustered observations

Syntax

expandcl $\left[=\right]exp$ $\left[if\right]$ $\left[in\right]$, <u>cl</u>uster(*varlist*) <u>gene</u>rate(*newvar*)

Menu

Data > Create or change data > Other variable-transformation commands > Duplicate clustered observations

Description

expandcl duplicates clusters of observations and generates a new variable that identifies the clusters uniquely.

expandcl replaces each cluster in the dataset with n copies of the cluster, where n is equal to the required expression rounded to the nearest integer. The expression is required to be constant within cluster. If the expression is less than 1 or equal to *missing*, it is interpreted as if it were 1, and the cluster is retained but not duplicated.

Options

cluster(*varlist*) is required and specifies the variables that identify the clusters before expanding the data.

generate(*newvar*) is required and stores unique identifiers for the duplicated clusters in *newvar*. *newvar* will identify the clusters by using consecutive integers starting from 1.

Remarks

▷ Example 1

We will show how expandcl works by using a small dataset with five clusters. In this dataset, cl identifies the clusters, x contains a unique value for each observation, and n identifies how many copies we want of each cluster.

```
. use http://www.stata-press.com/data/r12/expclxmpl

. list, sepby(cl)
```

	cl	x	n
1.	10	1	-1
2.	10	2	-1
3.	20	3	0
4.	20	4	0
5.	30	5	1
6.	30	6	1
7.	40	7	2.7
8.	40	8	2.7
9.	50	9	3
10.	50	10	3
11.	60	11	.
12.	60	12	.

```
. expandcl n, generate(newcl) cluster(cl)
(2 missing counts ignored; observations not deleted)
(2 noninteger counts rounded to integer)
(2 negative counts ignored; observations not deleted)
(2 zero counts ignored; observations not deleted)
(8 observations created)

. sort newcl cl x
```

. list, sepby(newcl)

	cl	x	n	newcl
1.	10	1	-1	1
2.	10	2	-1	1
3.	20	3	0	2
4.	20	4	0	2
5.	30	5	1	3
6.	30	6	1	3
7.	40	7	2.7	4
8.	40	8	2.7	4
9.	40	7	2.7	5
10.	40	8	2.7	5
11.	40	7	2.7	6
12.	40	8	2.7	6
13.	50	9	3	7
14.	50	10	3	7
15.	50	9	3	8
16.	50	10	3	8
17.	50	9	3	9
18.	50	10	3	9
19.	60	11	.	10
20.	60	12	.	10

The first three clusters were not replicated because n was less than or equal to 1. n is 2.7 in the fourth cluster, so expandcl created two replications (2.7 was rounded to 3) of this cluster, bringing the total number of clusters of this type to 3. expandcl created two replications of cluster 50 because n is 3. Finally, expandcl did not replicate the last cluster because n was missing.

◁

Methods and formulas

expandcl is implemented as an ado-file.

Also see

[R] **bsample** — Sampling with replacement

[D] **expand** — Duplicate observations

Title

---
export — Overview of exporting data from Stata

Description

This entry provides a quick reference for determining which method to use for exporting Stata data from memory to other formats.

Remarks

Remarks are presented under the following headings:

Summary of the different methods
    export excel
    outsheet
    odbc
    outfile
    export sasxport
    xmlsave

Summary of the different methods

export excel

- `export excel` creates Microsoft Excel worksheets in `.xls` and `.xlsx` files.
- Entire worksheets can be exported, or custom cell ranges can be overwritten.
- See [D] **import excel**.

outsheet

- `outsheet` creates comma-separated (CSV) or tab-delimited files that many other programs can read.
- A custom delimiter may also be specified.
- The first line of the file can optionally contain the names of the variables.
- See [D] **outsheet**.

odbc

- ODBC, an acronym for Open DataBase Connectivity, is a standard for exchanging data between programs. Stata supports the ODBC standard for exporting data via the `odbc` command and can write to any ODBC data source on your computer.
- See [D] **odbc**.

outfile

- ○ outfile creates text-format datasets.
- ○ The data can be written in space-separated or comma-separated format.
- ○ Alternatively, the data can be written in fixed-column format.
- ○ See [D] **outfile**.

export sasxport

- ○ export sasxport saves SAS XPORT Transport format files.
- ○ export sasxport can also write value label information to a formats.xpf XPORT file.
- ○ See [D] **import sasxport**.

xmlsave

- ○ xmlsave writes extensible markup language (XML) files—highly adaptable text-format files derived from the standard generalized markup language (SGML).
- ○ xmlsave can write either an Excel-format XML or a Stata-format XML file.
- ○ See [D] **xmlsave**.

Also see

[D] **import excel** — Import and export Excel files

[D] **import sasxport** — Import and export datasets in SAS XPORT format

[D] **outfile** — Export dataset in text format

[D] **outsheet** — Write spreadsheet-style dataset

[D] **odbc** — Load, write, or view data from ODBC sources

[D] **xmlsave** — Export or import dataset in XML format

[D] **import** — Overview of importing data into Stata

Title

> **filefilter** — Convert text or binary patterns in a file

Syntax

<u>filefil</u>ter *oldfile newfile* ,

$\Big\{$ <u>f</u>rom(*oldpattern*) <u>to</u>(*newpattern*) | ascii2ebcdic | ebcdic2ascii $\Big\}$ $\big[$ *options* $\big]$

where *oldpattern* and *newpattern* for ASCII characters are

"*string*" or *string*

string	:=	[*char*[*char*[*char*[...]]]]
char	:=	*regchar* | *code*
regchar	:=	ASCII 32–91, 93–128, 161–255; excludes '\'
code	:=	\BS backslash
		\r carriage return
		\n newline
		\t tab
		\M Classic Mac EOL, or \r
		\W Windows EOL, or \r\n
		\U Unix or Mac EOL, or \n
		\LQ left single quote, '
		\RQ right single quote, '
		\Q double quote, "
		\$ dollar sign, $
		\###d 3-digit [0–9] decimal ASCII
		\##h 2-digit [0–9, A–F] hexadecimal ASCII

options	Description
* <u>f</u>rom(*oldpattern*)	find *oldpattern* to be replaced
* <u>to</u>(*newpattern*)	use *newpattern* to replace occurrences of from()
* ascii2ebcdic	convert file from ASCII to EBCDIC
* ebcdic2ascii	convert file from EBCDIC to ASCII
<u>rep</u>lace	replace *newfile* if it already exists

* Both from(*oldpattern*) and to(*newpattern*) are required, or ascii2ebcdic or ebcdic2ascii is required.

Description

filefilter reads an input file, searching for *oldpattern*. Whenever a matching pattern is found, it is replaced with *newpattern*. All resulting data, whether matching or nonmatching, are then written to the new file.

Because of the buffering design of filefilter, arbitrarily large files can be converted quickly. filefilter is also useful when traditional editors cannot edit a file, such as when unprintable ASCII characters are involved. In fact, converting end-of-line characters between Mac, Classic Mac, Windows, and Unix is convenient with the EOL codes.

Unicode is not directly supported at this time, but you can attempt to operate on a Unicode file by breaking a 2-byte character into the corresponding two-character ASCII representation. However, this goes beyond the original design of the command and is technically unsupported. If you attempt to use `filefilter` in this manner, you might encounter problems with variable-length encoded Unicode.

Although it is not mandatory, you may want to use quotes to delimit a pattern, protecting the pattern from Stata's parsing routines. A pattern that contains blanks must be in quotes.

Options

`from(oldpattern)` specifies the pattern to be found and replaced. It is required unless `ascii2ebcdic` or `ebcdic2ascii` is specified.

`to(newpattern)` specifies the pattern used to replace occurrences of `from()`. It is required unless `ascii2ebcdic` or `ebcdic2ascii` is specified.

`ascii2ebcdic` specifies that characters in the file be converted from ASCII coding to EBCDIC coding. `from()`, `to()`, and `ebcdic2ascii` are not allowed with `ascii2ebcdic`.

`ebcdic2ascii` specifies that characters in the file be converted from EBCDIC coding to ASCII coding. `from()`, `to()`, and `ascii2ebcdic` are not allowed with `ebcdic2ascii`.

`replace` specifies that *newfile* be replaced if it already exists.

Remarks

Convert Classic Mac-style EOL characters to Windows-style

```
. filefilter macfile.txt winfile.txt, from(\M) to(\W) replace
```

Convert left quote (') characters to the string "left quote"

```
. filefilter auto1.csv auto2.csv, from(\LQ) to("left quote")
```

Convert the character with hexidecimal code 60 to the string "left quote"

```
. filefilter auto1.csv auto2.csv, from(\60h) to("left quote")
```

Convert the character with decimal code 96 to the string "left quote"

```
. filefilter auto1.csv auto2.csv, from(\096d) to("left quote")
```

Convert strings beginning with hexidecimal code 6B followed by "Text" followed by decimal character 100 followed by "Text" to an empty string (remove them from the file)

```
. filefilter file1.txt file2.txt, from("\6BhText\100dText") to("")
```

Convert file from EBCDIC to ASCII encoding

```
. filefilter ebcdicfile.txt asciifile.txt, ebcdic2ascii
```

Saved results

filefilter saves the following in r():

Scalars

r(occurrences)	number of *oldpattern* found
r(bytes_from)	# of bytes represented by *oldpattern*
r(bytes_to)	# of bytes represented by *newpattern*

Reference

Riley, A. R. 2008. Stata tip 60: Making fast and easy changes to files with filefilter. *Stata Journal* 8: 290–292.

Also see

[P] **file** — Read and write ASCII text and binary files

[D] **changeeol** — Convert end-of-line characters of text file

[D] **hexdump** — Display hexadecimal report on file

Title

fillin — Rectangularize dataset

Syntax

fillin *varlist*

Menu

Data > Create or change data > Other variable-transformation commands > Rectangularize dataset

Description

fillin adds observations with missing data so that all interactions of *varlist* exist, thus making a complete rectangularization of *varlist*. fillin also adds the variable _fillin to the dataset. _fillin is 1 for observations created by using fillin and 0 for previously existing observations.

Remarks

▷ Example 1

We have data on something by sex, race, and age group. We suspect that some of the combinations of sex, race, and age do not exist, but if so, we want them to exist with whatever remaining variables there are in the dataset set to missing. For example, rather than having a missing observation for black females aged 20–24, we want to create an observation that contains missing values:

```
. use http://www.stata-press.com/data/r12/fillin1
. list
```

	sex	race	age_gr~p	x1	x2
1.	female	white	20-24	20393	14.5
2.	male	white	25-29	32750	12.7
3.	female	black	30-34	39399	14.2

```
. fillin sex race age_group
```

```
. list, sepby(sex)
```

	sex	race	age_gr~p	x1	x2	_fillin
1.	female	white	20-24	20393	14.5	0
2.	female	white	25-29	.	.	1
3.	female	white	30-34	.	.	1
4.	female	black	20-24	.	.	1
5.	female	black	25-29	.	.	1
6.	female	black	30-34	39399	14.2	0
7.	male	white	20-24	.	.	1
8.	male	white	25-29	32750	12.7	0
9.	male	white	30-34	.	.	1
10.	male	black	20-24	.	.	1
11.	male	black	25-29	.	.	1
12.	male	black	30-34	.	.	1

◁

Methods and formulas

fillin is implemented as an ado-file.

References

Baum, C. F. 2009. *An Introduction to Stata Programming*. College Station, TX: Stata Press.

Cox, N. J. 2005. Stata tip 17: Filling in the gaps. *Stata Journal* 5: 135–136.

Also see

[D] **save** — Save Stata dataset

[D] **cross** — Form every pairwise combination of two datasets

[D] **expand** — Duplicate observations

[D] **joinby** — Form all pairwise combinations within groups

Title

> **format** — Set variables' output format

Syntax

Set formats

> <u>form</u>at *varlist* %*fmt*

> <u>form</u>at %*fmt varlist*

Set style of decimal point

> <u>se</u>t dp {<u>com</u>ma | <u>per</u>iod} [, <u>perman</u>ently]

Display long formats

> <u>form</u>at [*varlist*]

where %*fmt* can be a numerical, date, business calendar, or string format.

Numerical % *fmt*	Description	Example
right-justified		
%#.#g	general	%9.0g
%#.#f	fixed	%9.2f
%#.#e	exponential	%10.7e
%21x	hexadecimal	%21x
%16H	binary, hilo	%16H
%16L	binary, lohi	%16L
%8H	binary, hilo	%8H
%8L	binary, lohi	%8L
right-justified with commas		
%#.#gc	general	%9.0gc
%#.#fc	fixed	%9.2fc
right-justified with leading zeros		
%0#.#f	fixed	%09.2f
left-justified		
%-#.#g	general	%-9.0g
%-#.#f	fixed	%-9.2f
%-#.#e	exponential	%-10.7e
left-justified with commas		
%-#.#gc	general	%-9.0gc
%-#.#fc	fixed	%-9.2fc

You may substitute comma (,) for period (.) in any
of the above formats to make comma the decimal point. In
%9,2fc, 1000.03 is 1.000,03. Or you can set dp comma.

date % *fmt*	Description	Example
right-justified		
%tc	date/time	%tc
%tC	date/time	%tC
%td	date	%td
%tw	week	%tw
%tm	month	%tm
%tq	quarter	%tq
%th	half-year	%th
%ty	year	%ty
%tg	generic	%tg
left-justified		
%-tc	date/time	%-tc
%-tC	date/time	%-tC
%-td	date	%-td
etc.		

There are many variations allowed. See [D] **datetime display formats**.

business calendar % *fmt*	Description	Example
%tb*calname* $\big[$:*datetime-specifiers* $\big]$	a business calendar defined in *calname*.stbcal	%tbsimple

See [D] **datetime business calendars**.

string % *fmt*	Description	Example
right-justified %#s	string	%15s
left-justified %-#s	string	%-20s
centered %~#s	string	%~12s

The centered format is for use with display only.

Menu

Data > Variables Manager

Description

format *varlist* % *fmt* and format % *fmt varlist* are the same commands. They set the display format associated with the variables specified. The default formats are a function of the type of the variable:

byte	%8.0g
int	%8.0g
long	%12.0g
float	%9.0g
double	%10.0g
str#	%#s

set dp sets the symbol that Stata uses to represent the decimal point. The default is period, meaning that one and a half is displayed as 1.5.

format $\big[$ *varlist* $\big]$ displays the current formats associated with the variables. format by itself lists all variables that have formats too long to be listed in their entirety by describe. format *varlist* lists the formats for the specified variables regardless of their length. format * lists the formats for all the variables.

Option

permanently specifies that, in addition to making the change right now, the dp setting be remembered and become the default setting when you invoke Stata.

Remarks

Remarks are presented under the following headings:

> *Setting formats*
> *Setting European formats*
> *Details of formats*
>> *The %f format*
>> *The %fc format*
>> *The %g format*
>> *The %gc format*
>> *The %e format*
>> *The %21x format*
>> *The %16H and %16L formats*
>> *The %8H and %8L formats*
>> *The %t format*
>> *The %s format*
> *Other effects of formats*
> *Displaying current formats*

Setting formats

See [U] **12.5 Formats: Controlling how data are displayed** for an explanation of %*fmt*. To review: Stata's three numeric formats are denoted by a leading percent sign, %, followed by the string *w.d* (or *w,d* for European format), where *w* and *d* stand for two integers. The first integer, *w*, specifies the width of the format. The second integer, *d*, specifies the number of digits that are to follow the decimal point; *d* must be less than *w*. Finally, a character denoting the format type (e, f, or g) is appended. For example, %9.2f specifies the f format that is nine characters wide and has two digits following the decimal point. For f and g, a c may also be suffixed to indicate comma formats. Other "numeric" formats known collectively as the %t formats are used to display dates and times; see [D] **datetime display formats**. String formats are denoted by %*w*s, where *w* indicates the width of the format.

▷ Example 1

We have census data by region and state on median age and population in 1980.

```
. use http://www.stata-press.com/data/r12/census10
(1980 Census data by state)

. describe

Contains data from http://www.stata-press.com/data/r12/census10.dta
  obs:            50                          1980 Census data by state
  vars:            4                          9 Apr 2011 08:05
  size:         1,200

              storage   display    value
variable name   type    format     label      variable label

state          str14    %14s                   State
region         int      %8.0g      cenreg      Census region
pop            long     %11.0g                 Population
medage         float    %9.0g                  Median age

Sorted by:
```

```
. list in 1/8
```

	state	region	pop	medage
1.	Alabama	South	3893888	29.3
2.	Alaska	West	401851	26.1
3.	Arizona	West	2718215	29.2
4.	Arkansas	South	2286435	30.6
5.	California	West	23667902	29.9
6.	Colorado	West	2889964	28.6
7.	Connecticut	NE	3107576	32
8.	Delaware	South	594338	29.8

The `state` variable has a display format of %14s. To left-align the state data, we type

```
. format state %-14s
. list in 1/8
```

	state	region	pop	medage
1.	Alabama	South	3893888	29.3
2.	Alaska	West	401851	26.1
3.	Arizona	West	2718215	29.2
4.	Arkansas	South	2286435	30.6
5.	California	West	23667902	29.9
6.	Colorado	West	2889964	28.6
7.	Connecticut	NE	3107576	32
8.	Delaware	South	594338	29.8

Although it seems like `region` is a string variable, it is really a numeric variable with an attached value label. You do the same thing to left-align a numeric variable as you do a string variable: insert a negative sign.

```
. format region %-8.0g
. list in 1/8
```

	state	region	pop	medage
1.	Alabama	South	3893888	29.3
2.	Alaska	West	401851	26.1
3.	Arizona	West	2718215	29.2
4.	Arkansas	South	2286435	30.6
5.	California	West	23667902	29.9
6.	Colorado	West	2889964	28.6
7.	Connecticut	NE	3107576	32
8.	Delaware	South	594338	29.8

The pop variable would probably be easier to read if we inserted commas by appending a 'c':

```
. format pop %11.0gc
. list in 1/8
```

	state	region	pop	medage
1.	Alabama	South	3,893,888	29.3
2.	Alaska	West	401,851	26.1
3.	Arizona	West	2,718,215	29.2
4.	Arkansas	South	2,286,435	30.6
5.	California	West	23667902	29.9
6.	Colorado	West	2,889,964	28.6
7.	Connecticut	NE	3,107,576	32
8.	Delaware	South	594,338	29.8

Look at the value of pop for observation 5. There are no commas. This number was too large for Stata to insert commas and still respect the current width of 11. Let's try again:

```
. format pop %12.0gc
. list in 1/8
```

	state	region	pop	medage
1.	Alabama	South	3,893,888	29.3
2.	Alaska	West	401,851	26.1
3.	Arizona	West	2,718,215	29.2
4.	Arkansas	South	2,286,435	30.6
5.	California	West	23,667,902	29.9
6.	Colorado	West	2,889,964	28.6
7.	Connecticut	NE	3,107,576	32
8.	Delaware	South	594,338	29.8

Finally, medage would look better if the decimal points were vertically aligned.

```
. format medage %8.1f
. list in 1/8
```

	state	region	pop	medage
1.	Alabama	South	3,893,888	29.3
2.	Alaska	West	401,851	26.1
3.	Arizona	West	2,718,215	29.2
4.	Arkansas	South	2,286,435	30.6
5.	California	West	23,667,902	29.9
6.	Colorado	West	2,889,964	28.6
7.	Connecticut	NE	3,107,576	32.0
8.	Delaware	South	594,338	29.8

Display formats are permanently attached to variables by the format command. If we save the data, the next time we use it, state will still be formatted as %-14s, region will still be formatted as %-8.0g, etc.

◁

▷ Example 2

Suppose that we have an employee identification variable, empid, and that we want to retain the leading zeros when we list our data. format has a leading-zero option that allows this.

```
. use http://www.stata-press.com/data/r12/fmtxmpl
. describe empid

              storage  display   value
variable name   type   format    label      variable label

empid                   float    %9.0g
. list empid in 83/87
```

	empid
83.	98
84.	99
85.	100
86.	101
87.	102

```
. format empid %05.0f
. list empid in 83/87
```

	empid
83.	00098
84.	00099
85.	00100
86.	00101
87.	00102

◁

❑ Technical note

The syntax of the format command allows a *varlist* and not just one variable name. Thus you can attach the %9.2f format to the variables myvar, thisvar, and thatvar by typing

```
. format myvar thisvar thatvar %9.2f
```

❑

▷ Example 3

We have employee data that includes hiredate and login and logout times. hiredate is stored as a float, but we were careful to store login and logout as doubles. We need to attach a date format to these three variables.

```
. use http://www.stata-press.com/data/r12/fmtxmpl2
. format hiredate login logout

variable name  display format

hiredate       %9.0g
login          %10.0g
logout         %10.0g
```

```
. format login logout %tcDDmonCCYY_HH:MM:SS.ss
. list login logout in 1/5
```

	login	logout
1.	08nov2006 08:16:42.30	08nov2006 05:32:23.53
2.	08nov2006 08:07:20.53	08nov2006 05:57:13.40
3.	08nov2006 08:10:29.48	08nov2006 06:17:07.51
4.	08nov2006 08:30:02.19	08nov2006 05:42:23.17
5.	08nov2006 08:29:43.25	08nov2006 05:29:39.48

```
. format hiredate %td
. list hiredate in 1/5
```

	hiredate
1.	24jan1986
2.	10mar1994
3.	29sep2006
4.	14apr2006
5.	03dec1999

We remember that the project manager requested that hire dates be presented in the same form as they were previously.

```
. format hiredate %tdDD/NN/CCYY
. list hiredate in 1/5
```

	hiredate
1.	24/01/1986
2.	10/03/1994
3.	29/09/2006
4.	14/04/2006
5.	03/12/1999

◁

Setting European formats

Do you prefer that one and one half be written as 1,5 and that one thousand one and a half be written as 1.001,5? Stata will present numbers in that format if, when you set the format, you specify ',' rather than '.' as follows:

```
. use http://www.stata-press.com/data/r12/census10
(1980 Census data by state)
. format pop %12,0gc
. format medage %9,2f
```

```
. list in 1/8
```

	state	region	pop	medage
1.	Alabama	South	3.893.888	29,30
2.	Alaska	West	401.851	26,10
3.	Arizona	West	2.718.215	29,20
4.	Arkansas	South	2.286.435	30,60
5.	California	West	23.667.902	29,90
6.	Colorado	West	2.889.964	28,60
7.	Connecticut	NE	3.107.576	32,00
8.	Delaware	South	594.338	29,80

You can also leave the formats just as they were and instead type `set dp comma`. That tells Stata to interpret all formats as if you had typed the comma instead of the period:

```
. format pop %12.0gc                (put the formats back as they were)
. format medage %9.2f
. set dp comma                      (tell Stata to use European format)
. list in 1/8
(same output appears as above)
```

`set dp comma` affects all Stata output, so if you run a regression, display summary statistics, or make a table, commas will be used instead of periods in the output:

```
. tabulate region [fw=pop]
```

Census region	Freq.	Percent	Cum.
NE	49.135.283	21,75	21,75
N Cntrl	58.865.670	26,06	47,81
South	74.734.029	33,08	80,89
West	43.172.490	19,11	100,00
Total	225.907.472	100,00	

You can return to using periods by typing

```
. set dp period
```

Setting a variable's display format to European affects how the variable's values are displayed by `list` and in a few other places. Setting `dp` to `comma` affects every bit of Stata.

Also, `set dp comma` affects only how Stata displays output, not how it gets input. When you need to type one and a half, you must type `1.5` regardless of context.

❑ Technical note

`set dp comma` makes drastic changes inside Stata, and we mention this because some older, user-written programs may not be able to deal with those changes. If you are using an older, user-written program, you might `set dp comma` only to find that the program does not work and instead presents some sort of syntax error.

If, using any program, you get an unanticipated error, try setting `dp` back to `period`.

Even with `set dp comma`, you might still see some output with the decimal symbol shown as a period rather than a comma. There are two places in Stata where Stata ignores `set dp comma` because the features are generally used to produce what will be treated as input, and `set dp comma` does not affect how Stata inputs numbers. First,

```
local x = sqrt(2)
```

stores the string "1.414213562373095" in x and not "1,414213562373095", so if some program were to display 'x' as a string in the output, the period would be displayed. Most programs, however, would use 'x' in subsequent calculations or, at the least, when the time came to display what was in 'x', would display it as a number. They would code

```
display ... 'x' ...
```

and not

```
display ... "'x'" ...
```

so the output would be

```
... 1,4142135 ...
```

The other place where Stata ignores set dp comma is the string() function. If you type

```
. gen res = string(numvar)
```

new variable res will contain the string representation of numeric variable numvar, with the decimal symbol being a period, even if you have previously set dp comma. Of course, if you explicitly ask that string() use European format,

```
. gen res = string(numvar,"%9,0g")
```

then string() honors your request; string() merely ignores the global set dp comma.

❏

Details of formats

The %f format

In %*w*.d*f*, *w* is the total output width, including sign and decimal point, and *d* is the number of digits to appear to the right of the decimal point. The result is right-justified.

The number 5.139 in %12.2f format displays as

```
----+----1--
        5.14
```

When $d = 0$, the decimal point is not displayed. The number 5.14 in %12.0f format displays as

```
----+----1--
           5
```

%-*w*.d*f* works the same way, except that the output is left-justified in the field. The number 5.139 in %-12.2f displays as

```
----+----1--
5.14
```

The %fc format

%*w*.d*f*c works like %*w*.d*f* except that commas are inserted to make larger numbers more readable. *w* records the total width of the result, including commas.

The number 5.139 in %12.2fc format displays as

```
----+----1--
        5.14
```

The number 5203.139 in %12.2fc format displays as

```
----+----1--
    5,203.14
```

As with %f, if $d = 0$, the decimal point is not displayed. The number 5203.139 in %12.0fc format displays as

```
----+----1--
       5,203
```

As with %f, a minus sign may be inserted to left justify the output. The number 5203.139 in %-12.0fc format displays as

```
----+----1--
5,203
```

The %g format

In %$w.d$g, w is the overall width, and d is usually specified as 0, which leaves up to the format the number of digits to be displayed to the right of the decimal point. If $d \neq 0$ is specified, then not more than d digits will be displayed. As with %f, a minus sign may be inserted to left-justify results.

%g differs from %f in that 1) it decides how many digits to display to the right of the decimal point, and 2) it will switch to a %e format if the number is too large or too small.

The number 5.139 in %12.0g format displays as

```
----+----1--
       5.139
```

The number 5231371222.139 in %12.0g format displays as

```
----+----1--
  5231371222
```

The number 52313712223.139 displays as

```
----+----1--
 5.23137e+10
```

The number 0.0000029394 displays as

```
----+----1--
 2.93940e-06
```

The %gc format

%$w.d$gc is %$w.d$g with commas. It works in the same way as the %g and %fc formats.

The %e format

%$w.d$e displays numeric values in exponential format. w records the width of the format. d records the number of digits to be shown after the decimal place. w should be greater than or equal to $d+7$ or, if 3-digit exponents are expected, $d+8$.

The number 5.139 in %12.4e format is

```
----+----1--
 5.1390e+00
```

The number 5.139×10^{220} is

```
----+----1--
5.1390e+220
```

The %21x format

The %21x format is for those, typically programmers, who wish to analyze routines for numerical roundoff error. There is no better way to look at numbers than how the computer actually records them.

The number 5.139 in %21x format is

```
----+----1----+----2-
+1.48e5604189375X+002
```

The number 5.125 is

```
----+----1----+----2-
+1.4800000000000X+002
```

Reported is a signed, base-16 number with base-16 point, the letter X, and a signed, 3-digit base-16 integer. Call the two numbers f and e. The interpretation is $f \times 2^e$.

The %16H and %16L formats

The %16H and %16L formats show the value in the IEEE floating point, double-precision form. %16H shows the value in most-significant-byte-first (hilo) form. %16L shows the number in least-significant-byte-first (lohi) form.

The number 5.139 in %16H is

```
----+----1----+-
40148e5604189375
```

The number 5.139 in %16L is

```
----+----1----+-
75931804568e1440
```

The format is sometimes used by programmers who are simultaneously studying a hexadecimal dump of a binary file.

The %8H and %8L formats

%8H and %8L are similar to %16H and %16L but show the number in IEEE single-precision form.

The number 5.139 in %8H is

```
----+---
40a472b0
```

The number 5.139 in %8L is

```
----+---
b072a440
```

The %t format

The %t format displays numerical variables as dates and times. See [D] **datetime display formats**.

The %s format

The %*w*s format displays a string in a right-justified field of width *w*. %-*w*s displays the string left-justified.

"Mary Smith" in %16s format is

```
----+----1----+-
     Mary Smith
```

"Mary Smith" in %-16s format is

```
----+----1----+-
Mary Smith
```

Also, in some contexts, particularly display (see [P] **display**), %~*w*s is allowed, which centers the string. "Mary Smith" in %~16s format is

```
----+----1----+-
   Mary Smith
```

Other effects of formats

You have data on the age of employees, and you type summarize age to obtain the mean and standard deviation. By default, Stata uses its default g format to provide as much precision as possible.

```
. use http://www.stata-press.com/data/r12/fmtxmpl
. summarize age
```

Variable	Obs	Mean	Std. Dev.	Min	Max
age	204	30.18627	10.38067	18	66

If you attach a %9.2f format to the variable and specify the format option, Stata uses that specification to format the results:

```
. format age %9.2f
. summarize age, format
```

Variable	Obs	Mean	Std. Dev.	Min	Max
age	204	30.19	10.38	18.00	66.00

Displaying current formats

format *varlist* is not often used to display the formats associated with variables because using describe (see [D] **describe**) is easier and provides more information. The exceptions are date variables. Unless you use the default %tc, %tC, ... formats (and most people do), the format specifier itself can become very long, such as

```
. format admittime %tcDDmonCCYY_HH:MM:SS.sss
```

Such formats are too long for describe to display, so it gives up. In such cases, you can use format to display the format:

```
. format admittime
  variable name   display format
  ---------------------------------
  admittime       %tcDDmonCCYY_HH:MM:SS.sss
  ---------------------------------
```

Type format * to see the formats for all the variables.

References

Cox, N. J. 2011. Speaking Stata: MMXI and all that: Handling Roman numerals within Stata. *Stata Journal* 11: 126–142.

Gould, W. W. 2011a. How to read the %21x format. The Stata Blog: Not Elsewhere Classified. http://blog.stata.com/2011/02/02/how-to-read-the-percent-21x-format/

——. 2011b. How to read the %21x format, part 2. The Stata Blog: Not Elsewhere Classified. http://blog.stata.com/2011/02/02/how-to-read-the-percent-21x-format-part-2/

Linhart, J. M. 2008. Mata Matters: Overflow, underflow and the IEEE floating-point format. *Stata Journal* 8: 255–268.

Also see

[P] **display** — Display strings and values of scalar expressions

[D] **datetime business calendars** — Business calendars

[D] **datetime display formats** — Display formats for dates and times

[D] **list** — List values of variables

[D] **varmanage** — Manage variable labels, formats, and other properties

[U] **12.5 Formats: Controlling how data are displayed**

[U] **12.6 Dataset, variable, and value labels**

Title

> **functions** — Functions

Description

This entry describes the functions allowed by Stata. For information on Mata functions, see [M-4] **intro**.

A quick note about missing values: Stata denotes a numeric missing value by ., .a, .b, ..., or .z. A string missing value is denoted by "" (the empty string). Here any one of these may be referred to by *missing*. If a numeric value x is missing, then $x \geq .$ is true. If a numeric value x is not missing, then $x < .$ is true.

Functions are listed under the following headings:

> *Mathematical functions*
> *Probability distributions and density functions*
> *Random-number functions*
> *String functions*
> *Programming functions*
> *Date and time functions*
> *Selecting time spans*
> *Matrix functions returning a matrix*
> *Matrix functions returning a scalar*

Mathematical functions

abs(x)
> Domain: $-8\text{e}+307$ to $8\text{e}+307$
> Range: 0 to $8\text{e}+307$
> Description: returns the absolute value of x.

acos(x)
> Domain: -1 to 1
> Range: 0 to π
> Description: returns the radian value of the arccosine of x.

acosh(x)
> Domain: 1 to $8.9\text{e}+307$
> Range: 0 to 709.77
> Description: returns the inverse hyperbolic cosine of x, $\text{acosh}(x) = \ln(x + \sqrt{x^2 - 1})$.

asin(x)
> Domain: -1 to 1
> Range: $-\pi/2$ to $\pi/2$
> Description: returns the radian value of the arcsine of x.

asinh(x)
> Domain: $-8.9\text{e}+307$ to $8.9\text{e}+307$
> Range: -709.77 to 709.77
> Description: returns the inverse hyperbolic sine of x, $\text{asinh}(x) = \ln(x + \sqrt{x^2 + 1})$.

atan(x)
    Domain: $-8e+307$ to $8e+307$
    Range: $-\pi/2$ to $\pi/2$
    Description: returns the radian value of the arctangent of x.

atan2(y, x)
    Domain y: $-8e+307$ to $8e+307$
    Domain x: $-8e+307$ to $8e+307$
    Range: $-\pi$ to π
    Description: returns the radian value of the arctangent of y/x, where the signs of the parameters y and x are used to determine the quadrant of the answer.

atanh(x)
    Domain: -1 to 1
    Range: $-8e+307$ to $8e+307$
    Description: returns the inverse hyperbolic tangent of x, atanh(x) $= \frac{1}{2}\{\ln(1+x) - \ln(1-x)\}$.

ceil(x)
    Domain: $-8e+307$ to $8e+307$
    Range: integers in $-8e+307$ to $8e+307$
    Description: returns the unique integer n such that $n - 1 < x \le n$.
                   returns x (not ".") if x is missing, meaning that ceil(.a) = .a.

                   Also see floor(x), int(x), and round(x).

cloglog(x)
    Domain: 0 to 1
    Range: $-8e+307$ to $8e+307$
    Description: returns the complementary log-log of x,
                   cloglog(x) $= \ln\{-\ln(1-x)\}$.

comb(n,k)
    Domain n: integers 1 to 1e+305
    Domain k: integers 0 to n
    Range: 0 to 8e+307 and *missing*
    Description: returns the combinatorial function $n!/\{k!(n-k)!\}$.

cos(x)
    Domain: $-1e+18$ to $1e+18$
    Range: -1 to 1
    Description: returns the cosine of x, where x is in radians.

cosh(x)
    Domain: -709 to 709
    Range: 1 to 4.11e+307
    Description: returns the hyperbolic cosine of x, cosh(x) $= \{\exp(x) + \exp(-x)\}/2$.

digamma(x)
    Domain: $-1e+15$ to $8e+307$
    Range: $-8e+307$ to $8e+307$ and *missing*
    Description: returns the digamma() function, $d\ln\Gamma(x)/dx$. This is the derivative of lngamma(x).

                   The digamma(x) function is sometimes called the psi function, $\psi(x)$.

$\exp(x)$

    Domain: $-8e+307$ to 709

    Range: 0 to 8e+307

    Description: returns the exponential function e^x. This function is the inverse of $\ln(x)$.

$\text{floor}(x)$

    Domain: $-8e+307$ to 8e+307

    Range: integers in $-8e+307$ to 8e+307

    Description: returns the unique integer n such that $n \le x < n+1$.
                returns x (not ".") if x is missing, meaning that `floor(.a)` = `.a`.

                Also see `ceil(x)`, `int(x)`, and `round(x)`.

$\text{int}(x)$

    Domain: $-8e+307$ to 8e+307

    Range: integers in $-8e+307$ to 8e+307

    Description: returns the integer obtained by truncating x toward 0; thus,
                        `int(5.2)` $= 5$
                        `int(-5.8)` $= -5$
                returns x (not ".") if x is missing, meaning that `int(.a)` = `.a`.

                One way to obtain the closest integer to x is `int(x+sign(x)/2)`, which simplifies to `int(x+0.5)` for $x \ge 0$. However, use of the `round()` function is preferred. Also see `ceil(x)`, `int(x)`, and `round(x)`.

$\text{invcloglog}(x)$

    Domain: $-8e+307$ to 8e+307

    Range: 0 to 1 and *missing*

    Description: returns the inverse of the complementary log-log function of x,
$$\text{invcloglog}(x) = 1 - \exp\{-\exp(x)\}.$$

$\text{invlogit}(x)$

    Domain: $-8e+307$ to 8e+307

    Range: 0 to 1 and *missing*

    Description: returns the inverse of the logit function of x,
$$\text{invlogit}(x) = \exp(x)/\{1 + \exp(x)\}.$$

$\ln(x)$

    Domain: 1e–323 to 8e+307

    Range: -744 to 709

    Description: returns the natural logarithm, $\ln(x)$. This function is the inverse of $\exp(x)$.

                The logarithm of x in base b can be calculated via $\log_b(x) = \log_a(x)/\log_a(b)$. Hence,
$$\log_5(x) = \ln(x)/\ln(5) = \log(x)/\log(5) = \log10(x)/\log10(5)$$
$$\log_2(x) = \ln(x)/\ln(2) = \log(x)/\log(2) = \log10(x)/\log10(2)$$

                You can calculate $\log_b(x)$ by using the formula that best suits your needs.

lnfactorial(*n*)

Domain: integers 0 to 1e+305
Range: 0 to 8e+307
Description: returns the natural log of factorial = $\ln(n!)$.

To calculate $n!$, use round(exp(lnfactorial(*n*)),1) to ensure that the result is an integer. Logs of factorials are generally more useful than the factorials themselves because of overflow problems.

lngamma(*x*)

Domain: −2,147,483,648 to 1e+305 (excluding negative integers)
Range: −8e+307 to 8e+307
Description: returns $\ln\{\Gamma(x)\}$. Here the gamma function, $\Gamma(x)$, is defined by $\Gamma(x) = \int_0^\infty t^{x-1}e^{-t}dt$. For integer values of $x > 0$, this is $\ln((x-1)!)$.

lngamma(*x*) for $x < 0$ returns a number such that exp(lngamma(*x*)) is equal to the absolute value of the gamma function, $\Gamma(x)$. That is, lngamma(*x*) always returns a real (not complex) result.

log(*x*)

Domain: 1e–323 to 8e+307
Range: −744 to 709
Description: returns the natural logarithm, $\ln(x)$, which is a synonym for ln(*x*). Also see ln(*x*) for more information.

log10(*x*)

Domain: 1e–323 to 8e+307
Range: −323 to 308
Description: returns the base-10 logarithm of x.

logit(*x*)

Domain: 0 to 1 (exclusive)
Range: −8e+307 to 8e+307 and *missing*
Description: returns the log of the odds ratio of x,
               $$\text{logit}(x) = \ln\{x/(1-x)\}.$$

max(*x₁*,*x₂*,...,*xₙ*)

Domain *x₁*: −8e+307 to 8e+307 and *missing*
Domain *x₂*: −8e+307 to 8e+307 and *missing*
. . .
Domain *xₙ*: −8e+307 to 8e+307 and *missing*
Range: −8e+307 to 8e+307 and *missing*
Description: returns the maximum value of $x_1, x_2, \ldots, x_n$. Unless all arguments are *missing*, missing values are ignored.
               max(2,10,.,7) = 10
               max(.,.,.,.) = .

$\min(x_1, x_2, \ldots, x_n)$
    Domain x_1: $-8e+307$ to $8e+307$ and *missing*
    Domain x_2: $-8e+307$ to $8e+307$ and *missing*
    . . .
    Domain x_n: $-8e+307$ to $8e+307$ and *missing*
    Range: $-8e+307$ to $8e+307$ and *missing*
    Description: returns the minimum value of $x_1, x_2, \ldots, x_n$. Unless all arguments are *missing*,
                  missing values are ignored.
                  `min(2,10,.,7) = 2`
                  `min(.,.,.) = .`

$\mathrm{mod}(x, y)$
    Domain x: $-8e+307$ to $8e+307$
    Domain y: 0 to $8e+307$
    Range: 0 to $8e+307$
    Description: returns the modulus of x with respect to y.
                  $\mathrm{mod}(x, y) = x - y\ \mathtt{int}(x/y)$
                  $\mathrm{mod}(x, 0) = .$

$\mathrm{reldif}(x, y)$
    Domain x: $-8e+307$ to $8e+307$ and *missing*
    Domain y: $-8e+307$ to $8e+307$ and *missing*
    Range: $-8e+307$ to $8e+307$ and *missing*
    Description: returns the "relative" difference $|x - y|/(|y| + 1)$.
             returns 0 if both arguments are the same type of extended missing value.
             returns *missing* if only one argument is missing or if the two arguments are
                  two different types of *missing*.

$\mathrm{round}(x, y)$ or $\mathrm{round}(x)$
    Domain x: $-8e+307$ to $8e+307$
    Domain y: $-8e+307$ to $8e+307$
    Range: $-8e+307$ to $8e+307$
    Description: returns x rounded in units of y or x rounded to the nearest integer if the argument
                  y is omitted.
             returns x (not ".") if x is missing, meaning that `round(.a) = .a` and
                  `round(.a,`y`) = .a` if y is not missing; if y is missing, then "." is returned.

             For $y = 1$, or with y omitted, this amounts to the closest integer to x; `round(5.2,1)`
             is 5, as is `round(4.8,1)`; `round(-5.2,1)` is -5, as is `round(-4.8,1)`. The
             rounding definition is generalized for $y \neq 1$. With $y = 0.01$, for instance, x is
             rounded to two decimal places; `round(sqrt(2),.01)` is 1.41. y may also be larger
             than 1; `round(28,5)` is 30, which is 28 rounded to the closest multiple of 5.
             For $y = 0$, the function is defined as returning x unmodified. Also see
             `int(`x`)`, `ceil(`x`)`, and `floor(`x`)`.

$\mathrm{sign}(x)$
    Domain: $-8e+307$ to $8e+307$ and *missing*
    Range: -1, 0, 1 and *missing*
    Description: returns the sign of x: -1 if $x < 0$, 0 if $x = 0$, 1 if $x > 0$, and *missing*
                  if x is missing.

$\sin(x)$
  Domain: −1e+18 to 1e+18
  Range: −1 to 1
  Description: returns the sine of x, where x is in radians.

$\sinh(x)$
  Domain: −709 to 709
  Range: −4.11e+307 to 4.11e+307
  Description: returns the hyperbolic sine of x, $\sinh(x) = \{\exp(x) - \exp(-x)\}/2$.

$\mathrm{sqrt}(x)$
  Domain: 0 to 8e+307
  Range: 0 to 1e+154
  Description: returns the square root of x.

$\mathrm{sum}(x)$
  Domain: all real numbers and *missing*
  Range: −8e+307 to 8e+307 (excluding *missing*)
  Description: returns the running sum of x, treating missing values as zero.

              For example, following the command `generate y=sum(x)`, the jth observation on y contains the sum of the first through jth observations on x. See [D] **egen** for an alternative sum function, `total()`, that produces a constant equal to the overall sum.

$\tan(x)$
  Domain: −1e+18 to 1e+18
  Range: −1e+17 to 1e+17 and *missing*
  Description: returns the tangent of x, where x is in radians.

$\tanh(x)$
  Domain: −8e+307 to 8e+307
  Range: −1 to 1 and *missing*
  Description: returns the hyperbolic tangent of x,
                $\tanh(x) = \{\exp(x) - \exp(-x)\}/\{\exp(x) + \exp(-x)\}$.

$\mathrm{trigamma}(x)$
  Domain: −1e+15 to 8e+307
  Range: 0 to 8e+307 and *missing*
  Description: returns the second derivative of $\mathrm{lngamma}(x) = d^2 \ln\Gamma(x)/dx^2$. The `trigamma()` function is the derivative of $\mathrm{digammma}(x)$.

$\mathrm{trunc}(x)$ is a synonym for $\mathrm{int}(x)$.

❑ Technical note

The trigonometric functions are defined in terms of *radians*. There are 2π radians in a circle. If you prefer to think in terms of *degrees*, because there are also 360 degrees in a circle, you may convert degrees into radians by using the formula $r = d\pi/180$, where d represents degrees and r represents radians. Stata includes the built-in constant _pi, equal to π to machine precision. Thus, to calculate the sine of theta, where theta is measured in degrees, you could type

    sin(theta*_pi/180)

`atan()` similarly returns radians, not degrees. The arccotangent can be obtained as

    acot(x) =_pi/2 - atan(x)

❑

Probability distributions and density functions

The probability distributions and density functions are organized under the following headings:

Beta and noncentral beta distributions
Binomial distribution
Chi-squared and noncentral chi-squared distributions
Dunnett's multiple range distribution
F and noncentral F distributions
Gamma distribution
Hypergeometric distribution
Negative binomial distribution
Normal (Gaussian), log of the normal, and binormal distributions
Poisson distribution
Random-number functions
Student's t distribution
Tukey's Studentized range distribution

Beta and noncentral beta distributions

ibeta(a,b,x)

Domain a: 1e–10 to 1e+17
Domain b: 1e–10 to 1e+17
Domain x: −8e+307 to 8e+307
            Interesting domain is $0 \leq x \leq 1$
Range: 0 to 1
Description: returns the cumulative beta distribution with shape parameters a and b defined by

$$I_x(a,b) = \frac{\Gamma(a+b)}{\Gamma(a)\Gamma(b)} \int_0^x t^{a-1}(1-t)^{b-1}\, dt$$

returns 0 if $x < 0$.
returns 1 if $x > 1$.

ibeta() returns the regularized incomplete beta function, also known as the incomplete beta function ratio. The incomplete beta function without regularization is given by (gamma(a)*gamma(b)/gamma(a+b))*ibeta(a,b,x) or, better when a or b might be large, exp(lngamma(a)+lngamma(b)-lngamma(a+b))*ibeta(a,b,x).

Here is an example of the use of the regularized incomplete beta function. Although Stata has a cumulative binomial function (see binomial()), the probability that an event occurs k or fewer times in n trials, when the probability of one event is p, can be evaluated as cond(k==n,1,1-ibeta(k+1,n-k,p)). The reverse cumulative binomial (the probability that an event occurs k or more times) can be evaluated as cond(k==0,1,ibeta(k,n-k+1,p)). See Press et al. (2007, 270–273) for a more complete description and for suggested uses for this function.

betaden(a,b,x)

  Domain a: 1e–323 to 8e+307
  Domain b: 1e–323 to 8e+307
  Domain x: 1e–323 to 8e+307
          Interesting domain is $0 \le x \le 1$
  Range: 0 to 8e+307
  Description: returns the probability density of the beta distribution,

$$\text{betaden}(a,b,x) = \frac{x^{a-1}(1-x)^{b-1}}{\int_0^\infty t^{a-1}(1-t)^{b-1}dt} = \frac{\Gamma(a+b)}{\Gamma(a)\Gamma(b)}x^{a-1}(1-x)^{b-1}$$

          where a and b are the shape parameters.
          returns 0 if $x < 0$ or $x > 1$.

ibetatail(a,b,x)

  Domain a: 1e–10 to 1e+17
  Domain b: 1e–10 to 1e+17
  Domain x: −8e+307 to 8e+307
          Interesting domain is $0 \le x \le 1$
  Range: 0 to 1
  Description: returns the reverse cumulative (upper tail or survivor) beta distribution with shape
          parameters a and b defined by

$$\text{ibetatail}(a,b,x) = 1 - \text{ibeta}(a,b,x) = \int_x^1 \text{betaden}(a,b,t)\,dt$$

          returns 1 if $x < 0$.
          returns 0 if $x > 1$.

          ibetatail() is also known as the complement to the incomplete beta function
          (ratio).

invibeta(a,b,p)

  Domain a: 1e–10 to 1e+17
  Domain b: 1e–10 to 1e+17
  Domain p: 0 to 1
  Range: 0 to 1
  Description: returns the inverse cumulative beta distribution: if $\text{ibeta}(a,b,x) = p$,
          then $\text{invibeta}(a,b,p) = x$.

invibetatail(a,b,p)

  Domain a: 1e–10 to 1e+17
  Domain b: 1e–10 to 1e+17
  Domain p: 0 to 1
  Range: 0 to 1
  Description: returns the inverse reverse cumulative (upper tail or survivor) beta distribution:
          if $\text{ibetatail}(a,b,x) = p$, then $\text{invibetatail}(a,b,p) = x$.

nibeta(a,b,λ,x)

Domain a: 1e–323 to 8e+307
Domain b: 1e–323 to 8e+307
Domain λ: 0 to 1,000
Domain x: −8e+307 to 8e+307
          Interesting domain is $0 \le x \le 1$
Range: 0 to 1
Description: returns the cumulative noncentral beta distribution

$$I_x(a, b, \lambda) = \sum_{j=0}^{\infty} \frac{e^{-\lambda/2}(\lambda/2)^j}{\Gamma(j+1)} I_x(a+j, b)$$

where a and b are shape parameters, λ is the noncentrality parameter, x is the value of a beta random variable, and $I_x(a, b)$ is the cumulative beta distribution, ibeta().
returns 0 if $x < 0$.
returns 1 if $x > 1$.

nibeta(a,b,0,x) = ibeta(a,b,x), but ibeta() is the preferred function to use for the central beta distribution. nibeta() is computed using an algorithm described in Johnson, Kotz, and Balakrishnan (1995).

nbetaden(a,b,λ,x)

Domain a: 1e–323 to 8e+307
Domain b: 1e–323 to 8e+307
Domain λ: 0 to 1,000
Domain x: −8e+307 to 8e+307
          Interesting domain is $0 \le x \le 1$
Range: 0 to 8e+307
Description: returns the probability density function of the noncentral beta distribution,

$$\sum_{j=0}^{\infty} \frac{e^{-\lambda/2}(\lambda/2)^j}{\Gamma(j+1)} \left\{ \frac{\Gamma(a+b+j)}{\Gamma(a+j)\Gamma(b)} x^{a+j-1}(1-x)^{b-1} \right\}$$

where a and b are shape parameters, λ is the noncentrality parameter, and x is the value of a beta random variable.
returns 0 if $x < 0$ or $x > 1$.

nbetaden(a,b,0,x) = betaden(a,b,x), but betaden() is the preferred function to use for the central beta distribution. nbetaden() is computed using an algorithm described in Johnson, Kotz, and Balakrishnan (1995).

invnibeta(a,b,λ,p)

Domain a: 1e–323 to 8e+307
Domain b: 1e–323 to 8e+307
Domain λ: 0 to 1,000
Domain p: 0 to 1
Range: 0 to 1
Description: returns the inverse cumulative noncentral beta distribution:
          if nibeta(a,b,λ,x) = p, then invibeta(a,b,λ,p) = x.

Binomial distribution

`binomial(`n`,`k`,`θ`)`

    Domain n: 0 to 1e+17

    Domain k: $-$8e+307 to 8e+307

               Interesting domain is $0 \le k < n$

    Domain θ: 0 to 1

    Range: 0 to 1

    Description: returns the probability of observing `floor(`k`)` or fewer successes in `floor(`n`)` trials when the probability of a success on one trial is θ.

                returns 0 if $k < 0$.

                returns 1 if $k > n$.

`binomialp(`n`,`k`,`p`)`

    Domain n: 1 to 1e+6

    Domain k: 0 to n

    Domain p: 0 to 1

    Range: 0 to 1

    Description: returns the probability of observing `floor(`k`)` successes in `floor(`n`)` trials when the probability of a success on one trial is p.

`binomialtail(`n`,`k`,`θ`)`

    Domain n: 0 to 1e+17

    Domain k: $-$8e+307 to 8e+307

               Interesting domain is $0 \le k < n$

    Domain θ: 0 to 1

    Range: 0 to 1

    Description: returns the probability of observing `floor(`k`)` or more successes in `floor(`n`)` trials when the probability of a success on one trial is θ.

                returns 1 if $k < 0$.

                returns 0 if $k > n$.

`invbinomial(`n`,`k`,`p`)`

    Domain n: 1 to 1e+17

    Domain k: 0 to $n-1$

    Domain p: 0 to 1 (exclusive)

    Range: 0 to 1

    Description: returns the inverse of the cumulative binomial; that is, it returns θ (θ = probability of success on one trial) such that the probability of observing `floor(`k`)` or fewer successes in `floor(`n`)` trials is p.

`invbinomialtail(`n`,`k`,`p`)`

    Domain n: 1 to 1e+17

    Domain k: 1 to n

    Domain p: 0 to 1 (exclusive)

    Range: 0 to 1

    Description: returns the inverse of the right cumulative binomial; that is, it returns θ (θ = probability of success on one trial) such that the probability of observing `floor(`k`)` or more successes in `floor(`n`)` trials is p.

Chi-squared and noncentral chi-squared distributions

chi2(n,x)
    Domain n: 2e–10 to 2e+17 (may be nonintegral)
    Domain x: −8e+307 to 8e+307
                  Interesting domain is $x \geq 0$
    Range: 0 to 1
    Description: returns the cumulative χ^2 distribution with n degrees of freedom.
                  chi2(n,x) = gammap($n/2$,$x/2$).
                  returns 0 if $x < 0$.

chi2tail(n,x)
    Domain n: 2e–10 to 2e+17 (may be nonintegral)
    Domain x: −8e+307 to 8e+307
                  Interesting domain is $x \geq 0$
    Range: 0 to 1
    Description: returns the reverse cumulative (upper tail or survivor) χ^2 distribution with n degrees
                  of freedom. chi2tail(n,x) = $1 -$ chi2(n,x)
                  returns 1 if $x < 0$.

invchi2(n,p)
    Domain n: 2e–10 to 2e+17 (may be nonintegral)
    Domain p: 0 to 1
    Range: 0 to 8e+307
    Description: returns the inverse of chi2(): if chi2(n,x) = p, then invchi2(n,p) = x.

invchi2tail(n,p)
    Domain n: 2e–10 to 2e+17 (may be nonintegral)
    Domain p: 0 to 1
    Range: 0 to 8e+307
    Description: returns the inverse of chi2tail(): if chi2tail(n,x) = p, then
                  invchi2tail(n,p) = x.

nchi2(n,λ,x)
    Domain n: integers 1 to 200
    Domain λ: 0 to 1,000
    Domain x: −8e+307 to 8e+307
                  Interesting domain is $x \geq 0$
    Range: 0 to 1
    Description: returns the cumulative noncentral χ^2 distribution,

$$\int_0^x \frac{e^{-t/2}\, e^{-\lambda/2}}{2^{n/2}} \sum_{j=0}^{\infty} \frac{t^{n/2+j-1}\, \lambda^j}{\Gamma(n/2+j)\, 2^{2j}\, j!}\, dt$$

           where n denotes the degrees of freedom, λ is the noncentrality parameter, and
           x is the value of χ^2.
        returns 0 if $x < 0$.

        nchi2(n,0,x) = chi2(n,x), but chi2() is the preferred function to use for
        the central χ^2 distribution. nchi2() is computed using the algorithm of
        Haynam, Govindarajulu, and Leone (1970).

`invnchi2(`n`,`λ`,`p`)`
  Domain n: integers 1 to 200
  Domain λ: 0 to 1,000
  Domain p: 0 to 1
  Range: 0 to 8e+307
  Description: returns the inverse cumulative noncentral χ^2 distribution:
          if `nchi2(`n`,`λ`,`x`)` $= p$, then `invnchi2(`n`,`λ`,`p`)` $= x$; n must be an integer.

`npnchi2(`n`,`x`,`p`)`
  Domain n: integers 1 to 200
  Domain x: 0 to 8e+307
  Domain p: 1e–138 to $1 - 2^{-52}$
  Range: 0 to 1,000
  Description: returns the noncentrality parameter, λ, for noncentral χ^2:
          if `nchi2(`n`,`λ`,`x`)` $= p$, then `npnchi2(`n`,`x`,`p`)` $= \lambda$.

Dunnett's multiple range distribution

`dunnettprob(`k`,`df`,`x`)`
  Domain k: 2 to 1e+6
  Domain df: 2 to 1e+6
  Domain x: −8e+307 to 8e+307
          Interesting domain is $x \geq 0$
  Range: 0 to 1
  Description: returns the cumulative multiple range distribution that is used in Dunnett's
          multiple-comparison method with k ranges and df degrees of freedom.
          returns 0 if $x < 0$.

          `dunnettprob()` is computed using an algorithm described in Miller (1981).

`invdunnettprob(`k`,`df`,`p`)`
  Domain k: 2 to 1e+6
  Domain df: 2 to 1e+6
  Domain p: 0 to 1 (right exclusive)
  Range: 0 to 8e+307
  Description: returns the inverse cumulative multiple range distribution that is used in Dunnett's
          multiple-comparison method with k ranges and df degrees of freedom. If
          `dunnettprob(`k`,`df`,`x`)` $= p$, then `invdunnettprob(`k`,`df`,`p`)` $= x$.

          `invdunnettprob()` is computed using an algorithm described in Miller (1981).

F and noncentral F distributions

$F(n_1, n_2, f)$
 Domain n_1: 2e–10 to 2e+17 (may be nonintegral)
 Domain n_2: 2e–10 to 2e+17 (may be nonintegral)
 Domain f: −8e+307 to 8e+307
      Interesting domain is $f \geq 0$
 Range: 0 to 1
 Description: returns the cumulative F distribution with n_1 numerator and n_2 denominator
      degrees of freedom: $F(n_1, n_2, f) = \int_0^f \texttt{Fden}(n_1, n_2, t)\ dt$.
      returns 0 if $f < 0$.

$\texttt{Fden}(n_1, n_2, f)$
 Domain n_1: 1e–323 to 8e+307 (may be nonintegral)
 Domain n_2: 1e–323 to 8e+307 (may be nonintegral)
 Domain f: −8e+307 to 8e+307
      Interesting domain is $f \geq 0$
 Range: 0 to 8e+307
 Description: returns the probability density function of the F distribution with n_1 numerator
      and n_2 denominator degrees of freedom:

$$\texttt{Fden}(n_1, n_2, f) = \frac{\Gamma(\frac{n_1+n_2}{2})}{\Gamma(\frac{n_1}{2})\Gamma(\frac{n_2}{2})} \left(\frac{n_1}{n_2}\right)^{\frac{n_1}{2}} \cdot f^{\frac{n_1}{2}-1} \left(1 + \frac{n_1}{n_2} f\right)^{-\frac{1}{2}(n_1+n_2)}$$

      returns 0 if $f < 0$.

$\texttt{Ftail}(n_1, n_2, f)$
 Domain n_1: 2e–10 to 2e+17 (may be nonintegral)
 Domain n_2: 2e–10 to 2e+17 (may be nonintegral)
 Domain f: −8e+307 to 8e+307
      Interesting domain is $f \geq 0$
 Range: 0 to 1
 Description: returns the reverse cumulative (upper tail or survivor) F distribution with n_1 numerator
      and n_2 denominator degrees of freedom. $\texttt{Ftail}(n_1, n_2, f) = 1 - F(n_1, n_2, f)$.
      returns 1 if $f < 0$.

$\texttt{invF}(n_1, n_2, p)$
 Domain n_1: 2e–10 to 2e+17 (may be nonintegral)
 Domain n_2: 2e–10 to 2e+17 (may be nonintegral)
 Domain p: 0 to 1
 Range: 0 to 8e+307
 Description: returns the inverse cumulative F distribution: if $F(n_1, n_2, f) = p$,
      then $\texttt{invF}(n_1, n_2, p) = f$.

$\texttt{invFtail}(n_1, n_2, p)$
 Domain n_1: 2e–10 to 2e+17 (may be nonintegral)
 Domain n_2: 2e–10 to 2e+17 (may be nonintegral)
 Domain p: 0 to 1
 Range: 0 to 8e+307
 Description: returns the inverse reverse cumulative (upper tail or survivor) F distribution:
      if $\texttt{Ftail}(n_1, n_2, f) = p$, then $\texttt{invFtail}(n_1, n_2, p) = f$.

nFden(n_1,n_2,λ,f)

    Domain n_1:   1e–323 to 8e+307 (may be nonintegral)

    Domain n_2:   1e–323 to 8e+307 (may be nonintegral)

    Domain λ: 0 to 1,000

    Domain f: −8e+307 to 8e+307

                 Interesting domain is $f \geq 0$

    Range: 0 to 8e+307

    Description: returns the probability density function of the noncentral F distribution with n_1 numerator and n_2 denominator degrees of freedom and noncentrality parameter λ.

                 returns 0 if $f < 0$.

                 nFden(n_1,n_2,0,F) = Fden(n_1,n_2,F), but Fden() is the preferred function to use for the central F distribution.

                 Also, if F follows the noncentral F distribution with n_1 and n_2 degrees of freedom and noncentrality parameter λ, then

$$\frac{n_1 F}{n_2 + n_1 F}$$

                 follows a noncentral beta distribution with shape parameters $a = \nu_1/2$, $b = \nu_2/2$, and noncentrality parameter λ, as given in nbetaden(). nFden() is computed based on this relationship.

nFtail(n_1,n_2,λ,f)

    Domain n_1:   1e–323 to 8e+307 (may be nonintegral)

    Domain n_2:   1e–323 to 8e+307 (may be nonintegral)

    Domain λ: 0 to 1,000

    Domain f: −8e+307 to 8e+307

                 Interesting domain is $f \geq 0$

    Range: 0 to 1

    Description: returns the reverse cumulative (upper tail or survivor) noncentral F distribution with n_1 numerator and n_2 denominator degrees of freedom and noncentrality parameter λ.

                 returns 1 if $f < 0$.

                 nFtail() is computed using nibeta() based on the relationship between the noncentral beta and F distributions. See Johnson, Kotz, and Balakrishnan (1995) for more details.

invnFtail(n_1,n_2,λ,p)

    Domain n_1:   1e–323 to 8e+307 (may be nonintegral)

    Domain n_2:   1e–323 to 8e+307 (may be nonintegral)

    Domain λ: 0 to 1,000

    Domain p: 0 to 1

    Range: 0 to 8e+307

    Description: returns the inverse reverse cumulative (upper tail or survivor) noncentral F distribution:

                 if nFtail(n_1,n_2,λ,x) = p, then invnFtail(n_1,n_2,λ,p) = x.

Gamma distribution

gammap(a,x)

    Domain a: 1e–10 to 1e+17

    Domain x: −8e+307 to 8e+307

                 Interesting domain is $x \geq 0$

    Range: 0 to 1

    Description: returns the cumulative gamma distribution with shape parameter a defined by

$$\frac{1}{\Gamma(a)} \int_0^x e^{-t} t^{a-1}\, dt$$

          returns 0 if $x < 0$.

          The cumulative Poisson (the probability of observing k or fewer events if the expected is x) can be evaluated as 1-gammap(k+1,x). The reverse cumulative (the probability of observing k or more events) can be evaluated as gammap(k,x). See Press et al. (2007, 259–266) for a more complete description and for suggested uses for this function.

          gammap() is also known as the incomplete gamma function (ratio).

          Probabilities for the three-parameter gamma distribution (see gammaden()) can be calculated by shifting and scaling x; that is, gammap(a,$(x-g)/b$).

gammaden(a,b,g,x)

    Domain a: 1e–323 to 8e+307

    Domain b: 1e–323 to 8e+307

    Domain g: −8e+307 to 8e+307

    Domain x: −8e+307 to 8e+307

                 Interesting domain is $x \geq g$

    Range: 0 to 8e+307

    Description: returns the probability density function of the gamma distribution defined by

$$\frac{1}{\Gamma(a)b^a}(x-g)^{a-1} e^{-(x-g)/b}$$

          where a is the shape parameter, b is the scale parameter, and g is the location parameter.

        returns 0 if $x < g$.

gammaptail(a,x)
   Domain a: 1e–10 to 1e+17
   Domain x: −8e+307 to 8e+307
                  Interesting domain is $x \geq 0$
   Range: 0 to 1
   Description: returns the reverse cumulative (upper tail or survivor) gamma distribution with shape
                  parameter a defined by

$$\text{gammaptail}(a,x) = 1 - \text{gammap}(a,x) = \int_x^\infty \text{gammaden}(a,t)\, dt$$

                  returns 1 if $x < 0$.

                  gammaptail() is also known as the complement to the incomplete gamma function
                  (ratio).

invgammap(a,p)
   Domain a: 1e–10 to 1e+17
   Domain p: 0 to 1
   Range: 0 to 8e+307
   Description: returns the inverse cumulative gamma distribution: if gammap(a,x) $= p$,
                  then invgammap(a,p) $= x$.

invgammaptail(a,p)
   Domain a: 1e–10 to 1e+17
   Domain p: 0 to 1
   Range: 0 to 8e+307
   Description: returns the inverse reverse cumulative (upper tail or survivor) gamma distribution:
                  if gammaptail(a,x) $= p$, then invgammaptail(a,p) $= x$.

dgammapda(a,x)
   Domain a: 1e–7 to 1e+17
   Domain x: −8e+307 to 8e+307
                  Interesting domain is $x \geq 0$
   Range: −16 to 0
   Description: returns $\frac{\partial P(a,x)}{\partial a}$, where $P(a,x) = \text{gammap}(a,x)$.
                  returns 0 if $x < 0$.

dgammapdada(a,x)
   Domain a: 1e–7 to 1e+17
   Domain x: −8e+307 to 8e+307
                  Interesting domain is $x \geq 0$
   Range: −0.02 to 4.77e+5
   Description: returns $\frac{\partial^2 P(a,x)}{\partial a^2}$, where $P(a,x) = \text{gammap}(a,x)$.
                  returns 0 if $x < 0$.

dgammapdadx(a,x)
   Domain a: 1e–7 to 1e+17
   Domain x: −8e+307 to 8e+307
                  Interesting domain is $x \geq 0$
   Range: −0.04 to 8e+307
   Description: returns $\frac{\partial^2 P(a,x)}{\partial a \partial x}$, where $P(a,x) = \text{gammap}(a,x)$.
                  returns 0 if $x < 0$.

dgammapdx(a,x)

    Domain a: 1e–10 to 1e+17

    Domain x: -8e+307 to 8e+307

                 Interesting domain is $x \geq 0$

    Range: 0 to 8e+307

    Description: returns $\frac{\partial P(a,x)}{\partial x}$, where $P(a,x) = $ gammap(a,x).

                 returns 0 if $x < 0$.

dgammapdxdx(a,x)

    Domain a: 1e–10 to 1e+17

    Domain x: -8e+307 to 8e+307

                 Interesting domain is $x \geq 0$

    Range: 0 to 1e+40

    Description: returns $\frac{\partial^2 P(a,x)}{\partial x^2}$, where $P(a,x) = $ gammap(a,x).

                 returns 0 if $x < 0$.

Hypergeometric distribution

hypergeometric(N,K,n,k)

    Domain N: 2 to 1e+5

    Domain K: 1 to $N-1$

    Domain n: 1 to $N-1$

    Domain k: max(0,$n - N + K$) to min(K,n)

    Range: 0 to 1

    Description: returns the cumulative probability of the hypergeometric distribution. N is the population size, K is the number of elements in the population that have the attribute of interest, and n is the sample size. Returned is the probability of observing k or fewer elements from a sample of size n that have the attribute of interest.

hypergeometricp(N,K,n,k)

    Domain N: 2 to 1e+5

    Domain K: 1 to $N-1$

    Domain n: 1 to $N-1$

    Domain k: max(0,$n - N + K$) to min(K,n)

    Range: 0 to 1 (right exclusive)

    Description: returns the hypergeometric probability of k successes (where success is obtaining an element with the attribute of interest) out of a sample of size n, from a population of size N containing K elements that have the attribute of interest.

Negative binomial distribution

nbinomial(n,k,p)

    Domain n: 1e–10 to 1e+17 (can be nonintegral)

    Domain k: 0 to $2^{53} - 1$

    Domain p: 0 to 1 (left exclusive)

    Range: 0 to 1

    Description: returns the cumulative probability of the negative binomial distribution. n can be nonintegral. When n is an integer, nbinomial() returns the probability of observing k or fewer failures before the nth success, when the probability of a success on one trial is p.

              The negative binomial distribution function is evaluated using the ibeta() function.

nbinomialp(n,k,p)

    Domain n: 1e–10 to 1e+6 (can be nonintegral)

    Domain k: 0 to 1e+10

    Domain p: 0 to 1 (left exclusive)

    Range: 0 to 1

    Description: returns the negative binomial probability. When n is an integer, nbinomialp() returns the probability of observing exactly floor(k) failures before the nth success, when the probability of a success on one trial is p.

nbinomialtail(n,k,p)

    Domain n: 1e–10 to 1e+17 (can be nonintegral)

    Domain k: 0 to $2^{53} - 1$

    Domain p: 0 to 1 (left exclusive)

    Range: 0 to 1

    Description: returns the reverse cumulative probability of the negative binomial distribution. When n is an integer, nbinomialtail() returns the probability of observing k or more failures before the nth success, when the probability of a success on one trial is p.

              The reverse negative binomial distribution function is evaluated using the ibetatail() function.

invnbinomial(n,k,q)

    Domain n: 1e–10 to 1e+17 (can be nonintegral)

    Domain k: 0 to $2^{53} - 1$

    Domain q: 0 to 1 (exclusive)

    Range: 0 to 1

    Description: returns the value of the negative binomial parameter, p, such that $q = $ nbinomial(n,k,p).

              invnbinomial() is evaluated using invibeta().

`invnbinomialtail(`n`,`k`,`q`)`

Domain n: 1e–10 to 1e+17 (can be nonintegral)
Domain k: 1 to $2^{53} - 1$
Domain q: 0 to 1 (exclusive)
Range: 0 to 1 (exclusive)
Description: returns the value of the negative binomial parameter, p, such that
$$q = \texttt{nbinomialtail}(n,k,p).$$

`invnbinomialtail()` is evaluated using `invibetatail()`.

Normal (Gaussian), log of the normal, and binormal distributions

`binormal(`h`,`k`,`ρ`)`

Domain h: −8e+307 to 8e+307
Domain k: −8e+307 to 8e+307
Domain ρ: −1 to 1
Range: 0 to 1
Description: returns the joint cumulative distribution $\Phi(h, k, \rho)$ of bivariate normal
                with correlation ρ; cumulative over $(-\infty, h] \times (-\infty, k]$:

$$\Phi(h, k, \rho) = \frac{1}{2\pi\sqrt{1 - \rho^2}} \int_{-\infty}^{h} \int_{-\infty}^{k} \exp\left\{ -\frac{1}{2(1 - \rho^2)} \left(x_1^2 - 2\rho x_1 x_2 + x_2^2 \right) \right\} dx_1 \, dx_2$$

`normal(`z`)`

Domain: −8e+307 to 8e+307
Range: 0 to 1
Description: returns the cumulative standard normal distribution.
$$\texttt{normal}(z) = \int_{-\infty}^{z} \frac{1}{\sqrt{2\pi}} e^{-x^2/2} dx$$

`normalden(`z`)`

Domain: −8e+307 to 8e+307
Range: 0 to 0.39894 ...
Description: returns the standard normal density, $N(0, 1)$.

`normalden(`z`,`σ`)`

Domain z: −8e+307 to 8e+307
Domain σ: 1e–308 to 8e+307
Range: 0 to 8e+307
Description: returns the rescaled standard normal density, $N(0, \sigma^2)$.
$$\texttt{normalden}(z,1) = \texttt{normalden}(z)$$
$$\texttt{normalden}(z,\sigma) = \texttt{normalden}(z)/\sigma$$

`normalden(`x`,`μ`,`σ`)`

Domain x: −8e+307 to 8e+307
Domain μ: −8e+307 to 8e+307
Domain σ: 1e–308 to 8e+307
Range: 0 to 8e+307
Description: returns the normal density with mean μ and standard deviation σ, $N(\mu, \sigma^2)$:
`normalden(`x`,0,1) = normalden(`x`)` and
`normalden(`x`,`μ`,`σ`) = normalden((`x` − `μ`)/`σ`)/`σ
In general,

$$\texttt{normalden}(z,\mu,\sigma) = \frac{1}{\sigma\sqrt{2\pi}}e^{-\frac{1}{2}\left\{\frac{(z-\mu)}{\sigma}\right\}^2}$$

`invnormal(`p`)`

Domain: 1e–323 to $1 - 2^{-53}$
Range: −38.449394 to 8.2095362
Description: returns the inverse cumulative standard normal distribution:
if `normal(`z`) = `p, then `invnormal(`p`) = `z.

`lnnormal(`z`)`

Domain: −1e+99 to 8e+307
Range: −5e+197 to 0
Description: returns the natural logarithm of the cumulative standard normal distribution:

$$\texttt{lnnormal}(z) = \ln\left(\int_{-\infty}^{z}\frac{1}{\sqrt{2\pi}}e^{-x^2/2}dx\right)$$

`lnnormalden(`z`)`

Domain: −1e+154 to 1e+154
Range: −5e+307 to −0.91893853 = `lnnormalden(0)`
Description: returns the natural logarithm of the standard normal density, $N(0, 1)$.

`lnnormalden(`z`,`σ`)`

Domain z: −1e+154 to 1e+154
Domain σ: 1e–323 to 8e+307
Range: −5e+307 to 742.82799
Description: returns the natural logarithm of the rescaled standard normal density, $N(0, \sigma^2)$.
`lnnormalden(`z`,1) = lnnormalden(`z`)`
`lnnormalden(`z`,`σ`) = lnnormalden(`z`) − ln(`σ`)`

`lnnormalden(`x`,`μ`,`σ`)`

Domain x: −8e+307 to 8e+307
Domain μ: −8e+307 to 8e+307
Domain σ: 1e–323 to 8e+307
Range: 1e–323 to 8e+307
Description: returns the natural logarithm of the normal density with mean μ and standard deviation σ, $N(\mu, \sigma^2)$: `lnnormalden(`x`,0,1) = lnnormalden(`x`)` and
`lnnormalden(`x`,`μ`,`σ`) = lnnormalden((`x` − `μ`)/`σ`) − ln(`σ`)`. In general,

$$\texttt{lnnormalden}(z,\mu,\sigma) = \ln\left[\frac{1}{\sigma\sqrt{2\pi}}e^{-\frac{1}{2}\left\{\frac{(z-\mu)}{\sigma}\right\}^2}\right]$$

Poisson distribution

poisson(m,k)
  Domain m: 1e–10 to $2^{53} - 1$
  Domain k: 0 to $2^{53} - 1$
  Range: 0 to 1
  Description: returns the probability of observing floor(k) or fewer outcomes that are distributed as Poisson with mean m.

  The Poisson distribution function is evaluated using the gammaptail() function.

poissonp(m,k)
  Domain m: 1e–10 to 1e+8
  Domain k: 0 to 1e+9
  Range: 0 to 1
  Description: returns the probability of observing floor(k) outcomes that are distributed as Poisson with mean m.

  The Poisson probability function is evaluated using the gammaden() function.

poissontail(m,k)
  Domain m: 1e–10 to $2^{53} - 1$
  Domain k: 0 to $2^{53} - 1$
  Range: 0 to 1
  Description: returns the probability of observing floor(k) or more outcomes that are distributed as Poisson with mean m.

  The reverse cumulative Poisson distribution function is evaluated using the gammap() function.

invpoisson(k,p)
  Domain k: 0 to $2^{53} - 1$
  Domain p: 0 to 1 (exclusive)
  Range: 1.110e–16 to 2^{53}
  Description: returns the Poisson mean such that the cumulative Poisson distribution evaluated at k is p: if poisson(m,k) $= p$, then invpoisson(k,p) $= m$.

  The inverse Poisson distribution function is evaluated using the invgammaptail() function.

invpoissontail(k,q)
  Domain k: 0 to $2^{53} - 1$
  Domain q: 0 to 1 (exclusive)
  Range: 0 to 2^{53} (left exclusive)
  Description: returns the Poisson mean such that the reverse cumulative Poisson distribution evaluated at k is q: if poissontail(m,k) $= q$, then invpoissontail(k,q) $= m$.

  The inverse of the reverse cumulative Poisson distribution function is evaluated using the invgammap() function.

Random-number functions

`runiform()`
> Range: 0 to nearly 1 (0 to $1 - 2^{-32}$)
> Description: returns uniform random variates.
>
> > `runiform()` returns uniformly distributed random variates on the interval $[0, 1)$. `runiform()` takes no arguments, but the parentheses must be typed. `runiform()` can be seeded with the **set seed** command; see the technical note at the end of this subsection. (See *Matrix functions* for the related `matuniform()` matrix function.)
> >
> > To generate random variates over the interval $[a, b)$, use a+(b-a)*`runiform()`.
> >
> > To generate random integers over $[a, b]$, use a+`int(`((b-a+1)*`runiform()))`.

`rbeta(`a`,`b`)`
> Domain a: 0.05 to 1e+5
> Domain b: 0.15 to 1e+5
> Range: 0 to 1 (exclusive)
> Description: returns beta(a,b) random variates, where a and b are the beta distribution shape parameters.
>
> > Besides the standard methodology for generating random variates from a given distribution, `rbeta()` uses the specialized algorithms of Johnk (Gentle 2003), Atkinson and Whittaker (1970, 1976), Devroye (1986), and Schmeiser and Babu (1980).

`rbinomial(`n`,`p`)`
> Domain n: 1 to 1e+11
> Domain p: 1e–8 to $1 - 1$e–8
> Range: 0 to n
> Description: returns binomial(n,p) random variates, where n is the number of trials and p is the success probability.
>
> > Besides the standard methodology for generating random variates from a given distribution, `rbinomial()` uses the specialized algorithms of Kachitvichyanukul (1982), Kachitvichyanukul and Schmeiser (1988), and Kemp (1986).

`rchi2(`df`)`
> Domain df: 2e–4 to 2e+8
> Range: 0 to `c(maxdouble)`
> Description: returns chi-squared, with df degrees of freedom, random variates.

`rgamma(`a`,`b`)`
  Domain a: 1e–4 to 1e+8
  Domain b: `c(smallestdouble)` to `c(maxdouble)`
  Range: 0 to `c(maxdouble)`
  Description: returns gamma(a,b) random variates, where a is the gamma shape parameter and b
               is the scale parameter.

               Methods for generating gamma variates are taken from Ahrens and Dieter (1974),
               Best (1983), and Schmeiser and Lal (1980).

`rhypergeometric(`N`,`K`,`n`)`
  Domain N: 2 to 1e+6
  Domain K: 1 to $N-1$
  Domain n: 1 to $N-1$
  Range: $\max(0, n - N + K)$ to $\min(K, n)$
  Description: returns hypergeometric random variates. The distribution parameters are integer
               valued, where N is the population size, K is the number of elements in
               the population that have the attribute of interest, and n is the sample size.

               Besides the standard methodology for generating random variates from a given
               distribution, `rhypergeometric()` uses the specialized algorithms of
               Kachitvichyanukul (1982) and Kachitvichyanukul and Schmeiser (1985).

`rnbinomial(`n`,`p`)`
  Domain n: 0.1 to 1e+5
  Domain p: 1e–4 to $1-1$e–4
  Range: 0 to $2^{53} - 1$
  Description: returns negative binomial random variates. If n is integer valued, `rnbinomial()`
               returns the number of failures before the nth success, where the probability of
               success on a single trial is p. n can also be nonintegral.

`rnormal()`
  Range: `c(mindouble)` to `c(maxdouble)`
  Description: returns standard normal (Gaussian) random variates, that is, variates from a normal
               distribution with a mean of 0 and a standard deviation of 1.

`rnormal(`m`)`
  Domain m: `c(mindouble)` to `c(maxdouble)`
  Range: `c(mindouble)` to `c(maxdouble)`
  Description: returns normal(m,1) (Gaussian) random variates, where m is the mean and the
               standard deviation is 1.

rnormal(m,s)
   Domain m: c(mindouble) to c(maxdouble)
   Domain s: c(smallestdouble) to c(maxdouble)
   Range: c(mindouble) to c(maxdouble)
   Description: returns normal(m,s) (Gaussian) random variates, where m is the mean and s is the
                   standard deviation.

   The methods for generating normal (Gaussian) random variates are taken from
   Knuth (1998, 122–128); Marsaglia, MacLaren, and Bray (1964); and Walker (1977).

rpoisson(m)
   Domain m: 1e–6 to 1e+11
   Range: 0 to $2^{53} - 1$
   Description: returns Poisson(m) random variates, where m is the distribution mean.

   Poisson variates are generated using the probability integral transform methods
   of Kemp and Kemp (1990, 1991), as well as the method of Kachitvichyanukul (1982).

rt(df)
   Domain df: 1 to $2^{53} - 1$
   Range: c(mindouble) to c(maxdouble)
   Description: returns Student's t random variates, where df is the degrees of freedom.

   Student's t variates are generated using the method of Kinderman and Monahan
   (1977, 1980).

❑ Technical note

The uniform pseudorandom-number function, runiform(), is based on George Marsaglia's (G. Marsaglia, 1994, pers. comm.) 32-bit pseudorandom-number generator KISS (keep it simple stupid). The KISS generator is composed of two 32-bit pseudorandom-number generators and two 16-bit generators (combined to make one 32-bit generator). The four generators are defined by the recursions

$$x_n = 69069\, x_{n-1} + 1234567 \quad \mathrm{mod}\ 2^{32} \tag{1}$$

$$y_n = y_{n-1}(I + L^{13})(I + R^{17})(I + L^5) \tag{2}$$

$$z_n = 65184\bigl(z_{n-1}\ \mathrm{mod}\ 2^{16}\bigr) + \mathrm{int}\bigl(z_{n-1}/2^{16}\bigr) \tag{3}$$

$$w_n = 63663\bigl(w_{n-1}\ \mathrm{mod}\ 2^{16}\bigr) + \mathrm{int}\bigl(w_{n-1}/2^{16}\bigr) \tag{4}$$

In recursion (2), the 32-bit word y_n is viewed as a 1×32 binary vector; L is the 32×32 matrix that produces a left shift of one (L has 1s on the first left subdiagonal, 0s elsewhere); and R is L transpose, affecting a right shift by one. In recursions (3) and (4), int(x) is the integer part of x.

The KISS generator produces the 32-bit random number

$$R_n = x_n + y_n + z_n + 2^{16}w_n \quad \mathrm{mod}\ 2^{32}$$

runiform() takes the output from the KISS generator and divides it by 2^{32} to produce a real number on the interval $[0, 1)$.

All the nonuniform random-number generators rely on uniform random numbers that are also generated using this KISS algorithm.

The recursions (1)–(4) have, respectively, the periods

$$2^{32} \tag{1}$$
$$2^{32} - 1 \tag{2}$$
$$(65184 \cdot 2^{16} - 2)/2 \approx 2^{31} \tag{3}$$
$$(63663 \cdot 2^{16} - 2)/2 \approx 2^{31} \tag{4}$$

Thus the overall period for the KISS generator is

$$2^{32} \cdot (2^{32} - 1) \cdot (65184 \cdot 2^{15} - 1) \cdot (63663 \cdot 2^{15} - 1) \approx 2^{126}$$

When Stata first comes up, it initializes the four recursions in KISS by using the seeds

$$x_0 = 123456789 \tag{1}$$
$$y_0 = 521288629 \tag{2}$$
$$z_0 = 362436069 \tag{3}$$
$$w_0 = 2262615 \tag{4}$$

Successive calls to `runiform()` then produce the sequence

$$\frac{R_1}{2^{32}}, \frac{R_2}{2^{32}}, \frac{R_3}{2^{32}}, \cdots$$

Hence, `runiform()` gives the same sequence of random numbers in every Stata session (measured from the start of the session) unless you reinitialize the seed. The full seed is the set of four numbers (x, y, z, w), but you can reinitialize the seed by simply issuing the command

    . set seed #

where # is any integer between 0 and $2^{31} - 1$, inclusive. When this command is issued, the initial value x_0 is set equal to #, and the other three recursions are restarted at the seeds y_0, z_0, and w_0 given above. The first 100 random numbers are discarded, and successive calls to `runiform()` give the sequence

$$\frac{R'_{101}}{2^{32}}, \frac{R'_{102}}{2^{32}}, \frac{R'_{103}}{2^{32}}, \cdots$$

However, if the command

    . set seed 123456789

is given, the first 100 random numbers are not discarded, and you get the same sequence of random numbers that `runiform()` produces by default; also see [R] **set seed**.

❏

❏ Technical note

You may "capture" the current seed (x, y, z, w) by coding

```
. local curseed = "`c(seed)'"
```

and, later in your code, reestablish that seed by coding

```
. set seed `curseed'
```

When the seed is set this way, the first 100 random numbers are not discarded.

`c(seed)` contains a 30-plus long character string similar to

X075bcd151f123bb5159a55e50022865746ad

The string contains an encoding of the four numbers (x, y, z, w) along with checksums and redundancy to ensure that, at `set seed` time, it is valid.

      ❏

Student's t distribution

`tden(n,t)`
- Domain n: 1e–323 to 8e+307
- Domain t: −8e+307 to 8e+307
- Range: 0 to 0.39894 . . .
- Description: returns the probability density function of Student's t distribution:

$$\texttt{tden}(n,t) = \frac{\Gamma\{(n+1)/2\}}{\sqrt{\pi n}\Gamma(n/2)} \cdot \left(1 + t^2/n\right)^{-(n+1)/2}$$

`ttail(n,t)`
- Domain n: 2e–10 to 2e+17 (may be nonintegral)
- Domain t: −8e+307 to 8e+307
- Range: 0 to 1
- Description: returns the reverse cumulative (upper tail or survivor) Student's t distribution; it returns

the probability $T > t$:

$$\texttt{ttail}(n,t) = \int_t^\infty \frac{\Gamma((n+1)/2)}{\sqrt{\pi n}\Gamma(n/2)} \cdot \left(1 + x^2/n\right)^{-(n+1)/2} \, dx$$

`invttail(n,p)`
- Domain n: 2e–10 to 2e+17 (may be nonintegral)
- Domain p: 0 to 1
- Range: −8e+307 to 8e+307
- Description: returns the inverse reverse cumulative (upper tail or survivor) Student's t distribution:
  if $\texttt{ttail}(n,t) = p$, then $\texttt{invttail}(n,p) = t$.

Tukey's Studentized range distribution

tukeyprob(k,df,x)

    Domain k: 2 to 1e+6

    Domain df: 2 to 1e+6

    Domain x: -8e+307 to 8e+307

                   Interesting domain is $x \geq 0$

    Range: 0 to 1

    Description: returns the cumulative Tukey's Studentized range distribution with k ranges and df degrees of freedom. If df is a missing value, then the normal distribution is used instead of Student's t.

                 returns 0 if $x < 0$.

                 tukeyprob() is computed using an algorithm described in Miller (1981).

invtukeyprob(k,df,p)

    Domain k: 2 to 1e+6

    Domain df: 2 to 1e+6

    Domain p: 0 to 1

    Range: 0 to 8e+307

    Description: returns the inverse cumulative Tukey's Studentized range distribution with k ranges and df degrees of freedom. If df is a missing value, then the normal distribution is used instead of Student's t. If tukeyprob(k,df,x) $= p$, then invtukeyprob(k,df,p) $= x$.

                 invtukeyprob() is computed using an algorithm described in Miller (1981).

String functions

Stata includes the following *string functions*. In the display below, s indicates a string subexpression (a string literal, a string variable, or another string expression), n indicates a numeric subexpression (a number, a numeric variable, or another numeric expression), and re indicates a regular expression based on Henry Spencer's NFA algorithms and this is nearly identical to the POSIX.2 standard.

abbrev(s,n)

    Domain s: strings

    Domain n: 5 to 32

    Range: strings

    Description: returns name s, abbreviated to n characters.

                 If any of the characters of s are a period, ".", and $n < 8$, then the value of n defaults to a value of 8. Otherwise, if $n < 5$, then n defaults to a value of 5. If n is *missing*, abbrev() will return the entire string s. abbrev() is typically used with variable names and variable names with factor-variable or time-series operators (the period case). abbrev("displacement",8) is displa~t.

char(n)
    Domain: integers 1 to 255
    Range: ASCII characters
    Description: returns the character corresponding to ASCII code n.
                  returns "" if n is not in the domain.

indexnot(s_1,s_2)
    Domain s_1: strings (to be searched)
    Domain s_2: strings of individual characters (to search for)
    Range: integers 0 to 244
    Description: returns the position in s_1 of the first character of s_1 not found in s_2, or 0
                  if all characters of s_1 are found in s_2.

itrim(s)
    Domain: strings
    Range: strings with no multiple, consecutive internal blanks
    Description: returns s with multiple, consecutive internal blanks collapsed to one blank.
                  itrim("hello   there") = "hello there"

length(s)
    Domain: strings
    Range: integers 0 to 244
    Description: returns the length of s. length("ab") = 2

lower(s)
    Domain: strings
    Range: strings with lowercased characters
    Description: returns the lowercased variant of s. lower("THIS") = "this"

ltrim(s)
    Domain: strings
    Range: strings without leading blanks
    Description: returns s without leading blanks. ltrim(" this") = "this"

plural(n,s) or plural(n,s_1,s_2)
    Domain n: real numbers
    Domain s: strings
    Domain s_1: strings
    Domain s_2: strings
    Range: strings
    Description: returns the plural of s, or s_1 in the 3-argument case, if $n \neq \pm 1$.
                  The plural is formed by adding "s" to s if you called plural(n,s). If
                  you called plural(n,s_1,s_2) and s_2 begins with the character "+", the plural
                  is formed by adding the remainder of s_2 to s_1. If s_2 begins with the character
                  "–", the plural is formed by subtracting the remainder of s_2 from s_1. If s_2
                  begins with neither "+" nor "–", then the plural is formed by returning s_2.
             returns s, or s_1 in the 3-argument case, if $n = \pm 1$.

                  plural(1, "horse") = "horse"
                  plural(2, "horse") = "horses"
                  plural(2, "glass", "+es") = "glasses"
                  plural(1, "mouse", "mice") = "mouse"
                  plural(2, "mouse", "mice") = "mice"
                  plural(2, "abcdefg", "-efg") = "abcd"

proper(*s*)
  Domain: strings
  Range: strings
  Description: returns a string with the first letter capitalized, and capitalizes any other letters immediately following characters that are not letters; all other letters converted to lowercase.
      proper("mR. joHn a. sMitH") = "Mr. John A. Smith"
      proper("jack o'reilly") = "Jack O'Reilly"
      proper("2-cent's worth") = "2-Cent'S Worth"

real(*s*)
  Domain: strings
  Range: −8e+307 to 8e+307 and *missing*
  Description: returns *s* converted to numeric, or returns *missing*.
      real("5.2")+1 = 6.2
      real("hello") = .

regexm(*s*,*re*)
  Domain *s*:   strings
  Domain *re*: regular expression
  Range: strings
  Description: performs a match of a regular expression and evaluates to 1 if regular expression *re* is satisfied by the string *s*, otherwise returns 0. Regular expression syntax is based on Henry Spencer's NFA algorithm, and this is nearly identical to the POSIX.2 standard.

regexr(*s*₁,*re*,*s*₂)
  Domain s_1: strings
  Domain *re*: regular expression
  Domain s_2: strings
  Range: strings
  Description: replaces the first substring within s_1 that matches *re* with s_2 and returns the resulting string. If s_1 contains no substring that matches *re*, the unaltered s_1 is returned.

regexs(*n*)
  Domain: 0 to 9
  Range: strings
  Description: returns subexpression *n* from a previous regexm() match, where $0 \leq n < 10$. Subexpression 0 is reserved for the entire string that satisfied the regular expression.

reverse(*s*)
  Domain: strings
  Range: reversed strings
  Description: returns *s* reversed. reverse("hello") = "olleh"

rtrim(*s*)
  Domain: strings
  Range: strings without trailing blanks
  Description: returns *s* without trailing blanks. rtrim("this ") = "this"

soundex(*s*)

    Domain: strings

    Range: strings

    Description: returns the soundex code for a string, *s*. The soundex code consists of a letter followed by three numbers: the letter is the first letter of the name and the numbers encode the remaining consonants. Similar sounding consonants are encoded by the same number.

```
soundex("Ashcraft") = "A226"
soundex("Robert") = "R163"
soundex("Rupert") = "R163"
```

soundex_nara(*s*)

    Domain: strings

    Range: strings

    Description: returns the U.S. Census soundex code for a string, *s*. The soundex code consists of a letter followed by three numbers: the letter is the first letter of the name and the numbers encode the remaining consonants. Similar sounding consonants are encoded by the same number.

```
soundex_nara("Ashcraft") = "A261"
```

string(*n*)

    Domain: $-8e+307$ to $8e+307$ and *missing*

    Range: strings

    Description: returns *n* converted to a string.

```
string(4)+"F" = "4F"
string(1234567) = "1234567"
string(12345678) = "1.23e+07"
string(.) = "."
```

string(*n*,*s*)

    Domain *n*: $-8e+307$ to $8e+307$ and *missing*

    Domain *s*:   strings containing %*fmt* numeric display format

    Range: strings

    Description: returns *n* converted to a string.

```
string(4,"%9.2f") = "4.00"
string(123456789,"%11.0g") = "123456789"
string(123456789,"%13.0gc") = "123,456,789"
string(0,"%td") = "01jan1960"
string(225,"%tq") = "2016q2"
string(225,"not a format") = ""
```

strlen(*s*) is a synonym for length(*s*).

strlower(*x*) is a synonym for lower(*x*).

strltrim(*x*) is a synonym for ltrim(*x*).

strmatch(s_1,s_2)
    Domain s: strings
    Range: 0 or 1
    Description: returns 1 if s_1 matches the pattern s_2; otherwise, it returns 0.
                strmatch("17.4","1??4") returns 1. In s_2, "?" means that one character
                goes here, and "*" means that zero or more characters go here. Also see
                regexm(), regexr(), and regexs().

strofreal(n) is a synonym for string(n).

strofreal(n,s) is a synonym for string(n,s).

strpos(s_1,s_2)
    Domain s_1: strings (to be searched)
    Domain s_2: strings (to search for)
    Range: integers 0 to 244
    Description: returns the position in s_1 at which s_2 is first found; otherwise, it returns 0.
                strpos("this","is") = 3
                strpos("this","it") = 0

strproper(x) is a synonym for proper(x).

strreverse(x) is a synonym for reverse(x).

strrtrim(x) is a synonym for rtrim(x).

strtoname(s,p)
    Domain s: strings
    Domain p: 0 or 1
    Range: strings
    Description: returns s translated into a Stata name. Each character in s that is not allowed
                in a Stata name is converted to an underscore character, _. If the first character
                in s is a numeric character and p is not 0, then the result is prefixed with
                an underscore. The result is truncated to 32 characters.

                strtoname("name",1) = "name"
                strtoname("a name",1) = "a_name"
                strtoname("5",1) = "_5"
                strtoname("5:30",1) = "_5_30"
                strtoname("5",0) = "5"
                strtoname("5:30",0) = "5_30"

strtoname(s)
    Domain s: strings
    Range: strings
    Description: returns s translated into a Stata name. Each character in s that is not allowed
                in a Stata name is converted to an underscore character, _. If the first character
                in s is a numeric character, then the result is prefixed with
                an underscore. The result is truncated to 32 characters.

                strtoname("name") = "name"
                strtoname("a name") = "a_name"
                strtoname("5") = "_5"
                strtoname("5:30") = "_5_30"

strtrim(x) is a synonym for trim(x).

strupper(x) is a synonym for upper(x).

subinstr(s_1,s_2,s_3,n)
   Domain s_1:   strings (to be substituted into)
   Domain s_2:   strings (to be substituted from)
   Domain s_3:   strings (to be substituted with)
   Domain n: integers 0 to 244 and *missing*
   Range: strings
   Description: returns s_1, where the first n occurrences in s_1 of s_2 have been replaced
                  with s_3. If n is *missing*, all occurrences are replaced.
                  Also see regexm(), regexr(), and regexs().
                  subinstr("this is this","is","X",1) = "thX is this"
                  subinstr("this is this","is","X",2) = "thX X this"
                  subinstr("this is this","is","X",.) = "thX X thX"

subinword(s_1,s_2,s_3,n)
   Domain s_1:   strings (to be substituted for)
   Domain s_2:   strings (to be substituted from)
   Domain s_3:   strings (to be substituted with)
   Domain n: integers 0 to 244 and *missing*
   Range: strings
   Description: returns s_1, where the first n occurrences in s_1 of s_2 as a word have
                  been replaced with s_3. A word is defined as a space-separated token.
                  A token at the beginning or end of s_1 is considered space-separated.
                  If n is *missing*, all occurrences are replaced.
                  Also see regexm(), regexr(), and regexs().

                  subinword("this is this","is","X",1) = "this X this"
                  subinword("this is this","is","X",.) = "this X this"
                  subinword("this is this","th","X",.) = "this is this"

substr(s,n_1,n_2)
   Domain s: strings
   Domain n_1:   integers 1 to 244 and -1 to -244
   Domain n_2:   integers 1 to 244 and -1 to -244
   Range: strings
   Description: returns the substring of s, starting at column n_1, for a length of n_2.
                  If $n_1 < 0$, n_1 is interpreted as distance from the end of the string;
                  if $n_2 = $. (*missing*), the remaining portion of the string is returned.

                  substr("abcdef",2,3) = "bcd"
                  substr("abcdef",-3,2) = "de"
                  substr("abcdef",2,.) = "bcdef"
                  substr("abcdef",-3,.) = "def"
                  substr("abcdef",2,0) = ""
                  substr("abcdef",15,2) = ""

trim(s)
   Domain: strings
   Range: strings without leading or trailing blanks
   Description: returns s without leading and trailing blanks; equivalent to
                  ltrim(rtrim(s)). trim(" this ") = "this"

upper(*s*)

> Domain: strings
> Range: strings with uppercased characters
> Description: returns the uppercased variant of *s*. upper("this") = "THIS"

word(*s*, *n*)

> Domain *s*:   strings
> Domain *n*:   integers $\ldots, -2, -1, 0, 1, 2, \ldots$
> Range: strings
> Description: returns the *n*th word in *s*. Positive numbers count words from the beginning of *s*, and negative numbers count words from the end of *s*. (1 is the first word in *s*, and -1 is the last word in *s*.) Returns *missing* ("") if *n* is missing.

wordcount(*s*)

> Domain: strings
> Range: nonnegative integers 0, 1, 2, $\ldots$
> Description: returns the number of words in *s*. A word is a set of characters that start and terminate with spaces, start with the beginning of the string, or terminate with the end of the string.

Programming functions

autocode(*x*,*n*,x_0,x_1)

> Domain *x*: $-8e+307$ to $8e+307$
> Domain *n*:   integers 1 to $8e+307$
> Domain x_0: $-8e+307$ to $8e+307$
> Domain x_1: x_0 to $8e+307$
> Range: x_0 to x_1
> Description: partitions the interval from x_0 to x_1 into *n* equal-length intervals and returns the upper bound of the interval that contains *x*. This function is an automated version of recode() (see below).
> See [U] **25 Working with categorical data and factor variables** for an example.

> The algorithm for autocode() is
>> if $(n \geq . \,|\, x_0 \geq . \,|\, x_1 \geq . \,|\, n \leq 0 \,|\, x_0 \geq x_1)$
>>> then return *missing*
>>> if $x \geq .$, then return *x*
>> otherwise
>>> for $i = 1$ to $n - 1$
>>>> $xmap = x_0 + i * (x_1 - x_0)/n$
>>>> if $x \leq xmap$ then return *xmap*
>>> end
>> otherwise
>>> return x_1

`byteorder()`

    Range: 1 and 2

    Description: returns 1 if your computer stores numbers by using a hilo byte order and evaluates to 2 if your computer stores numbers by using a lohi byte order. Consider the number 1 written as a 2-byte integer. On some computers (called hilo), it is written as "00 01", and on other computers (called lohi), it is written as "01 00" (with the least significant byte written first). There are similar issues for 4-byte integers, 4-byte floats, and 8-byte floats. Stata automatically handles byte-order differences for Stata-created files. Users need not be concerned about this issue. Programmers producing customary binary files can use `byteorder()` to determine the native byte ordering; see [P] **file**.

`c(`*name*`)`

    Domain: names

    Range: real values, strings, and *missing*

    Description: returns the value of the system or constant result `c(`*name*`)`; see [P] **creturn**.
                 Referencing `c(`*name*`)` will return an error if the result does not exist.
               returns a scalar if the result is scalar.
               returns a string of the result containing the first 244 characters.

`_caller()`

    Range: 1 to 12

    Description: returns `version` of the program or session that invoked the currently running program; see [P] **version**. The current version at the time of this writing is 12, so 12 is the upper end of this range. If Stata 12.1 were the current version, 12.1 would be the upper end of this range, and likewise, if Stata 13 were the current version, 13 would be the upper end of this range. This is a function for use by programmers.

`chop(`x, ϵ`)`

    Domain x: $-8e{+}307$ to $8e{+}307$

    Domain ϵ: $-8e{+}307$ to $8e{+}307$

    Range: $-8e{+}307$ to $8e{+}307$

    Description: returns `round(`x`)` if `abs(`$x - $`round(`$x$`))` $< \epsilon$; otherwise, returns x.
               returns x if x is missing.

`clip(`x,a,b`)`

    Domain x: $-8e{+}307$ to $8e{+}307$

    Domain a: $-8e{+}307$ to $8e{+}307$

    Domain b: $-8e{+}307$ to $8e{+}307$

    Range: $-8e{+}307$ to $8e{+}307$

    Description: returns x if $a < x < b$, b if $x \geq b$, a if $x \leq a$, and *missing* if x is missing or if $a > b$. If a or b is missing, this is interpreted as $a = -\infty$ or $b = +\infty$, respectively.
               returns x if x is missing.

cond(x,a,b,c) or cond(x,a,b)

Domain x:	$-8e+307$ to $8e+307$ and *missing*; $0 \Rightarrow$ *false*, otherwise interpreted as *true*
Domain a:	numbers and strings
Domain b:	numbers if a is a number; strings if a is a string
Domain c:	numbers if a is a number; strings if a is a string
Range:	a, b, and c
Description:	returns a if x is *true* and nonmissing, b if x is *false*, and c if x is *missing*. returns a if c is not specified and x evaluates to *missing*.

Note that expressions such as $x > 2$ will never evaluate to *missing*.

> cond(x>2,50,70) returns 50 if x > 2 (includes x ≥ .)
> cond(x>2,50,70) returns 70 if x ≤ 2

If you need a case for missing values in the above examples, try

> cond(missing(x), ., cond(x>2,50,70)) returns . if x is *missing*, returns 50 if x > 2, and returns 70 if x ≤ 2

If the first argument is a scalar that may contain a missing value or a variable containing missing values, the fourth argument has an effect.

> cond(wage,1,0,.) returns 1 if wage is not zero and not missing
> cond(wage,1,0,.) returns 0 if wage is zero
> cond(wage,1,0,.) returns . if wage is *missing*

Caution: If the first argument to cond() is a logical expression, that is, cond(x>2,50,70,.), the fourth argument is never reached.

e(*name*)

Domain:	names
Range:	strings, scalars, matrices, and *missing*
Description:	returns the value of saved result e(*name*); see [U] **18.8 Accessing results calculated by other programs**

e(*name*) = scalar missing if the saved result does not exist
e(*name*) = specified matrix if the saved result is a matrix
e(*name*) = scalar numeric value if the saved result is a scalar
e(*name*) = a string containing the first 244 characters if the saved result is a string

e(sample)

Range:	0 and 1
Description:	returns 1 if the observation is in the estimation sample and 0 otherwise.

epsdouble()

Range:	a double-precision number close to 0
Description:	returns the machine precision of a double-precision number. If $d <$ epsdouble() and (double) $x = 1$, then $x + d =$ (double) 1. This function takes no arguments, but the parentheses must be included.

epsfloat()
    Range: a floating-point number close to 0
    Description: returns the machine precision of a floating-point number. If $d <$ epsfloat() and (float) $x = 1$, then $x + d =$ (float) 1. This function takes no arguments, but the parentheses must be included.

float(x)
    Domain: -1e+38 to 1e+38
    Range: -1e+38 to 1e+38
    Description: returns the value of x rounded to float precision.

                Although you may store your numeric variables as byte, int, long, float, or double, Stata converts all numbers to double before performing any calculations. Consequently, difficulties can arise in comparing numbers that have no finite binary representation.

                For example, if the variable x is stored as a float and contains the value 1.1 (a repeating "decimal" in binary), the expression x==1.1 will evaluate to *false* because the literal 1.1 is the double representation of 1.1, which is different from the float representation stored in x. (They differ by 2.384×10^{-8}.) The expression x==float(1.1) will evaluate to *true* because the float() function converts the literal 1.1 to its float representation before it is compared with x. (See [U] **13.11 Precision and problems therein** for more information.)

fmtwidth(*fmtstr*)
    Range: strings
    Description: returns the output length of the %*fmt* contained in *fmtstr*.
                returns *missing* if *fmtstr* does not contain a valid %*fmt*. For example, fmtwidth("%9.2f") returns 9 and fmtwidth("%tc") returns 18.

has_eprop(*name*)
    Domain: names
    Range: 0 or 1
    Description: returns 1 if *name* appears as a word in e(properties); otherwise, returns 0.

inlist($z,a,b,\ldots$)
    Domain: all reals or all strings
    Range: 0 or 1
    Description: returns 1 if z is a member of the remaining arguments; otherwise, returns 0. All arguments must be reals or all must be strings. The number of arguments is between 2 and 255 for reals and between 2 and 10 for strings.

inrange(z,a,b)
    Domain: all reals or all strings
    Range: 0 or 1
    Description: returns 1 if it is known that $a \le z \le b$; otherwise, returns 0. The following ordered rules apply:
                $z \ge$. returns 0.
                $a \ge$. and $b =$. returns 1.
                $a \ge$. returns 1 if $z \le b$; otherwise, it returns 0.
                $b \ge$. returns 1 if $a \le z$; otherwise, it returns 0.
                Otherwise, 1 is returned if $a \le z \le b$.
                If the arguments are strings, "." is interpreted as "".

irecode(x,x_1,x_2,x_3,...,x_n)

Domain x: $-8e+307$ to $8e+307$

Domain x_i: $-8e+307$ to $8e+307$

Range: nonnegative integers

Description: returns *missing* if x is missing or $x_1,...,x_n$ is not weakly increasing.

returns 0 if $x \leq x_1$.

returns 1 if $x_1 < x \leq x_2$.

returns 2 if $x_2 < x \leq x_3$.

...

returns n if $x > x_n$.

Also see autocode() and recode() for other styles of recode functions.

irecode(3, -10, -5, -3, -3, 0, 15, .) $= 5$

matrix(*exp*)

Domain: any valid expression

Range: evaluation of *exp*

Description: restricts name interpretation to scalars and matrices; see scalar() function below.

maxbyte()

Range: one integer number

Description: returns the largest value that can be stored in storage type byte. This function takes no arguments, but the parentheses must be included.

maxdouble()

Range: one double-precision number

Description: returns the largest value that can be stored in storage type double. This function takes no arguments, but the parentheses must be included.

maxfloat()

Range: one floating-point number

Description: returns the largest value that can be stored in storage type float. This function takes no arguments, but the parentheses must be included.

maxint()

Range: one integer number

Description: returns the largest value that can be stored in storage type int. This function takes no arguments, but the parentheses must be included.

maxlong()

Range: one integer number

Description: returns the largest value that can be stored in storage type long. This function takes no arguments, but the parentheses must be included.

mi(x_1,x_2,...,x_n) is a synonym for missing(x_1,x_2,...,x_n).

minbyte()

Range: one integer number

Description: returns the smallest value that can be stored in storage type byte. This function takes no arguments, but the parentheses must be included.

mindouble()

Range: one double-precision number

Description: returns the smallest value that can be stored in storage type double. This function takes no arguments, but the parentheses must be included.

`minfloat()`
    Range: one floating-point number
    Description: returns the smallest value that can be stored in storage type `float`. This function takes no arguments, but the parentheses must be included.

`minint()`
    Range: one integer number
    Description: returns the smallest value that can be stored in storage type `int`. This function takes no arguments, but the parentheses must be included.

`minlong()`
    Range: one integer number
    Description: returns the smallest value that can be stored in storage type `long`. This function takes no arguments, but the parentheses must be included.

`missing($x_1, x_2, \ldots, x_n$)`
    Domain x_i: any string or numeric expression
    Range: 0 and 1
    Description: returns 1 if any x_i evaluates to *missing*; otherwise, returns 0.

    Stata has two concepts of missing values: a numeric missing value (., .a, .b, ..., .z) and a string missing value (""). `missing()` returns 1 (meaning *true*) if any expression x_i evaluates to *missing*. If x is numeric, `missing($x$)` is equivalent to $x \geq$ .. If x is string, `missing($x$)` is equivalent to $x==$ "".

`r(`*name*`)`
    Domain: names
    Range: strings, scalars, matrices, and *missing*
    Description: returns the value of the saved result `r(`*name*`)`;
                    see [U] **18.8 Accessing results calculated by other programs**
                    `r(`*name*`)` = scalar missing if the saved result does not exist
                    `r(`*name*`)` = specified matrix if the saved result is a matrix
                    `r(`*name*`)` = scalar numeric value if the saved result is a scalar
                                    that can be interpreted as a number
                    `r(`*name*`)` = a string containing the first 244 characters
                                    if the saved result is a string

`recode($x, x_1, x_2, \ldots, x_n$)`
    Domain x: $-8e+307$ to $8e+307$ and *missing*
    Domain x_1: $-8e+307$ to $8e+307$
    Domain x_2: x_1 to $8e+307$
    ...
    Domain x_n: x_{n-1} to $8e+307$
    Range: $x_1, x_2, \ldots, x_n$ and *missing*
    Description: returns *missing* if $x_1, \ldots, x_n$ is not weakly increasing.
                  returns x if x is missing.
                  returns x_1 if $x \leq x_1$; x_2 if $x \leq x_2$, ...; otherwise,
                      x_n if $x > x_1, x_2, \ldots, x_{n-1}$.
                      $x_i \geq$. is interpreted as $x_i = +\infty$.

    Also see `autocode()` and `irecode()` for other styles of recode functions.

replay()
> Range: integers 0 and 1, meaning *false* and *true*, respectively
> Description: returns 1 if the first nonblank character of local macro '0' is a comma,
>   or if '0' is empty. This is a function for use by programmers writing
>   estimation commands; see [P] **ereturn**.

return(*name*)
> Domain: names
> Range: strings, scalars, matrices, and *missing*
> Description: returns the value of the to-be-saved result r(*name*);
>   see [P] **return**.
>   return(*name*) = scalar missing if the saved result does not exist
>   return(*name*) = specified matrix if the saved result is a matrix
>   return(*name*) = scalar numeric value if the saved result is a scalar
>   return(*name*) = a string containing the first 244 characters
>   if the saved result is a string

s(*name*)
> Domain: names
> Range: strings and *missing*
> Description: returns the value of saved result s(*name*);
>   see [U] **18.8 Accessing results calculated by other programs**
>   s(*name*) = . if the saved result does not exist
>   s(*name*) = a string containing the first 244 characters
>   if the saved result is a string

scalar(*exp*)
> Domain: any valid expression
> Range: evaluation of *exp*
> Description: restricts name interpretation to scalars and matrices.
>
>   Names in expressions can refer to names of variables in the dataset, names of
>   matrices, or names of scalars. Matrices and scalars can have the same names as
>   variables in the dataset. If names conflict, Stata assumes that you are referring to the
>   name of the variable in the dataset.
>
>   matrix() and scalar() explicitly state that you are referring to matrices and
>   scalars. matrix() and scalar() are the same function; scalars and matrices may
>   not have the same names and so cannot be confused. Typing scalar(x) makes it
>   clear that you are referring to the scalar or matrix named x and not the variable
>   named x, should there happen to be a variable of that name.

smallestdouble()
> Range: a double-precision number close to 0
> Description: returns the smallest double-precision number greater than zero. If
> $0 < d <$ smallestdouble(), then d does not have full double
>   precision; these are called the denormalized numbers. This function
>   takes no arguments, but the parentheses must be included.

Date and time functions

Stata's *date and time functions* are described with examples in [U] **24 Working with dates and times** and [D] **datetime**. What follows is a technical description. We use the following notation:

e_b	%tb business calendar date (days)
e_{tc}	%tc encoded datetime (ms. since 01jan1960 00:00:00.000)
e_{tC}	%tC encoded datetime (ms. with leap seconds since 01jan1960 00:00:00.000)
e_d	%td encoded date (days since 01jan1960)
e_w	%tw encoded weekly date (weeks since 1960w1)
e_m	%tm encoded monthly date (months since 1960m1)
e_q	%tq encoded quarterly date (quarters since 1960q1)
e_h	%th encoded half-yearly date (half-years since 1960h1)
e_y	%ty encoded yearly date (years)
M	month, 1–12
D	day of month, 1–31
Y	year, 0100–9999
h	hour, 0–23
m	minute, 0–59
s	second, 0–59 or 60 if leap seconds
W	week number, 1–52
Q	quarter number, 1–4
H	half-year number, 1 or 2

The date and time functions, where integer arguments are required, allow noninteger values and use the `floor()` of the value.

A Stata date-and-time (%t) variable is recorded as the milliseconds, days, weeks, etc., depending upon the units from 01jan1960; negative values indicate dates and times before 01jan1960. Allowable dates and times are those between 01jan0100 and 31dec9999, inclusive, but all functions are based on the Gregorian calendar, and values do not correspond to historical dates before Friday, 15oct1582.

bofd("*cal*",e_d)
 Domain *cal*: business calendar names and formats
 Domain e_d: %td as defined by business calendar named *cal*
 Range: as defined by business calendar named *cal*
 Description: returns the e_b business date corresponding to e_d.

Cdhms(e_d,h,m,s)
 Domain e_d: %td dates 01jan0100 to 31dec9999 (integers −679,350 to 2,936,549)
 Domain h: integers 0 to 23
 Domain m: integers 0 to 59
 Domain s: reals 0.000 to 60.999
 Range: datetimes 01jan0100 00:00:00.000 to 31dec9999 23:59:59.999
               (integers −58,695,840,000,000 to > 253,717,919,999,999) and *missing*
 Description: returns the e_{tC} datetime (ms. with leap seconds since 01jan1960 00:00:00.000)
               corresponding to e_d, h, m, s.

Chms(h,m,s)
    Domain h: integers 0 to 23
    Domain m: integers 0 to 59
    Domain s: reals 0.000 to 60.999
    Range: datetimes 01jan0100 00:00:00.000 to 31dec9999 23:59:59.999
            (integers $-58{,}695{,}840{,}000{,}000$ to $>253{,}717{,}919{,}999{,}999$) and *missing*
    Description: returns the e_{tC} datetime (ms. with leap seconds since 01jan1960 00:00:00.000)
                corresponding to h, m, s on 01jan1960.

Clock($s_1,s_2\big[\,Y\big]$)
    Domain s_1:   strings
    Domain s_2:   strings
    Domain Y: integers 1000 to 9998 (but probably 2001 to 2099)
    Range: datetimes 01jan0100 00:00:00.000 to 31dec9999 23:59:59.999
            (integers $-58{,}695{,}840{,}000{,}000$ to $>253{,}717{,}919{,}999{,}999$) and *missing*
    Description: returns the e_{tC} datetime (ms. with leap seconds since 01jan1960 00:00:00.000)
                corresponding to s_1 based on s_2 and Y.

        Function Clock() works the same as function clock() except that Clock() returns
        a leap second–adjusted %tC value rather than an unadjusted %tc value. Use
        Clock() only if original time values have been adjusted for leap seconds.

clock($s_1,s_2\big[\,Y\big]$)
    Domain s_1:   strings
    Domain s_2:   strings
    Domain Y: integers 1000 to 9998 (but probably 2001 to 2099)
    Range: datetimes 01jan0100 00:00:00.000 to 31dec9999 23:59:59.999
            (integers $-58{,}695{,}840{,}000{,}000$ to $253{,}717{,}919{,}999{,}999$) and *missing*
    Description: returns the e_{tc} datetime (ms. since 01jan1960 00:00:00.000) corresponding to
                s_1 based on s_2 and Y.

        s_1 contains the date, time, or both, recorded as a string, in virtually any
        format. Months can be spelled out, abbreviated (to three characters), or indicated as
        numbers; years can include or exclude the century; blanks and punctuation are allowed.

        s_2 is any permutation of M, D, [##]Y, h, m, and s, with their order defining the
        order that month, day, year, hour, minute, and second occur (and whether they
        occur) in s_1. ##, if specified, indicates the default century for two-digit years in s_1.
        For instance, $s_2 = $ "MD19Y hm" would translate $s_1 = $ "11/15/91 21:14" as
        15nov1991 21:14. The space in "MD19Y hm" was not significant and the string would
        have translated just as well with "MD19Yhm".

        Y provides an alternate way of handling two-digit years. Y specifies the largest
        year that is to be returned when a two-digit year is encountered; see function date()
        below. If neither ## nor Y is specified, clock() returns *missing* when it
        encounters a two-digit year.

Cmdyhms(M,D,Y,h,m,s)

Domain M: integers 1 to 12
Domain D: integers 1 to 31
Domain Y: integers 0100 to 9999 (but probably 1800 to 2100)
Domain h: integers 0 to 23
Domain m: integers 0 to 59
Domain s: reals 0.000 to 60.999
Range: datetimes 01jan0100 00:00:00.000 to 31dec9999 23:59:59.999
                (integers $-58,695,840,000,000$ to $>253,717,919,999,999$) and *missing*
Description: returns the e_{tC} datetime (ms. with leap seconds since 01jan1960 00:00:00.000)
                corresponding to M, D, Y, h, m, s.

Cofc(e_{tc})

Domain e_{tc}: datetimes 01jan0100 00:00:00.000 to 31dec9999 23:59:59.999
                (integers $-58,695,840,000,000$ to $253,717,919,999,999$)
Range: datetimes 01jan0100 00:00:00.000 to 31dec9999 23:59:59.999
                (integers $-58,695,840,000,000$ to $>253,717,919,999,999$)
Description: returns the e_{tC} datetime (ms. with leap seconds since 01jan1960 00:00:00.000)
                of e_{tc} (ms. without leap seconds since 01jan1960 00:00:00.000).

cofC(e_{tC})

Domain e_{tC}: datetimes 01jan0100 00:00:00.000 to 31dec9999 23:59:59.999
                (integers $-58,695,840,000,000$ to $>253,717,919,999,999$)
Range: datetimes 01jan0100 00:00:00.000 to 31dec9999 23:59:59.999
                (integers $-58,695,840,000,000$ to $253,717,919,999,999$)
Description: returns the e_{tc} datetime (ms. without leap seconds since 01jan1960 00:00:00.000)
                of e_{tC} (ms. with leap seconds since 01jan1960 00:00:00.000).

Cofd(e_d)

Domain e_d: %td dates 01jan0100 to 31dec9999 (integers $-679,350$ to 2,936,549)
Range: datetimes 01jan0100 00:00:00.000 to 31dec9999 23:59:59.999
                (integers $-58,695,840,000,000$ to $>253,717,919,999,999$)
Description: returns the e_{tC} datetime (ms. with leap seconds since 01jan1960 00:00:00.000)
                of date e_d at time 00:00:00.000.

cofd(e_d)

Domain e_d: %td dates 01jan0100 to 31dec9999 (integers $-679,350$ to 2,936,549)
Range: datetimes 01jan0100 00:00:00.000 to 31dec9999 23:59:59.999
                (integers $-58,695,840,000,000$ to $253,717,919,999,999$)
Description: returns the e_{tc} datetime (ms. since 01jan1960 00:00:00.000) of date e_d at time
                00:00:00.000.

date(s_1, s_2 [, Y])
   Domain s_1:   strings
   Domain s_2:   strings
   Domain Y: integers 1000 to 9998 (but probably 2001 to 2099)
   Range: %td dates 01jan0100 to 31dec9999 (integers $-679,350$ to $2,936,549$) and *missing*
   Description: returns the e_d date (days since 01jan1960) corresponding to s_1 based on s_2 and Y.

   s_1 contains the date, recorded as a string, in virtually any format. Months can be spelled out, abbreviated (to three characters), or indicated as numbers; years can include or exclude the century; blanks and punctuation are allowed.

   s_2 is any permutation of M, D, and [##]Y, with their order defining the order that month, day, and year occur in s_1. ##, if specified, indicates the default century for two-digit years in s_1. For instance, s_2 = "MD19Y" would translate s_1 = "11/15/91" as 15nov1991.

   Y provides an alternate way of handling two-digit years. When a two-digit year is encountered, the largest year, *topyear*, that does not exceed Y is returned.

   date("1/15/08","MDY",1999) = 15jan1908
   date("1/15/08","MDY",2019) = 15jan2008

   date("1/15/51","MDY",2000) = 15jan1951
   date("1/15/50","MDY",2000) = 15jan1950
   date("1/15/49","MDY",2000) = 15jan1949

   date("1/15/01","MDY",2050) = 15jan2001
   date("1/15/00","MDY",2050) = 15jan2000

   If neither ## nor Y is specified, date() returns *missing* when it encounters a two-digit year. See *Working with two-digit years* in [D] **datetime translation** for more information.

day(e_d)
   Domain e_d: %td dates 01jan0100 to 31dec9999 (integers $-679,350$ to $2,936,549$)
   Range: integers 1 to 31 and *missing*
   Description: returns the numeric day of the month corresponding to e_d.

dhms(e_d, h, m, s)
   Domain e_d: %td dates 01jan0100 to 31dec9999 (integers $-679,350$ to $2,936,549$)
   Domain h: integers 0 to 23
   Domain m: integers 0 to 59
   Domain s: reals 0.000 to 59.999
   Range: datetimes 01jan0100 00:00:00.000 to 31dec9999 23:59:59.999
                     (integers $-58,695,840,000,000$ to $253,717,919,999,999$) and *missing*
   Description: returns the e_{tc} datetime (ms. since 01jan1960 00:00:00.000) corresponding to e_d, h, m, and s.

dofb(e_b, "cal")
   Domain e_b: %tb as defined by business calendar named cal
   Domain cal: business calendar names and formats
   Range: as defined by business calendar named cal
   Description: returns the e_d datetime corresponding to e_b.

dofC(e_{tC})
   Domain e_{tC}: datetimes 01jan0100 00:00:00.000 to 31dec9999 23:59:59.999
                  (integers $-58{,}695{,}840{,}000{,}000$ to $> 253{,}717{,}919{,}999{,}999$)
   Range: %td dates 01jan0100 to 31dec9999 (integers $-679{,}350$ to $2{,}936{,}549$)
   Description: returns the e_d date (days since 01jan1960) of datetime e_{tC} (ms. with leap
                  seconds since 01jan1960 00:00:00.000).

dofc(e_{tc})
   Domain e_{tc}: datetimes 01jan0100 00:00:00.000 to 31dec9999 23:59:59.999
                  (integers $-58{,}695{,}840{,}000{,}000$ to $253{,}717{,}919{,}999{,}999$)
   Range: %td dates 01jan0100 to 31dec9999 (integers $-679{,}350$ to $2{,}936{,}549$)
   Description: returns the e_d date (days since 01jan1960) of datetime e_{tc} (ms. since 01jan1960
                  00:00:00.000).

dofh(e_h)
   Domain e_h: %th dates 0100h1 to 9999h2 (integers $-3{,}720$ to $16{,}079$)
   Range: %td dates 01jan0100 to 01jul9999 (integers $-679{,}350$ to $2{,}936{,}366$)
   Description: returns the e_d date (days since 01jan1960) of the start of half-year e_h.

dofm(e_m)
   Domain e_m: %tm dates 0100m1 to 9999m12 (integers $-22{,}320$ to $96{,}479$)
   Range: %td dates 01jan0100 to 01dec9999 (integers $-679{,}350$ to $2{,}936{,}519$)
   Description: returns the e_d date (days since 01jan1960) of the start of month e_m.

dofq(e_q)
   Domain e_q: %tq dates 0100q1 to 9999q4 (integers $-7{,}440$ to $32{,}159$)
   Range: %td dates 01jan0100 to 01oct9999 (integers $-679{,}350$ to $2{,}936{,}458$)
   Description: returns the e_d date (days since 01jan1960) of the start of quarter e_q.

dofw(e_w)
   Domain e_w: %tw dates 0100w1 to 9999w52 (integers $-96{,}720$ to $418{,}079$)
   Range: %td dates 01jan0100 to 24dec9999 (integers $-679{,}350$ to $2{,}936{,}542$)
   Description: returns the e_d date (days since 01jan1960) of the start of week e_w.

dofy(e_y)
   Domain e_y: %ty dates 0100 to 9999 (integers 0100 to 9999)
   Range: %td dates 01jan0100 to 01jan9999 (integers $-679{,}350$ to $2{,}936{,}185$)
   Description: returns the e_d date (days since 01jan1960) of 01jan in year e_y.

dow(e_d)
   Domain e_d: %td dates 01jan0100 to 31dec9999 (integers $-679{,}350$ to $2{,}936{,}549$)
   Range: integers 0 to 6 and *missing*
   Description: returns the numeric day of the week corresponding to date e_d;
                  $0 = $ Sunday, $1 = $ Monday, ..., $6 = $ Saturday.

doy(e_d)
   Domain e_d: %td dates 01jan0100 to 31dec9999 (integers $-679{,}350$ to $2{,}936{,}549$)
   Range: integers 1 to 366 and *missing*
   Description: returns the numeric day of the year corresponding to date e_d.

halfyear(e_d)
   Domain e_d: %td dates 01jan0100 to 31dec9999 (integers $-679{,}350$ to $2{,}936{,}549$)
   Range: integers 1, 2, and *missing*
   Description: returns the numeric half of the year corresponding to date e_d.

`halfyearly(`s_1`,`s_2`[,`Y`])`

    Domain s_1:   strings

    Domain s_2:   strings `"HY"` and `"YH"`; Y may be prefixed with `##`

    Domain Y: integers 1000 to 9998 (but probably 2001 to 2099)

    Range: `%th` dates 0100h1 to 9999h2 (integers $-3{,}720$ to $16{,}079$) and *missing*

    Description: returns the e_h half-yearly date (half-years since 1960h1) corresponding to s_1 based on s_2 and Y; Y specifies *topyear*; see `date()`.

`hh(`e_{tc}`)`

    Domain e_{tc}:   datetimes 01jan0100 00:00:00.000 to 31dec9999 23:59:59.999 (integers $-58{,}695{,}840{,}000{,}000$ to $253{,}717{,}919{,}999{,}999$)

    Range: integers 0 through 23, *missing*

    Description: returns the hour corresponding to datetime e_{tc} (ms. since 01jan1960 00:00:00.000).

`hhC(`e_{tC}`)`

    Domain e_{tC}:   datetimes 01jan0100 00:00:00.000 to 31dec9999 23:59:59.999 (integers $-58{,}695{,}840{,}000{,}000$ to $>253{,}717{,}919{,}999{,}999$)

    Range: integers 0 through 23, *missing*

    Description: returns the hour corresponding to datetime e_{tC} (ms. with leap seconds since 01jan1960 00:00:00.000).

`hms(`h`,`m`,`s`)`

    Domain h: integers 0 to 23

    Domain m: integers 0 to 59

    Domain s: reals 0.000 to 59.999

    Range: datetimes 01jan1960 00:00:00.000 to 01jan1960 23:59:59.999 (integers 0 to 86,399,999 and *missing*)

    Description: returns the e_{tc} datetime (ms. since 01jan1960 00:00:00.000) corresponding to h, m, s on 01jan1960.

`hofd(`e_d`)`

    Domain e_d: `%td` dates 01jan0100 to 31dec9999 (integers $-679{,}350$ to $2{,}936{,}549$)

    Range: `%th` dates 0100h1 to 9999h2 (integers $-3{,}720$ to $16{,}079$)

    Description: returns the e_h half-yearly date (half years since 1960h1) containing date e_d.

`hours(`ms`)`

    Domain ms: real; milliseconds

    Range: real and *missing*

    Description: returns $ms/3{,}600{,}000$.

`mdy(`M`,`D`,`Y`)`

    Domain M: integers 1 to 12

    Domain D: integers 1 to 31

    Domain Y: integers 0100 to 9999 (but probably 1800 to 2100)

    Range: `%td` dates 01jan0100 to 31dec9999 (integers $-679{,}350$ to $2{,}936{,}549$) and *missing*

    Description: returns the e_d date (days since 01jan1960) corresponding to M, D, Y.

mdyhms(M,D,Y,h,m,s)
  Domain M: integers 1 to 12
  Domain D: integers 1 to 31
  Domain Y: integers 0100 to 9999 (but probably 1800 to 2100)
  Domain h: integers 0 to 23
  Domain m: integers 0 to 59
  Domain s: reals 0.000 to 59.999
  Range: datetimes 01jan0100 00:00:00.000 to 31dec9999 23:59:59.999
                 (integers $-58{,}695{,}840{,}000{,}000$ to $253{,}717{,}919{,}999{,}999$) and *missing*
  Description: returns the e_{tc} datetime (ms. since 01jan1960 00:00:00.000) corresponding to
                    M, D, Y, h, m, s.

minutes(ms)
  Domain ms: real; milliseconds
  Range: real and *missing*
  Description: returns $ms/60{,}000$.

mm(e_{tc})
  Domain e_{tc}:   datetimes 01jan0100 00:00:00.000 to 31dec9999 23:59:59.999
                 (integers $-58{,}695{,}840{,}000{,}000$ to $253{,}717{,}919{,}999{,}999$)
  Range: integers 0 through 59, *missing*
  Description: returns the minute corresponding to datetime e_{tc} (ms. since 01jan1960 00:00:00.000).

mmC(e_{tC})
  Domain e_{tC}:   datetimes 01jan0100 00:00:00.000 to 31dec9999 23:59:59.999
                 (integers $-58{,}695{,}840{,}000{,}000$ to $>253{,}717{,}919{,}999{,}999$)
  Range: integers 0 through 59, *missing*
  Description: returns the minute corresponding to datetime e_{tC} (ms. with leap seconds since
                    01jan1960 00:00:00.000).

mofd(e_d)
  Domain e_d: %td dates 01jan0100 to 31dec9999 (integers $-679{,}350$ to 2,936,549)
  Range: %tm dates 0100m1 to 9999m12 (integers $-22{,}320$ to 96,479)
  Description: returns the e_m monthly date (months since 1960m1) containing date e_d.

month(e_d)
  Domain e_d: %td dates 01jan0100 to 31dec9999 (integers $-679{,}350$ to 2,936,549)
  Range: integers 1 to 12 and *missing*
  Description: returns the numeric month corresponding to date e_d.

monthly(s_1,$s_2$$\big[$,$Y$$\big]$)
  Domain s_1: strings
  Domain s_2: strings "MY" and "YM"; Y may be prefixed with ##
  Domain Y: integers 1000 to 9998 (but probably 2001 to 2099)
  Range: %tm dates 0100m1 to 9999m12 (integers $-22{,}320$ to 96,479) and *missing*
  Description: returns the e_m monthly date (months since 1960m1) corresponding to s_1 based on
                    s_2 and Y; Y specifies *topyear*; see date().

msofhours(h)
  Domain h: real; hours
  Range: real and *missing*; milliseconds
  Description: returns $h \times 3{,}600{,}000$.

msofminutes(m)
    Domain m: real; minutes
    Range: real and *missing*; milliseconds
    Description: returns $m \times 60{,}000$.

msofseconds(s)
    Domain s: real; seconds
    Range: real and *missing*; milliseconds
    Description: returns $s \times 1{,}000$.

qofd(e_d)
    Domain e_d: %td dates 01jan0100 to 31dec9999 (integers $-679{,}350$ to $2{,}936{,}549$)
    Range: %tq dates 0100q1 to 9999q4 (integers $-7{,}440$ to $32{,}159$)
    Description: returns the e_q quarterly date (quarters since 1960q1) containing date e_d.

quarter(e_d)
    Domain e_d: %td dates 01jan0100 to 31dec9999 (integers $-679{,}350$ to $2{,}936{,}549$)
    Range: integers 1 to 4 and *missing*
    Description: returns the numeric quarter of the year corresponding to date e_d.

quarterly(s_1,$s_2$$\left[\,,Y\,\right]$)
    Domain s_1:   strings
    Domain s_2:   strings "QY" and "YQ"; Y may be prefixed with ##
    Domain Y: integers 1000 to 9998 (but probably 2001 to 2099)
    Range: %tq dates 0100q1 to 9999q4 (integers $-7{,}440$ to $32{,}159$) and *missing*
    Description: returns the e_q quarterly date (quarters since 1960q1) corresponding to s_1 based on s_2 and Y; Y specifies *topyear*; see date().

seconds(ms)
    Domain ms: real; milliseconds
    Range: real and *missing*
    Description: returns $ms/1{,}000$.

ss(e_{tc})
    Domain e_{tc}: datetimes 01jan0100 00:00:00.000 to 31dec9999 23:59:59.999
                     (integers $-58{,}695{,}840{,}000{,}000$ to $253{,}717{,}919{,}999{,}999$)
    Range: real 0.000 through 59.999, *missing*
    Description: returns the second corresponding to datetime e_{tc} (ms. since 01jan1960 00:00:00.000).

ssC(e_{tC})
    Domain e_{tC}: datetimes 01jan0100 00:00:00.000 to 31dec9999 23:59:59.999
                     (integers $-58{,}695{,}840{,}000{,}000$ to $>253{,}717{,}919{,}999{,}999$)
    Range: real 0.000 through 60.999, *missing*
    Description: returns the second corresponding to datetime e_{tC} (ms. with leap seconds since 01jan1960 00:00:00.000).

tC(l)
    Domain l: datetime literal strings 01jan0100 00:00:00.000 to 31dec9999 23:59:59.999
    Range: datetimes 01jan0100 00:00:00.000 to 31dec9999 23:59:59.999
                     (integers $-58{,}695{,}840{,}000{,}000$ to $>253{,}717{,}919{,}999{,}999$)
    Description: convenience function to make typing dates and times in expressions easier; same as tc(), except returns leap second–adjusted values; for example, typing tc(29nov2007 9:15) is equivalent to typing 1511946900000, whereas tC(29nov2007 9:15) is 1511946923000.

`tc(`*l*`)`

Domain *l*:   datetime literal strings 01jan0100 00:00:00.000 to 31dec9999 23:59:59.999

Range: datetimes 01jan0100 00:00:00.000 to 31dec9999 23:59:59.999
(integers $-58{,}695{,}840{,}000{,}000$ to $253{,}717{,}919{,}999{,}999$)

Description: convenience function to make typing dates and times in expressions easier;
for example, typing `tc(2jan1960 13:42)` is equivalent to typing 135720000;
the date but not the time may be omitted, and then 01jan1960 is
assumed; the seconds portion of the time may be omitted and
is assumed to be 0.000; `tc(11:02)` is equivalent to typing 39720000.

`td(`*l*`)`

Domain *l*:   date literal strings 01jan0100 to 31dec9999

Range: %td dates 01jan0100 to 31dec9999 (integers $-679{,}350$ to $2{,}936{,}549$)

Description: convenience function to make typing dates in expressions easier;
for example, typing `td(2jan1960)` is equivalent to typing 1.

`th(`*l*`)`

Domain *l*:   half-year literal strings 0100h1 to 9999h2

Range: %th dates 0100h1 to 9999h2 (integers $-3{,}720$ to $16{,}079$)

Description: convenience function to make typing half-yearly dates in expressions easier;
for example, typing `th(1960h2)` is equivalent to typing 1.

`tm(`*l*`)`

Domain *l*:   month literal strings 0100m1 to 9999m12

Range: %tm dates 0100m1 to 9999m12 (integers $-22{,}320$ to $96{,}479$)

Description: convenience function to make typing monthly dates in expressions easier;
for example, typing `tm(1960m2)` is equivalent to typing 1.

`tq(`*l*`)`

Domain *l*:   quarter literal strings 0100q1 to 9999q4

Range: %tq dates 0100q1 to 9999q4 (integers $-7{,}440$ to $32{,}159$)

Description: convenience function to make typing quarterly dates in expressions easier;
for example, typing `tq(1960q2)` is equivalent to typing 1.

`tw(`*l*`)`

Domain *l*:   week literal strings 0100w1 to 9999w52

Range: %tw dates 0100w1 to 9999w52 (integers $-96{,}720$ to $418{,}079$)

Description: convenience function to make typing weekly dates in expressions easier;
for example, typing `tw(1960w2)` is equivalent to typing 1.

`week(`e_d`)`

Domain e_d: %td dates 01jan0100 to 31dec9999 (integers $-679{,}350$ to $2{,}936{,}549$)

Range: integers 1 to 52 and *missing*

Description: returns the numeric week of the year corresponding to date e_d
(the first week of a year is the first 7-day period of the year).

`weekly(`s_1`,`s_2`[`,Y`])`

Domain s_1:   strings

Domain s_2:   strings "WY" and "YW"; Y may be prefixed with ##

Domain Y: integers 1000 to 9998 (but probably 2001 to 2099)

Range: %tw dates 0100w1 to 9999w52 (integers $-96{,}720$ to $418{,}079$) and *missing*

Description: returns the e_w weekly date (weeks since 1960w1) corresponding to s_1 based on s_2
and Y; Y specifies *topyear*; see `date()`.

wofd(e_d)
  Domain e_d: %td dates 01jan0100 to 31dec9999 (integers −679,350 to 2,936,549)
  Range: %tw dates 0100w1 to 9999w52 (integers −96,720 to 418,079)
  Description: returns the e_w weekly date (weeks since 1960w1) containing date e_d.

year(e_d)
  Domain e_d: %td dates 01jan0100 to 31dec9999 (integers −679,350 to 2,936,549)
  Range: integers 0100 to 9999 (but probably 1800 to 2100)
  Description: returns the numeric year corresponding to date e_d.

yearly(s_1,$s_2$$\left[\,,Y\right]$)
  Domain s_1:   strings
  Domain s_2:   string "Y"; Y may be prefixed with ##
  Domain Y: integers 1000 to 9998 (but probably 2001 to 2099)
  Range: %ty dates 0100 to 9999 (integers 0100 to 9999) and *missing*
  Description: returns the e_y yearly date (year) corresponding to s_1 based on s_2 and Y;
                Y specifies *topyear*; see date().

yh(Y,H)
  Domain Y: integers 1000 to 9999 (but probably 1800 to 2100)
  Domain H: integers 1, 2
  Range: %th dates 1000h1 to 9999h2 (integers −1,920 to 16,079)
  Description: returns the e_h half-yearly date (half-years since 1960h1) corresponding to year Y,
                half-year H.

ym(Y,M)
  Domain Y: integers 1000 to 9999 (but probably 1800 to 2100)
  Domain M: integers 1 to 12
  Range: %tm dates 1000m1 to 9999m12 (integers −11,520 to 96,479)
  Description: returns the e_m monthly date (months since 1960m1) corresponding to year Y,
                month M.

yofd(e_d)
  Domain e_d: %td dates 01jan0100 to 31dec9999 (integers −679,350 to 2,936,549)
  Range: %ty dates 0100 to 9999 (integers 0100 to 9999)
  Description: returns the e_y yearly date (year) containing date e_d.

yq(Y,Q)
  Domain Y: integers 1000 to 9999 (but probably 1800 to 2100)
  Domain Q: integers 1 to 4
  Range: %tq dates 1000q1 to 9999q4 (integers −3,840 to 32,159)
  Description: returns the e_q quarterly date (quarters since 1960q1) corresponding to year Y,
                quarter Q.

yw(Y,W)
  Domain Y: integers 1000 to 9999 (but probably 1800 to 2100)
  Domain W: integers 1 to 52
  Range: %tw dates 1000w1 to 9999w52 (integers −49,920 to 418,079)
  Description: returns the e_w weekly date (weeks since 1960w1) corresponding to year Y,
                week W.

Selecting time spans

$\texttt{tin}(d_1, d_2)$
 Domain d_1: date or time literals recorded in units of t previously $\texttt{tsset}$
 Domain d_2: date or time literals recorded in units of t previously $\texttt{tsset}$
 Range: 0 and 1, 1 $\Rightarrow$ *true*
 Description: *true* if $d_1 \le t \le d_2$, where t is the time variable previously $\texttt{tsset}$.

You must have previously $\texttt{tsset}$ the data to use $\texttt{tin}()$; see [TS] **tsset**. When you $\texttt{tsset}$ the data, you specify a time variable, t, and the format on t states how it is recorded. You type d_1 and d_2 according to that format.

If t has a $\texttt{\%tc}$ format, you could type $\texttt{tin(5jan1992 11:15, 14apr2002 12:25)}$.

If t has a $\texttt{\%td}$ format, you could type $\texttt{tin(5jan1992, 14apr2002)}$.

If t has a $\texttt{\%tw}$ format, you could type $\texttt{tin(1985w1, 2002w15)}$.

If t has a $\texttt{\%tm}$ format, you could type $\texttt{tin(1985m1, 2002m4)}$.

If t has a $\texttt{\%tq}$ format, you could type $\texttt{tin(1985q1, 2002q2)}$.

If t has a $\texttt{\%th}$ format, you could type $\texttt{tin(1985h1, 2002h1)}$.

If t has a $\texttt{\%ty}$ format, you could type $\texttt{tin(1985, 2002)}$.

Otherwise, t is just a set of integers, and you could type $\texttt{tin(12, 38)}$.

The details of the $\texttt{\%t}$ format do not matter. If your t is formatted $\texttt{\%tdnn/dd/yy}$ so that 5jan1992 displays as 1/5/92, you would still type the date in day–month–year order: $\texttt{tin(5jan1992, 14apr2002)}$.

$\texttt{twithin}(d_1, d_2)$
 Domain d_1: date or time literals recorded in units of t previously $\texttt{tsset}$
 Domain d_2: date or time literals recorded in units of t previously $\texttt{tsset}$
 Range: 0 and 1, 1 $\Rightarrow$ *true*
 Description: *true* if $d_1 < t < d_2$, where t is the time variable previously $\texttt{tsset}$;
     see the $\texttt{tin}()$ function above; $\texttt{twithin}()$ is similar, except the range is exclusive.

Matrix functions returning a matrix

In addition to the functions listed below, see [P] **matrix svd** for singular value decomposition, [P] **matrix symeigen** for eigenvalues and eigenvectors of symmetric matrices, and [P] **matrix eigenvalues** for eigenvalues of nonsymmetric matrices.

`cholesky(`M`)`

    Domain: $n \times n$, positive-definite, symmetric matrices
    Range: $n \times n$ lower-triangular matrices
    Description: returns the Cholesky decomposition of the matrix:
               if $R = $ `cholesky(`S`)`, then $RR^T = S$.
               R^T indicates the transpose of R.
               Row and column names are obtained from M.

`corr(`M`)`

    Domain: $n \times n$ symmetric variance matrices
    Range: $n \times n$ symmetric correlation matrices
    Description: returns the correlation matrix of the variance matrix.
               Row and column names are obtained from M.

`diag(`v`)`

    Domain: $1 \times n$ and $n \times 1$ vectors
    Range: $n \times n$ diagonal matrices
    Description: returns the square, diagonal matrix created from the row or column vector.
               Row and column names are obtained from the column names of M if M is
               a row vector or from the row names of M if M is a column vector.

`get(`*systemname*`)`

    Domain: existing names of system matrices
    Range: matrices
    Description: returns a copy of Stata internal system matrix *systemname*.

               This function is included for backward compatibility with previous versions
               of Stata.

`hadamard(`M`,`N`)`

    Domain M: $m \times n$ matrices
    Domain N: $m \times n$ matrices
    Range: $m \times n$ matrices
    Description: returns a matrix whose i, j element is $M[i,j] \cdot N[i,j]$ (if M and N
               are not the same size, this function reports a conformability error).

`I(`n`)`

    Domain: real scalars 1 to `matsize`
    Range: identity matrices
    Description: returns an $n \times n$ identity matrix if n is an integer; otherwise, this function returns
               the `round(`n`)`$\times$`round(`n`)` identity matrix.

`inv(`M`)`

    Domain: $n \times n$ nonsingular matrices
    Range: $n \times n$ matrices
    Description: returns the inverse of the matrix M. If M is singular, this will result in an error.

               The function `invsym()` should be used in preference to `inv()` because `invsym()`
               is more accurate. The row names of the result are obtained from the column
               names of M, and the column names of the result are obtained from the row names
               of M.

invsym(M)
   Domain: $n \times n$ symmetric matrices
   Range: $n \times n$ symmetric matrices
   Description: returns the inverse of M if M is positive definite. If M is not positive definite, rows will be inverted until the diagonal terms are zero or negative; the rows and columns corresponding to these terms will be set to 0, producing a g2 inverse. The row names of the result are obtained from the column names of M, and the column names of the result are obtained from the row names of M.

J(r,c,z)
   Domain r: integer scalars 1 to matsize
   Domain c: integer scalars 1 to matsize
   Domain z: scalars $-8e+307$ to $8e+307$
   Range: $r \times c$ matrices
   Description: returns the $r \times c$ matrix containing elements z.

matuniform(r,c)
   Domain r: integer scalars 1 to matsize
   Domain c: integer scalars 1 to matsize
   Range: $r \times c$ matrices
   Description: returns the $r \times c$ matrices containing uniformly distributed pseudorandom numbers on the interval $[0, 1)$.

nullmat(*matname*)
   Domain: matrix names, existing and nonexisting
   Range: matrices including null if *matname* does not exist
   Description: nullmat() is for use with the row-join (,) and column-join (\) operators in programming situations. Consider the following code fragment, which is an attempt to create the vector $(1, 2, 3, 4)$:

```
forvalues i = 1/4 {
        mat v = (v, 'i')
}
```

The above program will not work because, the first time through the loop, v will not yet exist, and thus forming (v, 'i') makes no sense. nullmat() relaxes that restriction:

```
forvalues i = 1/4 {
        mat v = (nullmat(v), 'i')
}
```

The nullmat() function informs Stata that if v does not exist, the function row-join is to be generalized. Joining nothing with 'i' results in ('i'). Thus the first time through the loop, v = (1) is formed. The second time through, v does exist, so v = (1, 2) is formed, and so on.

nullmat() can be used only with the , and \ operators.

`sweep(M,i)`

    Domain M: $n \times n$ matrices

    Domain i: integer scalars 1 to n

    Range: $n \times n$ matrices

    Description: returns matrix M with ith row/column swept. The row and column names of the resultant matrix are obtained from M, except that the nth row and column names are interchanged. If $B = \texttt{sweep}(A,k)$, then

$$B_{kk} = \frac{1}{A_{kk}}$$

$$B_{ik} = -\frac{A_{ik}}{A_{kk}}, \qquad i \neq k$$

$$B_{kj} = \frac{A_{kj}}{A_{kk}}, \qquad j \neq k$$

$$B_{ij} = A_{ij} - \frac{A_{ik}A_{kj}}{A_{kk}}, \qquad i \neq k, j \neq k$$

`vec(M)`

    Domain: matrices

    Range: column vectors ($n \times 1$ matrices)

    Description: returns a column vector formed by listing the elements of M, starting with the first column and proceeding column by column.

`vecdiag(M)`

    Domain: $n \times n$ matrices

    Range: $1 \times n$ vectors

    Description: returns the row vector containing the diagonal of matrix M. `vecdiag()` is the opposite of `diag()`. The row name is set to `r1`; the column names are obtained from the column names of M.

Matrix functions returning a scalar

`colnumb(M,s)`

    Domain M: matrices

    Domain s: strings

    Range: integer scalars 1 to `matsize` and *missing*

    Description: returns the column number of M associated with column name s. returns *missing* if the column cannot be found.

`colsof(M)`

    Domain: matrices

    Range: integer scalars 1 to `matsize`

    Description: returns the number of columns of M.

`det(M)`

    Domain: $n \times n$ (square) matrices

    Range: scalars $-8\text{e}{+}307$ to $8\text{e}{+}307$

    Description: returns the determinant of matrix M.

`diag0cnt(`M`)`

    Domain: $n \times n$ (square) matrices
    Range: integer scalars 0 to n
    Description: returns the number of zeros on the diagonal of M.

`el(`s,i,j`)`

    Domain s: strings containing matrix name
    Domain i: scalars 1 to `matsize`
    Domain j: scalars 1 to `matsize`
    Range: scalars $-8e+307$ to $8e+307$ and *missing*
    Description: returns $s[$`floor`$(i),$`floor`$(j)]$, the i, j element of the matrix named s.
                returns *missing* if i or j are out of range or if matrix s does not exist.

`issymmetric(`M`)`

    Domain M: matrices
    Range: integers 0 and 1
    Description: returns 1 if the matrix is symmetric; otherwise, returns 0.

`matmissing(`M`)`

    Domain M: matrices
    Range: integers 0 and 1
    Description: returns 1 if any elements of the matrix are missing; otherwise, returns 0.

`mreldif(`X,Y`)`

    Domain X: matrices
    Domain Y: matrices with same number of rows and columns as X
    Range: scalars $-8e+307$ to $8e+307$
    Description: returns the relative difference of X and Y, where the relative difference is
                defined as $\max_{i,j}\big(|x_{ij} - y_{ij}|/(|y_{ij}| + 1)\big)$.

`rownumb(`M,s`)`

    Domain M: matrices
    Domain s: strings
    Range: integer scalars 1 to `matsize` and *missing*
    Description: returns the row number of M associated with row name s.
                returns *missing* if the row cannot be found.

`rowsof(`M`)`

    Domain: matrices
    Range: integer scalars 1 to `matsize`
    Description: returns the number of rows of M.

`trace(`M`)`

    Domain: $n \times n$ (square) matrices
    Range: scalars $-8e+307$ to $8e+307$
    Description: returns the trace of matrix M.

Acknowledgments

We thank George Marsaglia of Florida State University for providing his KISS (keep it simple stupid) random-number generator.

We thank John R. Gleason of Syracuse University (retired) for directing our attention to Wichura (1988) for calculating the cumulative normal density accurately, for sharing his experiences about techniques with us, and for providing C code to make the calculations.

Jacques Salomon Hadamard (1865–1963) was born in Versailles, France. He studied at the Ecole Normale Supérieure in Paris and obtained a doctorate in 1892 for a thesis on functions defined by Taylor series. Hadamard taught at Bordeaux for 4 years and in a productive period published an outstanding theorem on prime numbers, proved independently by Charles de la Vallée Poussin, and worked on what are now called Hadamard matrices. In 1897, he returned to Paris, where he held a series of prominent posts. In his later career, his interests extended from pure mathematics toward mathematical physics. Hadamard produced papers and books in many different areas. He campaigned actively against anti-Semitism at the time of the Dreyfus affair. After the fall of France in 1940, he spent some time in the United States and then Great Britain.

References

Abramowitz, M., and I. A. Stegun, ed. 1968. *Handbook of Mathematical Functions with Formulas, Graphs, and Mathematical Tables*. 7th ed. Washington, DC: National Bureau of Standards.

Ahrens, J. H., and U. Dieter. 1974. Computer methods for sampling from gamma, beta, Poisson, and binomial distributions. *Computing* 12: 223–246.

Atkinson, A. C., and J. Whittaker. 1970. Algorithm AS 134: The generation of beta random variables with one parameter greater than and one parameter less than 1. *Applied Statistics* 28: 90–93.

———. 1976. A switching algorithm for the generation of beta random variables with at least one parameter less than 1. *Journal of the Royal Statistical Society, Series A* 139: 462–467.

Best, D. J. 1983. A note on gamma variate generators with shape parameters less than unity. *Computing* 30: 185–188.

Cox, N. J. 2003. Stata tip 2: Building with floors and ceilings. *Stata Journal* 3: 446–447.

———. 2004. Stata tip 6: Inserting awkward characters in the plot. *Stata Journal* 4: 95–96.

Devroye, L. 1986. *Non-uniform Random Variate Generation*. New York: Springer.

Gentle, J. E. 2003. *Random Number Generation and Monte Carlo Methods*. 2nd ed. New York: Springer.

Haynam, G. E., Z. Govindarajulu, and F. C. Leone. 1970. Tables of the cumulative noncentral chi-square distribution. In Vol. 1 of *Selected Tables in Mathematical Statistics*, ed. H. L. Harter and D. B. Owen, 1–78. Providence, RI: American Mathematical Society.

Hilbe, J. M. 2010. Creating synthetic discrete-response regression models. *Stata Journal* 10: 104–124.

Hilbe, J. M., and W. Linde-Zwirble. 1995. sg44: Random number generators. *Stata Technical Bulletin* 28: 20–21. Reprinted in *Stata Technical Bulletin Reprints*, vol. 5, pp. 118–121. College Station, TX: Stata Press.

———. 1998. sg44.1: Correction to random number generators. *Stata Technical Bulletin* 41: 23. Reprinted in *Stata Technical Bulletin Reprints*, vol. 7, p. 166. College Station, TX: Stata Press.

Johnson, N. L., S. Kotz, and N. Balakrishnan. 1995. *Continuous Univariate Distributions, Vol. 2*. 2nd ed. New York: Wiley.

Kachitvichyanukul, V. 1982. Computer Generation of Poisson, Binomial, and Hypergeometric Random Variables. PhD thesis, Purdue University.

Kachitvichyanukul, V., and B. W. Schmeiser. 1985. Computer generation of hypergeometric random variates. *Journal of Statistical Computation and Simulation* 22: 127–145.

———. 1988. Binomial random variate generation. *Communications of the Association for Computing Machinery* 31: 216–222.

Kantor, D., and N. J. Cox. 2005. Depending on conditions: A tutorial on the cond() function. *Stata Journal* 5: 413–420.

Kemp, A. W., and C. D. Kemp. 1990. A composition-search algorithm for low-parameter Poisson generation. *Journal of Statistical Computation and Simulation* 35: 239–244.

Kemp, C. D. 1986. A modal method for generating binomial variates. *Communications in Statistics, Theory and Methods* 15: 805–813.

Kemp, C. D., and A. W. Kemp. 1991. Poisson random variate generation. *Applied Statistics* 40: 143–158.

Kinderman, A. J., and J. F. Monahan. 1977. Computer generation of random variables using the ratio of uniform deviates. *ACM Transactions on Mathematical Software* 3: 257–260.

———. 1980. New methods for generating Student's t and gamma variables. *Computing* 25: 369–377.

Knuth, D. 1998. *The Art of Computer Programming, Volume 2: Seminumerical Algorithms.* 3rd ed. Reading, MA: Addison–Wesley.

Marsaglia, G., M. D. MacLaren, and T. A. Bray. 1964. A fast procedure for generating normal random variables. *Communications of the Association for Computing Machinery* 7: 4–10.

Mazýa, V., and T. Shaposhnikova. 1998. *Jacques Hadamard, A Universal mathematician.* Providence, RI: American Mathematical Society.

Miller, R. G., Jr. 1981. *Simultaneous Statistical Inference.* 2nd ed. New York: Springer.

Moore, R. J. 1982. Algorithm AS 187: Derivatives of the incomplete gamma integral. *Applied Statistics* 31: 330–335.

Oldham, K. B., J. C. Myland, and J. Spanier. 2009. *An Atlas of Functions.* 2nd ed. New York: Springer.

Posten, H. O. 1993. An effective algorithm for the noncentral beta distribution function. *American Statistician* 47: 129–131.

Press, W. H., S. A. Teukolsky, W. T. Vetterling, and B. P. Flannery. 2007. *Numerical Recipes in C: The Art of Scientific Computing.* 3rd ed. Cambridge: Cambridge University Press.

Rising, B. 2010. Stata tip 86: The missing() function. *Stata Journal* 10: 303–304.

Schmeiser, B. W., and A. J. G. Babu. 1980. Beta variate generation via exponential majorizing functions. *Operations Research* 28: 917–926.

Schmeiser, B. W., and R. Lal. 1980. Squeeze methods for generating gamma variates. *Journal of the American Statistical Association* 75: 679–682.

Walker, A. J. 1977. An efficient method for generating discrete random variables with general distributions. *ACM Transactions on Mathematical Software* 3: 253–256.

Weiss, M. 2009. Stata tip 80: Constructing a group variable with specified group sizes. *Stata Journal* 9: 640–642.

Wichura, M. J. 1988. Algorithm AS241: The percentage points of the normal distribution. *Applied Statistics* 37: 477–484.

Also see

Title

> **generate** — Create or change contents of variable

Syntax

Create new variable

    <u>g</u>enerate $[\,type\,]$ *newvar* $[\,:lblname\,]$ =*exp* $[\,if\,]$ $[\,in\,]$

Replace contents of existing variable

    replace *oldvar* =*exp* $[\,if\,]$ $[\,in\,]$ $[\,$, <u>nop</u>romote $]$

Specify default storage type assigned to new variables

    set <u>type</u> $\{\,$ float $|$ double $\,\}$ $[\,$, <u>perm</u>anently $]$

where *type* is one of byte | int | long | float | double | str | str1 | str2 | ... | str244.

See *Description* below for an explanation of str. For the other types, see [U] **12 Data**.

by is allowed with generate and replace; see [D] **by**.

Menu

generate

Data > Create or change data > Create new variable

replace

Data > Create or change data > Change contents of variable

Description

generate creates a new variable. The values of the variable are specified by =*exp*.

If no *type* is specified, the new variable type is determined by the type of result returned by =*exp*. A float variable (or a double, according to set type) is created if the result is numeric, and a string variable is created if the result is a string. In the latter case, a str# variable is created, where # is the smallest string that will hold the result.

If a *type* is specified, the result returned by =*exp* must be a string or numeric according to whether *type* is string or numeric. If str is specified, a str# variable is created, where # is the smallest string that will hold the result.

See [D] **egen** for extensions to generate.

replace changes the contents of an existing variable. Because replace alters data, the command cannot be abbreviated.

set type specifies the default storage type assigned to new variables (such as those created by generate) when the storage type is not explicitly specified.

Options

nopromote prevents replace from promoting the variable type to accommodate the change. For instance, consider a variable stored as an integer type (byte, int, or long), and assume that you replace some values with nonintegers. By default, replace changes the variable type to a floating point (float or double) and thus correctly stores the changed values. Similarly, replace promotes byte and int variables to longer integers (int and long) if the replacement value is an integer but is too large in absolute value for the current storage type. replace promotes strings to longer strings. nopromote prevents replace from doing this; instead, the replacement values are truncated to fit into the current storage type.

permanently specifies that, in addition to making the change right now, the new limit be remembered and become the default setting when you invoke Stata.

Remarks

Remarks are presented under the following headings:

> *generate and replace*
> *set type*

generate and replace

generate and replace are used to create new variables and to modify the contents of existing variables, respectively. Although the commands do the same thing, they have different names so that you do not accidentally replace values in your data. Detailed descriptions of expressions are given in [U] **13 Functions and expressions**.

Also see [D] **edit**.

▷ Example 1

We have a dataset containing the variable age2, which we have previously defined as age^2 (that is, age^2). We have changed some of the age data and now want to correct age2 to reflect the new values:

```
. use http://www.stata-press.com/data/r12/genxmpl1
(Wages of women)
. generate age2=age^2
age2 already defined
r(110);
```

When we attempt to re-generate age2, Stata refuses, telling us that age2 is already defined. We could drop age2 and then re-generate it, or we could use the replace command:

```
. replace age2=age^2
(204 real changes made)
```

When we use replace, we are informed of the number of actual changes made to the dataset.

◁

You can explicitly specify the storage type of the new variable being created by putting the *type*, such as byte, int, long, float, double, or str8, in front of the variable name. For example, you could type generate double revenue = qty * price. Not specifying a type is equivalent to specifying float if the variable is numeric, or, more correctly, it is equivalent to specifying the default type set by the set type command; see below. If the variable is alphanumeric, not specifying a type is equivalent to specifying str#, where # is the length of the largest string in the variable.

You may also specify a value label to be associated with the new variable by including ":*lblname*" after the variable name. This is seldom done because you can always associate the value label later by using the label define command; see [U] **12.6.3 Value labels**.

▷ Example 2

Among the variables in our dataset is name, which contains the first and last name of each person. We wish to create a new variable called lastname, which we will then use to sort the data. name is a string variable.

```
. use http://www.stata-press.com/data/r12/genxmpl2, clear
. list name
```

	name
1.	Johanna Roman
2.	Dawn Mikulin
3.	Malinda Vela
4.	Kevin Crow
5.	Zachary Bimslager

```
. generate lastname=word(name,2)
. describe
Contains data from http://www.stata-press.com/data/r12/genxmpl2.dta
  obs:            5
  vars:           2                          18 Jan 2011 12:24
  size:         130
```

	storage	display	value	
variable name	type	format	label	variable label
name	str17	%17s		
lastname	str9	%9s		

```
Sorted by:
    Note:  dataset has changed since last saved
```

Stata is smart. Even though we did not specify the storage type in our generate statement, Stata knew to create a str9 lastname variable, because the longest last name is Bimslager, which has nine characters.

◁

▷ Example 3

We wish to create a new variable, age2, that represents the variable age squared. We realize that because age is an integer, age2 will also be an integer and will certainly be less than 32,740. We therefore decide to store age2 as an int to conserve memory:

```
. use http://www.stata-press.com/data/r12/genxmpl3, clear
. generate int age2=age^2
(9 missing values generated)
```

Preceding `age2` with `int` told Stata that the variable was to be stored as an `int`. After creating the new variable, Stata informed us that nine missing values were generated. `generate` informs us whenever it produces missing values.

◁

See [U] **13 Functions and expressions** and [U] **25 Working with categorical data and factor variables** for more information and examples. Also see [D] **recode** for a convenient way to recode categorical variables.

❏ Technical note

If you specify the `if` modifier or `in` *range*, the `=exp` is evaluated only for those observations that meet the specified condition or are in the specified range (or both, if both `if` and `in` are specified). The other observations of the new variable are set to missing:

```
. use http://www.stata-press.com/data/r12/genxmpl3, clear
. generate int age2=age^2 if age>30
(290 missing values generated)
```

❏

▷ Example 4

`replace` can be used to change just one value, as well as to make sweeping changes to our data. For instance, say that we enter data on the first five odd and even positive integers and then discover that we made a mistake:

```
. use http://www.stata-press.com/data/r12/genxmpl4, clear
. list
```

	odd	even
1.	1	2
2.	3	4
3.	-8	6
4.	7	8
5.	9	10

The third observation is wrong; the value of `odd` should be 5, not -8. We can use `replace` to correct the mistake:

```
. replace odd=5 in 3
(1 real change made)
```

We could also have corrected the mistake by typing `replace odd=5 if odd==-8`.

◁

set type

When you create a new numeric variable and do not specify the storage type for it, say, by typing generate y=x+2, the new variable is made a float if you have not previously issued the set type command. If earlier in your session you typed set type double, the new numeric variable would be made a double.

Methods and formulas

You can do anything with replace that you can do with generate. The only difference between the commands is that replace requires that the variable already exist, whereas generate requires that the variable be new. In fact, inside Stata, generate and replace have the same code. Because Stata is an interactive system, we force a distinction between replacing existing values and generating new ones so that you do not accidentally replace valuable data while thinking that you are creating a new piece of information.

References

Gleason, J. R. 1997a. dm50: Defining variables and recording their definitions. *Stata Technical Bulletin* 40: 9–10. Reprinted in *Stata Technical Bulletin Reprints*, vol. 7, pp. 48–49. College Station, TX: Stata Press.

——. 1997b. dm50.1: Update to defv. *Stata Technical Bulletin* 51: 2. Reprinted in *Stata Technical Bulletin Reprints*, vol. 9, pp. 14–15. College Station, TX: Stata Press.

Newson, R. 2004. Stata tip 13: generate and replace use the current sort order. *Stata Journal* 4: 484–485.

Weesie, J. 1997. dm43: Automatic recording of definitions. *Stata Technical Bulletin* 35: 6–7. Reprinted in *Stata Technical Bulletin Reprints*, vol. 6, pp. 18–20. College Station, TX: Stata Press.

Also see

[D] **compress** — Compress data in memory

[D] **rename** — Rename variable

[D] **corr2data** — Create dataset with specified correlation structure

[D] **drawnorm** — Draw sample from multivariate normal distribution

[D] **edit** — Browse or edit data with Data Editor

[D] **egen** — Extensions to generate

[D] **encode** — Encode string into numeric and vice versa

[D] **label** — Manipulate labels

[D] **recode** — Recode categorical variables

[U] **12 Data**

[U] **13 Functions and expressions**

Title

> **gsort** — Ascending and descending sort

Syntax

> gsort [+|-] *varname* [[+|-] *varname* ...] [, generate(*newvar*) mfirst]

Menu

Data > Sort > Ascending and descending sort

Description

gsort arranges observations to be in ascending or descending order of the specified variables and so differs from sort in that sort produces ascending-order arrangements only; see [D] **sort**.

Each *varname* can be numeric or a string.

The observations are placed in ascending order of *varname* if + or nothing is typed in front of the name and are placed in descending order if - is typed.

Options

generate(*newvar*) creates *newvar* containing 1, 2, 3, ... for each group denoted by the ordered data. This is useful when using the ordering in a subsequent by operation; see [U] **11.5 by varlist: construct** and examples below.

mfirst specifies that missing values be placed first in descending orderings rather than last.

Remarks

gsort is almost a plug-compatible replacement for sort, except that you cannot specify a general *varlist* with gsort. For instance, sort alpha-gamma means to sort the data in ascending order of alpha, within equal values of alpha; sort on the next variable in the dataset (presumably beta), within equal values of alpha and beta; etc. gsort alpha-gamma would be interpreted as gsort alpha -gamma, meaning to sort the data in ascending order of alpha and, within equal values of alpha, in descending order of gamma.

▷ Example 1

The difference in *varlist* interpretation aside, gsort can be used in place of sort. To list the 10 lowest-priced cars in the data, we might type

```
. use http://www.stata-press.com/data/r12/auto
. gsort price
. list make price in 1/10
```

283

or, if we prefer,

```
. gsort +price
. list make price in 1/10
```

To list the 10 highest-priced cars in the data, we could type

```
. gsort -price
. list make price in 1/10
```

gsort can also be used with string variables. To list all the makes in reverse alphabetical order, we might type

```
. gsort -make
. list make
```

◁

> Example 2

gsort can be used with multiple variables. Given a dataset on hospital patients with multiple observations per patient, typing

```
. use http://www.stata-press.com/data/r12/bp3
. gsort id time
. list id time bp
```

lists each patient's blood pressures in the order the measurements were taken. If we typed

```
. gsort id -time
. list id time bp
```

then each patient's blood pressures would be listed in reverse time order.

◁

❏ Technical note

Say that we wished to attach to each patient's records the lowest and highest blood pressures observed during the hospital stay. The easier way to achieve this result is with egen's min() and max() functions:

```
. egen lo_bp = min(bp), by(id)
. egen hi_bp = max(bp), by(id)
```

See [D] **egen**. Here is how we could do it with gsort:

```
. use http://www.stata-press.com/data/r12/bp3, clear
. gsort id bp
. by id: gen lo_bp = bp[1]
. gsort id -bp
. by id: gen hi_bp = bp[1]
. list, sepby(id)
```

This works, even in the presence of missing values of bp, because such missing values are placed last within arrangements, regardless of the direction of the sort.

❏

❏ Technical note

Assume that we have a dataset containing x for which we wish to obtain the forward and reverse cumulatives. The forward cumulative is defined as $F(X) =$ the fraction of observations such that $x \leq X$. Again let's ignore the easier way to obtain the forward cumulative, which would be to use Stata's cumul command,

```
. set obs 100
. generate x = rnormal()
. cumul x, gen(cum)
```

(see [R] **cumul**). Eschewing cumul, we could type

```
. sort x
. by x: gen cum = _N if _n==1
. replace cum = sum(cum)
. replace cum = cum/cum[_N]
```

That is, we first place the data in ascending order of x; we used sort but could have used gsort. Next, for each observed value of x, we generated cum containing the number of observations that take on that value (you can think of this as the discrete density). We summed the density, obtaining the distribution, and finally normalized it to sum to 1.

The reverse cumulative $G(X)$ is defined as the fraction of data such that $x \geq X$. To obtain this, we could try simply reversing the sort:

```
. gsort -x
. by x: gen rcum = _N if _n==1
. replace rcum = sum(rcum)
. replace rcum = rcum/rcum[_N]
```

This would work, except for one detail: Stata will complain that the data are not sorted in the second line. Stata complains because it does not understand descending sorts (gsort is an ado-file). To remedy this problem, gsort's generate() option will create a new grouping variable that is in ascending order (thus satisfying Stata's narrow definition) and that is, in terms of the groups it defines, identical to that of the true sort variables:

```
. gsort -x, gen(revx)
. by revx: gen rcum = _N if _n==1
. replace rcum = sum(rcum)
. replace rcum = rcum/rcum[_N]
```

❏

Methods and formulas

gsort is implemented as an ado-file.

Also see

[D] **sort** — Sort data

Title

hexdump — Display hexadecimal report on file

Syntax

hexdump *filename* [, *options*]

options	Description
<u>an</u>alyze	display a report on the dump rather than the dump itself
<u>tab</u>ulate	display a full tabulation of the ASCII characters in the analyze report
<u>noe</u>xtended	do not display printable extended ASCII characters
<u>res</u>ults	save results containing the frequency with which each character code was observed; programmer's option
<u>f</u>rom(#)	dump or analyze first byte of the file; default is to start at first byte, from(0)
<u>t</u>o(#)	dump or analyze last byte of the file; default is to continue to the end of the file

Description

hexdump displays a hexadecimal dump of a file or, optionally, a report analyzing the dump.

Options

analyze specifies that a report on the dump, rather than the dump itself, be presented.

tabulate specifies in the analyze report that a full tabulation of the ASCII characters also be presented.

noextended specifies that hexdump not display printable extended ASCII characters, characters in the range 161–254 or, equivalently, 0xa1–0xfe. (hexdump does not display characters 128–160 and 255.)

results is for programmers. It specifies that, in addition to other saved results, hexdump save r(c0), r(c1), ..., r(c255), containing the frequency with which each character code was observed.

from(#) specifies the first byte of the file to be dumped or analyzed. The default is to start at the first byte of the file, from(0).

to(#) specifies the last byte of the file to be dumped or analyzed. The default is to continue to the end of the file.

Remarks

hexdump is useful when you are having difficulty reading a file with infile, infix, or insheet. Sometimes, the reason for the difficulty is that the file does not contain what you think it contains, or that it does contain the format you have been told, and looking at the file in text mode is either not possible or not revealing enough.

Pretend that we have the file `myfile.raw` containing

```
Datsun 210      4589   35   5   1
VW Scirocco     6850   25   4   1
Merc. Bobcat    3829   22   4   0
Buick Regal     5189   20   3   0
VW Diesel       5397   41   5   1
Pont. Phoenix   4424   19   .   0
Merc. Zephyr    3291   20   3   0
Olds Starfire   4195   24   1   0
BMW 320i        9735   25   4   1
```

We will use `myfile.raw` with `hexdump` to produce output that looks like the following:

```
. hexdump myfile.raw
                                                          character
                          hex representation            representation
      address   0 1  2 3  4 5  6 7  8 9  a b  c d  e f   0123456789abcdef

            0  4461 7473 756e 2032 3130 2020 2020 2034   Datsun 210     4
           10  3538 3920 2033 3520 2035 2020 310a 5657   589  35  5  1.VW
           20  2053 6369 726f 6363 6f20 2020 2036 3835    Scirocco    685
           30  3020 2032 3520 2034 2020 310a 4d65 7263   0  25  4  1.Merc

           40  2e20 426f 6263 6174 2020 2033 3832 3920   . Bobcat    3829
           50  2032 3220 2034 2020 300a 4275 6963 6b20   22  4  0.Buick
           60  5265 6761 6c20 2020 2035 3138 3920 2032   Regal    5189  2
           70  3020 2033 2020 300a 5657 2044 6965 7365   0  3  0.VW Diese

           80  6c20 2020 2020 2035 3339 3720 2034 3120   l      5397  41
           90  2035 2020 310a 506f 6e74 2e20 5068 6f65    5  1.Pont. Phoe
           a0  6e69 7820 2034 3432 3420 2031 3920 202e   nix  4424  19   .
           b0  2020 300a 4d65 7263 2e20 5a65 7068 7972     0.Merc. Zephyr

           c0  2020 2033 3239 3120 2032 3020 2033 2020     3291  20  3
           d0  300a 4f6c 6473 2053 7461 7266 6972 6520   0.Olds Starfire
           e0  2034 3139 3520 2032 3420 2031 2020 300a    4195  24  1  0.
           f0  424d 5720 3332 3069 2020 2020 2020 2039   BMW 320i       9
          100  3733 3520 2032 3520 2034 2020 310a        735  25  4  1.
```

hexdump can also produce output that looks like the following:

```
.  hexdump myfile.raw, analyze
   Line-end characters                    Line length (tab=1)
      \r\n         (Windows)       0         minimum                    29
      \r by itself (Mac)           0         maximum                    29
      \n by itself (Unix)          9
   Space/separator characters             Number of lines               9
      [blank]                     99         EOL at EOF?                yes
      [tab]                        0
      [comma] (,)                  0      Length of first 5 lines
   Control characters                        Line 1                     29
      binary 0                     0         Line 2                     29
      CTL excl. \r, \n, \t         0         Line 3                     29
      DEL                          0         Line 4                     29
      Extended (128-159,255)       0         Line 5                     29
   ASCII printable
      A-Z                         20
      a-z                         61      File format              ASCII
      0-9                         77
      Special (!@#$ etc.)          4
      Extended (160-254)           0
                             _____
   Total                         270

   Observed were:
      \n blank . 0 1 2 3 4 5 6 7 8 9 B D M O P R S V W Z a b c d e f g h i k l
      n o p r s t u x y
```

Of the two forms of output, the second is often the more useful because it summarizes the file, and the length of the summary is not a function of the length of the file. Here is the summary for a file that is just over 4 MB long:

```
.  hexdump bigfile.raw, analyze
   Line-end characters                    Line length (tab=1)
      \r\n         (Windows)  147,456        minimum                    29
      \r by itself (Mac)           0         maximum                    30
      \n by itself (Unix)          2
   Space/separator characters             Number of lines         147,458
      [blank]              1,622,039         EOL at EOF?                yes
      [tab]                        0
      [comma] (,)                  0      Length of first 5 lines
   Control characters                        Line 1                     30
      binary 0                     0         Line 2                     30
      CTL excl. \r, \n, \t         0         Line 3                     30
      DEL                          0         Line 4                     30
      Extended (128-159,255)       0         Line 5                     30
   ASCII printable
      A-Z                    327,684
      a-z                    999,436      File format              ASCII
      0-9                  1,261,587
      Special (!@#$ etc.)     65,536
      Extended (160-254)           0
                          _____
   Total                  4,571,196

   Observed were:
      \n \r blank . 0 1 2 3 4 5 6 7 8 9 B D M O P R S V W Z a b c d e f g h i
      k l n o p r s t u x y
```

Here is the same file but with a subtle problem:

```
. hexdump badfile.raw, analyze
  Line-end characters                      Line length (tab=1)
    \r\n        (Windows)     147,456        minimum                  30
    \r by itself (Mac)              0        maximum                  90
    \n by itself (Unix)             0
  Space/separator characters              Number of lines         147,456
    [blank]                 1,622,016        EOL at EOF?                 yes
    [tab]                           0
    [comma] (,)                     0      Length of first 5 lines
  Control characters                        Line 1                   30
    binary 0                        8        Line 2                   30
    CTL excl. \r, \n, \t            4        Line 3                   30
    DEL                             0        Line 4                   30
    Extended (128-159,255)         24        Line 5                   30
  ASCII printable
    A-Z                       327,683
    a-z                       999,426      File format              BINARY
    0-9                     1,261,568
    Special (!@#$ etc.)        65,539
    Extended (160-254)             16
                          _____
  Total                     4,571,196

  Observed were:
    \0 ^C ^D ^G \n \r ^U blank & . 0 1 2 3 4 5 6 7 8 9 B D E M O P R S U V W
    Z a b c d e f g h i k l n o p r s t u v x y } ~ E^A E^C E^I E^M E^P
    ë é ö 255
```

In the above, the line length varies between 30 and 90 (we were told that each line would be 30 characters long). Also the file contains what hexdump, analyze labeled control characters. Finally, hexdump, analyze declared the file to be BINARY rather than ASCII.

We created the second file by removing two valid lines from bigfile.raw (60 characters) and substituting 60 characters of binary junk. We would defy you to find the problem without using hexdump, analyze. You would succeed, but only after much work. Remember, this file has 147,456 lines, and only two of them are bad. If you print 1,000 lines at random from the file, your chances of listing the bad part are only 0.013472. To have a 50% chance of finding the bad lines, you would have to list 52,000 lines, which is to say, review about 945 pages of output. On those 945 pages, each line would need to be drawn at random. More likely, you would list lines in groups, and that would greatly reduce your chances of encountering the bad lines.

The situation is not as dire as we make it out to be because, were you to read badfile.raw by using infile, it would complain, and here it would tell you exactly where it was complaining. Still, at that point you might wonder whether the problem was with how you were using infile or with the data. Moreover, our 60 bytes of binary junk experiment corresponds to transmission error. If the problem were instead that the person who constructed the file constructed two of the lines differently, infile might not complain, but later you would notice some odd values in your data (because obviously you would review the summary statistics, right?). Here hexdump, analyze might be the only way you could prove to yourself and others that the raw data need to be reconstructed.

❏ Technical note

In the full hexadecimal dump,

```
. hexdump myfile.raw
```

address			hex representation						character representation 0123456789abcdef
	0 1	2 3	4 5	6 7	8 9	a b	c d	e f	
0	4461	7473	756e	2032	3130	2020	2020	2034	Datsun 210 4
10	3538	3920	2033	3520	2035	2020	310d	0a56	589 35 5 1..V
20	5720	5363	6972	6f63	636f	2020	2020	3638	W Scirocco 68
30	3530	2020	3235	2020	3420	2031	0d0a	4d65	50 25 4 1..Me

(output omitted)

addresses (listed on the left) are listed in hexadecimal. Above, 10 means decimal 16, 20 means decimal 32, and so on. Sixteen characters are listed across each line.

In some other dump, you might see something like

```
. hexdump myfile2.raw
```

address			hex representation						character representation 0123456789abcdef
	0 1	2 3	4 5	6 7	8 9	a b	c d	e f	
0	4461	7473	756e	2032	3130	2020	2020	2034	Datsun 210 4
10	3538	3920	2033	3520	2035	2020	3120	2020	589 35 5 1
20	2020	2020	2020	2020	2020	2020	2020	2020	
*									
160	2020	2020	2020	0a56	5720	5363	6972	6f63	.VW Sciroc
170	636f	2020	2020	3638	3530	2020	3235	2020	co 6850 25

(output omitted)

The ∗ in the address field indicates that the previous line is repeated until we get to hexadecimal address 160 (decimal 352).

❏

Saved results

hexdump, analyze and hexdump, results save the following in r():

Scalars

r(Windows)	number of \r\n
r(Mac)	number of \r by itself
r(Unix)	number of \n by itself
r(blank)	number of blanks
r(tab)	number of tab characters
r(comma)	number of comma (,) characters
r(ctl)	number of binary 0s; A–Z, excluding \r, \n, \t; DELs; and 128–159, 255
r(uc)	number of A–Z
r(lc)	number of a–z
r(digit)	number of 0–9
r(special)	number of printable special characters (!@#, etc.)
r(extended)	number of printable extended characters (160–254)
r(filesize)	number of characters
r(lmin)	minimum line length
r(lmax)	maximum line length
r(lnum)	number of lines
r(eoleof)	1 if EOL at EOF, 0 otherwise
r(l1)	length of 1st line
r(l2)	length of 2nd line
r(l3)	length of 3rd line
r(l4)	length of 4th line
r(l5)	length of 5th line
r(c0)	number of binary 0s (results only)
r(c1)	number of binary 1s (^A) (results only)
r(c2)	number of binary 2s (^B) (results only)
. . .	. . .
r(c255)	number of binary 255s (results only)

Macros

r(format)	ASCII, EXTENDED ASCII, or BINARY

Also see

[D] **filefilter** — Convert text or binary patterns in a file

[D] **type** — Display contents of a file

Title

icd9 — ICD-9-CM diagnostic and procedure codes

Syntax

Verify that variable contains defined codes

{icd9 | icd9p} check *varname* [, any list generate(*newvar*)]

Verify and clean variable

{icd9 | icd9p} clean *varname* [, dots pad]

Generate new variable from existing variable

{icd9 | icd9p} generate *newvar* = *varname* , main

{icd9 | icd9p} generate *newvar* = *varname* , description [long end]

{icd9 | icd9p} generate *newvar* = *varname* , range(*icd9rangelist*)

Display code descriptions

{icd9 | icd9p} lookup *icd9rangelist*

Search for codes from descriptions

{icd9 | icd9p} search ["]*text*["] [["]*text*["] ...] [, or]

Display ICD-9 code source

{icd9 | icd9p} query

where *icd9rangelist* is

icd9code	(the particular code)
icd9code∗	(all codes starting with)
icd9code/*icd9code*	(the code range)

or any combination of the above, such as 001∗ 018/019 E∗ 018.02. *icd9codes* must be typed with leading zeros: 1 is an error; type 001 (diagnostic code) or 01 (procedure code).

icd9 is for use with ICD-9 *diagnostic* codes, and icd9p is for use with *procedure* codes. The two commands' syntaxes parallel each other.

Menu

{icd9 | icd9p} check

Data > Other utilities > ICD9 utilities > Verify variable is valid

{icd9 | icd9p} clean

Data > Other utilities > ICD9 utilities > Clean and verify variable

{icd9 | icd9p} generate

Data > Other utilities > ICD9 utilities > Generate new variable from existing

{icd9 | icd9p} lookup

Data > Other utilities > ICD9 utilities > Display code descriptions

{icd9 | icd9p} search

Data > Other utilities > ICD9 utilities > Search for codes from descriptions

{icd9 | icd9p} query

Data > Other utilities > ICD9 utilities > Display ICD-9 code source

Description

icd9 and icd9p help when working with ICD-9-CM codes.

ICD-9 codes come in two forms: diagnostic codes and procedure codes. In this system, 001 (cholera) and 941.45 (deep 3rd deg burn nose) are examples of diagnostic codes, although some people write (and datasets record) 94145 rather than 941.45. Also, 01 (incise-excis brain/skull) and 55.01 (nephrotomy) are examples of procedure codes, although some people write 5501 rather than 55.01. icd9 and icd9p understand both ways of recording codes.

Important note: What constitutes a valid ICD-9 code changes over time. For the rest of this entry, a *defined code* is any code that is either currently valid, was valid at some point since version V16 (effective October 1, 1998), or has meaning as a grouping of codes. Some examples would help. The diagnosis code 001, though not valid on its own, is useful because it denotes cholera. It is kept as a defined code whose description ends with an asterisk (*). The diagnosis code 645.01 was deleted between versions V16 and V18. It remains as a defined code, and its description ends with a hash mark (#).

icd9 and icd9p parallel each other; icd9 is for use with diagnostic codes, and icd9p is for use with procedure codes.

icd9[p] check verifies that existing variable *varname* contains defined ICD-9 codes. If not, icd9[p] check provides a full report on the problems. icd9[p] check is useful for tracking down problems when any of the other icd9[p] commands tell you that the "variable does not contain ICD-9 codes". icd9[p] check verifies that each recorded code actually exists in the defined code list.

icd9[p] clean also verifies that existing variable *varname* contains valid ICD-9 codes, and, if it does, icd9[p] clean modifies the variable to contain the codes in either of two standard formats. All icd9[p] commands work equally well with cleaned or uncleaned codes. There are many ways of writing the same ICD-9 code, and icd9[p] clean is designed to ensure consistency and to make subsequent output look better.

icd9[p] generate produces new variables based on existing variables containing (cleaned or uncleaned) ICD-9 codes. icd9[p] generate, main produces *newvar* containing the main code. icd9[p] generate, description produces *newvar* containing a textual description of the ICD-9 code. icd9[p] generate, range() produces numeric *newvar* containing 1 if *varname* records an ICD-9 code in the range listed and 0 otherwise.

icd9[p] lookup and icd9[p] search are utility routines that are useful interactively. icd9[p] lookup simply displays descriptions of the codes specified on the command line, so to find out what diagnostic E913.1 means, you can type icd9 lookup e913.1. The data that you have in memory are irrelevant—and remain unchanged—when you use icd9[p] lookup. icd9[p] search is similar to icd9[p] lookup, except that it turns the problem around; icd9[p] search looks for relevant ICD-9 codes from the description given on the command line. For instance, you could type icd9 search liver or icd9p search liver to obtain a list of codes containing the word "liver".

icd9[p] query displays the identity of the source from which the ICD-9 codes were obtained and the textual description that icd9[p] uses.

ICD-9 codes are commonly written in two ways: with and without periods. For instance, with diagnostic codes, you can write 001, 86221, E8008, and V822, or you can write 001., 862.21, E800.8, and V82.2. With procedure codes, you can write 01, 50, 502, and 5021, or 01., 50., 50.2, and 50.21. The icd9[p] command does not care which syntax you use or even whether you are consistent. Case also is irrelevant: v822, v82.2, V822, and V82.2 are all equivalent. Codes may be recorded with or without leading and trailing blanks.

icd9[p] works with V28, V27, V26, V25, V24, V22, V21, V19, V18, and V16 codes.

Options for icd9[p] check

any tells icd9[p] check to verify that the codes fit the format of ICD-9 codes but not to check whether the codes are actually defined. This makes icd9[p] check run faster. For instance, diagnostic code 230.52 (or 23052, if you prefer) looks valid, but there is no such ICD-9 code. Without the any option, 230.52 would be flagged as an error. With any, 230.52 is not an error.

list reports any invalid codes that were found in the data by icd9[p] check. For example, 1, 1.1.1, and perhaps 230.52, if any is not specified, are to be individually listed.

generate(*newvar*) specifies that icd9[p] check create new variable *newvar* containing, for each observation, 0 if the code is defined and a number from 1 to 10 otherwise. The positive numbers indicate the kind of problem and correspond to the listing produced by icd9[p] check. For instance, 10 means that the code could be valid, but it turns out not to be on the list of defined codes.

Options for icd9[p] clean

dots specifies whether periods are to be included in the final format. Do you want the diagnostic codes recorded, for instance, as 86221 or 862.21? Without the dots option, the 86221 format would be used. With the dots option, the 862.21 format would be used.

pad specifies that the codes are to be padded with spaces, front and back, to make the codes line up vertically in listings. Specifying pad makes the resulting codes look better when used with most other Stata commands.

Options for icd9[p] generate

`main`, `description`, and `range`(*icd9rangelist*) specify what `icd9[p] generate` is to calculate. *varname* always specifies a variable containing ICD-9 codes.

`main` specifies that the main code be extracted from the ICD-9 code. For procedure codes, the main code is the first two characters. For diagnostic codes, the main code is usually the first three or four characters (the characters before the dot if the code has dots). In any case, `icd9[p] generate` does not care whether the code is padded with blanks in front or how strangely it might be written; `icd9[p] generate` will find the main code and extract it. The resulting variable is itself an ICD-9 code and may be used with the other `icd9[p]` subcommands. This includes `icd9[p] generate, main`.

`description` creates *newvar* containing descriptions of the ICD-9 codes.

`long` is for use with `description`. It specifies that the new variable, in addition to containing the text describing the code, contain the code, too. Without `long`, *newvar* in an observation might contain "bronchus injury-closed". With `long`, it would contain "862.21 bronchus injury-closed".

`end` modifies `long` (specifying `end` implies `long`) and places the code at the end of the string: "bronchus injury-closed 862.21".

`range`(*icd9rangelist*) allows you to create indicator variables equal to 1 when the ICD-9 code is in the inclusive range specified.

Option for icd9[p] search

`or` specifies that ICD-9 codes be searched for entries that contain any word specified after `icd9[p] search`. The default is to list only entries that contain all the words specified.

Remarks

Let's begin with the diagnostic codes that `icd9` processes. The format of an ICD-9 diagnostic code is

$$\left[\,\texttt{blanks}\,\right]\left\{\texttt{0-9,V,v}\right\}\left\{\texttt{0-9}\right\}\left\{\texttt{0-9}\right\}\left[\,.\,\right]\left[\,\texttt{0--9}\left[\,\texttt{0--9}\,\right]\,\right]\left[\,\texttt{blanks}\,\right]$$

or

$$\left[\,\texttt{blanks}\,\right]\left\{\texttt{E,e}\right\}\left\{\texttt{0-9}\right\}\left\{\texttt{0-9}\right\}\left\{\texttt{0-9}\right\}\left[\,.\,\right]\left[\,\texttt{0--9}\,\right]\left[\,\texttt{blanks}\,\right]$$

icd9 can deal with ICD-9 diagnostic codes written in any of the ways that this format allows. Items in square brackets are optional. The code might start with some number of blanks. Braces, { }, indicate required items. The code then has a digit from 0 to 9, the letter V (uppercase or lowercase, first line), or the letter E (uppercase or lowercase, second line). After that, it has two or more digits, perhaps followed by a period, and then it may have up to two more digits (perhaps followed by more blanks).

All the following codes meet the above definition:

```
001
001.
    001
001.9
        0019
86222
862.22
E800.2
e8002
V82
v82.2
V822
```

Meeting the above definition does not make the code valid. There are 133,100 possible codes meeting the above definition, of which fewer than 20,000 are currently defined.

Examples of currently defined diagnostic codes include

Code	Description
001	cholera*
001.0	cholera d/t vib cholerae
001.1	cholera d/t vib el tor
001.9	cholera nos
. . .	
999	complic medical care nec*
. . .	
V01	communicable dis contact*
V01.0	cholera contact
V01.1	tuberculosis contact
V01.2	poliomyelitis contact
V01.3	smallpox contact
V01.4	rubella contact
V01.5	rabies contact
V01.6	venereal dis contact
V01.7	viral dis contact nec#
V01.71	varicella contact/exp
V01.79	viral dis contact nec
V01.8	communic dis contact nec#
V01.81	contact/exposure-anthrax
V01.82	exposure to sars
V01.83	e. coli contact/exp
V01.84	meningococcus contact
V01.89	communic dis contact nec
V01.9	communic dis contact nos
. . .	
E800	rr collision nos*
E800.0	rr collision nos-employ
E800.1	rr coll nos-passenger
E800.2	rr coll nos-pedestrian
E800.3	rr coll nos-ped cyclist
E800.8	rr coll nos-person nec
E800.9	rr coll nos-person nos
. . .	

The *main code* refers to the part of the code to the left of the period. 001, 002, . . . , 999; V01, . . . , V82; and E800, . . . , E999 are main codes.

The main code corresponding to a detailed code can be obtained by taking the part of the code to the left of the period, except for codes beginning with 176, 764, 765, V29, and V69. Those main codes are not defined, yet there are more detailed codes under them:

Code	Description
176	CODE DOES NOT EXIST:
176.0	skin - kaposi's sarcoma
176.1	sft tisue - kpsi's srcma
...	
764	CODE DOES NOT EXIST:
764.0	lt-for-dates w/o fet mal*
764.00	light-for-dates wtnos
...	
765	CODE DOES NOT EXIST:
765.0	extreme immaturity*
765.00	extreme immatur wtnos
...	
V29	CODE DOES NOT EXIST:
V29.0	nb obsrv suspct infect
V29.1	nb obsrv suspct neurlgcl
...	
V69	CODE DOES NOT EXIST:
V69.0	lack of physical exercise
V69.1	inapprt diet eat habits
...	

Our solution is to define five new codes:

Code	Description
176	kaposi's sarcoma (Stata)*
764	light-for-dates (Stata)*
765	immat & preterm (Stata)*
V29	nb suspct cnd (Stata)*
V69	lifestyle (Stata)*

Things are less confusing with respect to the procedure codes processed by icd9p. The format of ICD-9 procedure codes is

$$\big[\texttt{blanks}\big]\big\{0\text{--}9\big\}\big\{0\text{--}9\big\}\big[\,.\,\big]\big[0\text{--}9\big[0\text{--}9\big]\big]\big[\texttt{blanks}\big]$$

Thus there are 10,000 possible procedure codes, of which fewer than 5,000 are currently valid. The first two digits represent the main code, of which 100 are feasible and 98 are currently used (00 and 17 are not used).

Descriptions

The description given for each of the codes is as found in the original source. The procedure codes contain the addition of five new codes created by Stata. An asterisk on the end of a description indicates that the corresponding ICD-9 diagnostic code has subcategories. A hash mark (#) at the end of a description denotes a code that is not valid in the most current version but that was valid at some time between version V16 and the present version.

icd9[p] query reports the original source of the information on the codes:

```
. icd9 query
_dta:
   1.  ICD9 Diagnostic Code Mapping Data for use with Stata, History
   2.  ─────── V16 ──────────────────────────────────────────────────
   3.  Dataset obtained 24aug1999 from http://www.hcfa.gov/stats/pufiles.htm,
       file http://www.hcfa.gov/stats/icd9v16.exe
   4.  Codes 176, 764, 765, V29, and V69 defined by StataCorp: 176 [kaposi's
       sarcoma (Stata)*], 765 [immat & preterm (Stata)*], 764 [light-for-dates
       (Stata)*], V29 [nb suspct cnd (Stata)*], V69 [lifestyle (Stata)*]
   5.  ─────── V18 ──────────────────────────────────────────────────
 (output omitted )
  12.  ─────── V19 ──────────────────────────────────────────────────
  13.  Dataset obtained 3jan2002 from http://www.hcfa.gov/stats/pufiles.htm,
       file http://www.hcfa.gov/stats/icd9v19.zip, file 9v19diag.txt
  14.  27feb2002: V19 put into Stata distribution
 (output omitted )
. icd9p query
_dta:
   1.  ICD9 Procedure Code Mapping Data for use with Stata, History
   2.  ─────── V16 ──────────────────────────────────────────────────
   3.  Dataset obtained 24aug1999 from http://www.hcfa.gov/stats/pufiles.htm,
       file http://www.hcfa.gov/stats/icd9v16.exe
   4.  ─────── V18 ──────────────────────────────────────────────────
   5.  Dataset obtained 10may2001 from http://www.hcfa.gov/stats/pufiles.htm,
       file http://www.hcfa.gov/stats/icd9v18.zip, file V18SURG.TXT
   6.  11jun2001: V18 data put into Stata distribution
   7.  BETWEEN V16 and V18: 9 codes added: 3971 3979 4107 4108 4109 4697 6096
       6097 9975
 (output omitted )
```

▷ Example 1

We have a dataset containing up to three diagnostic codes and up to two procedures on a sample of 1,000 patients:

```
. use http://www.stata-press.com/data/r12/patients
. list in 1/10
```

	patid	diag1	diag2	diag3	proc1	proc2
1.	1	65450			9383	
2.	2	23v.6	37456		8383	17
3.	3	V10.02				
4.	4	102.6			629	
5.	5	861.01				
6.	6	38601	2969		9337	
7.	7	705			7309	8385
8.	8	v53.32			7878	951
9.	9	20200	7548	E8247	0479	
10.	10	464.11	20197		4641	

Do not try to make sense of these data because, in constructing this example, the diagnostic and procedure codes were randomly selected.

First, variable `diag1` is recorded sloppily—sometimes the dot notation is used and sometimes not, and sometimes there are leading blanks. That does not matter. We decide to begin by using `icd9 clean` to clean up this variable:

```
. icd9 clean diag1
diag1 contains invalid ICD-9 codes
r(459);
```

`icd9 clean` refused because there are invalid codes among the 1,000 observations. We can use `icd9 check` to find and flag the problem observations (or observation, as here):

```
. icd9 check diag1, gen(prob)
diag1 contains invalid codes:
       1.  Invalid placement of period                    0
       2.  Too many periods                               0
       3.  Code too short                                 0
       4.  Code too long                                  0
       5.  Invalid 1st char (not 0-9, E, or V)            0
       6.  Invalid 2nd char (not 0-9)                     0
       7.  Invalid 3rd char (not 0-9)                     1
       8.  Invalid 4th char (not 0-9)                     0
       9.  Invalid 5th char (not 0-9)                     0
      10.  Code not defined                               0
                                                    ──────────
           Total                                          1

. list patid diag1 prob if prob
```

	patid	diag1	prob
2.	2	23v.6	7

Let's assume that we go back to the patient records and determine that this should have been coded 230.6:

```
. replace diag1 = "230.6" if patid==2
(1 real change made)
. drop prob
```

We now try again to clean up the formatting of the variable:

```
. icd9 clean diag1
(643 changes made)
. list in 1/10
```

	patid	diag1	diag2	diag3	proc1	proc2
1.	1	65450			9383	
2.	2	2306	37456		8383	17
3.	3	V1002				
4.	4	1026			629	
5.	5	86101				
6.	6	38601	2969		9337	
7.	7	705			7309	8385
8.	8	V5332			7878	951
9.	9	20200	7548	E8247	0479	
10.	10	46411	20197		4641	

Perhaps we prefer the dot notation. `icd9 clean` can be used again on `diag1`, and now we will clean up `diag2` and `diag3`:

```
. icd9 clean diag1, dots
(936 changes made)
. icd9 clean diag2, dots
(551 changes made)
. icd9 clean diag3, dots
(100 changes made)
. list in 1/10
```

	patid	diag1	diag2	diag3	proc1	proc2
1.	1	654.50			9383	
2.	2	230.6	374.56		8383	17
3.	3	V10.02				
4.	4	102.6			629	
5.	5	861.01				
6.	6	386.01	296.9		9337	
7.	7	705			7309	8385
8.	8	V53.32			7878	951
9.	9	202.00	754.8	E824.7	0479	
10.	10	464.11	201.97		4641	

We now turn to cleaning the procedure codes. We use `icd9p` (emphasis on the *p*) to clean these codes:

```
. icd9p clean proc1, dots
(816 changes made)
. icd9p clean proc2, dots
(140 changes made)
. list in 1/10
```

	patid	diag1	diag2	diag3	proc1	proc2
1.	1	654.50			93.83	
2.	2	230.6	374.56		83.83	17
3.	3	V10.02				
4.	4	102.6			62.9	
5.	5	861.01				
6.	6	386.01	296.9		93.37	
7.	7	705			73.09	83.85
8.	8	V53.32			78.78	95.1
9.	9	202.00	754.8	E824.7	04.79	
10.	10	464.11	201.97		46.41	

Both `icd9 clean` and `icd9p clean` verify only that the variable being cleaned follows the construction rules for the code; it does not check that the code is itself valid. `icd9[p] check` does that:

```
. icd9p check proc1
(proc1 contains valid ICD-9 procedure codes; 168 missing values)

. icd9p check proc2

proc2 contains invalid codes:
      1.   Invalid placement of period               0
      2.   Too many periods                          0
      3.   Code too short                            0
      4.   Code too long                             0
      5.   Invalid 1st char (not 0-9)                0
      6.   Invalid 2nd char (not 0-9)                0
      7.   Invalid 3rd char (not 0-9)                0
      8.   Invalid 4th char (not 0-9)                0
     10.   Code not defined                          1
                                                  ———————
           Total                                     1
```

proc2 has an invalid code. We could find it by using icd9p check, generate(), just as we did above with icd9 check, generate().

icd9[p] can create new variables containing textual descriptions of our diagnostic and procedure codes:

```
. icd9 generate td1 = diag1, description

. sort patid

. list patid diag1 td1 in 1/10
```

	patid	diag1	td1
1.	1	654.50	cerv incompet preg-unsp
2.	2	230.6	ca in situ anus nos
3.	3	V10.02	hx-oral/pharynx malg nec
4.	4	102.6	yaws of bone & joint
5.	5	861.01	heart contusion-closed
6.	6	386.01	meniere dis cochlvestib
7.	7	705	disorders of sweat gland*
8.	8	V53.32	ftng autmtc dfibrillator
9.	9	202.00	ndlr lym unsp xtrndl org
10.	10	464.11	ac tracheitis w obstruct

icd9[p] generate, description does not preserve the sort order of the data (and neither does icd9[p] check, unless you specify the any option).

Procedure code proc2 had an invalid code. Even so, icd9p generate, description is willing to create a textual description variable:

```
. icd9p gen tp2 = proc2, description
(1 nonmissing value invalid and so could not be labeled)
. sort patid
. list patid proc2 tp2 in 1/10
```

	patid	proc2	tp2
1.	1		
2.	2	17	
3.	3		
4.	4		
5.	5		
6.	6		
7.	7	83.85	musc/tend lng change nec
8.	8	95.1	form & structur eye exam*
9.	9		
10.	10		

tp2 contains nothing when proc2 is 17 because 17 is not a valid procedure code.

icd9[p] generate can also create variables containing main codes:

```
. icd9 generate main1 = diag1, main
. list patid diag1 main1 in 1/10
```

	patid	diag1	main1
1.	1	654.50	654
2.	2	230.6	230
3.	3	V10.02	V10
4.	4	102.6	102
5.	5	861.01	861
6.	6	386.01	386
7.	7	705	705
8.	8	V53.32	V53
9.	9	202.00	202
10.	10	464.11	464

icd9p generate, main can similarly generate main procedure codes.

Sometimes we might merely be examining an observation:

```
. list diag* if patid==563
```

	diag1	diag2	diag3
563.	526.4		

If we wondered what 526.4 was, we could type

        . icd9 lookup 526.4
    1 match found:
        526.4   inflammation of jaw

icd9[p] lookup can list ranges of codes:

        . icd9 lookup 526/526.99
    15 matches found:
        526   jaw diseases*
        526.0   devel odontogenic cysts
        526.1   fissural cysts of jaw
        526.2   cysts of jaws nec
        526.3   cent giant cell granulom
        526.4   inflammation of jaw
        526.5   alveolitis of jaw
        526.61   perfor root canal space
        526.62   endodontic overfill
        526.63   endodontic underfill
        526.69   periradicular path nec
        526.8   other jaw diseases*
        526.81   exostosis of jaw
        526.89   jaw disease nec
        526.9   jaw disease nos

The same result could be found by typing

        . icd9 lookup 526*

icd9[p] search can find a code from the description:

        . icd9 search jaw disease
    4 matches found:
        526   jaw diseases*
        526.8   other jaw diseases*
        526.89   jaw disease nec
        526.9   jaw disease nos

◁

Saved results

icd9 check and icd9p check save the following in r():

Scalars
    r(e#)   number of errors of type #
    r(esum)   total number of errors

icd9 clean and icd9p clean save the following in r():

Scalars
    r(N)   number of changes

Methods and formulas

icd9 and icd9p are implemented as ado-files.

Reference

Gould, W. W. 2000. dm76: ICD-9 diagnostic and procedure codes. *Stata Technical Bulletin* 54: 8–16. Reprinted in *Stata Technical Bulletin Reprints*, vol. 9, pp. 77–87. College Station, TX: Stata Press.

Title

import — Overview of importing data into Stata

Description

This entry provides a quick reference for determining which method to use for reading non-Stata data into memory. See [U] **21 Inputting and importing data** for more details.

Remarks

Remarks are presented under the following headings:

Summary of the different methods
   import excel
   insheet
   odbc
   infile (free format)—infile without a dictionary
   infix (fixed format)
   infile (fixed format)—infile with a dictionary
   import sasxport
   haver (Windows only)
   xmluse
Examples

Summary of the different methods

import excel

○ `import excel` reads worksheets from Microsoft Excel (`.xls` and `.xlsx`) files.

○ Entire worksheets can be read, or custom cell ranges can be read.

○ See [D] **import excel**.

insheet

○ `insheet` reads text files created by a spreadsheet or a database program.

○ The data must be tab-separated or comma-separated, but not both simultaneously. A custom delimiter may also be specified.

○ An observation must be on only one line.

○ The first line in the file can optionally contain the names of the variables.

○ See [D] **insheet**.

odbc

○ ODBC, an acronym for Open DataBase Connectivity, is a standard for exchanging data between programs. Stata supports the ODBC standard for importing data via the `odbc` command and can read from any ODBC data source on your computer.

○ See [D] **odbc**.

infile (free format)—infile without a dictionary

○ The data can be space-separated, tab-separated, or comma-separated.

○ Strings with embedded spaces or commas must be enclosed in quotes (even if tab- or comma-separated).

○ An observation can be on more than one line, or there can even be multiple observations per line.

○ See [D] **infile (free format)**.

infix (fixed format)

○ The data must be in fixed-column format.

○ An observation can be on more than one line.

○ `infix` has simpler syntax than `infile` (fixed format).

○ See [D] **infix (fixed format)**.

infile (fixed format)—infile with a dictionary

○ The data may be in fixed-column format.

○ An observation can be on more than one line.

○ ASCII or EBCDIC data can be read.

○ `infile` (fixed format) has the most capabilities for reading data.

○ See [D] **infile (fixed format)**.

import sasxport

○ `import sasxport` reads SAS XPORT Transport format files.

○ `import sasxport` will also read value label information from a `formats.xpf` XPORT file, if available.

○ See [D] **import sasxport**.

haver (Windows only)

○ `haver` reads Haver Analytics (http://www.haver.com/) database files.

○ `haver` is available only for Windows and requires a corresponding DLL (DLXAPI32.DLL) available from Haver Analytics.

○ See [TS] **haver**.

xmluse

○ `xmluse` reads extensible markup language (XML) files—highly adaptable text-format files derived from the standard generalized markup language (SGML).

○ `xmluse` can read either an Excel-format XML or a Stata-format XML file into Stata.

○ See [D] **xmlsave**.

Examples

> **Example 1: Tab-separated data**

```
——————————————————————————————————————————— begin example1.raw ———————————
1       0       1       John Smith      m
0       0       1       Paul Lin        m
0       1       0       Jan Doe f
0       0       .       Julie McDonald  f
——————————————————————————————————————————— end example1.raw ———————————
```

contains tab-separated data. The `type` command with the `showtabs` option shows the tabs:

```
. type example1.raw, showtabs
1<T>0<T>1<T>John Smith<T>m
0<T>0<T>1<T>Paul Lin<T>m
0<T>1<T>0<T>Jan Doe<T>f
0<T>0<T>.<T>Julie McDonald<T>f
```

It could be read in by

```
. insheet a b c name gender using example1
```

◁

> **Example 2: Comma-separated data**

```
——————————————————————————————————————————— begin example2.raw ———————————
a,b,c,name,gender
1,0,1,John Smith,m
0,0,1,Paul Lin,m
0,1,0,Jan Doe,f
0,0,,Julie McDonald,f
——————————————————————————————————————————— end example2.raw ———————————
```

could be read in by

```
. insheet using example2
```

◁

> **Example 3: Tab-separated data with double-quoted strings**

```
——————————————————————————————————————————— begin example3.raw ———————————
1       0       1       "John Smith"    m
0       0       1       "Paul Lin"      m
0       1       0       "Jan Doe"       f
0       0       .       "Julie McDonald"        f
——————————————————————————————————————————— end example3.raw ———————————
```

contains tab-separated data with strings in double quotes.

```
. type example3.raw, showtabs
1<T>0<T>1<T>"John Smith"<T>m
0<T>0<T>1<T>"Paul Lin"<T>m
0<T>1<T>0<T>"Jan Doe"<T>f
0<T>0<T>.<T>"Julie McDonald"<T>f
```

It could be read in by

```
. infile byte (a b c) str15 name str1 gender using example3
```

or

>   . insheet a b c name gender using example3

or

>   . infile using dict3

where the dictionary `dict3.dct` contains

———————————————————————————————— begin dict3.dct ————————

```
infile dictionary using example3 {
        byte    a
        byte    b
        byte    c
        str15   name
        str1    gender
}
```

———————————————————————————————— end dict3.dct ————————

◁

▷ Example 4: Space-separated data with double-quoted strings

———————————————————————————————— begin example4.raw ————————

```
1 0 1 "John Smith" m
0 0 1 "Paul Lin" m
0 1 0 "Jan Doe" f
0 0 . "Julie McDonald" f
```

———————————————————————————————— end example4.raw ————————

could be read in by

>   . infile byte (a b c) str15 name str1 gender using example4

or

>   . infile using dict4

where the dictionary `dict4.dct` contains

———————————————————————————————— begin dict4.dct ————————

```
infile dictionary using example4 {
        byte    a
        byte    b
        byte    c
        str15   name
        str1    gender
}
```

———————————————————————————————— end dict4.dct ————————

◁

▷ Example 5: Fixed-column format

———————————————————————————————— begin example5.raw ————————

```
101mJohn Smith
001mPaul Lin
010fJan Doe
00 fJulie McDonald
```

———————————————————————————————— end example5.raw ————————

could be read in by

```
. infix a 1 b 2 c 3 str gender 4 str name 5-19 using example5
```

or

```
. infix using dict5a
```

where dict5a.dct contains

```
                                                        ─ begin dict5a.dct ─
infix dictionary using example5 {
                a       1
                b       2
                c       3
        str     gender  4
        str     name    5-19
}
                                                        ─ end dict5a.dct ─
```

or

```
. infile using dict5b
```

where dict5b.dct contains

```
                                                        ─ begin dict5b.dct ─
infile dictionary using example5 {
        byte    a       %1f
        byte    b       %1f
        byte    c       %1f
        str1    gender  %1s
        str15   name    %15s
}
                                                        ─ end dict5b.dct ─
```

◁

▷ Example 6: Fixed-column format with headings

```
                                                        ─ begin example6.raw ─
line 1 : a heading
There are a total of 4 lines of heading.
The next line contains a useful heading:
----+----1----+----2----+----3----+----4----+-
1       0       1       m       John Smith
0       0       1       m       Paul Lin
0       1       0       f       Jan Doe
0       0               f       Julie McDonald
                                                        ─ end example6.raw ─
```

could be read in by

```
. infile using dict6a
```

where `dict6a.dct` contains

```
──────────────────────────────────────────── begin dict6a.dct ───────
    infile dictionary using example6 {
    _firstline(5)
                byte    a
                byte    b
    _column(17) byte    c          %1f
                str1    gender
    _column(33) str15   name       %15s
    }
────────────────────────────────────────────── end dict6a.dct ───────
```

or could be read in by

```
. infix 5 first a 1 b 9 c 17 str gender 25 str name 33-46 using example6
```

or could be read in by

```
. infix using dict6b
```

where `dict6b.dct` contains

```
──────────────────────────────────────────── begin dict6b.dct ───────
    infix dictionary using example6 {
    5 first
                a          1
                b          9
                c          17
          str   gender     25
          str   name       33-46
    }
────────────────────────────────────────────── end dict6b.dct ───────
```

◁

▷ Example 7: Fixed-column format with observations spanning multiple lines

```
──────────────────────────────────────────── begin example7.raw ───────
    a b c gender name
    1 0 1
    m
    John Smith
    0 0 1
    m
    Paul Lin
    0 1 0
    f
    Jan Doe
    0 0
    f
    Julie McDonald
────────────────────────────────────────────── end example7.raw ───────
```

could be read in by

```
. infile using dict7a
```

where `dict7a.dct` contains

```
––––––––––––––––––––––––––––––––––––––––––––––––––––––– begin dict7a.dct –––––––––––
    infile dictionary using example7 {
    _firstline(2)
            byte    a
            byte    b
            byte    c
    _line(2)
            str1    gender
    _line(3)
            str15   name    %15s
    }
––––––––––––––––––––––––––––––––––––––––––––––––––––––– end dict7a.dct –––––––––––
```

or, if we wanted to include variable labels,

    . infile using dict7b

where `dict7b.dct` contains

```
––––––––––––––––––––––––––––––––––––––––––––––––––––––– begin dict7b.dct –––––––––––
    infile dictionary using example7 {
    _firstline(2)
            byte    a           "Question 1"
            byte    b           "Question 2"
            byte    c           "Question 3"
    _line(2)
            str1    gender      "Gender of subject"
    _line(3)
            str15   name    %15s
    }
––––––––––––––––––––––––––––––––––––––––––––––––––––––– end dict7b.dct –––––––––––
```

`infix` could also read these data,

    . infix 2 first 3 lines a 1 b 3 c 5 str gender 2:1 str name 3:1-15 using example7

or the data could be read in by

    . infix using dict7c

where `dict7c.dct` contains

```
––––––––––––––––––––––––––––––––––––––––––––––––––––––– begin dict7c.dct –––––––––––
    infix dictionary using example7 {
    2 first
                a       1
                b       3
                c       5
            str gender  2:1
            str name    3:1-15
    }
––––––––––––––––––––––––––––––––––––––––––––––––––––––– end dict7c.dct –––––––––––
```

or the data could be read in by

    . infix using dict7d

where `dict7d.dct` contains

```
                                                    begin dict7d.dct
infix dictionary using example7 {
2 first
                a     1
                b     3
                c     5
/
        str     gender 1
/
        str     name   1-15
}
                                                    end dict7d.dct
```

◁

Also see

[D] **edit** — Browse or edit data with Data Editor

[D] **export** — Overview of exporting data from Stata

[D] **infile (fixed format)** — Read text data in fixed format with a dictionary

[D] **infile (free format)** — Read unformatted text data

[D] **infix (fixed format)** — Read text data in fixed format

[D] **input** — Enter data from keyboard

[D] **insheet** — Read text data created by a spreadsheet

[D] **import excel** — Import and export Excel files

[D] **import sasxport** — Import and export datasets in SAS XPORT format

[D] **odbc** — Load, write, or view data from ODBC sources

[D] **xmlsave** — Export or import dataset in XML format

[TS] **haver** — Load data from Haver Analytics database

[U] **21 Inputting and importing data**

Title

> **import excel** — Import and export Excel files

Syntax

Load an Excel file

> import ex̲c̲el [using] *filename* [, *import_excel_options*]

Load subset of variables from an Excel file

> import ex̲c̲el *extvarlist* using *filename* [, *import_excel_options*]

Describe contents of an Excel file

> import ex̲c̲el [using] *filename*, de̲sc̲ribe

Save data in memory to an Excel file

> export ex̲c̲el [using] *filename* [*if*] [*in*] [, *export_excel_options*]

Save subset of variables in memory to an Excel file

> export ex̲c̲el [*varlist*] using *filename* [*if*] [*in*] [, *export_excel_options*]

import_excel_options	Description
sh̲eet("*sheetname*")	Excel worksheet to load
ce̲llrange([*start*] [:*end*])	Excel cell range to load
fi̲rstrow	treat first row of Excel data as variable names
al̲lstring	import all Excel data as strings
clear	replace data in memory

export_excel_options	Description
Main	
sh̲eet("*sheetname*")	save to Excel worksheet
cell(*start*)	start (upper-left) cell in Excel to begin saving to
sheetmod̲ify	modify Excel worksheet
sheetreplace	replace Excel worksheet
fi̲rstrow(va̲riables | va̲rlabels)	save variable names or variable labels to first row
nol̲abel	export values instead of value labels
replace	overwrite Excel file
Advanced	
da̲testring("*datetime_format*")	save dates as strings with a *datetime_format*
mi̲ssing("*repval*")	save missing values as *repval*

313

extvarlist specifies variable names of imported columns. An *extvarlist* is one or more of any of the following:

> *varname*
> *varname*=*columnname*

Example: `import excel make mpg weight price using auto.xlsx, clear` imports columns A, B, C, and D from the Excel file `auto.xlsx`.

Example: `import excel make=A mpg=B price=D using auto.xlsx, clear` imports columns A, B, and D from the Excel file `auto.xlsx`. Column C and any columns after D are skipped.

Menu

import excel

File > Import > Excel spreadsheet

export excel

File > Export > Excel spreadsheet

Description

`import excel` loads an Excel file, also known as a workbook, into Stata. `import excel` *filename*, `describe` lists available sheets and ranges of an Excel file. `export excel` saves data in memory to an Excel file. Excel 1997/2003 (`.xls`) files and Excel 2007/2010 (`.xlsx`) files can be imported, exported, and described using `import excel`, `export excel`, and `import excel, describe`.

`import excel` and `export excel` are supported on Windows, Mac, and Linux.

`import excel` and `export excel` look at the file extension, `.xls` or `.xlsx`, to determine which Excel format to read or write.

For performance, `import excel` imposes a size limit of 50 MB for Excel 2007/2010 (`.xlsx`) files. Be warned that importing large `.xlsx` files can severely affect your machine's performance.

`import excel auto` first looks for `auto.xls` and then looks for `auto.xlsx` if `auto.xls` is not found in the current directory.

The default file extension for `export excel` is `.xls` if a file extension is not specified.

Options for import excel

`sheet("`*sheetname*`")` imports the worksheet named *sheetname* in the workbook. The default is to import the first worksheet.

`cellrange(`[*start*][:*end*]`)` specifies a range of cells within the worksheet to load. *start* and *end* are specified using standard Excel cell notation, for example, A1, BC2000, and C23.

`firstrow` specifies that the first row of data in the Excel worksheet consists of variable names. This option cannot be used with *extvarlist*. `firstrow` uses the first row of the cell range for variable names if `cellrange()` is specified. `import excel` translates the names in the first row to valid Stata variable names. The original names in the first row are stored unmodified as variable labels.

allstring forces import excel to import all Excel data as string data.

clear clears data in memory before loading data from the Excel workbook.

Options for export excel

⌐Main⌐

sheet("*sheetname*") saves to the worksheet named *sheetname*. If there is no worksheet named *sheetname* in the workbook, a new sheet named *sheetname* is created. If this option is not specified, the first worksheet of the workbook is used.

cell(*start*) specifies the start (upper-left) cell in the Excel worksheet to begin saving to. By default, export excel saves starting in the first row and first column of the worksheet.

sheetmodify exports data to the worksheet without changing the cells outside the exported range. sheetmodify cannot be combined with sheetreplace or replace.

sheetreplace clears the worksheet before the data are exported to it. sheetreplace cannot be combined with sheetmodify or replace.

firstrow(variables | varlabels) specifies that the variable names or the variable labels be saved in the first row in the Excel worksheet. The variable name is used if there is no variable label for a given variable.

nolabel exports the underlying numeric values instead of the value labels.

replace overwrites an existing Excel workbook. replace cannot be combined with sheetmodify or sheetreplace.

⌐Advanced⌐

datestring("*datetime_format*") exports all datetime variables as strings formatted by *datetime_format*. See [D] **datetime display formats**.

missing("*repval*") exports missing values as *repval*. *repval* can be either string or numeric. Without specifying this option, export excel exports the missing values as empty cells.

Remarks

To demonstrate the use of import excel and export excel, we will first load auto.dta and export it as an Excel file named auto.xls:

```
. use http://www.stata-press.com/data/r12/auto
(1978 Automobile Data)

. export excel auto, firstrow(variables)
file auto.xls saved
```

Now we can import from the auto.xls file we just created, telling Stata to clear the current data from memory and to treat the first row of the worksheet in the Excel file as variable names:

```
. import excel auto.xls, firstrow clear

. describe

Contains data
  obs:            74
  vars:           12
  size:        3,922
```

variable name	storage type	display format	value label	variable label
make	str17	%17s		make
price	int	%10.0g		price
mpg	byte	%10.0g		mpg
rep78	byte	%10.0g		rep78
headroom	double	%10.0g		headroom
trunk	byte	%10.0g		trunk
weight	int	%10.0g		weight
length	int	%10.0g		length
turn	byte	%10.0g		turn
displacement	int	%10.0g		displacement
gear_ratio	double	%10.0g		gear_ratio
foreign	str8	%9s		foreign

```
Sorted by:
      Note:  dataset has changed since last saved
```

We can also import a subrange of the cells in the Excel file:

```
. import excel auto.xls, cellrange(:D70) firstrow clear

. describe

Contains data
  obs:            69
  vars:            4
  size:        1,449
```

variable name	storage type	display format	value label	variable label
make	str17	%17s		make
price	int	%10.0g		price
mpg	byte	%10.0g		mpg
rep78	byte	%10.0g		rep78

```
Sorted by:
      Note:  dataset has changed since last saved
```

Both .xls and .xlsx files are supported by import excel and export excel. If a file extension is not specified with export excel, .xls is assumed, because this format is more common and is compatible with more applications that also can read from Excel files. To save the data in memory as a .xlsx file, specify the extension:

```
. use http://www.stata-press.com/data/r12/auto, clear
(1978 Automobile Data)

. export excel auto.xlsx
file auto.xlsx saved
```

To export a subset of variables and overwrite the existing auto.xls Excel file, specify a variable list and the replace option:

```
. export excel make mpg weight using auto, replace
file auto.xls saved
```

❑ Technical note: Excel data size limits

For an Excel .xls-type workbook, the worksheet size limits are 65,536 rows by 256 columns. The string size limit is 255 characters.

For an Excel .xlsx-type workbook, the worksheet size limits are 1,048,576 rows by 16,384 columns. The string size limit is 32,767 characters.

❑

❑ Technical note: Dates and times

Excel has two different date systems, the "1900 Date System" and the "1904 Date System". Excel stores a date and time as an integer representing the number of days since a start date plus a fraction of a 24-hour day.

In the 1900 Date System, the start date is 00Jan1900; in the 1904 Date System, the start date is 01Jan1904. In the 1900 Date System, there is another artificial date, 29feb1900, besides 00Jan1900. import excel translates 29feb1900 to 28feb1900 and 00Jan1900 to 31dec1899.

See *Using dates and times from other software* in [D] **datetime** for a discussion of the relationship between Stata datetimes and Excel datetimes.

❑

❑ Technical note: Mixed data types

Because Excel's data type is cell based, import excel may encounter a column of cells with mixed data types. In such a case, the following rules are used to determine the variable type in Stata of the imported column.

- If the column contains at least one cell with nonnumerical text, the entire column is imported as a string variable.

- If an all-numerical column contains at least one cell formatted as a date or time, the entire column is imported as a Stata date or datetime variable. import excel imports the column as a Stata date if all date cells in Excel are dates only; otherwise, a datetime is used.

❑

Saved results

import excel *filename*, describe saves the following in r():

Macros
r(N_worksheet)	number of worksheets in the Excel workbook
r(worksheet_#)	name of worksheet # in the Excel workbook
r(range_#)	available cell range for worksheet # in the Excel workbook

Methods and formulas

import excel and export excel are implemented as ado-files.

Also see

[D] **import** — Overview of importing data into Stata

[D] **export** — Overview of exporting data from Stata

[D] **datetime** — Date and time values and variables

Title

> **import sasxport** — Import and export datasets in SAS XPORT format

Syntax

Import SAS XPORT Transport file into Stata

> import sasxport *filename* [, *import_options*]

Describe contents of SAS XPORT Transport file

> import sasxport *filename*, <u>d</u>escribe [<u>m</u>ember(*mbrname*)]

Export data in memory to a SAS XPORT Transport file

> export sasxport *filename* [*if*] [*in*] [, *export_options*]

> export sasxport *varlist* using *filename* [*if*] [*in*] [, *export_options*]

import_options	Description
Main	
clear	replace data in memory
<u>noval</u>labels	ignore accompanying formats.xpf file if it exists
<u>m</u>ember(*mbrname*)	member to use; seldom used

export_options	Description
Main	
<u>ren</u>ame	rename variables and value labels to meet SAS XPORT restrictions
replace	overwrite files if they already exist
<u>vall</u>abfile(xpf)	save value labels in formats.xpf
<u>vall</u>abfile(<u>sas</u>code)	save value labels in SAS command file
<u>vall</u>abfile(both)	save value labels in formats.xpf and in a SAS command file
<u>vall</u>abfile(none)	do not save value labels

Menu

import sasxport

File > Import > SAS XPORT

export sasxport

File > Export > SAS XPORT

Description

import sasxport and export sasxport convert datasets from and to SAS XPORT Transport format. The U.S. Food and Drug Administration uses SAS XPORT transport format as the format for datasets submitted with new drug and new device applications (NDAs).

To save the data in memory as a SAS XPORT Transport file, type

>   . export sasxport *filename*

although sometimes you will want to type

>   . export sasxport *filename*, rename

It never hurts to specify the rename option. In any case, Stata will create *filename*.xpt as an XPORT file containing the data and, if needed, will also create formats.xpf—an additional XPORT file—containing the value-label definitions. These files can be easily read into SAS.

To read a SAS XPORT Transport file into Stata, type

>   . import sasxport *filename*

Stata will read into memory the XPORT file *filename*.xpt containing the data and, if available, will also read the value-label definitions stored in formats.xpf or FORMATS.xpf.

import sasxport, describe describes the contents of a SAS XPORT Transport file. The display is similar to that produced by describe. To describe a SAS XPORT Transport file, type

>   . import sasxport *filename*, describe

If *filename* is specified without an extension, .xpt is assumed.

Options for import sasxport

clear permits the data to be loaded, even if there is a dataset already in memory and even if that dataset has changed since the data were last saved.

novallabels specifies that value-label definitions stored in formats.xpf or FORMATS.xpf not be looked for or loaded. By default, if variables are labeled in *filename*.xpt, then import sasxport looks for formats.xpf to obtain and load the value-label definitions. If the file is not found, Stata looks for FORMATS.xpf. If that file is not found, a warning message is issued.

import sasxport can use only a formats.xpf or FORMATS.xpf file to obtain value-label definitions. import sasxport cannot understand value-label definitions from a SAS command file.

member(*mbrname*) is a rarely specified option indicating which member of the .xpt file is to be loaded. It is not used much anymore, but the original XPORT definition allowed multiple datasets to be placed in one file. The member() option allows you to read these old files. You can obtain a list of member names using import sasxport, describe. If member() is not specified—and it usually is not—import sasxport reads the first (and usually only) member.

Option for import sasxport, describe

┌─── Main ──

member(*mbrname*) is a rarely specified option indicating which member of the .xpt file is to be described. See the description of the member() option for import sasxport directly above. If member() is not specified, all members are described, one after the other. It is rare for an XPORT file to have more than one member.

Options for export sasxport

rename specifies that `export sasxport` may rename variables and value labels to meet the SAS XPORT restrictions, which are that names be no more than eight characters long and that there be no distinction between uppercase and lowercase letters.

We recommend specifying the `rename` option. If this option is specified, any name violating the restrictions is changed to a different but related name in the file. The name changes are listed. The new names are used only in the file; the names of the variables and value labels in memory remain unchanged.

If `rename` is not specified and one or more names violate the XPORT restrictions, an error message will be issued and no file will be saved. The alternative to the `rename` option is that you can rename variables yourself with the `rename` command:

```
. rename mylongvariablename myname
```

See [D] **rename**. Renaming value labels yourself is more difficult. The easiest way to rename value labels is to use `label save`, edit the resulting file to change the name, execute the file by using `do`, and reassign the new value label to the appropriate variables by using `label values`:

```
. label save mylongvaluelabel using myfile.do
. doedit myfile.do     (change mylongvaluelabel to, say, mlvlab)
. do myfile.do
. label values myvar mlvlab
```

See [D] **label** and [R] **do** for more information about renaming value labels.

replace permits `export sasxport` to overwrite existing *filename*.xpt, formats.xpf, and *filename*.sas files.

vallabfile(xpf | sascode | both | none) specifies whether and how value labels are to be stored. SAS XPORT Transport files do not really have value labels. Value-label definitions can be preserved in one of two ways:

1. In an additional SAS XPORT Transport file whose data contain the value-label definitions

2. In a SAS command file that will create the value labels

`export sasxport` can create either or both of these files.

vallabfile(xpf), the default, specifies that value labels be written into a separate SAS XPORT Transport file named formats.xpf. Thus `export sas xport` creates two files: *filename*.xpt, containing the data, and formats.xpf, containing the value labels. No formats.xpf file is created if there are no value labels.

SAS users can easily use the resulting .xpt and .xpf XPORT files.
See http://www.sas.com/govedu/fda/macro.html for SAS-provided macros for reading the XPORT files. The SAS macro fromexp() reads the XPORT files into SAS. The SAS macro toexp() creates XPORT files. When obtaining the macros, remember to save the macros at SAS's webpage as a plain-text file and to remove the examples at the bottom.

If the SAS macro file is saved as C:\project\macros.mac and the files mydat.xpt formats.xpf created by `export sasxport` are in C:\project\, the following SAS commands would create the corresponding SAS dataset and format library and list the data:

```
─────────────────── SAS commands ───────────────────
%include "C:\project\macros.mac" ;
%fromexp(C:\project, C:\project) ;
libname library 'C:\project' ;
data _null_ ; set library.mydat ; put _all_ ; run ;
proc print data = library.mydat ;
quit ;
```

`vallabfile(sascode)` specifies that the value labels be written into a SAS command file, *filename*.sas, containing SAS `proc format` and related commands. Thus `export sasxport` creates two files: *filename*.xpt, containing the data, and *filename*.sas, containing the value labels. SAS users may wish to edit the resulting *filename*.sas file to change the "libname datapath" and "libname xptfile xport" lines at the top to correspond to the location that they desire. `export sasxport` sets the location to the current working directory at the time `export sasxport` was issued. No .sas file will be created if there are no value labels.

`vallabfile(both)` specifies that both the actions described above be taken and that three files be created: *filename*.xpt, containing the data; `formats.xpf`, containing the value labels in XPORT format; and *filename*.sas, containing the value labels in SAS command-file format.

`vallabfile(none)` specifies that value-label definitions not be saved. Only one file is created: *filename*.xpt, which contains the data.

Remarks

All users, of course, may use these commands to transfer data between SAS and Stata, but there are limitations in the SAS XPORT Transport format, such as the eight-character limit on the names of variables (specifying `export sasxport`'s `rename` option works around that). For a complete listing of limitations and issues concerning the SAS XPORT Transport format, and an explanation of how `export sasxport` and `import sasxport` work around these limitations, see *Technical appendix* below. You may find it more convenient to use translation packages such as Stat/Transfer; see http://www.stata.com/products/transfer.html.

Remarks are presented under the following headings:

Saving XPORT files for transferring to SAS
Determining the contents of XPORT files received from SAS
Using XPORT files received from SAS

Saving XPORT files for transferring to SAS

▷ Example 1

To save the current dataset in `mydata.xpt` and the value labels in `formats.xpf`, type

```
. export sasxport mydata
```

To save the data as above but automatically rename variable names and value labels that are too long or are case sensitive, type

```
. export sasxport mydata, rename
```

To allow the replacement of any preexisting files, type

```
. export sasxport mydata, rename replace
```

To save the current dataset in `mydata.xpt` and the value labels in SAS command file `mydata.sas` and to automatically rename variable names and value labels, type

    . export sasxport mydata, rename vallab(sas)

To save the data as above but save the value labels in both `formats.xpf` and `mydata.sas`, type

    . export sasxport mydata, rename vallab(both)

To not save the value labels at all, thus creating only `mydata.xpt`, type

    . export sasxport mydata, rename vallab(none)

◁

Determining the contents of XPORT files received from SAS

▷ Example 2

To determine the contents of `testdata.xpt`, you might type

    . import sasxport testdata, describe

◁

Using XPORT files received from SAS

▷ Example 3

To read data from `testdata.xpt` and obtain value labels from `formats.xpf` (or FORMATS.xpf), if the file exists, you would type

    . import sasxport testdata

To read the data as above and discard any data in memory, type

    . import sasxport testdata, clear

◁

Saved results

import sasxport, describe saves the following in r():

Scalars
    r(N)   number of observations   r(size)   size of data
    r(k)   number of variables   r(n_members)   number of members

Macros
    r(members) names of members

Technical appendix

Technical details concerning the SAS XPORT Transport format and how `export sasxport` and `import sasxport` handle issues regarding the format are presented under the following headings:

> A1. Overview of SAS XPORT Transport format
> A2. Implications for writing XPORT datasets from Stata
> A3. Implications for reading XPORT datasets into Stata

A1. Overview of SAS XPORT Transport format

A SAS XPORT Transport file may contain one or more separate datasets, known as members. It is rare for a SAS XPORT Transport file to contain more than one member. See http://support.sas.com/techsup/technote/ts140.html for the SAS technical document describing the layout of the SAS XPORT Transport file.

A SAS XPORT dataset (member) is subject to certain restrictions:

1. The dataset may contain only 9,999 variables.

2. The names of the variables and value labels may not be longer than eight characters and are case insensitive; for example, `myvar`, `Myvar`, `MyVar`, and `MYVAR` are all the same name.

3. Variable labels may not be longer than 40 characters.

4. The contents of a variable may be numeric or string:

    a. Numeric variables may be integer or floating but may not be smaller than 5.398e–79 or greater than 9.046e+74, absolutely. Numeric variables may contain missing, which may be `.`, `._`, `.a`, `.b`, ..., `.z`.

    b. String variables may not exceed 200 characters. String variables are recorded in a "padded" format, meaning that, when variables are read, it cannot be determined whether the variable had trailing blanks.

5. Value labels are *not* written in the XPORT dataset. Suppose that you have variable `sex` in the data with values 0 and 1, and the values are labeled for gender (0=male, and 1=female). When the dataset is written in SAS XPORT Transport format, you can record that the variable label `gender` is associated with the `sex` variable, but you cannot record the association with the value labels male and female.

    Value-label definitions are typically stored in a second XPORT dataset or in a text file containing SAS commands. You can use the `vallabfile()` option of `export sasxport` to produce these datasets or files.

    Value labels and formats are recorded in the same position in an XPORT file, meaning that names corresponding to formats used in SAS cannot be used. Thus value labels may not be named

> `best`, `binary`, `comma`, `commax`, `d`, `date`, `datetime`, `dateampm`, `day`, `ddmmyy`, `dollar`, `dollarx`, `downname`, `e`, `eurdfdd`, `eurdfde`, `eurdfdn`, `eurdfdt`, `eurdfdwn`, `eurdfmn`, `eurdfmy`, `eurdfwdx`, `eurdfwkx`, `float`, `fract`, `hex`, `hhmm`, `hour`, `ib`, `ibr`, `ieee`, `julday`, `julian`, `percent`, `minguo`, `mmddyy`, `mmss`, `mmyy`, `monname`, `month`, `monyy`, `negparen`, `nengo`, `numx`, `octal`, `pd`, `pdjulg`, `pdjuli`, `pib`, `pibr`, `pk`, `pvalue`, `qtr`, `qtrr`, `rb`, `roman`, `s370ff`, `s370fib`, `s370fibu`, `s370fpd`, `s370fpdu`, `s370fpib`, `s370frb`, `s370fzd`, `s370fzdl`, `s370fzds`, `s370fzdt`, `s370fzdu`, `ssn`, `time`, `timeampm`, `tod`, `weekdate`, `weekdatx`, `weekday`, `worddate`, `worddatx`, `wordf`, `words`, `year`, `yen`, `yymm`, `yymmdd`, `yymon`, `yyq`, `yyqr`, `z`, `zd`, or any uppercase variation of these.

We refer to this as the "Known Reserved Word List" in this documentation. Other words may also be reserved by SAS; the technical documentation for the SAS XPORT Transport format provides no guidelines. This list was created by examining the formats defined in *SAS Language Reference: Dictionary, Version 8*. If SAS adds new formats, the list will grow.

6. A flaw in the XPORT design can make it impossible, in rare instances, to determine the exact number of observations in a dataset. This problem can occur only if 1) all variables in the dataset are string and 2) the sum of the lengths of all the string variables is less than 80. Actually, the above is the restriction, assuming that the code for reading the dataset is written well. If it is not, the flaw could occur if 1) the last variable or variables in the dataset are string and 2) the sum of the lengths of all variables is less than 80.

To prevent stumbling over this flaw, make sure that the last variable in the dataset is not a string variable. This is always sufficient to avoid the problem.

7. There is no provision for saving the Stata concepts `notes` and `characteristics`.

A2. Implications for writing XPORT datasets from Stata

Stata datasets for the most part fit well into the SAS XPORT Transport format. With the same numbering scheme as above,

1. Stata refuses to write the dataset if it contains more than 9,999 variables.

2. Stata issues an error message if any variable or label name violates the naming restrictions, or if the `rename` option is specified, Stata fixes any names that violate the restrictions.

   Whether or not `rename` is specified, names will be recorded case insensitively: you do not have to name all your variables with all lowercase or all uppercase letters. Stata verifies that ignoring case does not lead to problems, complaining or, if option `rename` is specified, fixing them.

3. Stata truncates variable labels to 40 characters to fit within the XPORT limit.

4. Stata treats variable contents as follows:

   a. If a numeric variable records a value greater than 9.046e+74 in absolute value, Stata issues an error message. If a variable records a value less than 5.398e–79 in absolute value, 0 is written.

   b. If you have string variables longer than 200 characters, Stata issues an error message. Also, if any string variable has trailing blanks, Stata issues an error message. To remove trailing blanks from string variable s, you can type

      ```
      . replace s = rtrim(s)
      ```

      To remove leading and trailing blanks, type

      ```
      . replace s = trim(s)
      ```

5. Value-label names are written in the XPORT dataset. The contents of the value label are not written in the same XPORT dataset. By default, `formats.xpf`, a second XPORT dataset, is created containing the value-label definitions.

   SAS recommends creating a `formats.xpf` file containing the value-label definitions (what SAS calls format definitions). They have provided SAS macros, making the reading of `.xpt` and `formats.xpf` files easy. See http://www.sas.com/govedu/fda/macro.html for details.

Alternatively, a SAS command file containing the value-label definitions can be produced. The `vallabfile()` option of `export sasxport` is used to indicate which, if any, of the formats to use for recording the value-label definitions.

If a value-label name matches a name on the Known Reserved Word List, and the `rename` option is not specified, Stata issues an error message.

If a variable has no value label, the following format information is recorded:

Stata format	SAS format
%td...	MMDDYY10.
%-td...	MMDDYY10.
%#s	$CHAR#.
%-#s	$CHAR#.
% #s	$CHAR#.
all other	BEST12.

6. If you have a dataset that could provoke the XPORT design flaw, a warning message is issued. Remember, the best way to avoid this flaw is to ensure that the last variable in the dataset is numeric. This is easily done. You could, for instance, type

   ```
   . gen ignoreme = 0
   . export sasxport ...
   ```

7. Because the XPORT file format does not support notes and characteristics, Stata ignores them when it creates the XPORT file. You may wish to incorporate important notes into the documentation that you provide to the user of your XPORT file.

A3. Implications for reading XPORT datasets into Stata

Reading SAS XPORT Transport format files into Stata is easy, but sometimes there are issues to consider:

1. If there are too many variables, Stata issues an error message. If you are using Stata/MP or Stata/SE, you can increase the maximum number of variables with the `set maxvar` command; see [D] **memory**.

2. The XPORT format variable naming restrictions are more restrictive than those of Stata, so no problems should arise. However, Stata reserves the following names:

   _all, _b, byte, _coef, _cons, double, float, if, in, int, long, _n, _N, _pi, _pred, _rc, _skip, str#, using, with

   If the XPORT file contains variables with any of these names, Stata issues an error message. Also, the error message

   ```
   . import sasxport ...
   _____ already defined
   r(110);
   ```

   indicates that the XPORT file was incorrectly prepared by some other software and that two or more variables share the same name.

3. The XPORT variable-label-length limit is more restrictive than that of Stata, so no problems can arise.

4. Variable contents may cause problems:

   a. The range of numeric variables in an XPORT dataset is a subset of that allowed by Stata, so no problems can arise. All variables are brought back as `doubles`; we recommend that you run `compress` after loading the dataset:

      ```
      . import sasxport ...
      . compress
      ```

      See [D] **compress**.

      Stata has no missing-value code corresponding to `. _`. If any value records `. _`, then `.u` is stored.

   b. String variables are brought back as recorded but with all trailing blanks stripped.

5. Value-label names are read directly from the XPORT dataset. Any value-label definitions are obtained from a separate XPORT dataset, if available. If a value-label name matches any in the Known Reserved Word List, no value-label name is recorded, and instead, the variable display format is set to `%9.0g`, `%10.0g`, or `%td`.

   The `%td` Stata format is used when the following SAS formats are encountered:

   DATE, EURDFDN, JULDAY, MONTH, QTRR, YEAR, DAY, EURDFDWN, JULIAN, MONYY, WEEKDATE, YYMM, DDMMYY, EURDFMN, MINGUO, NENGO, WEEKDATX, YYMMDD, DOW-NAME, EURDFMY, MMDDYY, PDJULG, WEEKDAY, YYMON, EURDFDD, EURDFWDX, MMYY, PDJULI, WORDDATE, YYQ, EURDFDE, EURDFWKX, MONNAME, QTR, WORDDATX, YYQR

   If the XPORT file indicates that one or more variables have value labels, `import sasxport` looks for the value-label definitions in `formats.xpf`, another XPORT file. If it does not find this file, it looks for `FORMATS.xpf`. If this file is not found, `import sasxport` issues a warning message unless the `novallabels` option is specified.

   Stata does not allow value-label ranges or string variables with value labels. If the `.xpt` file or `formats.xpf` file contains any of these, an error message is issued. The `novallabels` option allows you to read the data, ignoring all value labels.

6. If a dataset is read that provokes the all-strings XPORT design flaw, the dataset with the minimum number of possible observations is returned, and a warning message is issued. This duplicates the behavior of SAS.

7. SAS XPORT format does not allow notes or characteristics, so no issues can arise.

Also see

[D] **export** — Overview of exporting data from Stata

[D] **import** — Overview of importing data into Stata

Title

---
infile (fixed format) — Read text data in fixed format with a dictionary
---

Syntax

> <u>inf</u>ile using *dfilename* [*if*] [*in*] [, *options*]

options	Description
Main	
<u>us</u>ing(*filename*)	text dataset filename
clear	replace data in memory
Options	
<u>a</u>utomatic	create value labels from nonnumeric data
ebcdic	treat text dataset as EBCDIC

A dictionary is a text file that is created with the Do-file Editor or an editor outside Stata. This file specifies how Stata should read fixed-format data from a text file. The syntax for a dictionary is

--- begin dictionary file -------------

```
[infile] dictionary [using filename] {
        * comments may be included freely
        _lrecl(#)
        _firstlineoffile(#)
        _lines(#)

        _line(#)
        _newline[(#)]
        _column(#)
        _skip[(#)]

        [type] varname [:lblname] [% infmt] ["variable label"]
}
(your data might appear here)
```

--- end dictionary file -------------

where % *infmt* is { %[#[.#]]{f|g|e} | %[#]s | %[#]S }

Menu

File > Import > Text data in fixed format with a dictionary

Description

infile using reads a dataset that is stored in either ASCII or EBCDIC text form. infile using does this by first reading *dfilename*—a "dictionary" that describes the format of the data file—and then reads the file containing the data. The dictionary is a file you create with the Do-file Editor or an editor outside Stata. If *dfilename* is specified without an extension, .dct is assumed.

If using *filename* is not specified, the data are assumed to begin on the line following the closing brace. If using *filename* is specified, the data are assumed to be located in *filename*. If *filename* is specified without an extension, .raw is assumed.

If *dfilename* or *filename* contains embedded spaces, remember to enclose it in double quotes.

The data may be in the same file as the dictionary or in another file. If ebcdic is specified, the data will be converted from EBCDIC to ASCII as they are imported. The dictionary in all cases must be ASCII.

Another variation on infile omits the intermediate dictionary; see [D] **infile (free format)**. This variation is easier to use but will not read fixed-format files. On the other hand, although infile with a dictionary will read free-format files, infile without a dictionary is even better at it.

An alternative to infile using for reading fixed-format files is infix; see [D] **infix (fixed format)**. infix provides fewer features than infile using but is easier to use.

Stata has other commands for reading data. If you are not certain that infile using will do what you are looking for, see [D] **import** and [U] **21 Inputting and importing data**.

Options

using(*filename*) specifies the name of a file containing the data. If using() is not specified, the data are assumed to follow the dictionary in *dfilename*, or if the dictionary specifies the name of some other file, that file is assumed to contain the data. If using(*filename*) is specified, *filename* is used to obtain the data, even if the dictionary says otherwise. If *filename* is specified without an extension, .raw is assumed.

If *filename* contains embedded spaces, remember to enclose it in double quotes.

clear specifies that it is okay for the new data to replace what is currently in memory. To ensure that you do not lose something important, infile using will refuse to read new data if other data are already in memory. clear allows infile using to replace the data in memory. You can also drop the data yourself by typing drop _all before reading new data.

automatic causes Stata to create value labels from the nonnumeric data it reads. It also automatically widens the display format to fit the longest label.

ebcdic specifies that the data be stored using EBCDIC character encoding rather than ASCII, the default, and be converted from EBCDIC to ASCII as they are imported. In all cases, *dfilename*, the dictionary, must be specified using ASCII.

Dictionary directives

* marks comment lines. Wherever you wish to place a comment, begin the line with a *. Comments can appear many times in the same dictionary.

_lrecl(#) is used only for reading datasets that do not have end-of-line delimiters (carriage return, line feed, or some combination of these). Such files are often produced by mainframe computers and are either coded in EBCDIC or have been translated from EBCDIC into ASCII. _lrecl() specifies the logical record length. _lrecl() requests that infile act as if a line ends every # characters.

_lrecl() appears only once, and typically not at all, in a dictionary.

_firstlineoffile(#) (abbreviation _first()) is also rarely specified. It states the line of the file where the data begin. You do not need to specify _first() when the data follow the dictionary; Stata can figure that out for itself. However, you might specify _first() when reading data from another file in which the first line does not contain data because of headers or other markers.

_first() appears only once, and typically not at all, in a dictionary.

_lines(#) states the number of lines per observation in the file. Simple datasets typically have _lines(1). Large datasets often have many lines (sometimes called records) per observation. _lines() is optional, even when there is more than one line per observation because infile can sometimes figure it out for itself. Still, if _lines(1) is not right for your data, it is best to specify the correct number through _lines(#).

_lines() appears only once in a dictionary.

_line(#) tells infile to jump to line # of the observation. _line() is not the same as _lines(). Consider a file with _lines(4), meaning four lines per observation. _line(2) says to jump to the second line of the observation. _line(4) says to jump to the fourth line of the observation. You may jump forward or backward. infile does not care, and there is no inefficiency in going forward to _line(3), reading a few variables, jumping back to _line(1), reading another variable, and jumping forward again to _line(3).

You need not ensure that, at the end of your dictionary, you are on the last line of the observation. infile knows how to get to the next observation because it knows where you are and it knows _lines(), the total number of lines per observation.

_line() may appear many times in a dictionary.

_newline[(#)] is an alternative to _line(). _newline(1), which may be abbreviated _newline, goes forward one line. _newline(2) goes forward two lines. We do not recommend using _newline() because _line() is better. If you are currently on line 2 of an observation and want to get to line 6, you could type _newline(4), but your meaning is clearer if you type _line(6).

_newline() may appear many times in a dictionary.

_column(#) jumps to column # on the current line. You may jump forward or backward within a line. _column() may appear many times in a dictionary.

_skip[(#)] jumps forward # columns on the current line. _skip() is just an alternative to _column(). _skip() may appear many times in a dictionary.

[*type*] *varname* [:*lblname*] [% *infmt*] ["*variable label*"] instructs infile to read a variable. The simplest form of this instruction is the variable name itself: *varname*.

At all times, infile is on some column of some line of an observation. infile starts on column 1 of line 1, so pretend that is where we are. Given the simplest directive, '*varname*', infile goes through the following logic:

If the current column is blank, it skips forward until there is a nonblank column (or until the end of the line). If it just skipped all the way to the end of the line, it stores a missing value in *varname*. If it skipped to a nonblank column, it begins collecting what is there until it comes to a blank column or the end of the line. These are the data for *varname*. Then it sets the current column to wherever it is.

The logic is a bit more complicated. For instance, when skipping forward to find the data, infile might encounter a quote. If so, it then collects the characters for the data by skipping forward until it finds the matching quote. If you specified a % *infmt*, then infile skips the skipping-forward step and simply collects the specified number of characters. If you specified a %S *infmt*, then infile does not skip leading or trailing blanks. Nevertheless, the general logic is (optionally) skip, collect, and reset.

Remarks

Remarks are presented under the following headings:

> *Introduction*
> *Reading free-format files*
> *Reading fixed-format files*
> *Numeric formats*
> *String formats*
> *Specifying column and line numbers*
> *Examples of reading fixed-format files*
> *Reading fixed-block files*
> *Reading EBCDIC files*

Introduction

`infile using` follows a two-step process to read your data. You type something like `infile using descript`, and

1. `infile using` reads the file `descript.dct`, which tells `infile` about the format of the data; and

2. `infile using` then reads the data according to the instructions recorded in `descript.dct`.

`descript.dct` (the file could be named anything) is called a dictionary, and `descript.dct` is just a text file that you create with the Do-file Editor or an editor outside Stata.

As for the data, they can be in the same file as the dictionary or in a different file. It does not matter.

Reading free-format files

Another variation of `infile` for reading free-format files is described in [D] **infile (free format)**. We will refer to this variation as `infile` without a dictionary. The distinction between the two variations is in the treatment of line breaks. `infile` without a dictionary does not consider them significant. `infile` with a dictionary does.

A line, also known as a record, physical record, or physical line (as opposed to observations, logical records, or logical lines), is a string of characters followed by the line terminator. If you were to type the file, a line is what would appear on your screen if your screen were infinitely wide. Your screen would have to be infinitely wide so that there would be no possibility that one line could take more than one line of your screen, thus fooling you into thinking that there are multiple lines when there is only one.

A logical line, on the other hand, is a sequence of one or more physical lines that represent one observation of your data. `infile` with a dictionary does not spontaneously go to new physical lines; it goes to a new line only between observations and when you tell it to. `infile` without a dictionary, on the other hand, goes to a new line whenever it needs to, which can be right in the middle of an observation. Thus consider the following little bit of data, which is for three variables:

```
5 4
1 9 3
2
```

How do you interpret these data?

Here is one interpretation: There are 3 observations. The first is 5, 4, and missing. The second is 1, 9, and 3. The third is 2, missing, and missing. That is the interpretation that `infile` with a dictionary makes.

Here is another interpretation: There are 2 observations. The first is 5, 4, and 1. The second is 9, 3, and 2. That is the interpretation that `infile` without a dictionary makes.

Which is right? You would have to ask the person who entered these data. The question is, are the line breaks significant? Do they mean anything? If the line breaks are significant, you use `infile` with a dictionary. If the line breaks are not significant, you use `infile` without a dictionary.

The other distinction between the two `infile`s is that `infile` with a dictionary does not process comma-separated–value format. If your data are comma-separated, tab-separated, or otherwise delimited, see [D] **insheet** or [D] **infile (free format)**.

▷ Example 1

Outside Stata, we have typed into the file `highway.dct` information on the accident rate per million vehicle miles along a stretch of highway, the speed limit on that highway, and the number of access points (on-ramps and off-ramps) per mile. Our file contains

```
───────────────────────────────────── begin highway.dct, example 1 ──────────
    infile dictionary {
            acc_rate   spdlimit acc_pts
    }
    4.58 55 4.6
    2.86  60 4.4
    1.61  . 2.2
    3.02 60 4.7
──────────────────────────────────────── end highway.dct, example 1 ─────────
```

This file can be read by typing `infile using highway`. Stata displays the dictionary and reads the data:

```
. infile using highway
infile dictionary {
        acc_rate   spdlimit acc_pts
}
(4 observations read)
. list
```

	acc_rate	spdlimit	acc_pts
1.	4.58	55	4.6
2.	2.86	60	4.4
3.	1.61	.	2.2
4.	3.02	60	4.7

◁

▷ Example 2

We can include variable labels in a dictionary so that after we `infile` the data, the data will be fully labeled. We could change `highway.dct` to read

```
                                              begin highway.dct, example 2
infile dictionary {
* This is a comment and will be ignored by Stata
* You might type the source of the data here.
        acc_rate   "Acc. Rate/Million Miles"
        spdlimit   "Speed Limit (mph)"
        acc_pts    "Access Pts/Mile"
}
4.58 55 4.6
2.86  60 4.4
1.61  . 2.2
3.02 60 4.7
                                              end highway.dct, example 2
```

Now when we type `infile using highway`, Stata not only reads the data but also labels the variables.

◁

▷ Example 3

We can indicate the variable types in the dictionary. For instance, if we wanted to store `acc_rate` as a `double` and `spdlimit` as a `byte`, we could change `highway.dct` to read

```
                                              begin highway.dct, example 3
infile dictionary {
* This is a comment and will be ignored by Stata
* You might type the source of the data here.
 double acc_rate   "Acc. Rate/Million Miles"
 byte    spdlimit   "Speed Limit (mph)"
        acc_pts    "Access Pts/Mile"
}
4.58 55 4.6
2.86  60 4.4
1.61  . 2.2
3.02 60 4.7
                                              end highway.dct, example 3
```

Because we do not indicate the variable type for `acc_pts`, it is given the default variable type `float` (or the type specified by the `set type` command).

◁

▷ Example 4

By specifying the types, we can read string variables as well as numeric variables. For instance,

```
                                              begin emp.dct
infile dictionary {
* data on employees
  str20 name       "Name"
        age        "Age"
    int sex        "Sex coded 0 male 1 female"
}
"Lisa Gilmore" 25 1
Branton 32 1
'Bill Ross' 27 0
                                              end emp.dct
```

The strings can be delimited by single or double quotes, and quotes may be omitted altogether if the string contains no blanks or other special characters.

◁

▷ Example 5

You may attach value labels to variables in the dictionary by using the colon notation:

```
───────────────────────────────────────────── begin emp2.dct ──────────
infile dictionary {
* data on name, sex, and age
  str16 name      "Name"
        sex:sexlbl  "Sex"
    int age        "Age"
}
"Arthur Doyle" Male 22
"Mary Hope" Female 37
"Guy Fawkes" Male 48
"Karen Cain" Female 25
────────────────────────────────────────────── end emp2.dct ──────────
```

If you want the value labels to be created automatically, you must specify the automatic option on the infile command. These data could be read by typing infile using emp2, automatic, assuming the dictionary and data are stored in the file emp2.dct.

◁

▷ Example 6

The data need not be in the same file as the dictionary. We might leave the highway data in highway.raw and write a dictionary called highway.dct describing the data:

```
────────────────────────────────────── begin highway.dct, example 4 ──────
infile dictionary using highway {
* This dictionary reads the file highway.raw.  If the
* file were called highway.txt, the first line would
* read "dictionary using highway.txt"
        acc_rate  "Acc. Rate/Million Miles"
        spdlimit  "Speed Limit (mph)"
        acc_pts   "Access Pts/Mile"
}
─────────────────────────────────────── end highway.dct, example 4 ──────
```

◁

▷ Example 7

The firstlineoffile() directive allows us to ignore lines at the top of the file. Consider the following raw dataset:

```
─────────────────────────────────────────── begin mydata.raw ──────
The following data was entered by Marsha Martinez.  It was checked by
Helen Troy.
id income educ sex age
1024 25000 HS Male 28
1025 27000 C Female 24
──────────────────────────────────────────── end mydata.raw ──────
```

Our dictionary might read

```
——————————————————————————————————— begin mydata.dct ———————————
infile dictionary using mydata {
        _first(4)
        int id "Identification Number"
        income "Annual income"
        str2 educ "Highest educ level"
        str6 sex
        byte age
}
——————————————————————————————————— end mydata.dct ———————————
```

◁

▷ Example 8

The _line() and _lines() directives tell Stata how to read our data when there are multiple records per observation. We have the following in mydata2.raw:

```
———————————————————————————————— begin mydata2.raw ———————————
id income educ sex age
1024 25000 HS
Male
28
1025 27000 C
Female
24
1035 26000 HS
Male
32
1036 25000 C
Female
25
———————————————————————————————— end mydata2.raw ———————————
```

We can read this with a dictionary mydata2.dct, which we will just let Stata list as it simultaneously reads the data:

```
. infile using mydata2, clear
infile dictionary using mydata2 {
    _first(2)                              * Begin reading on line 2
    _lines(3)                              * Each observation takes 3 lines.
    int id "Identification Number"         * Since _line is not specified, Stata
    income "Annual income"                 * assumes that it is 1.
    str2 educ "Highest educ level"
    _line(2)                               * Go to line 2 of the observation.
    str6 sex                               * (values for sex are located on line 2)
    _line(3)                               * Go to line 3 of the observation.
    int age                                * (values for age are located on line 3)
}
(4 observations read)
. list
```

	id	income	educ	sex	age
1.	1024	25000	HS	Male	28
2.	1025	27000	C	Female	24
3.	1035	26000	HS	Male	32
4.	1036	25000	C	Female	25

Here is the really good part: we read these variables in order, but that was not necessary. We could just as well have used the dictionary:

```
                                                    begin mydata2p.dct
    infile dictionary using mydata2 {
            _first(2)
            _lines(3)

            _line(1)   int    id      "Identification number"
                              income  "Annual income"
                       str2   educ    "Highest educ level"
            _line(3)   int    age
            _line(2)   str6   sex
    }
                                                     end mydata2p.dct
```

We would have obtained the same results just as quickly, the only difference being that our variables in the final dataset would be in the order specified: id, income, educ, age, and sex.

◁

❑ Technical note

You can use _newline to specify where breaks occur, if you prefer:

```
                                                    begin highway.dct, example 5
    infile dictionary {
            acc_rate   "Acc. Rate/Million Miles"
            spdlimit   "Speed Limit (mph)"
    _newline acc_pts   "Access Pts/Mile"
    }
    4.58 55
    4.6
    2.86   60
     4.4
    1.61 .
    2.2
    3.02 60
     4.7
                                                     end highway.dct, example 5
```

The line reading '1.61 .' could have been read 1.61 (without the period), and the results would have been unchanged. Because dictionaries do not go to new lines automatically, a missing value is assumed for all values not found in the record.

❑

Reading fixed-format files

Values in formatted data are sometimes packed one against the other with no intervening blanks. For instance, the highway data might appear as

```
                                                    begin highway.raw, example 6
    4.58554.6
    2.86604.4
    1.61  2.2
    3.02604.7
                                                     end highway.raw, example 6
```

The first four columns of each record represent the accident rate; the next two columns, the speed limit; and the last three columns, the number of access points per mile.

To read these data, you must specify the % *infmt* in the dictionary. Numeric % *infmt*s are denoted by a leading percent sign (%) followed optionally by a string of the form w or $w.d$, where w and d stand for two integers. The first integer, w, specifies the width of the format. The second integer, d, specifies the number of digits that are to follow the decimal point. d must be less than or equal to w. Finally, a character denoting the format type (f, g, or e) is appended. For example, %9.2f specifies an f format that is nine characters wide and has two digits following the decimal point.

Numeric formats

The f format indicates that `infile` is to attempt to read the data as a number. When you do not specify the % *infmt* in the dictionary, `infile` assumes the %f format. The width, w, being missing means that `infile` is to attempt to read the data in free format.

As it starts reading each observation, `infile` reads a record into its buffer and sets a column pointer to 1, indicating that it is currently on the first column. When `infile` processes a %f format, it moves the column pointer forward through white space. It then collects the characters up to the next occurrence of white space and attempts to interpret those characters as a number. The column pointer is left at the first occurrence of white space following those characters. If the next variable is also free format, the logic repeats.

When you explicitly specify the field width w, as in %wf, `infile` does not skip leading white space. Instead, it collects the next w characters starting at the column pointer and attempts to interpret the result as a number. The column pointer is left at the old value of the column pointer plus w, that is, on the first character following the specified field.

▷ Example 9

If the data above were stored in `highway.raw`, we could create the following dictionary to read the data:

```
                                          begin highway.dct, example 6
     infile dictionary using highway {
            acc_rate   %4f   "Acc. Rate/Million Miles"
            spdlimit   %2f   "Speed Limit (mph)"
            acc_pts    %3f   "Access Pts/Mile
     }
                                          end highway.dct, example 6
```

When we explicitly indicate the field width, `infile` does not skip intervening characters. The first four columns are used for the variable `acc_rate`, the next two for `spdlimit`, and the last three for `acc_pts`.

◁

❏ Technical note

The d specification in the %$w.d$f indicates the number of *implied* decimal places in the data. For instance, the string 212 read in a %3.2f format represents the number 2.12. Do *not* specify d unless your data have elements of this form. The w alone is sufficient to tell `infile` how to read data in which the decimal point is explicitly indicated.

When you specify d, Stata takes it only as a suggestion. If the decimal point is explicitly indicated in the data, that decimal point always overrides the d specification. Decimal points are also not implied if the data contain an E, e, D, or d, indicating scientific notation.

Fields are right-justified before implying decimal points. Thus '2   ', ' 2 ', and ' 2' are all read as 0.2 by the %3.1f format.

❏

❏ Technical note

The g and e formats are the same as the f format. You can specify any of these letters interchangeably. The letters g and e are included as a convenience to those familiar with Fortran, in which the e format indicates scientific notation. For example, the number 250 could be indicated as 2.5E+02 or 2.5D+02. Fortran programmers would refer to this as an E7.5 format, and in Stata, this format would be indicated as %7.5e. In Stata, however, you need specify only the field width w, so you could read this number by using %7f, %7g, or %7e.

The g format is really a Fortran output format that indicates a freer format than f. In Stata, the two formats are identical.

Throughout this section, you may freely substitute the g or e formats for the f format.

❏

❏ Technical note

Be careful to distinguish between %*fmts* and %*infmts*. %*fmts* are also known as *display* formats—they describe how a variable is to look when it is displayed; see [U] **12.5 Formats: Controlling how data are displayed**. %*infmts* are also known as *input* formats—they describe how a variable looks when you input it. For instance, there is an output date format, %td, but there is no corresponding input format. (See [U] **24 Working with dates and times** for recommendations on how to read dates.) For the other formats, we have attempted to make the input and output definitions as similar as possible. Thus we include g, e, and f %*infmts*, even though they all mean the same thing, because g, e, and f are also %*fmts*.

❏

String formats

The s and S formats are used for reading strings. The syntax is %ws or %wS, where the w is optional. If you do not specify the field width, your strings must either be enclosed in quotes (single or double) or not contain any characters other than letters, numbers, and "_".

This may surprise you, but the s format can be used for reading numeric variables, and the f format can be used for reading string variables! When you specify the field width, w, in the %wf format, all embedded blanks in the field are removed before the result is interpreted. They are not removed by the %ws format.

For instance, the %3f format would read "- 2", "-2 ", or " -2" as the number -2. The %3s format would not be able to read "- 2" as a number, because the sign is separated from the digit, but it could read " -2" or "-2 ". The %wf format removes blanks; datasets written by some Fortran programs separate the sign from the number.

There are, however, some side effects of this practice. The string "2 2" will be read as 22 by a %3f format. Most Fortran compilers would read this number as 202. The %3s format would issue a warning and store a *missing* value.

Now consider reading the string "a b" into a string variable. Using a %3s format, Stata will store it as it appears: a b. Using a %3f format, however, it will be stored as ab—the middle blank will be removed.

%wS is a special case of %ws. A string read with %ws will have leading and trailing blanks removed, but a string read with %wS will not have them removed.

Examples using the %s format are provided below, after we discuss specifying column and line numbers.

Specifying column and line numbers

_column() jumps to the specified column. For instance, the documentation of some dataset indicates that the variable age is recorded as a two-digit number in column 47. You could read this by coding

```
_column(47) age %2f
```

After typing this, you are now at column 49, so if immediately following age there were a one-digit number recording sex as 0 or 1, you could code

```
_column(47) age %2f
            sex %1f
```

or, if you wanted to be explicit about it, you could instead code

```
_column(47) age %2f
_column(49) sex %1f
```

It makes no difference. If at column 50 there were a one-digit code for race and you wanted to read it but skip reading the sex code, you could code

```
_column(47) age %2f
_column(50) race %1f
```

You could equivalently skip forward using _skip():

```
_column(47) age %2f
_skip(1)    race %1f
```

One advantage of column() over _skip is that it lets you jump forward or backward in a record. If you wanted to read race and then age, you could code

```
_column(50) race %1f
_column(47) age %2f
```

If the data you are reading have multiple lines per observation (sometimes said as multiple records per observation), you can tell infile how many lines per record there are by using _lines():

```
_lines(4)
```

_lines() appears only once in a dictionary. Good style says that it should be placed near the top of the dictionary, but Stata does not care.

When you want to go to a particular line, include the _line() directive. In our example, let's assume that race, sex, and age are recorded on the second line of each observation:

```
_lines(4)
_line(2)
    _column(47) age %2f
    _column(50) race %1f
```

Let's assume that id is recorded on line 1.

```
_lines(4)
_line(1)
    _column(1)  id  %4f
_line(2)
    _column(47) age %2f
    _column(50) race %1f
```

_line() works like _column() in that you can jump forward or backward, so these data could just as well be read by

```
_lines(4)
_line(2)
    _column(47) age %2f
    _column(50) race %1f
_line(1)
    _column(1)  id  %4f
```

Remember that this dataset has four lines per observation, and yet we have never referred to line(3) or line(4). That is okay. Also, at the end of our dictionary, we are on line 1, not line 4. That is okay, too. infile will still get to the next observation correctly.

❏ Technical note

Another way to move between records is _newline(). _newline() is to _line() as _skip() is to _column(), which is to say, _newline() can only go forward. There is one difference: _skip() has its uses, whereas _newline() is useful only for backward capability with older versions of Stata.

_skip() has its uses because sometimes we think in columns and sometimes we think in widths. Some data documentation might include the sentence, "At column 54 are recorded the answers to the 25 questions, with one column allotted to each." If we want to read the answers to questions 1 and 5, it would indeed be natural to code

```
_column(54) q1 %1f
_skip(3)
            q5 %1f
```

Nobody has ever read data documentation with the statement, "Demographics are recorded on record 2, and two records after that are the income values." The documentation would instead say, "Record 2 contains the demographic information and record 4, income." The _newline() way of thinking is based on what is convenient for the computer, which does, after all, have to move past a certain number of records. That, however, is no reason for making you think that way.

Before that thought occurred to us, Stata users specified _newline() to go forward a number of records. They still can, so their old dictionaries will work. When you use _newline() and do not specify _lines(), you must move past the correct number of records so that, at the end of the dictionary, you are on the last record. In this mode, when Stata reexecutes the dictionary to process the next observation, it goes forward one record.

❏

Examples of reading fixed-format files

▷ Example 10

In this example, each observation occupies two lines. The first 2 observations in the dataset are

```
John Dunbar                      10001   101 North 42nd Street
1010111111
Sam K. Newey Jr.                 10002   15663 Roustabout Boulevard
0101000000
```

The first observation tells us that the name of the respondent is John Dunbar; that his ID is 10001; that his address is 101 North 42nd Street; and that his answers to questions 1–10 were yes, no, yes, no, yes, yes, yes, yes, yes, and yes.

The second observation tells us that the name of the respondent is Sam K. Newey Jr.; that his ID is 10002; that his address is 15663 Roustabout Boulevard; and that his answers to questions 1–10 were no, yes, no, yes, no, no, no, no, no, and no.

To see the layout within the file, we can temporarily add two rulers to show the appropriate columns:

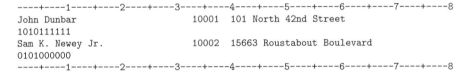

```
----+----1----+----2----+----3----+----4----+----5----+----6----+----7----+----8
John Dunbar                      10001   101 North 42nd Street
1010111111
Sam K. Newey Jr.                 10002   15663 Roustabout Boulevard
0101000000
----+----1----+----2----+----3----+----4----+----5----+----6----+----7----+----8
```

Each observation in the data appears in two physical lines within our text file. We had to check in our editor to be sure that there really were new-line characters (for example, "hard returns") after the address. This is important because some programs will wrap output for you so that one line may appear as many lines. The two seemingly identical files will differ in that one has a hard return and the other has a soft return added only for display purposes.

In our data, the name occupies columns 1–32; a person identifier occupies columns 33–37; and the address occupies columns 40–80. Our worksheet revealed that the widest address ended in column 80.

The text file containing these data is called `fname.txt`. Our dictionary file looks like this:

```
─────────────────────────────────────────────────── begin fname.dct ───────
infile dictionary using fname.txt {
*
* Example reading in data where observations extend across more
* than one line.  The next line tells infile there are 2 lines/obs:
*
_lines(2)
*
                str50    name    %32s       "Name of respondent"
_column(33)     long     id      %5f        "Person id"
_skip(2)        str50    addr    %41s       "Address"
_line(2)
_column(1)      byte     q1      %1f        "Question 1"
                byte     q2      %1f        "Question 2"
                byte     q3      %1f        "Question 3"
                byte     q4      %1f        "Question 4"
                byte     q5      %1f        "Question 5"
                byte     q6      %1f        "Question 6"
                byte     q7      %1f        "Question 7"
                byte     q8      %1f        "Question 8"
                byte     q9      %1f        "Question 9"
                byte     q10     %1f        "Question 10"
}
─────────────────────────────────────────────────────── end fname.dct ───────
```

Up to five pieces of information may be supplied in the dictionary for each variable: the location of the data, the storage type of the variable, the name of the variable, the input format, and the variable label.

Thus the `str50` line says that the first variable is to be given a storage type of `str50`, called `name`, and is to have the variable label "Name of respondent". The `%32s` is the input format, which tells Stata how to read the data. The `s` tells Stata not to remove any embedded blanks; the 32 tells Stata to go across 32 columns when reading the data.

The next line says that the second variable is to be assigned a storage type of `long`, named `id`, and be labeled "Person id". Stata should start reading the information for this variable in column 33. The `f` tells Stata to remove any embedded blanks, and the 5 says to read across five columns.

The third variable is to be given a storage type of `str50`, called `addr`, and be labeled "Address". The `_skip(2)` directs Stata to skip two columns before beginning to read the data for this variable, and the `%41s` instructs Stata to read across 41 columns and not to remove embedded blanks.

`line(2)` instructs Stata to go to line 2 of the observation.

The remainder of the data is 0/1 coded, indicating the answers to the questions. It would be convenient if we could use a shorthand to specify this portion of the dictionary, but we must supply explicit directives.

◁

□ Technical note

In the preceding example, there were two pieces of information about location: where the data begin for each variable (the `_column()`, `_skip()`, `_line()`) and how many columns the data span (the `%32s`, `%5f`, `%41s`, `%1f`). In our dictionary, some of this information was redundant. After reading `name`, Stata had finished with 32 columns of information. Unless instructed otherwise, Stata would proceed to the next column—column 33—to begin reading information about `id`. The `_column(33)` was unnecessary.

The _skip(2) was necessary, however. Stata had read 37 columns of information and was ready to look at column 38. Although the address information does not begin until column 40, columns 38 and 39 contain blanks. Because these are leading blanks instead of embedded blanks, Stata would just ignore them without any trouble. The problem is with the %41s. If Stata begins reading the address information from column 38 and reads 41 columns, Stata would stop reading in column 78 ($78 - 41 + 1 = 38$), but the widest address ends in column 80. We could have omitted the _skip(2) if we had specified an input format of %43s.

The _line(2) was necessary, although we could have read the second line by coding _newline instead.

The _column(1) could have been omitted. After the _line(), Stata begins in column 1.

See the next example for a dataset in which both pieces of location information are required.

❑

▷ Example 11

The following file contains six variables in a variety of formats. In the dictionary, we read the variables fifth and sixth out of order by forcing the column pointer.

```
──────────────────────────────────────────────── begin example.dct ──────────
infile dictionary {
                          first    %3f
              double   second   %2.1f
                          third    %6f
     _skip(2)     str4    fourth   %4s
     _column(21)          sixth %4.1f
     _column(18)          fifth %2f
}
1.2125.7e+252abcd 1 .232
1.3135.7    52efgh2    5
1.41457    52abcd 3 100.
1.5155.7D+252efgh04 1.7
16 16 .57  52abcd 5 1.71
──────────────────────────────────────────────── end example.dct ──────────
```

Assuming that the above is stored in a file called example.dct, we can infile and list it by typing

```
. infile using example
infile dictionary {
                          first    %3f
              double   second   %2.1f
                          third    %6f
     _skip(2)     str4    fourth   %4s
     _column(21)          sixth %4.1f
     _column(18)          fifth %2f
}
(5 observations read)
```

```
. list
```

	first	second	third	fourth	sixth	fifth
1.	1.2	1.2	570	abcd	.232	1
2.	1.3	1.3	5.7	efgh	.5	2
3.	1.4	1.4	57	abcd	100	3
4.	1.5	1.5	570	efgh	1.7	4
5.	16	1.6	.57	abcd	1.71	5

◁

Reading fixed-block files

❏ Technical note

The $\_lrecl(\#)$ directive is used for reading datasets that do not have end-of-line delimiters (carriage return, line feed, or some combination of these). Such datasets are typical of IBM mainframes, where they are known as fixed block, or FB. The abbreviation LRECL is IBM mainframe jargon for logical record length.

In a fixed-block dataset, each $\#$ characters are to be interpreted as a record. For instance, consider the data

```
1 21
2 42
3 63
```

In fixed-block format, these data might be recorded as

――――――――――――――――――――――――――――――――― begin mydata.ibm ―――――――
```
1 212 423 63
```
――――――――――――――――――――――――――――――――― end mydata.ibm ―――――――

and you would be told, on the side, that the LRECL is 4. If you then pass along that information to infile, it can read the data:

――――――――――――――――――――――――――――――――― begin mydata.dct ―――――――
```
infile dictionary using mydata.ibm {
        _lrecl(4)
        int     id
        int     age
}
```
――――――――――――――――――――――――――――――――― end mydata.dct ―――――――

When you do not specify the $\_lrecl(\#)$ directive, infile assumes that each line ends with the standard ASCII delimiter (which can be a line feed, a carriage return, a line feed followed by a carriage return, or a carriage return followed by a line feed). When you specify $\_lrecl(\#)$, infile reads the data in blocks of $\#$ characters and then acts as if that is a line.

A common mistake in processing fixed-block datasets is to use an incorrect LRECL value, such as 160 when it is really 80. To understand what can happen, pretend that you thought the LRECL in your data was 6 rather than 4. Taking the characters in groups of 6, the data appear as

```
1 212
423 63
```

Stata cannot verify that you have specified the correct LRECL, so if the data appear incorrect, verify that you have the correct number.

The maximum LRECL `infile` allows is 524,275.

❏

Reading EBCDIC files

In the previous section, we discussed the `_lrecl(#)` directive that is often necessary for files that originated on mainframes and do not have end-of-line delimiters.

Such files sometimes are not even ASCII files, which are commonly known just as a plain-text file. Sometimes, these files have an alternate character encoding known as extended binary coded decimal interchange code (EBCDIC). The EBCDIC encoding was created in the 1960s by IBM for its mainframes.

Because EBCDIC is a different character encoding, we cannot even show you a printed example; it would be unreadable. Nevertheless, Stata can convert EBCDIC files to ASCII (see [D] **filefilter**) and can read data from EBCDIC files.

If you have a data file encoded with EBCDIC, you undoubtedly also have a description of it from which you can create a dictionary that includes the LRECL of the file (EBCDIC files do not typically have end-of-line delimiters) and the character positions of the fields in the file. You create a dictionary for an EBCDIC file just as you would for an ASCII file, using the Do-file Editor or another text editor, and being sure to use the `_lrecl()` directive in the dictionary to specify the LRECL. You then simply specify the `ebcdic` option for `infile`, and Stata will convert the characters in the file from EBCDIC to ASCII on the fly:

```
. infile using mydict, ebcdic
```

References

Gleason, J. R. 1998. dm54: Capturing comments from data dictionaries.i *Stata Technical Bulletin* 42: 3–4. Reprinted in *Stata Technical Bulletin Reprints*, vol. 7, pp. 55–57. College Station, TX: Stata Press.

Gould, W. W. 1992. dm10: Infiling data: Automatic dictionary creation. *Stata Technical Bulletin* 9: 4–8. Reprinted in *Stata Technical Bulletin Reprints*, vol. 2, pp. 28–34. College Station, TX: Stata Press.

Nash, J. D. 1994. dm19: Merging raw data and dictionary files. *Stata Technical Bulletin* 20: 3–5. Reprinted in *Stata Technical Bulletin Reprints*, vol. 4, pp. 22–25. College Station, TX: Stata Press.

Also see

[D] **infile (free format)** — Read unformatted text data

[D] **infix (fixed format)** — Read text data in fixed format

[D] **export** — Overview of exporting data from Stata

[D] **import** — Overview of importing data into Stata

[U] **21 Inputting and importing data**

Title

> **infile (free format)** — Read unformatted text data

Syntax

> infile *varlist* [_skip[(#)]] [*varlist* [_skip[(#)] ...]]] using *filename* [*if*] [*in*]
>
> [, *options*]

options	Description
Main	
clear	replace data in memory
Options	
automatic	create value labels from nonnumeric data
byvariable(#)	organize external file by variables; # is number of observations

Menu

File > Import > Unformatted text data

Description

infile reads into memory from a disk a dataset that is not in Stata format. If *filename* is specified without an extension, .raw is assumed.

Note for Stata for Mac and Stata for Windows users: If your *filename* contains embedded spaces, remember to enclose it in double quotes.

Here we discuss using infile to read free-format data, meaning datasets in which Stata does not need to know the formatting information. Another variation on infile allows reading fixed-format data; see [D] **infile (fixed format)**. Yet another alternative is insheet, which is easier to use if your data are tab- or comma-separated and contain 1 observation per line. Stata has other commands for reading data, too. If you are not certain that infile will do what you are looking for, see [D] **import** and [U] **21 Inputting and importing data**.

After the data are read into Stata, they can be saved in a Stata-format dataset; see [D] **save**.

Options

> **Main**

clear specifies that it is okay for the new data to replace the data that are currently in memory. To ensure that you do not lose something important, infile will refuse to read new data if data are already in memory. clear allows infile to replace the data in memory. You can also drop the data yourself by typing drop _all before reading new data.

346

`automatic` causes Stata to create value labels from the nonnumeric data it reads. It also automatically widens the display format to fit the longest label.

`byvariable(#)` specifies that the external data file is organized by variables rather than by observations. All the observations on the first variable appear, followed by all the observations on the second variable, and so on. Time-series datasets sometimes come in this format.

Remarks

This section describes `infile` features for reading data in free or comma-separated–value format.

Remarks are presented under the following headings:

> *Reading free-format data*
> *Reading comma-separated data*
> *Specifying variable types*
> *Reading string variables*
> *Skipping variables*
> *Skipping observations*
> *Reading time-series data*

Reading free-format data

In free format, data are separated by one or more white-space characters—blanks, tabs, or new lines (carriage return, line feed, or carriage-return/line feed combinations). Thus one observation may span any number of lines.

Numeric missing values are indicated by single periods (".").

▷ Example 1

In the file `highway.raw`, we have information on the accident rate per million vehicle miles along a stretch of highway, the speed limit on that highway, and the number of access points (on-ramps and off-ramps) per mile. Our file contains

```
──────────────────────────────────── begin highway.raw, example 1 ────────
4.58 55 4.6
2.86  60 4.4
1.61 . 2.2
3.02 60
4.7
──────────────────────────────────── end highway.raw, example 1 ────────
```

We can read these data by typing

```
. infile acc_rate spdlimit acc_pts using highway
(4 observations read)

. list
```

	acc_rate	spdlimit	acc_pts
1.	4.58	55	4.6
2.	2.86	60	4.4
3.	1.61	.	2.2
4.	3.02	60	4.7

The spacing of the numbers in the original file is irrelevant.

◁

❏ Technical note

Missing values need not be indicated by one period. The third observation on the speed limit is missing in example 1. The raw data file indicates this by recording one period. Let's assume, instead, that the missing value was indicated by the word unknown. Thus the raw data file appears as

―― begin highway.raw, example 2 ――――――――――
```
4.58 55 4.6
2.86  60 4.4
1.61 unknown 2.2
3.02 60
4.7
```
―― end highway.raw, example 2 ――――――――――

Here is the result of infiling these data:

```
. infile acc_rate spdlimit acc_pts using highway
'unknown' cannot be read as a number for spdlimit[3]
(4 observations read)
```

infile warned us that it could not read the word unknown, stored a *missing*, and then continued to read the rest of the dataset. Thus aside from the warning message, results are unchanged.

Because not all packages indicate missing data in the same way, this feature can be useful when reading data. Whenever infile sees something that it does not understand, it warns you, records a *missing*, and continues. If, on the other hand, the missing values were recorded not as unknown but as, say, 99, Stata would have had no difficulty reading the number, but it would also have stored 99 rather than missing. To convert such coded missing values to true missing values, see [D] **mvencode**.

❏

Reading comma-separated data

In comma-separated–value format, data are separated by commas. You may mix comma-separated–value and free formats. Missing values are indicated either by single periods or by multiple commas that serve as placeholders, or both. As with free format, 1 observation may span any number of input lines.

▷ Example 2

We can modify the format of highway.raw used in example 1 without affecting infile's ability to read it. The dataset can be read with the same command, and the results would be the same if the file instead contained

―― begin highway.raw, example 3 ――――――――――
```
4.58,55 4.6
2.86, 60,4.4
1.61,,2.2
3.02,60
4.7
```
―― end highway.raw, example 3 ――――――――――

◁

Specifying variable types

The variable names you type after the word `infile` are new variables. The syntax for a new variable is

$$[\mathit{type}]\ \mathit{new\_varname}[\,\mathbf{:}\mathit{label\_name}]$$

A full discussion of this syntax can be found in [U] **11.4 varlists**. As a quick review, new variables are, by default, of type `float`. This default can be overridden by preceding the variable name with a storage type (`byte`, `int`, `long`, `float`, `double`, or `str#`) or by using the `set type` command. A list of variables placed in parentheses will be given the same type. For example,

double(*first_var second_var* ... *last_var*)

causes *first_var second_var* ... *last_var* to all be of type `double`.

There is also a shorthand syntax for variable names with numeric suffixes. The varlist `var1-var4` is equivalent to specifying `var1 var2 var3 var4`.

▷ Example 3

In the highway example, we could `infile` the data `acc_rate`, `spdlimit`, and `acc_pts` and force the variable `spdlimit` to be of type `int` by typing

```
. infile acc_rate int spdlimit acc_pts using highway, clear
(4 observations read)
```

We could force all variables to be of type `double` by typing

```
. infile double(acc_rate spdlimit acc_pts) using highway, clear
(4 observations read)
```

We could call the three variables `v1`, `v2`, and `v3` and make them all of type `double` by typing

```
. infile double(v1-v3) using highway, clear
(4 observations read)
```

◁

Reading string variables

By explicitly specifying the types, you can read string variables, as well as numeric variables.

▷ Example 4

Typing `infile str20 name age sex using myfile` would read

─── begin myfile.raw ───────────
```
    "Sherri Holliday" 25 1
    Branton 32 1
    "Bill Ross" 27,0
```
─── begin myfile.raw ───────────

or even

─── begin myfile.raw, variation 2 ───────────
```
    'Sherri Holliday' 25,1 "Branton" 32
    1,'Bill Ross', 27,0
```
─── end myfile.raw, variation 2 ───────────

The spacing is irrelevant, and either single or double quotes may be used to delimit strings. The quotes do not count when calculating the length of strings. Quotes may be omitted altogether if the string contains no blanks or other special characters (anything other than letters, numbers, or underscores).

Typing

```
. infile str20 name age sex using myfile, clear
(3 observations read)
```

makes `name` a `str20` and `age` and `sex` `floats`. We might have typed

```
. infile str20 name age int sex using myfile, clear
(3 observations read)
```

to make `sex` an `int` or

```
. infile str20 name int(age sex) using myfile, clear
(3 observations read)
```

to make both `age` and `sex` `ints`.   ◁

❏ Technical note

`infile` can also handle nonnumeric data by using *value labels*. We will briefly review value labels, but you should see [U] **12.6.3 Value labels** for a complete description.

A value label is a mapping from the set of integers to words. For instance, if we had a variable called `sex` in our data that represented the sex of the individual, we might code 0 for male and 1 for female. We could then just remember that every time we see a value of 0 for `sex`, that observation refers to a male, whereas 1 refers to a female.

Even better, we could inform Stata that 0 represents males and 1 represents females by typing

```
. label define sexfmt 0 "Male" 1 "Female"
```

Then we must tell Stata that this coding scheme is to be associated with the variable `sex`. This is typically done by typing

```
. label values sex sexfmt
```

Thereafter, Stata will print `Male` rather than 0 and `Female` rather than 1 for this variable.

Stata has the ability to turn a value label around. Not only can it go from numeric codes to words such as "Male" and "Female", it can also go from the words to the numeric code. We tell `infile` the value label that goes with each variable by placing a colon (`:`) after the variable name and typing the name of the value label. Before we do that, we use the `label define` command to inform Stata of the coding.

Let's assume that we wish to `infile` a dataset containing the words `Male` and `Female` and that we wish to store numeric codes rather than the strings themselves. This will result in considerable data compression, especially if we store the numeric code as a `byte`. We have a dataset named `persons.raw` that contains name, sex, and age:

```
————————————————————————————————————— begin persons.raw ———————
  "Arthur Doyle" Male 22
  "Mary Hope" Female 37
  "Guy Fawkes" Male 48
  "Carrie House" Female 25
————————————————————————————————————— end persons.raw ————————
```

Here is how we read and encode it at the same time:

```
. label define sexfmt 0 "Male" 1 "Female"
. infile str16 name sex:sexfmt age using persons
(4 observations read)
. list
```

	name	sex	age
1.	Arthur Doyle	Male	22
2.	Mary Hope	Female	37
3.	Guy Fawkes	Male	48
4.	Carrie House	Female	25

The `str16` in the `infile` command applies only to the `name` variable; `sex` is a numeric variable, which we can prove by typing

```
. list, nolabel
```

	name	sex	age
1.	Arthur Doyle	0	22
2.	Mary Hope	1	37
3.	Guy Fawkes	0	48
4.	Carrie House	1	25

❑

❑ Technical note

When `infile` is directed to use a value label and it finds an entry in the file that does not match any of the codings recorded in the label, it prints a warning message and stores *missing* for the observation. By specifying the `automatic` option, you can instead have `infile` automatically add new entries to the value label.

Say that we have a dataset containing three variables. The first, region of the country, is a character string; the remaining two variables, which we will just call `var1` and `var2`, contain numbers. We have stored the data in a file called `geog.raw`:

```
──────────────────────────────────────────────── begin geog.raw ───────────
   "NE"      31.23      87.78
   'NCntrl'  29.52      98.92
   South     29.62     114.69
   West      28.28     218.92
   NE        17.50      44.33
   NCntrl    22.51      55.21
──────────────────────────────────────────────────── end geog.raw ─────────
```

The easiest way to read this dataset is to type

```
. infile str6 region var1 var2 using geog
```

making `region` a string variable. We do not want to do this, however, because we are practicing for reading a dataset like this containing 20,000 observations. If `region` were numerically encoded and stored as a `byte`, there would be a 5-byte saving per observation, reducing the size of the data by 100,000 bytes. We also do not want to bother with first creating the value label. Using the `automatic` option, `infile` creates the value label automatically as it encounters new regions.

```
. infile byte region:regfmt var1 var2 using geog, automatic clear
(6 observations read)
. list, sep(0)
```

	region	var1	var2
1.	NE	31.23	87.78
2.	NCntrl	29.52	98.92
3.	South	29.62	114.69
4.	West	28.28	218.92
5.	NE	17.5	44.33
6.	NCntrl	22.51	55.21

infile automatically created and defined a new value label called regfmt. We can use the label list command to view its contents:

```
. label list regfmt
regfmt:
           1 NE
           2 NCntrl
           3 South
           4 West
```

The value label need not be undefined before we use infile with the automatic option. If the value label regfmt had been previously defined as

```
. label define regfmt 2 "West"
```

the result of label list after the infile would have been

```
regfmt:
           2 West
           3 NE
           4 NCntrl
           5 South
```

The automatic option is convenient, but there is one reason for using it. Suppose that we had a dataset containing, among other things, information about an individual's sex. We know that the sex variable is supposed to be coded male and female. If we read the data by using the automatic option and if one of the records contains fmlae, then infile will blindly create a third sex rather than print a warning.

❑

Skipping variables

Specifying _skip instead of a variable name directs infile to ignore the variable in that location. This feature makes it possible to extract manageable subsets from large disk datasets. A number of contiguous variables can be skipped by specifying _skip(#), where # is the number of variables to ignore.

▷ Example 5

In the highway example from example 1, the data file contained three variables: acc_rate, spdlimit, and acc_pts. We can read the first two variables by typing

```
. infile acc_rate spdlimit _skip using highway
(4 observations read)
```

We can read the first and last variables by typing

```
. infile acc_rate _skip acc_pts using highway, clear
(4 observations read)
```

We can read the first variable by typing

```
. infile acc_rate _skip(2) using highway, clear
(4 observations read)
```

_skip may be specified more than once. If we had a dataset containing four variables—say, a, b, c, and d—and we wanted to read just a and c, we could type `infile a _skip c _skip using` *filename*.

◁

Skipping observations

Subsets of observations can be extracted by specifying if *exp*, which also makes it possible to extract manageable subsets from large disk datasets. Do not, however, use the *_variable* _N in *exp*. Use the in *range* modifier to refer to observation numbers within the disk dataset.

▷ Example 6

Again referring to the highway example, if we type

```
. infile acc_rate spdlimit acc_pts if acc_rate>3 using highway, clear
(2 observations read)
```

only observations for which `acc_rate` is greater than 3 will be infiled. We can type

```
. infile acc_rate spdlimit acc_pts in 2/4 using highway, clear
(eof not at end of obs)
(3 observations read)
```

to read only the second, third, and fourth observations.

◁

Reading time-series data

If you are dealing with time-series data, you may receive datasets organized by variables rather than by observations. All the observations on the first variable appear, followed by all the observations on the second variable, and so on. The `byvariable(#)` option specifies that the external data file is organized in this way. You specify the number of observations in the parentheses, because `infile` needs to know that number to read the data properly. You can also mark the end of one variable's data and the beginning of another's data by placing a semicolon (";") in the raw data file. You may then specify a number larger than the number of observations in the dataset and leave it to `infile` to determine the actual number of observations. This method can also be used to read unbalanced data.

▷ Example 7

We have time-series data on 4 years recorded in the file `time.raw`. The dataset contains information on year, amount, and cost, and is organized by variable:

```
                                                    begin time.raw
1980  1981  1982  1983
14  17  25  30
120  135  150
180
                                                    end time.raw
```

We can read these data by typing

```
. infile year amount cost using time, byvariable(4) clear
(4 observations read)

. list
```

	year	amount	cost
1.	1980	14	120
2.	1981	17	135
3.	1982	25	150
4.	1983	30	180

If the data instead contained semicolons marking the end of each series and had no information for amount in 1983, the raw data might appear as

```
1980  1981  1982  1983 ;
14  17  25 ;
120  135  150
180 ;
```

We could read these data by typing

```
. infile year amount cost using time, byvariable(100) clear
(4 observations read)

. list
```

	year	amount	cost
1.	1980	14	120
2.	1981	17	135
3.	1982	25	150
4.	1983	.	180

◁

Also see

[D] **infile (fixed format)** — Read text data in fixed format with a dictionary

[D] **import** — Overview of importing data into Stata

[D] **export** — Overview of exporting data from Stata

[U] **21 Inputting and importing data**

Title

> **infix (fixed format)** — Read text data in fixed format

Syntax

> infix using *dfilename* [*if*] [*in*] [, using(*filename₂*) clear]

> infix *specifications* using *filename* [*if*] [*in*] [, clear]

where *dfilename*, if it exists, contains

——————————————————————————————— begin dictionary file ————————

    infix dictionary [using *filename*] {
         * *comments preceded by asterisk may appear freely*
         specifications
    }
    (*your data might appear here*)

——————————————————————————————— end dictionary file ————————
---

and where *specifications* is

  # firstlineoffile
  # lines
  #:
  /
  [byte | int | float | long | double | str] *varlist* [*#:*]*#*[*-#*]

Menu

File > Import > Text data in fixed format

Description

infix reads into memory from a disk dataset that is *not* in Stata format. infix requires that the data be in fixed-column format.

If *dfilename* is specified without an extension, .dct is assumed. If *filename* is specified without an extension, .raw is assumed. If *dfilename* contains embedded spaces, remember to enclose it in double quotes.

In the first syntax, if using *filename₂* is not specified on the command line and using *filename* is not specified in the dictionary, the data are assumed to begin on the line following the closing brace.

infile and insheet are alternatives to infix. infile can also read data in fixed format—see [D] **infile (fixed format)**—and it can read data in free format—see [D] **infile (free format)**. Most people think that infix is easier to use for reading fixed-format data, but infile has more features. If your data are not fixed format, you can use insheet; see [D] **insheet**. If you are not certain that infix will do what you are looking for, see [D] **import** and [U] **21 Inputting and importing data**.

In its first syntax, infix reads the data in a two-step process. You first create a disk file describing how the data are recorded. You tell infix to read that file—called a dictionary—and from there, infix reads the data. The data can be in the same file as the dictionary or in a different file.

In its second syntax, you tell infix how to read the data right on the command line with no intermediate file.

Options

_____| Main |_____

using(*filename₂*) specifies the name of a file containing the data. If using() is not specified, the data are assumed to follow the dictionary in *dfilename*, or if the dictionary specifies the name of some other file, that file is assumed to contain the data. If using(*filename₂*) is specified, *filename₂* is used to obtain the data, even if the dictionary says otherwise. If *filename₂* is specified without an extension, .raw is assumed. If *filename₂* contains embedded spaces, remember to enclose it in double quotes.

clear specifies that it is okay for the new data to replace what is currently in memory. To ensure that you do not lose something important, infix will refuse to read new data if data are already in memory. clear allows infix to replace the data in memory. You can also drop the data yourself by typing drop _all before reading new data.

Specifications

firstlineoffile (abbreviation first) is rarely specified. It states the line of the file at which the data begin. You need not specify first when the data follow the dictionary; infix can figure that out for itself. You can specify first when only the data appear in a file and the first few lines of that file contain headers or other markers.

first appears only once in the specifications.

lines states the number of lines per observation in the file. Simple datasets typically have "1 lines". Large datasets often have many lines (sometimes called records) per observation. lines is optional, even when there is more than one line per observation, because infix can sometimes figure it out for itself. Still, if 1 lines is not right for your data, it is best to specify the appropriate number of lines.

lines appears only once in the specifications.

#: tells infix to jump to line # of the observation. Consider a file with 4 lines, meaning four lines per observation. 2: says to jump to the second line of the observation. 4: says to jump to the fourth line of the observation. You may jump forward or backward: infix does not care, and there is no inefficiency in going forward to 3:, reading a few variables, jumping back to 1:, reading another variable, and jumping back again to 3:.

You need not ensure that, at the end of your specification, you are on the last line of the observation. infix knows how to get to the next observation because it knows where you are and it knows lines, the total number of lines per observation.

#: may appear many times in the specifications.

/ is an alternative to #:. / goes forward one line. // goes forward two lines. We do not recommend using / because #: is better. If you are currently on line 2 of an observation and want to get to line 6, you could type ////, but your meaning is clearer if you type 6:.

/ may appear many times in the specifications.

[byte | int | float | long | double | str] *varlist* [#:]#[-#] instructs infix to read a variable or, sometimes, more than one.

The simplest form of this is *varname #*, such as sex 20. That says that variable *varname* be read from column # of the current line; that variable sex be read from column 20; and that here, sex is a one-digit number.

varname #-#, such as age 21-23, says that *varname* be read from the column range specified; that age be read from columns 21 through 23; and that here, age is a three-digit number.

You can prefix the variable with a storage type. str name 25-44 means to read the string variable name from columns 25 through 44. If you do not specify str, the variable is assumed to be numeric. You can specify the numeric subtype if you wish.

You can specify more than one variable, with or without a type. byte q1-q5 51-55 means read variables q1, q2, q3, q4, and q5 from columns 51 through 55 and store the five variables as bytes.

Finally, you can specify the line on which the variable(s) appear. age 2:21-23 says that age is to be obtained from the second line, columns 21 through 23. Another way to do this is to put together the #: directive with the input-variable directive: 2: age 21-23. There is a difference, but not with respect to reading the variable age. Let's consider two alternatives:

```
1:  str name 25-44     age 2:21-23   q1-q5 51-55
1:  str name 25-44  2: age 21-23     q1-q5 51-55
```

The difference is that the first directive says that variables q1 through q5 are on line 1, whereas the second says that they are on line 2.

When the colon is put in front, it indicates the line on which variables are to be found when we do not explicitly say otherwise. When the colon is put inside, it applies only to the variable under consideration.

Remarks

Remarks are presented under the following headings:

> *Two ways to use infix*
> *Reading string variables*
> *Reading data with multiple lines per observation*
> *Reading subsets of observations*

Two ways to use infix

There are two ways to use infix. One is to type the specifications that describe how to read the fixed-format data on the command line:

```
. infix acc_rate 1-4  spdlimit 6-7  acc_pts 9-11  using highway.raw
```

The other is to type the specifications into a file,

```
──────────────────────────────────────── begin highway.dct, example 1 ────────
  infix dictionary using highway.raw {
        acc_rate 1-4
        spdlimit 6-7
        acc_pts  9-11
  }
──────────────────────────────────────── end highway.dct, example 1 ────────
```

and then, in Stata, type

```
. infix using highway.dct
```

The method you use makes no difference to Stata. The first method is more convenient if there are only a few variables, and the second method is less prone to error if you are reading a big, complicated file.

The second method allows two variations, the one we just showed—where the data are in another file—and one where the data are in the same file as the dictionary:

```
──────────────────────────────────────────── begin highway.dct, example 2 ────────
    infix dictionary {
            acc_rate  1-4
            spdlimit  6-7
            acc_pts   9-11
    }
    4.58 55 .46
    2.86 60 4.4
    1.61    2.2
    3.02 60 4.7
──────────────────────────────────────────────── end highway.dct, example 2 ────────
```

Note that in the first example, the top line of the file read `infix dictionary using highway.raw`, whereas in the second, the line reads simply `infix dictionary`. When you do not say where the data are, Stata assumes that the data follow the dictionary.

▷ Example 1

So, let's complete the example we started. We have a dataset on the accident rate per million vehicle miles along a stretch of highway, the speed limit on that highway, and the number of access points per mile. We have created the dictionary file, `highway.dct`, which contains the dictionary and the data:

```
──────────────────────────────────────────── begin highway.dct, example 2 ────────
    infix dictionary {
            acc_rate  1-4
            spdlimit  6-7
            acc_pts   9-11
    }
    4.58 55 .46
    2.86 60 4.4
    1.61    2.2
    3.02 60 4.7
──────────────────────────────────────────────── end highway.dct, example 2 ────────
```

We created this file outside Stata by using an editor or word processor. In Stata, we now read the data. `infix` lists the dictionary so that we will know the directives it follows:

```
. infix using highway
infix dictionary {
        acc_rate  1-4
        spdlimit  6-7
        acc_pts   9-11
}
(4 observations read)

. list
```

	acc_rate	spdlimit	acc_pts
1.	4.58	55	.46
2.	2.86	60	4.4
3.	1.61	.	2.2
4.	3.02	60	4.7

We simply typed `infix using highway` rather than `infix using highway.dct`. When we do not specify the file extension, `infix` assumes that we mean `.dct`.

◁

Reading string variables

When you do not say otherwise in your specification—either in the command line or in the dictionary—`infix` assumes that variables are numeric. You specify that a variable is a string by placing `str` in front of its name:

```
. infix id 1-6  str name 7-36  age 38-39  str sex 41  using employee.raw
```

or

```
                                                  ─────── begin employee.dct ────────
infix dictionary using employee.raw {
        id          1-6
        str name   7-36
        age        38-39
        str sex      40
}
                                                  ─────── end employee.dct ────────
```

Reading data with multiple lines per observation

When a dataset has multiple lines per observation—sometimes called multiple records per observation—you specify the number of lines per observation by using `lines`, and you specify the line on which the elements appear by using `#:`. For example,

```
. infix  2 lines  1: id 1-6  str name 7-36  2: age 1-2  str sex 4  using emp2.raw
```

or

```
                                                ─────── begin emp2.dct ────────
infix dictionary using emp2.raw {
    2 lines
    1:
        id          1-6
        str name   7-36
    2:
        age          1-2
        str sex        4
}
                                                ─────── end emp2.dct ────────
```

There are many different ways to do the same thing.

▷ Example 2

Consider the following raw data:

```
                                                     ── begin mydata.raw ─────
id income educ / sex age / rcode, answers to questions 1-5
1024 25000 HS
      Male   28
      1 1 9 5 0 3
1025 27000 C
      Female 24
      0 2 2 1 1 3
1035 26000 HS
      Male   32
      1 1 0 3 2 1
1036 25000 C
      Female 25
      1 3 1 2 3 2
──────────────────────────────────────────────── end mydata.raw ─────
```

This dataset has three lines per observation, and the first line is just a comment. One possible method for reading these data is

```
                                                     ── begin mydata1.dct ─────
infix dictionary using mydata {
    2 first
    3 lines
    1:     id        1-4
           income    6-10
           str educ 12-13
    2:     str sex   6-11
           int age  13-14
    3:     rcode     6
           q1-q5     7-16
}
──────────────────────────────────────────────── end mydata1.dct ─────
```

although we prefer

```
                                                     ── begin mydata2.dct ─────
infix dictionary using mydata {
    2 first
    3 lines
           id        1: 1-4
           income    1: 6-10
           str educ 1:12-13
           str sex   2: 6-11
           age       2:13-14
           rcode     3: 6
           q1-q5     3: 7-16
}
──────────────────────────────────────────────── end mydata2.dct ─────
```

Either method will read these data, so we will use the first and then explain why we prefer the second.

```
. infix using mydata1
infix dictionary using mydata {
    2 first
    3 lines
    1:    id        1-4
          income    6-10
          str educ 12-13
    2:    str sex   6-11
          int age  13-14
    3:    rcode      6
          q1-q5     7-16
}
(4 observations read)
. list in 1/2
```

	id	income	educ	sex	age	rcode	q1	q2	q3	q4	q5
1.	1024	25000	HS	Male	28	1	1	9	5	0	3
2.	1025	27000	C	Female	24	0	2	2	1	1	3

What is better about the second is that the location of each variable is completely documented on each line—the line number and column. Because infix does not care about the order in which we read the variables, we could take the dictionary and jumble the lines, and it would still work. For instance,

──────────────────────────────────── begin mydata3.dct ────────

```
infix dictionary using mydata {
    2 first
    3 lines
        str sex   2: 6-11
        rcode     3: 6
        str educ  1:12-13
        age       2:13-14
        id        1: 1-4
        q1-q5     3: 7-16
        income    1: 6-10
}
```

──────────────────────────────────── end mydata3.dct ────────

will also read these data even though, for each observation, we start on line 2, go forward to line 3, jump back to line 1, and end up on line 1. It is not inefficient to do this because infix does not really jump to record 2, then record 3, then record 1 again, etc. infix takes what we say and organizes it efficiently. The order in which we say it makes no difference, except that the order of the variables in the resulting Stata dataset will be the order we specify.

Here the reordering is senseless, but in real datasets, reordering variables is often desirable. Moreover, we often construct dictionaries, realize that we omitted a variable, and then go back and modify them. By making each line complete, we can add new variables anywhere in the dictionary and not worry that, because of our addition, something that occurs later will no longer read correctly.

◁

Reading subsets of observations

If you wanted to read only the information about males from some raw data file, you might type

```
. infix id 1-6  str name 7-36  age 38-39  str sex 41  using employee.raw
> if sex=="M"
```

If your specification was instead recorded in a dictionary, you could type

```
. infix using employee.dct if sex=="M"
```

In another dataset, if you wanted to read just the first 100 observations, you could type

```
. infix 2 lines  1: id 1-6  str name 7-36  2: age 1-2  str sex 4  using emp2.raw
> in 1/100
```

or if the specification was instead recorded in a dictionary and you wanted observations 101–573, you could type

```
. infix using emp2.dct in 101/573
```

Also see

[D] **infile (fixed format)** — Read text data in fixed format with a dictionary

[D] **export** — Overview of exporting data from Stata

[D] **import** — Overview of importing data into Stata

[U] **21 Inputting and importing data**

Title

input — Enter data from keyboard

Syntax

<u>inp</u>ut [*varlist*] [, <u>a</u>utomatic <u>l</u>abel]

Description

input allows you to type data directly into the dataset in memory. See also [D] **edit** for a windowed alternative to input.

Options

automatic causes Stata to create value labels from the nonnumeric data it encounters. It also automatically widens the display format to fit the longest label. Specifying automatic implies label, even if you do not explicitly type the label option.

label allows you to type the labels (strings) instead of the numeric values for variables associated with value labels. New value labels are not automatically created unless automatic is specified.

Remarks

If no data are in memory, you must specify a *varlist* when you type input. Stata will then prompt you to enter the new observations until you type end.

▷ Example 1

We have data on the accident rate per million vehicle miles along a stretch of highway, along with the speed limit on that highway. We wish to type these data directly into Stata:

```
. input
nothing to input
r(104);
```

Typing input by itself does not provide enough information about our intentions. Stata needs to know the names of the variables we wish to create.

```
. input acc_rate spdlimit

     acc_rate   spdlimit
 1. 4.58 55
 2. 2.86 60
 3. 1.61 .
 4. end

.  _
```

We typed `input acc_rate spdlimit`, and Stata responded by repeating the variable names and prompting us for the first observation. We entered the values for the first two observations, pressing *Return* after each value was entered. For the third observation, we entered the accident rate (1.61), but we entered a period (.) for missing because we did not know the corresponding speed limit for the highway. After entering data for the fourth observation, we typed `end` to let Stata know that there were no more observations.

We can now `list` the data to verify that we have entered the data correctly:

```
. list
```

```
     ┌──────────────────────┐
     │ acc_rate    spdlimit │
     ├──────────────────────┤
 1.  │    4.58          55  │
 2.  │    2.86          60  │
 3.  │    1.61           .  │
     └──────────────────────┘
```

◁

If you have data in memory and type `input` without a *varlist*, you will be prompted to enter more information on *all* the variables. This continues until you type `end`.

▷ Example 2: Adding observations

We now have another observation that we wish to add to the dataset. Typing `input` by itself tells Stata that we wish to add new observations:

```
. input
        acc_rate    spdlimit
 4. 3.02 60
 5. end

. _
```

Stata reminded us of the names of our variables and prompted us for the fourth observation. We entered the numbers 3.02 and 60 and pressed *Return*. Stata then prompted us for the fifth observation. We could add as many new observations as we wish. Because we needed to add only 1 observation, we typed `end`. Our dataset now has 4 observations.

◁

You may add new variables to the data in memory by typing `input` followed by the names of the new variables. Stata will begin by prompting you for the first observation, then the second, and so on, until you type `end` or enter the last observation.

▷ Example 3: Adding variables

In addition to the accident rate and speed limit, we now obtain data on the number of access points (on-ramps and off-ramps) per mile along each stretch of highway. We wish to enter the new data.

```
. input acc_pts
        acc_pts
 1. 4.6
 2. 4.4
 3. 2.2
 4. 4.7

. _
```

When we typed input acc_pts, Stata responded by prompting us for the first observation. There are 4.6 access points per mile for the first highway, so we entered 4.6. Stata then prompted us for the second observation, and so on. We entered each of the numbers. When we entered the final observation, Stata automatically stopped prompting us—we did not have to type end. Stata knows that there are 4 observations in memory, and because we are adding a new variable, it stops automatically.

We can, however, type end anytime we wish, and Stata fills the remaining observations on the new variables with *missing*. To illustrate this, we enter one more variable to our data and then list the result:

```
. input junk

           junk
  1. 1
  2. 2
  3. end
. list
```

	acc_rate	spdlimit	acc_pts	junk
1.	4.58	55	4.6	1
2.	2.86	60	4.4	2
3.	1.61	.	2.2	.
4.	3.02	60	4.7	.

◁

You can input string variables by using input, but you must remember to indicate explicitly that the variables are strings by specifying the type of the variable before the variable's name.

▷ Example 4: Inputting string variables

String variables are indicated by the types str#, where # represents the storage length, or maximum length, of the variable. For instance, a str4 variable has a maximum length of 4, meaning that it can contain the strings a, ab, abc, and abcd, but not abcde. Strings shorter than the maximum length can be stored in the variable, but strings longer than the maximum length cannot. You can create variables up to str244.

Although a str80 variable can store strings shorter than 80 characters, you should not make all your string variables str80 because Stata allocates space for strings on the basis of their *maximum* length. Thus doing so would waste the computer's memory.

Let's assume that we have no data in memory and wish to enter the following data:

```
. input str16 name age str6 sex

                name        age        sex
  1. "Arthur Doyle" 22 male
  2. "Mary Hope" 37 "female"
  3. Guy Fawkes 48 male
'Fawkes' cannot be read as a number
  3. "Guy Fawkes" 48 male
  4. "Kriste Yeager" 25 female
  5. end

  . _
```

We first typed input str16 name age str6 sex, meaning that name is to be a str16 variable and sex a str6 variable. Because we did not specify anything about age, Stata made it a numeric variable.

Stata then prompted us to enter our data. On the first line, the name is Arthur Doyle, which we typed in double quotes. The double quotes are not really part of the string; they merely delimit the beginning and end of the string. We followed that with Mr. Doyle's age, 22, and his sex, male. We did not bother to type double quotes around the word `male` because it contained no blanks or special characters. For the second observation, we typed the double quotes around `female`; it changed nothing.

In the third observation, we omitted the double quotes around the name, and Stata informed us that Fawkes could not be read as a number and reprompted us for the observation. When we omitted the double quotes, Stata interpreted `Guy` as the name, `Fawkes` as the age, and 48 as the sex. This would have been okay with Stata, except for one problem: `Fawkes` looks nothing like a number, so Stata complained and gave us another chance. This time, we remembered to put the double quotes around the name.

Stata was satisfied, and we continued. We entered the fourth observation and typed `end`. Here is our dataset:

```
. list
```

	name	age	sex
1.	Arthur Doyle	22	male
2.	Mary Hope	37	female
3.	Guy Fawkes	48	male
4.	Kriste Yeager	25	female

◁

▷ Example 5: Specifying numeric storage types

Just as we indicated the string variables by placing a storage type in front of the variable name, we can indicate the storage type of our numeric variables as well. Stata has five numeric storage types: `byte`, `int`, `long`, `float`, and `double`. When you do not specify the storage type, Stata assumes that the variable is a `float`. See the definitions of numbers in [U] **12 Data**.

There are two reasons for explicitly specifying the storage type: to induce more precision or to conserve memory. The default type `float` has plenty of precision for most circumstances because Stata performs all calculations in double precision, no matter how the data are stored. If you were storing nine-digit Social Security numbers, however, you would want to use a different storage type, or the last digit would be rounded. `long` would be the best choice; `double` would work equally well, but it would waste memory.

Sometimes you do not need to store a variable as `float`. If the variable contains only integers between −32,767 and 32,740, it can be stored as an `int` and would take only half the space. If a variable contains only integers between −127 and 100, it can be stored as a `byte`, which would take only half again as much space. For instance, in example 4 we entered data for `age` without explicitly specifying the storage type; hence, it was stored as a `float`. It would have been better to store it as a `byte`. To do that, we would have typed

```
. input str16 name byte age str6 sex
                 name        age        sex
  1. "Arthur Doyle" 22 male
  2. "Mary Hope" 37 "female"
  3. "Guy Fawkes" 48 male
  4. "Kriste Yeager" 25 female
  5. end

.  _
```

Stata understands several shorthands. For instance, typing

. input int(a b) c

allows you to input three variables—a, b, and c—and makes both a and b ints and c a float. Remember, typing

. input int a b c

would make a an int but both b and c floats. Typing

. input a long b double(c d) e

would make a a float, b a long, c and d doubles, and e a float.

Stata has a shorthand for variable names with numeric suffixes. Typing v1-v4 is equivalent to typing v1 v2 v3 v4. Thus typing

. input int(v1-v4)

inputs four variables and stores them as ints.

◁

❏ Technical note

The rest of this section deals with using input with value labels. If you are not familiar with value labels, see [U] **12.6.3 Value labels**.

Value labels map numbers into words and vice versa. There are two aspects to the process. First, we must define the association between numbers and words. We might tell Stata that 0 corresponds to male and 1 corresponds to female by typing label define sexlbl 0 "male" 1 "female". The correspondences are named, and here we have named the 0↔male 1↔female correspondence sexlbl.

Next we must associate this value label with a variable. If we had already entered the data and the variable were called sex, we would do this by typing label values sex sexlbl. We would have entered the data by typing 0s and 1s, but at least now when we list the data, we would see the words rather than the underlying numbers.

We can do better than that. After defining the value label, we can associate the value label with the variable at the time we input the data and tell Stata to use the value label to interpret what we type:

```
. label define sexlbl 0 "male" 1 "female"
. input str16 name byte(age sex:sexlbl), label
                 name         age        sex
  1. "Arthur Doyle" 22 male
  2. "Mary Hope" 37 "female"
  3. "Guy Fawkes" 48 male
  4. "Kriste Yeager" 25 female
  5. end

  . _
```

After defining the value label, we typed our input command. We added the label option at the end of the command, and we typed sex:sexlbl for the name of the sex variable. The byte(...) around age and sex:sexlbl was not really necessary; it merely forced both age and sex to be stored as bytes.

Let's first decipher sex:sexlbl. sex is the name of the variable we want to input. The :sexlbl part tells Stata that the new variable is to be associated with the value label named sexlbl. The label option tells Stata to look up any strings we type for labeled variables in their corresponding value label and substitute the number when it stores the data. Thus when we entered the first observation of our data, we typed male for Mr. Doyle's sex, even though the corresponding variable is numeric. Rather than complaining that '"male" could not be read as a number", Stata accepted what we typed, looked up the number corresponding to male, and stored that number in the data.

That Stata has actually stored a number rather than the words male or female is almost irrelevant. Whenever we list the data or make a table, Stata will use the words male and female just as if those words were actually stored in the dataset rather than their numeric codings:

```
. list
```

	name	age	sex
1.	Arthur Doyle	22	male
2.	Mary Hope	37	female
3.	Guy Fawkes	48	male
4.	Kriste Yeager	25	female

```
. tabulate sex
```

sex	Freq.	Percent	Cum.
male	2	50.00	50.00
female	2	50.00	100.00
Total	4	100.00	

It is only almost irrelevant because we can use the underlying numbers in statistical analyses. For instance, if we were to ask Stata to calculate the mean of sex by typing summarize sex, Stata would report 0.5. We would interpret that to mean that one-half of our sample is female.

Value labels are permanently associated with variables, so once we associate a value label with a variable, we never have to do so again. If we wanted to add another observation to these data, we could type

```
. input, label
             name         age        sex
5. "Mark Esman" 26 male
6. end

. _
```
❑

❑ Technical note

The automatic option automates the definition of the value label. In the previous example, we informed Stata that male corresponds to 0 and female corresponds to 1 by typing label define sexlbl 0 "male" 1 "female". It was not necessary to explicitly specify the mapping. Specifying the automatic option tells Stata to interpret what we type as follows:

First, see if the value is a number. If so, store that number and be done with it. If it is not a number, check the value label associated with the variable in an attempt to interpret it. If an interpretation exists, store the corresponding numeric code. If one does not exist, add a new numeric code corresponding to what was typed. Store that new number and update the value label so that the new correspondence is never forgotten.

We can use these features to reenter our age and sex data. Before reentering the data, we drop _all and label drop _all to prove that we have nothing up our sleeve:

```
. drop _all

. label drop _all

. input str16 name byte(age sex:sexlbl), automatic
                   name        age        sex
  1. "Arthur Doyle" 22 male
  2. "Mary Hope" 37 "female"
  3. "Guy Fawkes" 48 male
  4. "Kriste Yeager" 25 female
  5. end

  . _
```

We previously defined the value label sexlbl so that male corresponded to 0 and female corresponded to 1. The label that Stata automatically created is slightly different but is just as good:

```
. label list sexlbl
sexlbl:
              1 male
              2 female
```

❏

Reference

Kohler, U. 2005. Stata tip 16: Using input to generate variables. *Stata Journal* 5: 134.

Also see

[D] **save** — Save Stata dataset

[D] **edit** — Browse or edit data with Data Editor

[D] **import** — Overview of importing data into Stata

[U] **21 Inputting and importing data**

Title

insheet — Read text data created by a spreadsheet

Syntax

insheet [*varlist*] using *filename* [, *options*]

options	Description
[no]double	override default storage type
tab	tab-delimited data
comma	comma-delimited data
delimiter("*char*")	use *char* as delimiter
clear	replace data in memory
case	preserve variable name's case
[no]names	variable names are included on the first line of the file

[no]names does not appear in the dialog box.

Menu

File > Import > Text data created by a spreadsheet

Description

insheet reads into memory from a disk a dataset that is not in Stata format. insheet is intended for reading files created by a spreadsheet or database program. Regardless of the creator of the file, insheet reads text (ASCII) files in which there is 1 observation per line and the values are separated by tabs or commas. Also the first line of the file can contain the variable names. If you type

. insheet using *filename*

insheet reads your data; that is all there is to it.

If *filename* is specified without an extension, .raw is assumed. If your *filename* contains embedded spaces, remember to enclose it in double quotes.

Stata has other commands for reading data. If you are not sure that insheet will do what you are looking for, see [D] **import** and [U] **21 Inputting and importing data**. If you want to save your data in spreadsheet-style text format, see [D] **outsheet**. However, export excel may be a better option; see [D] **import excel**.

Options

[no]double affects the way Stata handles the storage of floating-point variables. If the default storage type (see [D] **generate**) is set to float, specifying the double option forces Stata to store floating-point variables as doubles rather than floats. If the default storage type has been set to double, you must specify nodouble to have floating-point variables stored as floats rather than doubles; see [U] **12.2.2 Numeric storage types**.

tab tells Stata that the values are tab-separated. Specifying this option will speed insheet's processing, assuming that you are right. insheet can determine for itself whether the separation character is a tab or a comma.

comma tells Stata that the values are comma-separated. Specifying this option will speed insheet's processing, assuming that you are right. insheet can determine for itself whether the separation character is a comma or a tab.

delimiter("*char*") allows you to specify other separation characters. For instance, if values in the file are separated by a semicolon, specify delimiter(";").

clear specifies that it is okay for the new data to replace the data that are currently in memory. To ensure that you do not lose something important, insheet will refuse to read new data if data are already in memory. clear allows insheet to replace the data in memory. You can also drop the data yourself by typing drop _all before reading new data.

case preserves the variable name's case. By default, all variable names are imported as lowercase.

The following option is available with insheet but is not shown in the dialog box:

[no]names informs Stata whether variable names are included on the first line of the file. Specifying this option will speed insheet's processing, assuming that you are right. insheet can determine for itself whether the file includes variable names.

Remarks

insheet is easy. You type

```
. insheet using filename
```

and insheet reads your data. That is, it reads your data if

1. it can find the file and

2. the file meets insheet's expectations as to its format.

Assuring 1 is easy enough; just realize that if you type insheet using myfile, Stata interprets this as an instruction to read myfile.raw. If your file is called myfile.txt, type insheet using myfile.txt.

As for the file's format, most spreadsheets and some database programs write data in the form insheet expects. It is easy enough to look—as we will show you—and it is even easier simply to try and see what happens. If typing

```
. insheet using filename
```

does not produce the desired result, try one of Stata's other infile commands; see [D] **import**.

▷ Example 1

We have a raw data file on automobiles called auto.raw. This file was saved by a spreadsheet and can be read by typing

```
. insheet using auto
(5 vars, 10 obs)

. _
```

That done, we can now look at what we just loaded:

```
. describe
Contains data
    obs:            10
    vars:            5
    size:          270

                  storage   display    value
variable name     type      format     label        variable label

make              str13     %13s
price             int       %8.0g
mpg               byte      %8.0g
rep78             byte      %8.0g
foreign           str10     %10s

Sorted by:
    Note:  dataset has changed since last saved

. list
```

	make	price	mpg	rep78	foreign
1.	AMC Concord	4099	22	3	Domestic
2.	AMC Pacer	4749	17	3	Domestic
3.	AMC Spirit	3799	22	.	Domestic
4.	Buick Century	4816	20	3	Domestic
5.	Buick Electra	7827	15	4	Domestic
6.	Buick LeSabre	5788	18	3	Domestic
7.	Buick Opel	4453	26	.	Domestic
8.	Buick Regal	5189	20	3	Domestic
9.	Buick Riviera	10372	16	3	Domestic
10.	Buick Skylark	4082	19	3	Domestic

These data contain a combination of string and numeric variables. insheet figured all that out by itself.

◁

❑ Technical note

Now let's back up and look at the auto.raw file. Stata's type command will display files on the screen:

```
. type auto.raw
make      price   mpg   rep78   foreign
AMC Concord   4099   22   3       Domestic
AMC Pacer     4749   17   3       Domestic
AMC Spirit    3799   22   .       Domestic
Buick Century 4816   20   3       Domestic
Buick Electra 7827   15   4       Domestic
Buick LeSabre 5788   18   3       Domestic
Buick Opel    4453   26   .       Domestic
Buick Regal   5189   20   3       Domestic
Buick Riviera 10372  16   3       Domestic
Buick Skylark 4082   19   3       Domestic
```

These data have tab characters between values. Tab characters are invisible and are indistinguishable from blanks. type's showtabs option makes the tabs visible:

```
. type auto.raw, showtabs
make<T>price<T>mpg<T>rep78<T>foreign
AMC Concord<T>4099<T>22<T>3<T>Domestic
AMC Pacer<T>4749<T>17<T>3<T>Domestic
AMC Spirit<T>3799<T>22<T>.<T>Domestic
Buick Century<T>4816<T>20<T>3<T>Domestic
Buick Electra<T>7827<T>15<T>4<T>Domestic
Buick LeSabre<T>5788<T>18<T>3<T>Domestic
Buick Opel<T>4453<T>26<T>.<T>Domestic
Buick Regal<T>5189<T>20<T>3<T>Domestic
Buick Riviera<T>10372<T>16<T>3<T>Domestic
Buick Skylark<T>4082<T>19<T>3<T>Domestic
```

This is an example of the kind of data that `insheet` is willing to read. The first line contains the variable names, although that is not necessary. What is necessary is that the data values have tab characters between them.

`insheet` would be just as happy if the data values were separated by commas. Here is another variation on `auto.raw` that `insheet` can read:

```
. type auto2.raw
make,price,mpg,rep78,foreign
AMC Concord,4099,22,3,Domestic
AMC Pacer,4749,17,3,Domestic
AMC Spirit,3799,22,,Domestic
Buick Century,4816,20,3,Domestic
Buick Electra,7827,15,4,Domestic
Buick LeSabre,5788,18,3,Domestic
Buick Opel,4453,26,,Domestic
Buick Regal,5189,20,3,Domestic
Buick Riviera,10372,16,3,Domestic
Buick Skylark,4082,19,3,Domestic
```

It is easier for us human beings to see the commas rather than the tabs, but computers do not care one way or the other.

❏

▷ Example 2

The file does not have to contain variable names. Here is another variation on `auto.raw` without the first line, this time with commas rather than tabs separating the values:

```
. type auto3.raw
AMC Concord,4099,22,3,Domestic
AMC Pacer,4749,17,3,Domestic
 (output omitted )
Buick Skylark,4082,19,3,Domestic
```

Here is what happens when we read it:

```
. insheet using auto3
you must start with an empty dataset
r(18);

. _
```

Oops! We still have the data from the last example in memory. We need to clear the old data before reading the new data.

```
. insheet using auto3, clear
(5 vars, 10 obs)

. describe

Contains data
  obs:            10
  vars:            5
  size:          270
```

variable name	storage type	display format	value label	variable label
v1	str13	%13s		
v2	int	%8.0g		
v3	byte	%8.0g		
v4	byte	%8.0g		
v5	str10	%10s		

```
Sorted by:
      Note:  dataset has changed since last saved

. list
```

	v1	v2	v3	v4	v5
1.	AMC Concord	4099	22	3	Domestic
2.	AMC Pacer	4749	17	3	Domestic
	(output omitted)				
10.	Buick Skylark	4082	19	3	Domestic

The only difference in this dataset is that rather than the variables being nicely named make, price, mpg, rep78, and foreign, they are named v1, v2, ..., v5. We could now give our variables nicer names:

```
. rename v1 make

. rename v2 price

. _
```

We can also specify the variable names when reading the data:

```
. insheet make price mpg rep78 foreign using auto3, clear
(5 vars, 10 obs)

. list
```

	make	price	mpg	rep78	foreign
1.	AMC Concord	4099	22	3	Domestic
2.	AMC Pacer	4749	17	3	Domestic
	(output omitted)				
10.	Buick Skylark	4082	19	3	Domestic

If we use this approach, we must not specify too few variables,

```
. insheet make price mpg rep78 using auto3, clear
too few variables specified
error in line 11 of file
r(102);
```

or too many:

```
. insheet make price mpg rep78 foreign weight using auto3, clear
too many variables specified
error in line 11 of file
r(103);
```

We recommend typing

```
. insheet using filename
```

It is not difficult to rename your variables afterward, should that be necessary.

◁

> ## Example 3

The data may not always be appropriate for reading by insheet. Here is yet another version of the automobile data:

```
. type auto4.raw, showtabs
"AMC Concord"    4099  22  3  Domestic
"AMC Pacer"      4749  17  3  Domestic
"AMC Spirit"     3799  22  .  Domestic
"Buick Century"  4816  20  3  Domestic
"Buick Electra"  7827  15  4  Domestic
"Buick LeSabre"  5788  18  3  Domestic
"Buick Opel"     4453  26  .  Domestic
"Buick Regal"    5189  20  3  Domestic
"Buick Riviera" 10372  16  3  Domestic
"Buick Skylark"  4082  19  3  Domestic
```

We specified type's showtabs option, and no tabs are shown. These data are not tab-delimited or comma-delimited and are not the kind of data that insheet is designed to read. Let's try insheet anyway:

```
. insheet using auto4, clear
(1 var, 10 obs)
. describe
Contains data
  obs:            10
  vars:            1
  size:          390
```

variable name	storage type	display format	value label	variable label
v1	str39	%39s		

```
Sorted by:
     Note:  dataset has changed since last saved
. list
```

	v1
1.	AMC Concord 4099 22 3 Domestic
2.	AMC Pacer 4749 17 3 Domestic
	(output omitted)
10.	Buick Skylark 4082 19 3 Domestic

When `insheet` tries to read data that have no tabs or commas, it is fooled into thinking that the data contain just one variable. If we had these data, we would have to read the data with one of Stata's other commands, such as `infile` (free format).

◁

Also see

[D] **export** — Overview of exporting data from Stata

[D] **import** — Overview of importing data into Stata

[D] **rename** — Rename variable

[U] **21 Inputting and importing data**

Title

inspect — Display simple summary of data's attributes

Syntax

<u>insp</u>ect [*varlist*] [*if*] [*in*]

by is allowed; see [D] **by**.

Menu

Data > Describe data > Inspect variables

Description

The `inspect` command provides a quick summary of a numeric variable that differs from the summary provided by `summarize` or `tabulate`. It reports the number of negative, zero, and positive values; the number of integers and nonintegers; the number of unique values; and the number of *missing*; and it produces a small histogram. Its purpose is not analytical but is to allow you to quickly gain familiarity with unknown data.

Remarks

Typing `inspect` by itself produces an inspection for all the variables in the dataset. If you specify a *varlist*, an inspection of just those variables is presented.

▷ Example 1

`inspect` is not a replacement or substitute for `summarize` and `tabulate`. It is instead a data-management or information tool that lets us quickly gain insight into the values stored in a variable.

For instance, we receive data that purport to be on automobiles, and among the variables in the dataset is one called mpg. Its variable label is `Mileage (mpg)`, which is surely suggestive. We inspect the variable,

```
. use http://www.stata-press.com/data/r12/auto
(1978 Automobile Data)
. inspect mpg
```

mpg: Mileage (mpg)   Number of Observations

							Total	Integers	Nonintegers
		#				Negative	–	–	–
		#				Zero	–	–	–
		#				Positive	74	74	–
	#	#							
	#	#	#			Total	74	74	–
	#	#	#	#	.	Missing	–		

```
 12                41                                  74
    (21 unique values)
```

and we discover that the variable is never *missing*; all 74 observations in the dataset have some value for mpg. Moreover, the values are all positive and are all integers, as well. Among those 74 observations are 21 unique (different) values. The variable ranges from 12 to 41, and we are provided with a small histogram that suggests that the variable appears to be what it claims.

◁

▷ Example 2

Bob, a coworker, presents us with some census data. Among the variables in the dataset is one called region, which is labeled Census Region and is evidently a numeric variable. We inspect this variable:

```
. use http://www.stata-press.com/data/r12/bobsdata
(1980 Census data by state)
. inspect region
region:  Census region                       Number of Observations

                                       Total    Integers   Nonintegers
              #                Negative    -          -           -
              #    #           Zero        -          -           -
         #    #    #           Positive   50         50           -
    #    #    #    #                     ─────     ─────       ─────
    #    #    #    #           Total      50         50           -
    #    #    #    #    .       Missing    -

1                   5                       50
     (5 unique values)
          region is labeled but 1 value is NOT documented in the label.
```

In this dataset something may be wrong. region takes on five unique values. The variable has a value label, however, and one of the observed values is not documented in the label. Perhaps there is a typographical error.

◁

▷ Example 3

There was indeed an error. Bob fixes it and returns the data to us. Here is what inspect produces now:

```
. use http://www.stata-press.com/data/r12/census
(1980 Census data by state)
. inspect region
region:  Census region                       Number of Observations

                                       Total    Integers   Nonintegers
              #                Negative    -          -           -
              #                Zero        -          -           -
         #    #    #           Positive   50         50           -
    #    #    #    #                     ─────     ─────       ─────
    #    #    #    #           Total      50         50           -
    #    #    #    #           Missing     -

1                   4                       50
     (4 unique values)
          region is labeled and all values are documented in the label.
```

◁

▷ Example 4

We receive data on the climate in 956 U.S. cities. The variable `tempjan` records the Average January `temperature` in degrees Fahrenheit. The results of `inspect` are

```
. use http://www.stata-press.com/data/r12/citytemp
(City Temperature Data)
. inspect tempjan
tempjan:  Average January temperature              Number of Observations

                                             Total   Integers   Nonintegers
          #                   Negative          -          -            -
          #                   Zero              -          -            -
          #                   Positive        954         78          876
          #   #   #                         _____    _____       _____
          #   #   #           Total           954         78          876
     .    #   #   #    .       Missing           2
          +--------------------              _____
        2.2           72.6                    956
        (More than 99 unique values)
```

In two of the 956 observations, `tempjan` is *missing*. Of the 954 cities that have a recorded `tempjan`, all are positive, and 78 of them are integer values. `tempjan` varies between 2.2 and 72.6. There are more than 99 unique values of `tempjan` in the dataset. (Stata stops counting unique values after 99.)

◁

Saved results

`inspect` saves the following in `r()`:

Scalars
`r(N)`	number of observations
`r(N_neg)`	number of negative observations
`r(N_0)`	number of observations equal to 0
`r(N_pos)`	number of positive observations
`r(N_negint)`	number of negative integer observations
`r(N_posint)`	number of positive integer observations
`r(N_unique)`	number of unique values or . if more than 99
`r(N_undoc)`	number of undocumented values or . if not labeled

Also see

[D] **codebook** — Describe data contents

[D] **compare** — Compare two variables

[D] **describe** — Describe data in memory or in file

[D] **ds** — List variables matching name patterns or other characteristics

[D] **isid** — Check for unique identifiers

[R] **lv** — Letter-value displays

[R] **summarize** — Summary statistics

[R] **table** — Tables of summary statistics

[R] **tabulate oneway** — One-way tables of frequencies

[R] **tabulate, summarize()** — One- and two-way tables of summary statistics

[R] **tabulate twoway** — Two-way tables of frequencies

Title

ipolate — Linearly interpolate (extrapolate) values

Syntax

ipolate *yvar* *xvar* [*if*] [*in*] , generate(*newvar*) [epolate]

by is allowed; see [D] **by**.

Menu

Data > Create or change data > Other variable-creation commands > Linearly interpolate/extrapolate values

Description

ipolate creates in *newvar* a linear interpolation of *yvar* on *xvar* for missing values of *yvar*.

Because interpolation requires that *yvar* be a function of *xvar*, *yvar* is also interpolated for tied values of *xvar*. When *yvar* is not missing and *xvar* is neither missing nor repeated, the value of *newvar* is just *yvar*.

Options

generate(*newvar*) is required and specifies the name of the new variable to be created.

epolate specifies that values be both interpolated and extrapolated. Interpolation only is the default.

Remarks

▷ Example 1

We have data points on y and x, although sometimes the observations on y are missing. We believe that y is a function of x, justifying filling in the missing values by linear interpolation:

```
. use http://www.stata-press.com/data/r12/ipolxmpl1
. list, sep(0)
```

	x	y
1.	0	.
2.	1	3
3.	1.5	.
4.	2	6
5.	3	.
6.	3.5	.
7.	4	18

```
. ipolate y x, gen(y1)
(1 missing value generated)
. ipolate y x, gen(y2) epolate
```

```
. list, sep(0)
```

	x	y	y1	y2
1.	0	.	.	0
2.	1	3	3	3
3.	1.5	.	4.5	4.5
4.	2	6	6	6
5.	3	.	12	12
6.	3.5	.	15	15
7.	4	18	18	18

◁

▷ Example 2

We have a dataset of circulations for 10 magazines from 1980 through 2003. The identity of the magazines is recorded in `magazine`, circulation is recorded in `circ`, and the year is recorded in `year`. In a few of the years, the circulation is not known, so we want to fill it in by linear interpolation.

```
. use http://www.stata-press.com/data/r12/ipolxmpl2, clear
. by magazine: ipolate circ year, gen(icirc)
```

When the `by` prefix is specified, interpolation is performed separately for each group.

◁

Methods and formulas

`ipolate` is implemented as an ado-file.

The value y at x is found by finding the closest points (x_0, y_0) and (x_1, y_1), such that $x_0 < x$ and $x_1 > x$ where y_0 and y_1 are observed, and calculating

$$y = \frac{y_1 - y_0}{x_1 - x_0}(x - x_0) + y_0$$

If `epolate` is specified and if (x_0, y_0) and (x_1, y_1) cannot be found on both sides of x, the two closest points on the same side of x are found, and the same formula is applied.

If there are multiple observations with the same value for x_0, then y_0 is taken as the average of the corresponding y values for those observations. (x_1, y_1) is handled in the same way.

Also see

[R] **lowess** — Lowess smoothing

[MI] **mi impute** — Impute missing values

Title

> **isid** — Check for unique identifiers

Syntax

isid *varlist* [using *filename*] [, <u>s</u>ort <u>m</u>issok]

Menu

Data > Data utilities > Check for unique identifiers

Description

isid checks whether the specified variables uniquely identify the observations.

Options

sort specifies that the dataset be sorted by *varlist*.

missok indicates that missing values are permitted in *varlist*.

Remarks

▷ Example 1

Suppose that we want to check whether the mileage ratings (mpg) uniquely identify the observations in our auto dataset.

```
. use http://www.stata-press.com/data/r12/auto
(1978 Automobile Data)
. isid mpg
variable mpg does not uniquely identify the observations
r(459);
```

isid returns an error and reports that there are multiple observations with the same mileage rating. We can locate those observations manually:

```
. sort mpg
. by mpg: generate nobs = _N
. list make mpg if nobs >1, sepby(mpg)
```

	make	mpg
1.	Linc. Mark V	12
2.	Linc. Continental	12
	(output omitted)	
68.	Dodge Colt	30
69.	Mazda GLC	30
72.	Datsun 210	35
73.	Subaru	35

◁

▷ Example 2

isid is useful for checking a time-series panel dataset. For this type of dataset, we usually need two variables to identify the observations: one that labels the individual IDs and another that labels the periods. Before we set the data using tsset, we want to make sure that there are no duplicates with the same panel ID and time. Suppose that we have a dataset that records the yearly gross investment of 10 companies for 20 years. The panel and time variables are company and year.

```
. use http://www.stata-press.com/data/r12/grunfeld, clear
. isid company year
```

isid reports no error, so the two variables company and year uniquely identify the observations. Therefore, we should be able to tsset the data successfully:

```
. tsset company year
        panel variable:  company (strongly balanced)
         time variable:  year, 1935 to 1954
                 delta:  1 year
```

◁

❏ Technical note

The sort option is a convenient shortcut, especially when combined with using. The command

```
. isid patient_id date using newdata, sort
```

is equivalent to

```
. preserve
. use newdata, clear
. sort patient_id date
. isid patient_id date
. save, replace
. restore
```

❏

Methods and formulas

isid is implemented as an ado-file.

Also see

[D] **describe** — Describe data in memory or in file

[D] **ds** — List variables matching name patterns or other characteristics

[D] **duplicates** — Report, tag, or drop duplicate observations

[D] **lookfor** — Search for string in variable names and labels

[D] **codebook** — Describe data contents

[D] **inspect** — Display simple summary of data's attributes

Title

> **joinby** — Form all pairwise combinations within groups

Syntax

joinby [*varlist*] using *filename* [, *options*]

options	Description
When observations match:	
update	replace missing data in memory with values from *filename*
replace	replace all data in memory with values from *filename*
When observations do not match:	
unmatched(none)	ignore all; the default
unmatched(both)	include from both datasets
unmatched(master)	include from data in memory
unmatched(using)	include from data in *filename*
_merge(*varname*)	*varname* marks source of resulting observation; default is _merge
nolabel	do not copy value-label definitions from *filename*

Menu

Data > Combine datasets > Form all pairwise combinations within groups

Description

joinby joins, within groups formed by *varlist*, observations of the dataset in memory with *filename*, a Stata-format dataset. By *join* we mean to form all pairwise combinations. *filename* is required to be sorted by *varlist*. If *filename* is specified without an extension, .dta is assumed.

If *varlist* is not specified, joinby takes as *varlist* the set of variables common to the dataset in memory and in *filename*.

Observations unique to one or the other dataset are ignored unless unmatched() specifies differently. Whether you load one dataset and join the other or vice versa makes no difference in the number of resulting observations.

If there are common variables between the two datasets, however, the combined dataset will contain the values from the master data for those observations. This behavior can be modified with the update and replace options.

Options

┌─ Options └──

update varies the action that joinby takes when an observation is matched. By default, values from the master data are retained when the same variables are found in both datasets. If update is specified, however, the values from the using dataset are retained where the master dataset contains missing.

replace, allowed with update only, specifies that nonmissing values in the master dataset be replaced with corresponding values from the using dataset. A nonmissing value, however, will never be replaced with a missing value.

unmatched(none | both | master | using) specifies whether observations unique to one of the datasets are to be kept, with the variables from the other dataset set to missing. Valid values are

none	ignore all unmatched observations (default)
both	include unmatched observations from the master and using data
master	include unmatched observations from the master data
using	include unmatched observations from the using data

_merge(*varname*) specifies the name of the variable that will mark the source of the resulting observation. The default name is _merge(_merge). To preserve compatibility with earlier versions of joinby, _merge is generated only if unmatched is specified.

nolabel prevents Stata from copying the value-label definitions from the dataset on disk into the dataset in memory. Even if you do not specify this option, label definitions from the disk dataset do not replace label definitions already in memory.

Remarks

The following, admittedly artificial, example illustrates joinby.

▷ Example 1

We have two datasets: child.dta and parent.dta. Both contain a family_id variable, which identifies the people who belong to the same family.

```
. use http://www.stata-press.com/data/r12/child
(Data on Children)

. describe
Contains data from http://www.stata-press.com/data/r12/child.dta
  obs:            5                          Data on Children
  vars:           4                          11 Dec 2010 21:08
  size:          30
```

variable name	storage type	display format	value label	variable label
family_id	int	%8.0g		Family ID number
child_id	byte	%8.0g		Child ID number
x1	byte	%8.0g		
x2	int	%8.0g		

```
Sorted by:  family_id
```

```
. list
```

	family~d	child_id	x1	x2
1.	1025	3	11	320
2.	1025	1	12	300
3.	1025	4	10	275
4.	1026	2	13	280
5.	1027	5	15	210

```
. use http://www.stata-press.com/data/r12/parent
(Data on Parents)

. describe

Contains data from http://www.stata-press.com/data/r12/parent.dta
  obs:             6                      Data on Parents
  vars:            4                      11 Dec 2010 03:06
  size:           84
```

	storage	display	value	
variable name	type	format	label	variable label
family_id	int	%8.0g		Family ID number
parent_id	float	%9.0g		Parent ID number
x1	float	%9.0g		
x3	float	%9.0g		

```
Sorted by:
. list, sep(0)
```

	family~d	parent~d	x1	x3
1.	1030	10	39	600
2.	1025	11	20	643
3.	1025	12	27	721
4.	1026	13	30	760
5.	1026	14	26	668
6.	1030	15	32	684

We want to join the information for the parents and their children. The data on parents are in memory, and the data on children are posted at http://www.stata-press.com. child.dta has been sorted by family_id, but parent.dta has not, so first we sort the parent data on family_id:

```
. sort family_id

. joinby family_id using http://www.stata-press.com/data/r12/child

. describe
```

Contains data

obs:	8	Data on Parents
vars:	6	
size:	136	

variable name	storage type	display format	value label	variable label
family_id	int	%8.0g		Family ID number
parent_id	float	%9.0g		Parent ID number
x1	float	%9.0g		
x3	float	%9.0g		
child_id	byte	%8.0g		Child ID number
x2	int	%8.0g		

Sorted by:
    Note: dataset has changed since last saved

```
. list, sepby(family_id) abbrev(12)
```

	family_id	parent_id	x1	x3	child_id	x2
1.	1025	12	27	721	1	300
2.	1025	12	27	721	4	275
3.	1025	12	27	721	3	320
4.	1025	11	20	643	4	275
5.	1025	11	20	643	1	300
6.	1025	11	20	643	3	320
7.	1026	13	30	760	2	280
8.	1026	14	26	668	2	280

1. family_id of 1027, which appears only in child.dta, and family_id of 1030, which appears only in parent.dta, are not in the combined dataset. Observations for which the matching variables are not in both datasets are omitted.

2. The x1 variable is in both datasets. Values for this variable in the joined dataset are the values from parent.dta—the dataset in memory when we issued the joinby command. If we had child.dta in memory and parent.dta on disk when we requested joinby, the values for x1 would have been those from child.dta. Values from the dataset in memory take precedence over the dataset on disk.

◁

Methods and formulas

joinby is implemented as an ado-file.

Acknowledgment

joinby was written by Jeroen Weesie, Department of Sociology, Utrecht University, The Netherlands.

Reference

Baum, C. F. 2009. *An Introduction to Stata Programming*. College Station, TX: Stata Press.

Also see

[D] **save** — Save Stata dataset

[D] **append** — Append datasets

[D] **cross** — Form every pairwise combination of two datasets

[D] **fillin** — Rectangularize dataset

[D] **merge** — Merge datasets

[U] **22 Combining datasets**

Title

> **label** — Manipulate labels

Syntax

Label dataset

> <u>la</u>bel <u>data</u> [*"label"*]

Label variable

> <u>la</u>bel <u>var</u>iable *varname* [*"label"*]

Define value label

> <u>la</u>bel <u>def</u>ine *lblname* # *"label"* [# *"label"* ...] [, <u>a</u>dd modify replace nofix]

Assign value label to variables

> <u>la</u>bel <u>val</u>ues *varlist* [*lblname* | .] [, nofix]

List names of value labels

> <u>la</u>bel <u>dir</u>

List names and contents of value labels

> <u>la</u>bel <u>l</u>ist [*lblname* [*lblname* ...]]

Copy value labels

> <u>la</u>bel copy *lblname* *lblname* [, replace]

Drop value labels

> <u>la</u>bel drop { *lblname* [*lblname* ...] | _all }

Save value labels in do-file

> <u>la</u>bel save [*lblname* [*lblname* ...]] using *filename* [, replace]

where *#* is an integer or an extended missing value (.a, .b, ..., .z).

Menu

label data

Data > Data utilities > Label utilities > Label dataset

390

label variable

Data > Variables Manager

label define

Data > Variables Manager

label values

Data > Variables Manager

label list

Data > Data utilities > Label utilities > List value labels

label copy

Data > Data utilities > Label utilities > Copy value labels

label drop

Data > Variables Manager

label save

Data > Data utilities > Label utilities > Save value labels as do-file

Description

label data attaches a label (up to 80 characters) to the dataset in memory. Dataset labels are displayed when you use the dataset and when you describe it. If no label is specified, any existing label is removed.

label variable attaches a label (up to 80 characters) to a variable. If no label is specified, any existing variable label is removed.

label define defines a list of up to 65,536 (1,000 for Small Stata) associations of integers and text called value labels. Value labels are attached to variables by label values.

label values attaches a value label to *varlist*. If . is specified instead of *lblname*, any existing value label is detached from that *varlist*. The value label, however, is not deleted. The syntax label values *varname* (that is, nothing following the *varname*) acts the same as specifying the .. Value labels may be up to 32,000 characters long.

label dir lists the names of value labels stored in memory.

label list lists the names and contents of value labels stored in memory.

label copy makes a copy of an existing value label.

label drop eliminates value labels.

label save saves value labels in a do-file. This is particularly useful for value labels that are not attached to a variable because these labels are not saved with the data.

See [D] **label language** for information on the label language command.

Options

add allows you to add # ↔ *label* correspondences to *lblname*. If add is not specified, you may create only new *lblnames*. If add is specified, you may create new *lblnames* or add new entries to existing *lblnames*.

modify allows you to modify or delete existing # ↔ *label* correspondences and add new correspondences. Specifying modify implies add, even if you do not type the add option.

replace, with label define, allows an existing value label to be redefined. replace, with label copy, allows an existing value label to be copied over. replace, with label save, allows *filename* to be replaced.

nofix prevents display formats from being widened according to the maximum length of the value label. Consider label values myvar mylab, and say that myvar has a %9.0g display format right now. Say that the maximum length of the strings in mylab is 12 characters. label values would change the format of myvar from %9.0g to %12.0g. nofix prevents this.

> nofix is also allowed with label define, but it is relevant only when you are modifying an existing value label. Without the nofix option, label define finds all the variables that use this value label and considers widening their display formats. nofix prevents this.

Remarks

See [U] **12.6 Dataset, variable, and value labels** for a complete description of labels. This entry deals only with details not covered there.

label dir lists the names of all defined value labels. label list displays the contents of a value label.

▷ Example 1

Although describe shows the names of the value labels, those value labels may not exist. Stata does not consider it an error to label the values of a variable with a nonexistent label. When this occurs, Stata still shows the association on describe but otherwise acts as if the variable's values are unlabeled. This way, you can associate a value label name with a variable before creating the corresponding label. Similarly, you can define labels that you have not yet used.

```
. use http://www.stata-press.com/data/r12/hbp4
. describe
Contains data from http://www.stata-press.com/data/r12/hbp4.dta
  obs:         1,130
  vars:            7                          22 Jan 2011 11:12
  size:       19,210
```

variable name	storage type	display format	value label	variable label
id	str10	%10s		Record identification number
city	byte	%8.0g		
year	int	%8.0g		
age_grp	byte	%8.0g		
race	byte	%8.0g		
hbp	byte	%8.0g		
female	byte	%8.0g	sexlbl	

```
Sorted by:
```

The dataset is using the value label sexlbl. Let's define the value label yesno:

```
. label define yesno 0 "no" 1 "yes"
```

label dir shows you the labels that you have actually defined:

```
. label dir
yesno
sexlbl
```

We have two value labels stored in memory: yesno and sexlbl.

We can display the contents of a value label with the label list command:

```
. label list yesno
yesno:
           0 no
           1 yes
```

The value label yesno labels the values 0 as no and 1 as yes.

If you do not specify the name of the value label on the label list command, Stata lists all the value labels:

```
. label list
yesno:
           0 no
           1 yes
sexlbl:
           0 male
           1 female
```

◁

❏ Technical note

Because Stata can have more value labels stored in memory than are actually used in the dataset, you may wonder what happens when you save the dataset. Stata stores only those value labels actually associated with variables.

When you use a dataset, Stata eliminates all the value labels stored in memory before loading the dataset.

❏

You can add new codings to an existing value label by using the add option with the label define command. You can modify existing codings by using the modify option. You can redefine a value label by specifying the replace option.

▷ Example 2

The label yesno codes 0 as no and 1 as yes. You might wish later to add a third coding: 2 as maybe. Typing label define with no options results in an error:

```
. label define yesno 2 maybe
label yesno already defined
r(110);
```

If you do not specify the add, modify, or replace options, label define can be used only to create *new* value labels. The add option lets you add codings to an existing label:

```
. label define yesno 2 maybe, add
. label list yesno
yesno:
          0 no
          1 yes
          2 maybe
```

Perhaps you have accidentally mislabeled a value. For instance, 2 may not mean "maybe" but may instead mean "don't know". add does not allow you to change an existing label:

```
. label define yesno 2 "don't know", add
invalid attempt to modify label
r(180);
```

Instead, you would specify the modify option:

```
. label define yesno 2 "don't know", modify
. label list yesno
yesno:
          0 no
          1 yes
          2 don't know
```

In this way, Stata attempts to protect you from yourself. If you type label define with no options, you can only create a new value label—you cannot accidentally change an existing one. If you specify the add option, you can add new labels to a label, but you cannot accidentally change any existing label. If you specify the modify option, which you may not abbreviate, you can change any existing label.

You can even use the modify option to eliminate existing labels. To do this, you map the numeric code to a *null string*, that is, "":

```
. label define yesno 2 "", modify
. label list yesno
yesno:
          0 no
          1 yes
```
◁

You can eliminate entire value labels by using the label drop command.

▷ Example 3

We currently have two value labels stored in memory—sexlbl and yesno—as shown by the label dir command:

```
. label dir
yesno
sexlbl
```

The dataset that we have in memory uses only one of the labels—sexlbl. describe reports that yesno is not being used:

```
. describe
Contains data from http://www.stata-press.com/data/r12/hbp4.dta
  obs:         1,130
  vars:            7                           22 Jan 2011 11:12
  size:       19,210

              storage   display    value
variable name   type    format     label      variable label

id            str10    %10s                   Record identification number
city          byte     %8.0g
year          int      %8.0g
age_grp       byte     %8.0g
race          byte     %8.0g
hbp           byte     %8.0g
female        byte     %8.0g      sexlbl

Sorted by:
```

We can eliminate the yesno label by typing

```
. label drop yesno
. label dir
sexlbl
```

We could eliminate *all* the value labels in memory by typing

```
. label drop _all
. label dir
. _
```

The value label sexlbl, which no longer exists, was associated with the variable female. Even after dropping the value label, sexlbl is still associated with the variable:

```
. describe
Contains data from http://www.stata-press.com/data/r12/hbp4.dta
  obs:         1,130
  vars:            7                           22 Jan 2011 11:12
  size:       19,210

              storage   display    value
variable name   type    format     label      variable label

id            str10    %10s                   Record identification number
city          byte     %8.0g
year          int      %8.0g
age_grp       byte     %8.0g
race          byte     %8.0g
hbp           byte     %8.0g
female        byte     %8.0g      sexlbl

Sorted by:
```

Stata does not mind if a nonexistent value label is associated with a variable. When Stata uses such a variable, it simply acts as if the variable is not labeled:

```
. list in 1/4
```

	id	city	year	age_grp	race	hbp	female
1.	8008238923	1	1993	2	2	0	1
2.	8007143470	1	1992	5	.	0	.
3.	8000468015	1	1988	4	2	0	0
4.	8006167153	1	1991	4	2	0	0

◁

The `label save` command creates a *do-file* containing `label define` commands for each label you specify. If you do not specify the *lblnames*, all value labels are stored in the file. If you do not specify the extension for *filename*, `.do` is assumed.

▷ Example 4

`label copy` is useful when you want to create a new value label that is similar to an existing value label. For example, assume that we currently have the value label `yesno` in memory:

```
. label list yesno
yesno:
           1 yes
           2 no
```

Assume that we have some variables in our dataset coded with 1 and 2 for "yes" and "no" and that we have some other variables coded with 1 for "yes", 2 for "no", and 3 for "maybe".

We could make a copy of label `yesno` and then add the new coding to that copy:

```
. label copy yesno yesnomaybe
. label define yesnomaybe 3 "maybe", add
. label list
yesnomaybe:
           1 yes
           2 no
           3 maybe
yesno:
           1 yes
           2 no
```

◁

▷ Example 5

Labels are automatically stored with your dataset when you `save` it. Conversely, the `use` command drops all labels before loading the new dataset. You may occasionally wish to move a value label from one dataset to another. The `label save` command allows you to do this.

For example, assume that we currently have the value label `yesnomaybe` in memory:

```
. label list yesnomaybe
yesnomaybe:
           1 yes
           2 no
           3 maybe
```

We have a dataset stored on disk called `survey.dta` to which we wish to add this value label. We might use `survey` and then retype the `label define yesnomaybe` command. Retyping the label would not be too tedious here but if the value label in memory mapped, say, the 50 states of the union, retyping it would be irksome. `label save` provides an alternative:

```
. label save yesnomaybe using ynfile
file ynfile.do saved
```

Typing `label save yesnomaybe using ynfile` caused Stata to create a do-file called `ynfile.do` containing the definition of the `yesnomaybe` label.

To see the contents of the file, we can use the `type` command:

```
. type ynfile.do
label define yesnomaybe 1 '"yes"', modify
label define yesnomaybe 2 '"no"', modify
label define yesnomaybe 3 '"maybe"', modify
```

We can now use our new dataset, `survey.dta`:

```
. use survey
(Household survey data)
. label dir

. _
```

Using the new dataset causes Stata to eliminate all value labels stored in memory. The label `yesnomaybe` is now gone. Because we saved it in the file `ynfile.do`, however, we can get it back by typing either `do ynfile` or `run ynfile`. If we type `do`, we will see the commands in the file execute. If we type `run`, the file will execute silently:

```
. run ynfile
. label dir
yesnomaybe
```

The label is now restored just as if we had typed it from the keyboard.

◁

❑ Technical note

You can also use the `label save` command to more easily edit value labels. You can save a label in a file, leave Stata and use your word processor or editor to edit the label, and then return to Stata. Using `do` or `run`, you can load the edited values.

❑

Saved results

`label list` saves the following in `r()`:

Scalars
`r(k)`	number of mapped values, including missings
`r(min)`	minimum nonmissing value label
`r(max)`	maximum nonmissing value label
`r(hasemiss)`	1 if extended missing values labeled, 0 otherwise

`label dir` saves the following in `r()`:

Macros
    r(names)   names of value labels

References

Gleason, J. R. 1998a. dm56: A labels editor for Windows and Macintosh. *Stata Technical Bulletin* 43: 3–6. Reprinted in *Stata Technical Bulletin Reprints*, vol. 8, pp. 5–10. College Station, TX: Stata Press.

——. 1998b. dm56.1: Update to labedit. *Stata Technical Bulletin* 51: 2. Reprinted in *Stata Technical Bulletin Reprints*, vol. 9, p. 15. College Station, TX: Stata Press.

Long, J. S. 2009. *The Workflow of Data Analysis Using Stata*. College Station, TX: Stata Press.

Weesie, J. 1997. dm47: Verifying value label mappings. *Stata Technical Bulletin* 37: 7–8. Reprinted in *Stata Technical Bulletin Reprints*, vol. 7, pp. 39–40. College Station, TX: Stata Press.

——. 2005a. Value label utilities: labeldup and labelrename. *Stata Journal* 5: 154–161.

——. 2005b. Multilingual datasets. *Stata Journal* 5: 162–187.

Also see

[D] **label language** — Labels for variables and values in multiple languages

[D] **labelbook** — Label utilities

[D] **encode** — Encode string into numeric and vice versa

[D] **varmanage** — Manage variable labels, formats, and other properties

[U] **12.6 Dataset, variable, and value labels**

Title

label language — Labels for variables and values in multiple languages

Syntax

List defined languages

> <u>la</u>bel <u>lan</u>guage

Change labels to specified language name

> <u>la</u>bel <u>lan</u>guage *languagename*

Create new set of labels with specified language name

> <u>la</u>bel <u>lan</u>guage *languagename*, new [copy]

Rename current label set

> <u>la</u>bel <u>lan</u>guage *languagename*, <u>ren</u>ame

Delete specified label set

> <u>la</u>bel <u>lan</u>guage *languagename*, delete

Menu

Data > Data utilities > Label utilities > Set label language

Description

label language lets you create and use datasets that contain different sets of data, variable, and value labels. A dataset might contain one set in English, another in German, and a third in Spanish. A dataset may contain up to 100 sets of labels.

We will write about the different sets as if they reflect different spoken languages, but you need not use the multiple sets in this way. You could create a dataset with one set of long labels and another set of shorter ones.

One set of labels is in use at any instant, but a dataset may contain multiple sets. You can choose among the sets by typing

> . label language *languagename*

When other Stata commands produce output (such as describe and tabulate), they use the currently set language. When you define or modify the labels by using the other label commands (see [D] **label**), you modify the current set.

label language (without arguments)
>   lists the available languages and the name of the current one. The current language refers to the labels you will see if you used, say, describe or tabulate. The available languages refer to the names of the other sets of previously created labels. For instance, you might currently be using the labels in en (English), but labels in de (German) and es (Spanish) may also be available.

label language *languagename*
>   changes the labels to those of the specified language. For instance, if label language revealed that en, de, and es were available, typing label language de would change the current language to German.

label language *languagename*, new
>   allows you to create a new set of labels and collectively name them *languagename*. You may name the set as you please, as long as the name does not exceed 24 characters. If the labels correspond to spoken languages, we recommend that you use the language's ISO 639-1 two-letter code, such as en for English, de for German, and es for Spanish. A list of codes for popular languages is listed in the appendix below. For a complete list, see http://lcweb.loc.gov/standards/iso639-2/iso639jac.html.

label language *languagename*, rename
>   changes the name of the label set currently in use. If the label set in use were named default and you now wanted to change that to en, you could type label language en, rename.
>
>   Our choice of the name default in the example was not accidental. If you have not yet used label language to create a new language, the dataset will have one language, named default.

label language *languagename*, delete
>   deletes the specified label set. If *languagename* is also the current language, one of the other available languages becomes the current language.

Option

>   copy is used with label language, new and copies the labels from the current language to the new language.

Remarks

>   Remarks are presented under the following headings:
>
>   *Creating labels in the first language*
>   *Creating labels in the second and subsequent languages*
>   *Creating labels from a clean slate*
>   *Creating labels from a previously existing language*
>   *Switching languages*
>   *Changing the name of a language*
>   *Deleting a language*
>   *Appendix: Selected ISO 639-1 two-letter codes*

Creating labels in the first language

You can begin by ignoring the `label language` command. You create the data, variable, and value labels just as you would ordinarily; see [D] **label**.

```
. label data "1978 Automobile Data"
. label variable foreign "Car type"
. label values foreign origin
. label define origin  0 "Domestic"  1 "Foreign"
```

At some point—at the beginning, the middle, or the end—rename the language appropriately. For instance, if the labels you defined were in English, type

```
. label language en, rename
```

`label language, rename` simply changes the name of the currently set language. You may change the name as often as you wish.

Creating labels in the second and subsequent languages

After creating the first language, you can create a new language by typing

```
. label language newlanguagename, new
```

or by typing the two commands

```
. label language existinglanguagename
. label language newlanguagename, new copy
```

In the first case, you start with a clean slate: no data, variable, or value labels are defined. In the second case, you start with the labels from *existinglanguagename*, and you can make the changes from there.

Creating labels from a clean slate

To create new labels in the language named `de`, type

```
. label language de, new
```

If you were now to type `describe`, you would find that there are no data, variable, or value labels. You can define new labels in the usual way:

```
. label data "1978 Automobil Daten"
. label variable foreign "Art Auto"
. label values foreign origin_de
. label define origin_de  0 "Innen"  1 "Ausländisch"
```

Creating labels from a previously existing language

It is sometimes easier to start with the labels from a previously existing language, which you can then translate:

```
. label language en
. label language de, new copy
```

If you were now to type describe, you would see the English-language labels, even though the new language is named de. You can then work to translate the labels:

```
. label data "1978 Automobil Daten"
. label variable foreign "Art Auto"
```

Typing describe, you might also discover that the variable foreign has the value label origin. Do not change the contents of the value label. Instead, create a new value label:

```
. label define origin_de  0 "Innen"  1 "Ausländisch"
. label values foreign origin_de
```

Creating value labels with the copy option is no different from creating them from a clean slate, except that you start with an existing set of labels from another language. Using describe can make it easier to translate them.

Switching languages

You can discover the names of the previously defined languages by typing

```
. label language
```

You can switch to a previously defined language—say, to en—by typing

```
. label language en
```

Changing the name of a language

To change the name of a previously defined language make it the current language and then specify the rename option:

```
. label language de
. label language German, rename
```

You may rename a language as often as you wish:

```
. label language de, rename
```

Deleting a language

To delete a previously defined language, such as de, type

```
. label language de, delete
```

The delete option deletes the specified language and, if the language was also the currently set language, resets the current language to one of the other languages or to default if there are none.

Appendix: Selected ISO 639-1 two-letter codes

You may name languages as you please. You may name German labels Deutsch, German, Aleman, or whatever else appeals to you. For consistency across datasets, if the language you are creating is a spoken language, we suggest that you use the ISO 639-1 two-letter codes. Some of them are listed below, and the full list can be found at http://lcweb.loc.gov/standards/iso639-2/iso639jac.html.

Two-letter code	English name of language
ar	Arabic
cs	Czech
cy	Welsh
de	German
el	Greek
en	English
es	Spanish; Castillian
fa	Persian
fi	Finnish
fr	French
ga	Irish
he	Hebrew
hi	Hindi
is	Icelandic
it	Italian
ja	Japanese
kl	Kalaallisut; Greenlandic
lt	Lithuanian
lv	Latvian
nl	Dutch; Flemish
no	Norwegian
pl	Polish
pt	Portuguese
ro	Romanian; Moldavian
ru	Russian
sk	Slovak
sr	Serbian
sv	Swedish
tr	Turkish
uk	Ukrainian
uz	Uzbek
zh	Chinese

Saved results

`label language` without arguments saves the following in `r()`:

Scalars
    `r(k)`   number of languages defined

Macros
    `r(languages)` list of languages, listed one after the other
    `r(language)` name of current language

Methods and formulas

This section is included for programmers who wish to access or extend the services label language provides.

Language sets are implemented using [P] **char**. The names of the languages and the name of the current language are stored in

$$\_dta\big[\_lang\_list\big] \qquad \text{list of defined languages}$$

$$\_dta\big[\_lang\_c\big] \qquad \text{currently set language}$$

If these characteristics are undefined, results are as if each contained the word "default". Do not change the contents of the above two macros except by using label language.

For each language *languagename* except the current language, data, variable, and value labels are stored in

$$\_dta\big[\_lang\_v\_languagename\big] \qquad \text{data label}$$

$$\textit{varname}\big[\_lang\_v\_languagename\big] \qquad \text{variable label}$$

$$\textit{varname}\big[\_lang\_l\_languagename\big] \qquad \text{value-label name}$$

References

Golbe, D. L. 2010. Stata tip 83: Merging multilingual datasets. *Stata Journal* 10: 152–156.

Weesie, J. 2005. Multilingual datasets. *Stata Journal* 5: 162–187.

Also see

[D] **label** — Manipulate labels

[D] **codebook** — Describe data contents

[D] **labelbook** — Label utilities

Title

> **labelbook** — Label utilities

Syntax

Produce a codebook describing value labels

> labelbook [*lblname-list*] [, *labelbook_options*]

Prefix numeric values to value labels

> numlabel [*lblname-list*] , { <u>a</u>dd | <u>r</u>emove } [*numlabel_options*]

Make dataset containing value-label information

> uselabel [*lblname-list*] [using *filename*] [, clear <u>v</u>ar]

labelbook_options	Description
<u>a</u>lpha	alphabetize label mappings
<u>l</u>ength(*#*)	check if value labels are unique to length *#*; default is length(12)
<u>l</u>ist(*#*)	list maximum of *#* mappings; default is list(32000)
<u>p</u>roblems	describe potential problems in a summary report
<u>d</u>etail	do not suppress detailed report on variables or value labels

numlabel_options	Description
* <u>a</u>dd	prefix numeric values to value labels
* <u>r</u>emove	remove numeric values from value labels
<u>m</u>ask(*str*)	mask for formatting numeric labels; default mask is "#. "
force	force adding or removing of numeric labels
<u>d</u>etail	provide details about value labels, where some labels are prefixed with numbers and others are not

* Either add or remove must be specified.

Menu

labelbook

Data > Data utilities > Label utilities > Produce codebook of value labels

numlabel

Data > Data utilities > Label utilities > Prepend values to value labels

uselabel

Data > Data utilities > Label utilities > Create dataset from value labels

Description

labelbook displays information for the value labels specified or, if no labels are specified, all the labels in the data.

For multilingual datasets (see [D] **label language**), labelbook lists the variables to which value labels are attached in all defined languages.

numlabel prefixes numeric values to value labels. For example, a value mapping of 2 -> "catholic" will be changed to 2 -> "2. catholic". See option mask() for the different formats. Stata commands that display the value labels also show the associated numeric values. Prefixes are removed with the remove option.

uselabel is a programmer's command that reads the value-label information from the currently loaded dataset or from an optionally specified filename.

uselabel creates a dataset in memory that contains only that value-label information. The new dataset has four variables named label, lname, value, and trunc; is sorted by lname value; and has 1 observation per mapping. Value labels can be longer than the maximum string length in Stata; see help limits. The new variable trunc contains 1 if the value label is truncated to fit in a string variable in the dataset created by uselabel.

uselabel complements label, save, which produces a text file of the value labels in a format that allows easy editing of the value-label texts.

Specifying no list or _all is equivalent to specifying all value labels. Value-label names may not be abbreviated or specified with wildcards.

Options for labelbook

alpha specifies that the list of value-label mappings be sorted alphabetically on label. The default is to sort the list on value.

length(#) specifies the minimum length that labelbook checks to determine whether shortened value labels are still unique. It defaults to 12, the width used by most Stata commands. labelbook also reports whether value labels are unique at their full length.

list(#) specifies the maximum number of value-label mappings to be listed. If a value label defines more mappings, a random subset of # mappings is displayed. By default, labelbook displays all mappings. list(0) suppresses the listing of the value-label definitions.

problems specifies that a summary report be produced describing potential problems that were diagnosed:

1. Value label has gaps in mapped values (for example, values 0 and 2 are labeled, while 1 is not)

2. Value label strings contain leading or trailing blanks

3. Value label contains duplicate labels, that is, there are different values that map into the same string

4. Value label contains duplicate labels at length 12

5. Value label contains numeric → numeric mappings

6. Value label contains numeric → null string mappings

7. Value label is not used by variables

detail may be specified only with problems. It specifies that the detailed report on the variables or value labels not be suppressed.

Options for numlabel

add specifies that numeric values be prefixed to value labels. Value labels that are already numlabeled (using the same mask) are not modified.

remove specifies that numeric values be removed from the value labels. If you added numeric values by using a nondefault mask, you must specify the same mask to remove them. Value labels that are not numlabeled or are numlabeled using a different mask are not modified.

mask(*str*) specifies a mask for formatting the numeric labels. In the mask, # is replaced by the numeric label. The default mask is "#. " so that numeric value 3 is shown as "3. ". Spaces are relevant. For the mask "[#]", numeric value 3 would be shown as "[3]".

force specifies that adding or removing numeric labels be performed, even if some value labels are numlabeled using the mask and others are not. Here only labels that are not numlabeled will be modified.

detail specifies that details be provided about the value labels that are sometimes, but not always, numlabeled using the mask.

Options for uselabel

clear permits the dataset to be created, even if the dataset already in memory has changed since it was last saved.

var specifies that the varlists using value label *vl* be returned in r(*vl*).

Remarks

Remarks are presented under the following headings:

> *labelbook*
> *Diagnosing problems*
> *numlabel*
> *uselabel*

labelbook

labelbook produces a detailed report of the value labels in your data. You can restrict the report to a list of labels, meaning that no abbreviations or wildcards will be allowed. labelbook is a companion command to [D] **codebook**, which describes the data, focusing on the variables.

For multilingual datasets (see [D] **label language**), labelbook lists the variables to which value labels are attached in any of the languages.

▷ Example 1

We request a labelbook report for value labels in a large dataset on the internal organization of households. We restrict output to three value labels: agree5 (used for five-point Likert-style items), divlabor (division of labor between husband and wife), and noyes for simple no-or-yes questions.

```
. use http://www.stata-press.com/data/r12/labelbook1

. labelbook agree5 divlabor noyes
```

---

value label agree5

---

values		**labels**	
range:	[1,5]	string length:	[8,11]
N:	5	unique at full length:	yes
gaps:	no	unique at length 12:	yes
missing .*:	0	null string:	no
		leading/trailing blanks:	no
		numeric -> numeric:	no

definition
```
           1    -- disagree
           2    - disagree
           3    indifferent
           4    + agree
           5    ++ agree
```
variables: rs056 rs057 rs058 rs059 rs060 rs061 rs062 rs063 rs064 rs065
             rs066 rs067 rs068 rs069 rs070 rs071 rs072 rs073 rs074 rs075
             rs076 rs077 rs078 rs079 rs080 rs081

---

value label divlabor

---

values		**labels**	
range:	[1,7]	string length:	[7,16]
N:	7	unique at full length:	yes
gaps:	no	unique at length 12:	yes
missing .*:	0	null string:	no
		leading/trailing blanks:	yes
		numeric -> numeric:	no

definition
```
           1    wife only
           2    wife >> husband
           3    wife > husband
           4    equally
           5    husband > wife
           6    husband >> wife
           7    husband only
```
variables: hm01_a hm01_b hm01_c hm01_d hm01_e hn19 hn21 hn25_a hn25_b
             hn25_c hn25_d hn25_e hn27_a hn27_b hn27_c hn27_d hn27_e hn31
             hn36 hn38 hn42 hn46_a hn46_b hn46_c hn46_d hn46_e ho01_a ho01_b
             ho01_c ho01_d ho01_e

```
value label noyes
```

values		**labels**	
range:	[1,2]	string length:	[2,16]
N:	4	unique at full length:	yes
gaps:	yes	unique at length 12:	yes
missing .*:	2	null string:	no
		leading/trailing blanks:	no
		numeric -> numeric:	no

```
definition
        1    no
        2    yes
       .a    not applicable
       .b    ambiguous answer
variables:   hb12 hd01_a hd01_b hd03 hd04_a hd04_b he03_a he03_b hlat hn09_b
             hn24_a hn34 hn49 hu05_a hu06_1c hu06_2c hx07_a hx08 hlat2
             hfinish rh02 rj10_01 rk16_a rk16_b rl01 rl03 rl08_a rl08_b
             rl09_a rs047 rs048 rs049 rs050 rs051 rs052 rs053 rs054 rs093
             rs095 rs096 rs098
```

The report is largely self-explanatory. Extended missing values are denoted by ".*". In the definition of the mappings, the leading 12 characters of longer value labels are underlined to make it easier to check that the value labels still make sense after truncation. The following example emphasizes this feature. The option alpha specifies that the value-label mappings be sorted in alphabetical order by the label strings rather than by the mapped values.

```
. use http://www.stata-press.com/data/r12/labelbook2

. labelbook sports, alpha
```

```
value label sports
```

values		**labels**	
range:	[1,5]	string length:	[16,23]
N:	4	unique at full length:	yes
gaps:	yes	unique at length 12:	no
missing .*:	0	null string:	no
		leading/trailing blanks:	no
		numeric -> numeric:	no

```
definition
        5    college baseball
        4    college basketball
        2    professional baseball
        1    professional basketball
variables:   active passive
```

The report includes information about potential problems in the data. These are discussed in greater detail in the next section.

◁

Diagnosing problems

labelbook can diagnose a series of potential problems in the value-label mappings. labelbook produces warning messages for a series of problems:

1. Gaps in the labeled values (for example, values 0 and 2 are labeled, whereas 1 is not) may occur when value labels of the intermediate values have not been defined.

2. Leading or trailing blanks in the value labels may distort Stata output.

3. Stata allows you to define blank labels, that is, the mapping of a number to the empty string. Below we give you an example of the unexpected output that may result. Blank labels are most often the result of a mistaken value-label definition, for instance, the expansion of a nonexisting macro in the definition of a value label.

4. Stata does not require that the labels within each value label consist of *unique* strings, that is, that different values be mapped into different strings. For instance, you might accidentally define the value label `gender` as

   ```
   label define gender 1 female 2 female
   ```

   You will probably catch most of the problems, but in more complicated value labels, it is easy to miss the error. `labelbook` finds such problems and displays a warning.

5. Stata allows long value labels (32,000 characters), so labels can be long. However, some commands may need to display truncated value labels, typically at length 12. Consequently, even if the value labels are unique, the truncated value labels may not be, which can cause problems. `labelbook` warns you for value labels that are not unique at length 12.

6. Stata allows value labels that can be interpreted as numbers. This is sometimes useful, but it can cause highly misleading output. Think about tabulating a variable for which the associated value label incorrectly maps 1 into "2", 2 into "3", and 3 into "1". `labelbook` looks for such problematic labels and warns you if they are found.

7. In Stata, value labels are defined as separate objects that can be associated with more than one variable:

   ```
   label define labname # str # str ....
   label value varname1 labname
   label value varname2 labname
   ...
   ```

   If you forget to associate a variable label with a variable, Stata considers the label unused and drops its definition. `labelbook` reports unused value labels so that you may fix the problem.

   The related command `codebook` reports on two other potential problems concerning value labels:

   a. A variable is value labeled, but some values of the variable are not labeled. You may have forgotten to define a mapping for some values, or you generated a variable incorrectly; for example, your `sex` variable has an unlabeled value 3, and you are not working in experimental genetics!

   b. A variable has been associated with an undefined value label.

   `labelbook` can also be invoked with the `problems` option, specifying that only a report on potential problems be displayed without the standard detailed description of the value labels.

❑ Technical note

The following two examples demonstrate some features of value labels that may be difficult to understand. In the first example, we `encode` a string variable with blank strings of various sizes; that is, we turn a string variable into a value-labeled numeric variable. Then we tabulate the generated variable.

```
. clear all
. set obs 5
obs was 0, now 5
. generate str10 horror = substr("        ", 1, _n)
. encode horror, gen(Ihorror)
. tabulate horror
```

horror	Freq.	Percent	Cum.
	1	20.00	20.00
	1	20.00	40.00
	1	20.00	60.00
	1	20.00	80.00
	1	20.00	100.00
Total	5	100.00	

It may look as if you have discovered a bug in Stata because there are no value labels in the first column of the table. This happened because we encoded a variable with only blank strings, so the associated value label maps integers into blank strings.

```
. label list Ihorror
Ihorror:
            1
            2
            3
            4
            5
```

In the first column of the table, `tabulate` displayed the value-label texts, just as it should. Because these texts are all blank, the first column is empty. As illustrated below, `labelbook` would have warned you about this odd value label.

Our second example illustrates what could go wrong with numeric values stored as string values. We want to turn this into a numeric variable, but we incorrectly `encode` the variable rather than using the appropriate command, `destring`.

```
. generate str10 horror2 = string(_n+1)
. encode horror2, gen(Ihorror2)
. tabulate Ihorror2
```

Ihorror2	Freq.	Percent	Cum.
2	1	20.00	20.00
3	1	20.00	40.00
4	1	20.00	60.00
5	1	20.00	80.00
6	1	20.00	100.00
Total	5	100.00	

```
. tabulate Ihorror2, nolabel
```

Ihorror2	Freq.	Percent	Cum.
1	1	20.00	20.00
2	1	20.00	40.00
3	1	20.00	60.00
4	1	20.00	80.00
5	1	20.00	100.00
Total	5	100.00	

```
. label list Ihorror2
Ihorror2:
           1 2
           2 3
           3 4
           4 5
           5 6
```
                                                                           ❏

labelbook skips the detailed descriptions of the value labels and reports only the potential
problems in the value labels if the problems option is specified. This report would have alerted you
to the problems with the value labels we just described.

```
. use http://www.stata-press.com/data/r12/data_in_trouble, clear
. labelbook, problem

  Potential problems in dataset    http://www.stata-press.com/data/r12/
> data_in_trouble.dta

              potential problem    value labels

            numeric -> numeric    Ihorror2
  leading or trailing blanks       Ihorror
         not used by variables     unused
```

Running labelbook, problems and codebook, problems on new data might catch a series of
annoying problems.

numlabel

The numlabel command allows you to prefix numeric codes to value labels. The reason you
might want to do this is best seen in an example using the automobile data. First, we create a value
label for the variable rep78 (repair record in 1978),

```
. use http://www.stata-press.com/data/r12/auto
(1978 Automobile Data)
. label define repair 1 "very poor" 2 "poor" 3 "medium" 4 good 5 "very good"
. label values rep78 repair
```

and tabulate it.

```
. tabulate rep78
```

Repair Record 1978	Freq.	Percent	Cum.
very poor	2	2.90	2.90
poor	8	11.59	14.49
medium	30	43.48	57.97
good	18	26.09	84.06
very good	11	15.94	100.00
Total	69	100.00	

Suppose that we want to recode the variable by joining the categories *poor* and *very poor*. To do
this, we need the numerical codes of the categories, not the value labels. However, Stata does not
display both the numeric codes and the value labels. We could redisplay the table with the nolabel
option. The numlabel command provides a simple alternative: it modifies the value labels so that
they also contain the numeric codes.

```
. numlabel, add

. tabulate rep78
```

Repair Record 1978	Freq.	Percent	Cum.
1. very poor	2	2.90	2.90
2. poor	8	11.59	14.49
3. medium	30	43.48	57.97
4. good	18	26.09	84.06
5. very good	11	15.94	100.00
Total	69	100.00	

If you do not like the way the numeric codes are formatted, you can use `numlabel` to change the formatting. First, we remove the numeric codes again:

```
. numlabel repair, remove
```

In this example, we specified the name of the label. If we had not typed it, `numlabel` would have removed the codes from all the value labels. We can include the numeric codes while specifying a mask:

```
. numlabel, add mask("[#]  ")

. tabulate rep78
```

Repair Record 1978	Freq.	Percent	Cum.
[1] very poor	2	2.90	2.90
[2] poor	8	11.59	14.49
[3] medium	30	43.48	57.97
[4] good	18	26.09	84.06
[5] very good	11	15.94	100.00
Total	69	100.00	

`numlabel` prefixes rather than postfixes the value labels with numeric codes. Because value labels can be fairly long (up to 80 characters), Stata usually displays only the first 12 characters.

uselabel

uselabel is of interest primarily to programmers. Here we briefly illustrate it with the auto dataset.

▷ Example 2

```
. use http://www.stata-press.com/data/r12/auto
(1978 Automobile Data)
. uselabel
. describe
Contains data
  obs:              2
  vars:             4
  size:            32
```

variable name	storage type	display format	value label	variable label
lname	str6	%9s		
value	byte	%9.0g		
label	str8	%9s		
trunc	byte	%9.0g		

```
Sorted by:  lname   value
     Note:  dataset has changed since last saved
. list
```

	lname	value	label	trunc
1.	origin	0	Domestic	0
2.	origin	1	Foreign	0

uselabel created a dataset containing the labels and values for the value label origin.

The maximum length of the text associated with a value label is 32,000 characters, whereas the maximum length of a string variable in a Stata dataset is 244. uselabel uses only the first 244 characters of the label. The trunc variable will record a 1 if the text was truncated for this reason.

◁

Saved results

labelbook saves the following in r():

Macros
r(names)	*lblname-list*	
r(gaps)	gaps in mapped values	
r(blanks)	leading or trailing blanks	
r(null)	name of value label containing null strings	
r(nuniq)	duplicate labels	
r(nuniq_sh)	duplicate labels at length 12	
r(ntruniq)	duplicate labels at maximum string length	
r(notused)	not used by any of the variables	
r(numeric)	name of value label containing mappings to numbers	

uselabel saves the following in r():

Macros
    r(*lblname*)   list of variables that use value label *lblname* (only when var option is specified)

Methods and formulas

labelbook, numlabel, and uselabel are implemented as ado-files.

Acknowledgments

labelbook and numlabel were written by Jeroen Weesie, Department of Sociology, Utrecht University. A command similar to numlabel was written by J. M. Lauritsen (2001).

References

Lauritsen, J. M. 2001. dm84: labjl: Adding numerical codes to value labels. *Stata Technical Bulletin* 59: 6–7. Reprinted in *Stata Technical Bulletin Reprints*, vol. 10, pp. 35–37. College Station, TX: Stata Press.

Weesie, J. 1997. dm47: Verifying value label mappings. *Stata Technical Bulletin* 37: 7–8. Reprinted in *Stata Technical Bulletin Reprints*, vol. 7, pp. 39–40. College Station, TX: Stata Press.

Also see

[D] **codebook** — Describe data contents

[D] **describe** — Describe data in memory or in file

[D] **ds** — List variables matching name patterns or other characteristics

[D] **encode** — Encode string into numeric and vice versa

[D] **label** — Manipulate labels

[U] **12.6 Dataset, variable, and value labels**

[U] **15 Saving and printing output—log files**

Title

list — List values of variables

Syntax

$\underline{l}$ist [*varlist*] [*if*] [*in*] [, *options*]

$\underline{f}$list is equivalent to list with the fast option.

options	Description
Main	
$\underline{compress}$	compress width of columns in both table and display formats
$\underline{noc}$ompress	use display format of each variable
fast	synonym for nocompress; no delay in output of large datasets
$\underline{abbreviate}$(#)	abbreviate variable names to # characters; default is ab(8)
$\underline{string}$(#)	truncate string variables to # characters; default is string(10)
$\underline{noo}$bs	do not list observation numbers
fvall	display all levels of factor variables
Options	
$\underline{table}$	force table format
$\underline{display}$	force display format
$\underline{header}$	display variable header once; default is table mode
$\underline{noh}$eader	suppress variable header
$\underline{h}$eader(#)	display variable header every # lines
clean	force table format with no divider or separator lines
$\underline{divider}$	draw divider lines between columns
$\underline{separator}$(#)	draw a separator line every # lines; default is separator(5)
$\underline{sepby}$(*varlist₂*)	draw a separator line whenever *varlist₂* values change
$\underline{nol}$abel	display numeric codes rather than label values
Summary	
mean [(*varlist₂*)]	add line reporting the mean for the (specified) variables
sum [(*varlist₂*)]	add line reporting the sum for the (specified) variables
N [(*varlist₂*)]	add line reporting the number of nonmissing values for the (specified) variables
$\underline{labvar}$(*varname*)	substitute Mean, Sum, or N for value of *varname* in last row of table
Advanced	
$\underline{constant}$ [(*varlist₂*)]	separate and list variables that are constant only once
$\underline{notrim}$	suppress string trimming
$\underline{absolute}$	display overall observation numbers when using by *varlist*:
nodotz	display numerical values equal to .z as field of blanks
$\underline{subvarname}$	substitute characteristic for variable name in header
$\underline{linesize}$(#)	columns per line; default is linesize(79)

varlist may contain factor variables; see [U] **11.4.3 Factor variables**.

varlist may contain time-series operators; see [U] **11.4.4 Time-series varlists**.

by is allowed with list; see [D] **by**.

Menu

Data > Describe data > List data

Description

list displays the values of variables. If no *varlist* is specified, the values of all the variables are displayed. Also see browse in [D] **edit**.

Options

────┐ Main └──

compress and nocompress change the width of the columns in both table and display formats. By default, list examines the data and allocates the needed width to each variable. For instance, a variable might be a string with a %18s format, and yet the longest string will be only 12 characters long. Or a numeric variable might have a %9.0g format, and yet, given the values actually present, the widest number needs only four columns.

nocompress prevents list from examining the data. Widths will be set according to the display format of each variable. Output generally looks better when nocompress is not specified, but for very large datasets (say, 1,000,000 observations or more), nocompress can speed up the execution of list.

compress allows list to engage in a little more compression than it otherwise would by telling list to abbreviate variable names to fewer than eight characters.

fast is a synonym for nocompress. fast may be of interest to those with very large datasets who wish to see output appear without delay.

abbreviate(#) is an alternative to compress that allows you to specify the minimum abbreviation of variable names to be considered. For example, you could specify abbreviate(16) if you never wanted variables abbreviated to less than 16 characters.

string(#) specifies that when string variables are listed, they be truncated to # characters in the output. Any value that is truncated will be appended with "..." to indicate the truncation. string() is useful for displaying just a part of long strings.

noobs suppresses the listing of the observation numbers.

fvall specifies that the entire dataset be used to determine how many levels are in any factor variables specified in *varlist*. The default is to determine the number of levels by using only the observations in the if and in qualifiers.

────┐ Options └──

table and display determine the style of output. By default, list determines whether to use table or display on the basis of the width of your screen and the linesize() option, if you specify it.

table forces table format. Forcing table format when list would have chosen otherwise generally produces impossible-to-read output because of the linewraps. However, if you are logging output in SMCL format and plan to print the output on wide paper later, specifying table can be a reasonable thing to do.

display forces display format.

header, noheader, and header(#) specify how the variable header is to be displayed.

header is the default in table mode and displays the variable header once, at the top of the table.

noheader suppresses the header altogether.

header(#) redisplays the variable header every # observations. For example, header(10) would display a new header every 10 observations.

The default in display mode is to display the variable names interweaved with the data:

	make	price	mpg	rep78	headroom	trunk	weight	length
1.	AMC Concord	4,099	22	3	2.5	11	2,930	186

	turn	displa~t		gear_r~o		foreign	
	40	121		3.58		Domestic	

However, if you specify header, the header is displayed once, at the top of the table:

	make	price	mpg	rep78	headroom	trunk	weight	length
	turn	displa~t		gear_r~o		foreign	
1.	AMC Concord	4,099	22	3	2.5	11	2,930	186
	40		121		3.58		Domestic	

clean is a better alternative to table when you want to force table format and your goal is to produce more readable output on the screen. clean implies table, and it removes all dividing and separating lines, which is what makes wrapped table output nearly impossible to read.

divider, separator(#), and sepby(*varlist₂*) specify how dividers and separator lines should be displayed. These three options affect only table format.

divider specifies that divider lines be drawn between columns. The default is nodivider.

separator(#) and sepby(*varlist₂*) indicate when separator lines should be drawn between rows.

separator(#) specifies how often separator lines should be drawn between rows. The default is separator(5), meaning every 5 observations. You may specify separator(0) to suppress separators altogether.

sepby(*varlist₂*) specifies that a separator line be drawn whenever any of the variables in sepby(*varlist₂*) change their values; up to 10 variables may be specified. You need not make sure the data were sorted on sepby(*varlist₂*) before issuing the list command. The variables in sepby(*varlist₂*) also need not be among the variables being listed.

nolabel specifies that numeric codes be displayed rather than the label values.

mean, sum, N, mean(*varlist*$_2$), sum(*varlist*$_2$), and N(*varlist*$_2$) all specify that lines be added to the output reporting the mean, sum, or number of nonmissing values for the (specified) variables. If you do not specify the variables, all numeric variables in the *varlist* following list are used.

labvar(*varname*) is for use with mean$\big[()\big]$, sum$\big[()\big]$, and N$\big[()\big]$. list displays Mean, Sum, or N where the observation number would usually appear to indicate the end of the table—where a row represents the calculated mean, sum, or number of observations.

labvar(*varname*) changes that. Instead, Mean, Sum, or N is displayed where the value for *varname* would be displayed. For instance, you might type

. list group costs profits, sum(costs profits) labvar(group)

	group	costs	profits
1.	1	47	5
2.	2	123	10
3.	3	22	2
	Sum	192	17

and then also specify the noobs option to suppress the observation numbers.

constant and constant(*varlist*$_2$) specify that variables that do not vary observation by observation be separated out and listed only once.

constant specifies that list determine for itself which variables are constant.

constant(*varlist*$_2$) allows you to specify which of the constant variables you want listed separately. list verifies that the variables you specify really are constant and issues an error message if they are not.

constant and constant() respect if *exp* and in *range*. If you type

. list if group==3

variable x might be constant in the selected observations, even though the variable varies in the entire dataset.

notrim affects how string variables are listed. The default is to trim strings at the width implied by the widest possible column given your screen width (or linesize(), if you specified that). notrim specifies that strings not be trimmed. notrim implies clean (see above) and, in fact, is equivalent to the clean option, so specifying either makes no difference.

absolute affects output only when list is prefixed with by *varlist*:. Observation numbers are displayed, but the overall observation numbers are used rather than the observation numbers within each by-group. For example, if the first group had 4 observations and the second had 2, by default the observations would be numbered 1, 2, 3, 4 and 1, 2. If absolute is specified, the observations will be numbered 1, 2, 3, 4 and 5, 6.

nodotz is a programmer's option that specifies that numerical values equal to .z be listed as a field of blanks rather than as .z.

subvarname is a programmer's option. If a variable has the characteristic *var*$\big[$varname$\big]$ set, then the contents of that characteristic will be used in place of the variable's name in the headers.

linesize(#) specifies the width of the page to be used for determining whether table or display format should be used and for formatting the resulting table. Specifying a value of linesize() that is wider than your screen width can produce truly ugly output on the screen, but that output can nevertheless be useful if you are logging output and plan to print the log later on a wide printer.

Remarks

list, typed by itself, lists all the observations and variables in the dataset. If you specify *varlist*, only those variables are listed. Specifying one or both of in *range* and if *exp* limits the observations listed.

list respects line size. That is, if you resize the Results window (in windowed versions of Stata) before running list, it will take advantage of the available horizontal space. Stata for Unix(console) users can instead use the set linesize command to take advantage of this feature; see [R] **log**.

list may not display all the large strings. You have two choices: 1) you can specify the clean option, which makes a different, less attractive listing, or 2) you can increase line size, as discussed above.

▷ Example 1

list has two output formats, known as table and display. The table format is suitable for listing a few variables, whereas the display format is suitable for listing an unlimited number of variables. Stata chooses automatically between those two formats:

```
. use http://www.stata-press.com/data/r12/auto
(1978 Automobile Data)
. list in 1/2
```

1.	make AMC Concord	price 4,099	mpg 22	rep78 3	headroom 2.5	trunk 11	weight 2,930	length 186
	turn 40	displa~t 121			gear_r~o 3.58		foreign Domestic	

2.	make AMC Pacer	price 4,749	mpg 17	rep78 3	headroom 3.0	trunk 11	weight 3,350	length 173
	turn 40	displa~t 258			gear_r~o 2.53		foreign Domestic	

```
. list make mpg weight displ rep78 in 1/5
```

	make	mpg	weight	displa~t	rep78
1.	AMC Concord	22	2,930	121	3
2.	AMC Pacer	17	3,350	258	3
3.	AMC Spirit	22	2,640	121	.
4.	Buick Century	20	3,250	196	3
5.	Buick Electra	15	4,080	350	4

The first case is an example of display format; the second is an example of table format. The table format is more readable and takes less space, but it is effective only if the variables can fit on one line across the screen. Stata chose to list all 12 variables in display format, but when the *varlist* was restricted to five variables, Stata chose table format.

If you are dissatisfied with Stata's choice, you can decide for yourself. You can specify the display option to force display format and the nodisplay option to force table format.

◁

❏ Technical note

If you have long string variables in your data—say, str75 or longer—by default, list displays only the first 70 or so characters of each; the exact number is determined by the width of your Results window. The first 70 or so characters will be shown followed by "...". If you need to see the entire contents of the string, you can

1. specify the clean option, which makes a different (and uglier) style of list, or

2. make your Results window wider [Stata for Unix(console) users: increase set linesize].

❏

❏ Technical note

Among the things that determine the widths of the columns, the variable names play a role. Left to itself, list will never abbreviate variable names to fewer than eight characters. You can use the compress option to abbreviate variable names to fewer characters than that.

❏

❏ Technical note

When Stata lists a string variable in table output format, the variable is displayed right-justified by default.

When Stata lists a string variable in display output format, it decides whether to display the variable right-justified or left-justified according to the display format for the string variable; see [U] **12.5 Formats: Controlling how data are displayed**. In our previous example, make has a display format of %-18s.

```
. describe make
```

variable name	storage type	display format	value label	variable label
make	str18	%-18s		Make and Model

The negative sign in the %-18s instructs Stata to left-justify this variable. If the display format had been %18s, Stata would have right-justified the variable.

The foreign variable appears to be string, but if we describe it, we see that it is not:

```
. describe foreign
```

variable name	storage type	display format	value label	variable label
foreign	byte	%8.0g	origin	Car type

foreign is stored as a byte, but it has an associated value label named origin; see [U] **12.6.3 Value labels**. Stata decides whether to right-justify or left-justify a numeric variable with an associated value label by using the same rule used for string variables: it looks at the display format of the variable. Here the display format of %8.0g tells Stata to right-justify the variable. If the display format had been %-8.0g, Stata would have left-justified this variable.

❑

❑ Technical note

You can list the variables in any order. When you specify the *varlist*, list displays the variables in the order you specify. You may also include variables more than once in the *varlist*.

❑

▷ Example 2

Sometimes you may wish to suppress the observation numbers. You do this by specifying the noobs option:

```
. list make mpg weight displ foreign in 46/55, noobs
```

make	mpg	weight	displa~t	foreign
Plym. Volare	18	3,330	225	Domestic
Pont. Catalina	18	3,700	231	Domestic
Pont. Firebird	18	3,470	231	Domestic
Pont. Grand Prix	19	3,210	231	Domestic
Pont. Le Mans	19	3,200	231	Domestic
Pont. Phoenix	19	3,420	231	Domestic
Pont. Sunbird	24	2,690	151	Domestic
Audi 5000	17	2,830	131	Foreign
Audi Fox	23	2,070	97	Foreign
BMW 320i	25	2,650	121	Foreign

After seeing the table, we decide that we want to separate the "Domestic" observations from the "Foreign" observations, so we specify sepby(foreign).

```
. list make mpg weight displ foreign in 46/55, noobs sepby(foreign)
```

make	mpg	weight	displa~t	foreign
Plym. Volare	18	3,330	225	Domestic
Pont. Catalina	18	3,700	231	Domestic
Pont. Firebird	18	3,470	231	Domestic
Pont. Grand Prix	19	3,210	231	Domestic
Pont. Le Mans	19	3,200	231	Domestic
Pont. Phoenix	19	3,420	231	Domestic
Pont. Sunbird	24	2,690	151	Domestic
Audi 5000	17	2,830	131	Foreign
Audi Fox	23	2,070	97	Foreign
BMW 320i	25	2,650	121	Foreign

◁

▷ Example 3

We want to add vertical lines in the table to separate the variables, so we specify the `divider` option. We also want to draw a horizontal line after every 2 observations, so we specify `separator(2)`.

. list make mpg weight displ foreign in 46/55, divider separator(2)

	make	mpg	weight	displa~t	foreign
46.	Plym. Volare	18	3,330	225	Domestic
47.	Pont. Catalina	18	3,700	231	Domestic
48.	Pont. Firebird	18	3,470	231	Domestic
49.	Pont. Grand Prix	19	3,210	231	Domestic
50.	Pont. Le Mans	19	3,200	231	Domestic
51.	Pont. Phoenix	19	3,420	231	Domestic
52.	Pont. Sunbird	24	2,690	151	Domestic
53.	Audi 5000	17	2,830	131	Foreign
54.	Audi Fox	23	2,070	97	Foreign
55.	BMW 320i	25	2,650	121	Foreign

After seeing the table, we decide that we do not want to abbreviate `displacement`, so we specify `abbreviate(12)`.

. list make mpg weight displ foreign in 46/55, divider sep(2) abbreviate(12)

	make	mpg	weight	displacement	foreign
46.	Plym. Volare	18	3,330	225	Domestic
47.	Pont. Catalina	18	3,700	231	Domestic
48.	Pont. Firebird	18	3,470	231	Domestic
49.	Pont. Grand Prix	19	3,210	231	Domestic
50.	Pont. Le Mans	19	3,200	231	Domestic
51.	Pont. Phoenix	19	3,420	231	Domestic
52.	Pont. Sunbird	24	2,690	151	Domestic
53.	Audi 5000	17	2,830	131	Foreign
54.	Audi Fox	23	2,070	97	Foreign
55.	BMW 320i	25	2,650	121	Foreign

◁

❑ Technical note

You can suppress the use of value labels by specifying the `nolabel` option. For instance, the `foreign` variable in the examples above really contains numeric codes, with 0 meaning Domestic and 1 meaning Foreign. When we `list` the variable, however, we see the corresponding value labels rather than the underlying numeric code:

```
. list foreign in 51/55
```

	foreign
51.	Domestic
52.	Domestic
53.	Foreign
54.	Foreign
55.	Foreign

Specifying the `nolabel` option displays the underlying numeric codes:

```
. list foreign in 51/55, nolabel
```

	foreign
51.	0
52.	0
53.	1
54.	1
55.	1

❑

References

Harrison, D. A. 2006. Stata tip 34: Tabulation by listing. *Stata Journal* 6: 425–427.

Lauritsen, J. M. 2001. dm84: labjl: Adding numerical codes to value labels. *Stata Technical Bulletin* 59: 6–7. Reprinted in *Stata Technical Bulletin Reprints*, vol. 10, pp. 35–37. College Station, TX: Stata Press.

Riley, A. R. 1993. dm15: Interactively list values of variables. *Stata Technical Bulletin* 16: 2–6. Reprinted in *Stata Technical Bulletin Reprints*, vol. 3, pp. 37–41. College Station, TX: Stata Press.

Royston, P., and P. Sasieni. 1994. dm16: Compact listing of a single variable. *Stata Technical Bulletin* 17: 7–8. Reprinted in *Stata Technical Bulletin Reprints*, vol. 3, pp. 41–43. College Station, TX: Stata Press.

Weesie, J. 1999. dm68: Display of variables in blocks. *Stata Technical Bulletin* 50: 3–4. Reprinted in *Stata Technical Bulletin Reprints*, vol. 9, pp. 27–29. College Station, TX: Stata Press.

Also see

[D] **edit** — Browse or edit data with Data Editor

[P] **display** — Display strings and values of scalar expressions

[P] **tabdisp** — Display tables

[R] **table** — Tables of summary statistics

Title

lookfor — Search for string in variable names and labels

Syntax

lookfor *string* $\left[\, string\, [\ldots]\,\right]$

Description

lookfor helps you find variables by searching for *string* among all variable names and labels. If multiple *strings* are specified, lookfor will search for each of them separately. You may search for a phrase by enclosing *string* in double quotes.

Remarks

▷ Example 1

lookfor finds variables by searching for *string*, ignoring case, among the variable names and labels.

```
. use http://www.stata-press.com/data/r12/nlswork
(National Longitudinal Survey.  Young Women 14-26 years of age in 1968)
. lookfor code
```

variable name	storage type	display format	value label	variable label
idcode	int	%8.0g		NLS ID
ind_code	byte	%8.0g		industry of employment
occ_code	byte	%8.0g		occupation

Three variable names contain the word code.

```
. lookfor married
```

variable name	storage type	display format	value label	variable label
msp	byte	%8.0g		1 if married, spouse present
nev_mar	byte	%8.0g		1 if never married

Two variable labels contain the word married.

```
. lookfor gnp
```

variable name	storage type	display format	value label	variable label
ln_wage	float	%9.0g		ln(wage/GNP deflator)

lookfor ignores case, so lookfor gnp found GNP in a variable label.

◁

▷ Example 2

If multiple strings are specified, all variable names or labels containing any of the strings are listed.

. lookfor code married

variable name	storage type	display format	value label	variable label
idcode	int	%8.0g		NLS ID
msp	byte	%8.0g		1 if married, spouse present
nev_mar	byte	%8.0g		1 if never married
ind_code	byte	%8.0g		industry of employment
occ_code	byte	%8.0g		occupation

◁

To search for a phrase, enclose *string* in double quotes.

. lookfor "never married"

variable name	storage type	display format	value label	variable label
nev_mar	byte	%8.0g		1 if never married

Saved results

lookfor saves the following in r():

Macros
    r(varlist)   the varlist of found variables

Methods and formulas

lookfor is implemented as an ado-file.

Also see

[D] **describe** — Describe data in memory or in file

[D] **ds** — List variables matching name patterns or other characteristics

Title

memory — Memory management

Syntax

Display memory usage report

    memory

Display memory settings

    query memory

Modify memory settings

    set maxvar   # [, permanently]

    set niceness   # [, permanently]

    set min_memory *amt* [, permanently]

    set max_memory *amt* [, permanently]

    set segmentsize *amt* [, permanently]

where *amt* is #[b | k | m | g], and the default unit is b.

Parameter	Default	Minimum	Maximum	
maxvar	5000	2048	32767	(MP and SE)
	2048	2048	2048	(IC)
	99	99	99	(Small)
niceness	5	0	10	
min_memory	0	0	max_memory	
max_memory	.	2×segmentsize	.	
segmentsize	32m	1m	32g	(64-bit)
	16m	1m	1g	(32-bit)

Notes:

1. The maximum number of variables in your dataset is limited to maxvar, which is 2,048 by default. set maxvar is allowed with Stata/SE and Stata/MP only. In Stata/IC, maxvar is fixed at 2,048. In Small Stata, maxvar is fixed at 99.

2. Most users do not need to read beyond this point. Stata's memory management is completely automatic. If, however, you are using the Linux operating system, see *Serious bug in Linux OS* under *Remarks* below.

3. The maximum number of observations is fixed at 2,147,483,647 regardless of computer size or memory settings. Depending on the amount of memory on your computer, you may face a lower practical limit.

4. `max_memory` specifies the maximum amount of memory Stata can use to store your data. The default of missing (.) means all the memory the operating system is willing to supply. There are three reasons to change the value from missing to a finite number.

    1. You are a Linux user; see *Serious bug in Linux OS* under *Remarks* below.

    2. You wish to reduce the chances of accidents, such as typing `expand 100000` with a large dataset in memory and actually having Stata do it. You would rather see an insufficient-memory error message. Set `max_memory` to the amount of physical memory on your computer or more than that if you are willing to use virtual memory.

    3. You are a system administrator; see *Notes for system administrators* under *Remarks* below.

5. The remaining memory parameters—`niceness`, `min_memory`, and `segment_size`—affect efficiency only; they do not affect the size of datasets you can analyze.

6. Memory amounts for `min_memory`, `max_memory`, and `segmentsize` may be specified in bytes, kilobytes, megabytes, or gigabytes; suffix b, k, m, or g to the end of the number. The following are equivalent ways of specifying 1 gigabyte:

    ```
    1073741824
     1048576k
        1024m
           1g
    ```

Suffix k is defined as (multiply by) 1024, m is defined as 1024^2, and g is defined as 1024^3.

7. 64-bit computers can theoretically provide up to 18,446,744,073,709,551,616 bytes of memory, equivalent to 17,179,869,184 gigabytes, 16,777,216 terabytes, 16,384 petabytes, 16 exabytes. Real computers have less.

8. 32-bit computers can theoretically provide up to 4,294,967,296 bytes of memory, equivalent to 4,194,304 kilobytes, 4,096 megabytes, or 4 gigabytes. Most 32-bit operating systems limit Stata to half that.

9. Stata allocates memory for data in units of `segmentsize`. Smaller values of `segmentsize` can result in more efficient use of available memory but require Stata to jump around more. The default provides a good balance. We recommend resetting `segmentsize` only if your computer has large amounts of memory.

10. If you have large amounts of memory and you use it to process large datasets, you may wish to increase `segmentsize`. Suggested values are

memory	segmentsize
32g	64m
64g	128m
128g	256m
256g	512m
512g	1g
1024g	2g

11. `niceness` affects how soon Stata gives back unused segments to the operating system. If Stata releases them too soon, it often needs to turn around and get them right back. If Stata waits too long, Stata is consuming memory that it is not using. One reason to give memory back is to be nice to other users on multiuser systems or to be nice to yourself if you are running other processes.

The default value of 5 is defined to provide good performance. Waiting times are currently defined as

niceness	waiting time (m:s)
10	0:00.000
9	0:00.125
8	0:00.500
7	0:01
6	0:30
5	1:00
4	5:00
3	10:00
2	15:00
1	20:00
0	30:00

Niceness 10 corresponds to being totally nice. Niceness 0 corresponds to being an inconsiderate, self-centered, totally selfish jerk.

12. `min_memory` specifies an amount of memory Stata will not fall below. For instance, you have a long do-file. You know that late in the do-file, you will need 8 gigabtyes. You want to ensure that the memory will be available later. At the start of your do-file, you `set min_memory 8g`.

13. Concerning `min_memory` and `max_memory`, be aware that Stata allocates memory in segmentsize blocks. Both `min_memory` and `max_memory` are rounded down. Thus the actual minimum memory Stata will reserve will be

    `segmentsize*trunc(min_memory/segmentsize)`

The effective maximum memory is calculated similarly. (Stata does not round up `min_memory` because some users set `min_memory` equal to `max_memory`.)

Description

Memory usage and settings are described here.

`memory` displays a report on Stata's current memory usage.

`query memory` displays the current values of Stata's memory settings.

`set maxvar`, `set niceness`, `set min_memory`, `set max_memory`, and `set segmentsize` change the values of the memory settings.

If you are a Unix user, see *Serious bug in Linux OS* under *Remarks* below.

Options

permanently specifies that, in addition to making the change right now, the new limit be remembered and become the default setting when you invoke Stata.

once is not shown in the syntax diagram but is allowed with set niceness, set min_memory, set max_memory, and set segmentsize. It is for use by system administrators; see *Notes for system administrators* under *Remarks* below.

Remarks

Remarks are presented under the following headings:

> *Examples*
> *Serious bug in Linux OS*
> *Notes for system administrators*

Examples

Here is our memory-usage report after we load auto.dta that comes with Stata using Stata/MP:

```
. sysuse auto
(1978 Automobile Data)

. memory
```

Memory usage

	used	allocated
data (incl. buffers)	3,225	33,554,432
var. names, %fmts, ...	1,739	25,609
overhead	1,064,964	1,065,360
Stata matrices	0	0
ado-files	4,518	4,518
saved results	0	0
Mata matrices	0	0
Mata functions	0	0
set maxvar usage	1,391,728	1,391,728
other	1,409	1,409
total	2,466,215	36,043,056

We could then obtain the current memory-settings report by typing

```
. query memory
```

Memory settings

set maxvar	5000	2048-32767; max. vars allowed
set matsize	400	10-11000; max. # vars in models
set niceness	5	0-10
set min_memory	0	0-0
set max_memory	.	16m-0 or .
set segmentsize	16m	1m-1g

Serious bug in Linux OS

If you use Linux OS, we strongly suggest that you set `max_memory`. Here's why:

"By default, Linux follows an optimistic memory allocation strategy. This means that when malloc() returns non-NULL there is no guarantee that the memory really is available. This is a really bad bug. In case it turns out that the system is out of memory, one or more processes will be killed by the infamous OOM killer. In case Linux is employed under circumstances where it would be less desirable to suddenly lose some randomly picked processes, and moreover the kernel version is sufficiently recent, one can switch off this overcommitting behavior using [. . .]"

— Output from Unix command `man malloc`.

What this means is that Stata requests memory from Linux, Linux says yes, and then later when Stata uses that memory, the memory might not be available and Linux crashes Stata, or worse. The Linux documentation writer exercised admirable restraint. This bug can cause Linux itself to crash. It is easy.

The proponents of this behavior call it "optimistic memory allocation". We will, like the documentation writer, refer to it as a bug.

The bug is fixable. Type `man malloc` at the Unix prompt for instructions. Note that `man malloc` is an instruction of Unix, not Stata. If the bug is not mentioned, perhaps it has been fixed. Before assuming that, we suggest using a search engine to search for "linux optimistic memory allocation".

Alternatively, Stata can live with the bug if you set `max_memory`. Find out how much physical memory is on your computer and `set max_memory` to that. If you want to use virtual memory, you might set it larger, just make sure your Linux system can provide the requested memory. Specify the option `permanently` so you only need to do this once. For example,

```
. set max_memory 16g, permanently
```

Doing this does not guarantee that the bug does not bite, but it makes it unlikely.

Notes for system administrators

System administrators can set `max_memory`, `min_memory`, and `niceness` so that Stata users cannot change them. You may want to do this on shared computers to prevent individual users from hogging resources.

There is no reason you would want to do this on users' personal computers.

You can also set `segmentsize`, but there is no reason to do this even on shared systems.

The instructions are to create (or edit) the text file `sysprofile.do` in the directory where the Stata executable resides. Add the lines

```
set min_memory 0, once
set max_memory 16g, once
set niceness 5, once
```

The file must be plain text, and there must be end-of-line characters at the end of each line, including the last line. Blank lines at the end are recommended.

The 16g on `set max_memory` is merely for example. Choose an appropriate number.

The values of 0 for `min_memory` and 5 for `niceness` are recommended.

Saved results

memory saves all reported numbers in r(). StataCorp may change what memory reports, and you should not expect the same r() results to exist in future versions of Stata. To see the saved results from memory, type return list, all.

Reference

Sasieni, P. 1997. ip20: Checking for sufficient memory to add variables. *Stata Technical Bulletin* 40: 13. Reprinted in *Stata Technical Bulletin Reprints*, vol. 7, p. 86. College Station, TX: Stata Press.

Also see

[R] **query** — Display system parameters

[P] **creturn** — Return c-class values

[R] **matsize** — Set the maximum number of variables in a model

[U] **6 Managing memory**

Title

merge — Merge datasets	

Syntax

One-to-one merge on specified key variables

    <u>merge</u> 1:1 *varlist* using *filename* [, *options*]

Many-to-one merge on specified key variables

    <u>merge</u> m:1 *varlist* using *filename* [, *options*]

One-to-many merge on specified key variables

    <u>merge</u> 1:m *varlist* using *filename* [, *options*]

Many-to-many merge on specified key variables

    <u>merge</u> m:m *varlist* using *filename* [, *options*]

One-to-one merge by observation

    <u>merge</u> 1:1 _n using *filename* [, *options*]

options	Description
Options	
<u>keepus</u>ing(*varlist*)	variables to keep from using data; default is all
<u>g</u>enerate(*newvar*)	name of new variable to mark merge results; default is _merge
<u>no</u>generate	do not create _merge variable
<u>nol</u>abel	do not copy value-label definitions from using
<u>non</u>otes	do not copy notes from using
update	update missing values of same-named variables in master with values from using
replace	replace all values of same-named variables in master with nonmissing values from using (requires update)
<u>nore</u>port	do not display match result summary table
force	allow string/numeric variable type mismatch without error
Results	
<u>ass</u>ert(*results*)	specify required match results
keep(*results*)	specify which match results to keep
<u>sorted</u>	do not sort; dataset already sorted

sorted does not appear in the dialog box.

433

Menu

Data > Combine datasets > Merge two datasets

Description

merge joins corresponding observations from the dataset currently in memory (called the master dataset) with those from *filename*.dta (called the using dataset), matching on one or more key variables. merge can perform match merges (one-to-one, one-to-many, many-to-one, and many-to-many), which are often called *joins* by database people. merge can also perform sequential merges, which have no equivalent in the relational database world.

merge is for adding new variables from a second dataset to existing observations. You use merge, for instance, when combining hospital patient and discharge datasets. If you wish to add new observations to existing variables, then see [D] **append**. You use append, for instance, when adding current discharges to past discharges.

By default, merge creates a new variable, _merge, containing numeric codes concerning the source and the contents of each observation in the merged dataset. These codes are explained below in the match results table.

If *filename* is specified without an extension, then .dta is assumed.

Options

_____ Options ____

keepusing(*varlist*) specifies the variables from the using dataset that are kept in the merged dataset. By default, all variables are kept. For example, if your using dataset contains 2,000 demographic characteristics but you want only sex and age, then type merge ... , keepusing(sex age) ....

generate(*newvar*) specifies that the variable containing match results information should be named *newvar* rather than _merge.

nogenerate specifies that _merge not be created. This would be useful if you also specified keep(match), because keep(match) ensures that all values of _merge would be 3.

nolabel specifies that value-label definitions from the using file be ignored. This option should be rare, because definitions from the master are already used.

nonotes specifies that notes in the using dataset not be added to the merged dataset; see [D] **notes**.

update and replace both perform an update merge rather than a standard merge. In a standard merge, the data in the master are the authority and inviolable. For example, if the master and using datasets both contain a variable age, then matched observations will contain values from the master dataset, while unmatched observations will contain values from their respective datasets.

If update is specified, then matched observations will update missing values from the master dataset with values from the using dataset. Nonmissing values in the master dataset will be unchanged.

If replace is specified, then matched observations will contain values from the using dataset, unless the value in the using dataset is missing.

Specifying either update or replace affects the meanings of the match codes. See *Treatment of overlapping variables* for details.

noreport specifies that merge not present its summary table of match results.

force allows string/numeric variable type mismatches, resulting in missing values from the using dataset. If omitted, merge issues an error; if specified, merge issues a warning.

⌐ Results ⌐

assert(*results*) specifies the required match results. The possible *results* are

Numeric code	Equivalent word (*results*)	Description
1	master	observation appeared in master only
2	using	observation appeared in using only
3	match	observation appeared in both
4	match_update	observation appeared in both, missing values updated
5	match_conflict	observation appeared in both, conflicting nonmissing values

Codes 4 and 5 can arise only if the update option is specified. If codes of both 4 and 5 could pertain to an observation, then 5 is used.

Numeric codes and words are equivalent when used in the assert() or keep() options.

The following synonyms are allowed: masters for master, usings for using, matches and matched for match, match_updates for match_update, and match_conflicts for match_conflict.

Using assert(match master) specifies that the merged file is required to include only matched master or using observations and unmatched master observations, and may not include unmatched using observations. Specifying assert() results in merge issuing an error if there are match results among those observations you allowed.

The order of the words or codes is not important, so all the following assert() specifications would be the same:

> assert(match master)
>
> assert(master matches)
>
> assert(1 3)

When the match results contain codes other than those allowed, return code 9 is returned, and the merged dataset with the unanticipated results is left in memory to allow you to investigate.

keep(*results*) specifies which observations are to be kept from the merged dataset. Using keep(match master) specifies keeping only matched observations and unmatched master observations after merging.

keep() differs from assert() because it selects observations from the merged dataset rather than enforcing requirements. keep() is used to pare the merged dataset to a given set of observations when you do not care if there are other observations in the merged dataset. assert() is used to verify that only a given set of observations is in the merged dataset.

You can specify both assert() and keep(). If you require matched observations and unmatched master observations but you want only the matched observations, then you could specify assert(match master) keep(match).

assert() and keep() are convenience options whose functionality can be duplicated using _merge directly.

> . merge ..., assert(match master) keep(match)

is identical to

```
. merge ...
. assert _merge==1 | _merge==3
. keep if _merge==3
```

The following option is available with `merge` but is not shown in the dialog box:

`sorted` specifies that the master and using datasets are already sorted by *varlist*. If the datasets are already sorted, then `merge` runs a little more quickly; the difference is hardly detectable, so this option is of interest only where speed is of the utmost importance.

Remarks

Remarks are presented under the following headings:

> *Overview*
> *Basic description*
> *1:1 merges*
> *m:1 merges*
> *1:m merges*
> *m:m merges*
> *Sequential merges*
> *Treatment of overlapping variables*
> *Sort order*
> *Troubleshooting m:m merges*
> *Examples*

Overview

`merge 1:1` *varlist* ... specifies a one-to-one match merge. *varlist* specifies variables common to both datasets that together uniquely identify single observations in both datasets. For instance, suppose you have a dataset of customer information, called `customer.dta`, and have a second dataset of other information about roughly the same customers, called `other.dta`. Suppose further that both datasets identify individuals by using the `pid` variable, and there is only one observation per individual in each dataset. You would merge the two datasets by typing

```
. use customer
. merge 1:1 pid using other
```

Reversing the roles of the two files would be fine. Choosing which dataset is the master and which is the using matters only if there are overlapping variable names. `1:1` merges are less common than `1:m` and `m:1` merges.

`merge 1:m` and `merge m:1` specify one-to-many and many-to-one match merges, respectively. To illustrate the two choices, suppose you have a dataset containing information about individual hospitals, called `hospitals.dta`. In this dataset, each observation contains information about one hospital, which is uniquely identified by the `hospitalid` variable. You have a second dataset called `discharges.dta`, which contains information on individual hospital stays by many different patients. `discharges.dta` also identifies hospitals by using the `hospitalid` variable. You would like to join all the information in both datasets. There are two ways you could do this.

`merge 1:m` *varlist* ... specifies a one-to-many match merge.

```
. use hospitals
. merge 1:m hospitalid using discharges
```

would join the discharge data to the hospital data. This is a 1:m merge because hospitalid uniquely identifies individual observations in the dataset in memory (hospitals), but could correspond to many observations in the using dataset.

merge m:1 *varlist* ... specifies a many-to-one match merge.

```
. use discharges
. merge m:1 hospitalid using hospitals
```

would join the hospital data to the discharge data. This is an m:1 merge because hospitalid can correspond to many observations in the master dataset, but uniquely identifies individual observations in the using dataset.

merge m:m *varlist* ... specifies a many-to-many match merge. This is allowed for completeness, but it is difficult to imagine an example of when it would be useful. For an m:m merge, *varlist* does not uniquely identify the observations in either dataset. Matching is performed by combining observations with equal values of *varlist*; within matching values, the first observation in the master dataset is matched with the first matching observation in the using dataset; the second, with the second; and so on. If there is an unequal number of observations within a group, then the last observation of the shorter group is used repeatedly to match with subsequent observations of the longer group. Use of merge m:m is not encouraged.

merge 1:1 _n performs a sequential merge. _n is not a variable name; it is Stata syntax for observation number. A sequential merge performs a one-to-one merge on observation number. The first observation of the master dataset is matched with the first observation of the using dataset; the second, with the second; and so on. If there is an unequal number of observations, the remaining observations are unmatched. Sequential merges are dangerous, because they require you to rely on sort order to know that observations belong together. Use this merge at your own risk.

Basic description

Think of merge as being *master* + *using* = *merged result*.

Call the dataset in memory the *master* dataset, and the dataset on disk the *using* dataset. This way we have general names that are not dependent on individual datasets.

Suppose we have two datasets,

master in memory   on disk in file *filename*

id	age
1	22
2	56
5	17

id	wgt
1	130
2	180
4	110

We would like to join together the age and weight information. We notice that the id variable identifies unique observations in both datasets: if you tell me the id number, then I can tell you the one observation that contains information about that id. This is true for both the master and the using datasets.

Because id uniquely identifies observations in both datasets, this is a 1:1 merge. We can bring in the dataset from disk by typing

. merge 1:1 id using *filename*

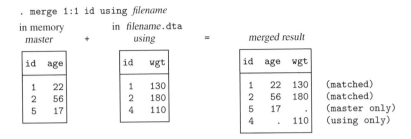

The original data in memory are called the master data. The data in *filename*.dta are called the using data. After merge, the merged result is left in memory. The id variable is called the key variable. Stata jargon is that the datasets were merged on id.

Observations for id==1 existed in both the master and using datasets and so were combined in the merged result. The same occurred for id==2. For id==5 and id==4, however, no matches were found and thus each became a separate observation in the merged result. Thus each observation in the merged result came from one of three possible sources:

Numeric code	Equivalent word	Description
1	master	originally appeared in master only
2	using	originally appeared in using only
3	match	originally appeared in both

merge encodes this information into new variable _merge, which merge adds to the merged result:

```
in memory          in filename.dta
  master      +        using       =        merged result

id  age            id   wgt            id  age  wgt  _merge

 1   22             1   130             1   22  130      3
 2   56             2   180             2   56  180      3
 5   17             4   110             5   17    .      1
                                        4    .  110      2
```

Note: Above we show the master and using data sorted by id before merging; this was for illustrative purposes. The dataset resulting from a 1:1 merge will have the same data, regardless of the sort order of the master and using datasets.

The formal definition for merge behavior is the following: Start with the first observation of the master. Find the corresponding observation in the using data, if there is one. Record the matched or unmatched result. Proceed to the next observation in the master dataset. When you finish working through the master dataset, work through unused observations from the using data. By default, unmatched observations are kept in the merged data, whether they come from the master dataset or the using dataset.

Remember this formal definition. It will serve you well.

1:1 merges

The example shown above is called a 1:1 merge, because the key variable uniquely identified each observation in each of the datasets.

A variable or variable list uniquely identifies the observations if each distinct value of the variable(s) corresponds to one observation in the dataset.

In some datasets, multiple variables are required to identify the observations. Imagine data obtained by observing patients at specific points in time so that variables pid and time, taken together, identify the observations. Below we have two such datasets and run a 1:1 merge on pid and time,

. merge 1:1 pid time using *filename*

master				using				merged result				
pid	time	x1		pid	time	x2		pid	time	x1	x2	_merge
14	1	0		14	1	7		14	1	0	7	3
14	2	0		14	2	9		14	2	0	9	3
14	4	0		16	1	2		14	4	0	.	1
16	1	1		16	2	3		16	1	1	2	3
16	2	1		17	1	5		16	2	1	3	3
17	1	0		17	2	2		17	1	0	5	3
								17	2	.	2	2

This is a 1:1 merge because the combination of the values of pid and time uniquely identifies observations in both datasets.

By default, there is nothing about a 1:1 merge that implies that all, or even any of, the observations match. Above five observations matched, one observation was only in the master (subject 14 at time 4), and another was only in the using (subject 17 at time 2).

m:1 merges

In an m:1 merge, the key variable or variables uniquely identify the observations in the using data, but not necessarily in the master data. Suppose you had person-level data within regions and you wished to bring in regional data. Here is an example:

. merge m:1 region using *filename*

master				using				merged result				
id	region	a		region	x			id	region	a	x	_merge
1	2	26		1	15			1	2	26	13	3
2	1	29		2	13			2	1	29	15	3
3	2	22		3	12			3	2	22	13	3
4	3	21		4	11			4	3	21	12	3
5	1	24						5	1	24	15	3
6	5	20						6	5	20	.	1
								.	4	.	11	2

To bring in the regional information, we need to merge on region. The values of region identify individual observations in the using data, but it is not an identifier in the master data.

We show the merged dataset sorted by id because this makes it easier to see how the merged dataset was constructed. For each observation in the master data, merge finds the corresponding observation in the using data. merge combines the values of the variables in the using dataset to the observations in the master dataset.

1:m merges

1:m merges are similar to m:1, except that now the key variables identify unique observations in the master dataset. Any datasets that can be merged using an m:1 merge may be merged using a 1:m merge by reversing the roles of the master and using datasets. Here is the same example as used previously, with the master and using datasets reversed:

```
. merge 1:m region using filename
```

| master | + | using | = | merged result |

region	x
1	15
2	13
3	12
4	11

id	region	a
1	2	26
2	1	29
3	2	22
4	3	21
5	1	24
6	5	20

region	x	id	a	_merge
1	15	2	29	3
1	15	5	24	3
2	13	1	26	3
2	13	3	22	3
3	12	4	21	3
4	11	.	.	1
5	.	6	20	2

This merged result is identical to the merged result in the previous section, except for the sort order and the contents of _merge. This time, we show the merged result sorted by region rather than id. Reversing the roles of the files causes a reversal in the 1s and 2s for _merge: where _merge was previously 1, it is now 2, and vice versa. These exchanged _merge values reflect the reversed roles of the master and using data.

For each observation in the master data, merge found the corresponding observation(s) in the using data and then wrote down the matched or unmatched result. Once the master observations were exhausted, merge wrote down any observations from the using data that were never used.

m:m merges

m:m specifies a many-to-many merge and is a bad idea. In an m:m merge, observations are matched within equal values of the key variable(s), with the first observation being matched to the first; the second, to the second; and so on. If the master and using have an unequal number of observations within the group, then the last observation of the shorter group is used repeatedly to match with subsequent observations of the longer group. Thus m:m merges are dependent on the current sort order—something which should never happen.

Because m:m merges are such a bad idea, we are not going to show you an example. If you think that you need an m:m merge, then you probably need to work with your data so that you can use a 1:m or m:1 merge. Tips for this are given in *Troubleshooting m:m merges* below.

Sequential merges

In a *sequential* merge, there are no key variables. Observations are matched solely on their observation number:

```
. merge 1:1 _n using filename
```

master   +   using   =   merged result

x1
10
30
20
5

x2
7
2
1
9
3

x1	x2	_merge
10	7	3
30	2	3
20	1	3
5	9	3
.	3	2

In the example above, the using data are longer than the master, but that could be reversed. In most cases where sequential merges are appropriate, the datasets are expected to be of equal length, and you should type

```
. merge 1:1 _n using filename, assert(match) nogenerate
```

Sequential merges, like m:m merges, are dangerous. Both depend on the current sort order of the data.

Treatment of overlapping variables

When performing merges of any type, the master and using datasets may have variables in common other than the key variables. We will call such variables overlapping variables. For instance, if the variables in the master and using datasets are

master: id, region, sex, age, race

using: id, sex, bp, race

and id is the key variable, then the overlapping variables are sex and race.

By default, merge treats values from the master as inviolable. When observations match, it is the master's values of the overlapping variables that are recorded in the merged result.

If you specify the update option, however, then all missing values of overlapping variables in matched observations are replaced with values from the using data. Because of this new behavior, the merge codes change somewhat. Codes 1 and 2 keep their old meaning. Code 3 splits into codes 3, 4, and 5. Codes 3, 4, and 5 are filtered according to the following rules; the first applicable rule is used.

5 corresponds to matched observations where at least one overlapping variable had conflicting nonmissing values.

4 corresponds to matched observations where at least one missing value was updated, but there were no conflicting nonmissing values.

3 means observations matched, and there were neither updated missing values nor conflicting nonmissing values.

If you specify both the update and replace options, then the _merge==5 cases are updated with values from the using data.

Sort order

As we have mentioned, in the 1:1, 1:m, and m:1 match merges, the sort orders of the master and using datasets do not affect the data in the merged dataset. This is not the case of m:m, which we recommend you never use.

Sorting is used by merge internally for efficiency, so the merged result can be produced most quickly when the master and using datasets are already sorted by the key variable(s) before merging. You are not required to have the dataset sorted before using merge, however, because merge will sort behind the scenes, if necessary. If the using dataset is not sorted, then a temporary copy is made and sorted to ensure that the current sort order on disk is not affected.

All this is to reassure you that 1) your datasets on disk will not be modified by merge and 2) despite the fact that our discussion has ignored sort issues, merge is, in fact, efficient behind the scenes.

It hardly makes any difference in run times, but if you know that the master and using data are already sorted by the key variable(s), then you can specify the sorted option. All that will be saved is the time merge would spend discovering that fact for itself.

The merged result produced by merge orders the variables and observations in a special and sometimes useful way. If you think of datasets as tables, then the columns for the new variables appear to the right of what was the master. If the master data originally had k variables, then the new variables will be the $(k + 1)$st, $(k + 2)$nd, and so on. The new observations are similarly ordered so that they all appear at the end of what was the master. If the master originally had N observations, then the new observations, if any, are the $(N + 1)$st, $(N + 2)$nd, and so on. Thus the original master data can be found from the merged result by extracting the first k variables and first N observations. If merge with the update option was specified, however, then be aware that the extracted master may have some updated values.

The merged result is unsorted except for a 1:1 merge, where there are only matched observations. Here the dataset is sorted by the key variables.

Troubleshooting m:m merges

First, if you think you need to perform an m:m merge, then we suspect you are wrong. If you would like to match every observation in the master to every observation in the using with the same values of the key variable(s), then you should be using joinby; see [D] **joinby**.

If you still want to use merge, then it is likely that you have forgotten one or more key variables that could be used to identify observations within groups. Perhaps you have panel data with 4 observations on each subject, and you are thinking that what you need to do is

    . merge m:m subjectid using *filename*

Ask yourself if you have a variable that identifies observation within panel, such as a sequence number or a time. If you have, say, a time variable, then you probably should try something like

    . merge 1:m subjectid time using *filename*

(You might need a 1:1 or m:1 merge; 1:m was arbitrarily chosen for the example.)

If you do not have a time or time-like variable, then ask yourself if there is a meaning to matching the first observations within subject, the second observations within subject, and so on. If so, then there is a concept of sequence within subject.

Suppose you do indeed have a sequence concept, but in your dataset it is recorded via the ordering of the observations. Here you are in a dangerous situation because any kind of sorting would lose the identity of the first, second, and nth observation within subject. Your first goal should be to fix this problem by creating an explicit sequence variable from the current ordering—your merge can come later.

Start with your master data. Type

```
. sort subjectid, stable
. by subjectid: gen seqnum = _n
```

Do not omit sort's stable option. That is what will keep the observations in the same order within subject. Save the data. Perform these same three steps on your using data.

After fixing the datasets, you can now type

```
. merge 1:m subjectid seqnum using filename
```

If you do not think there is a meaning to being the first, second, and nth observation within subject, then you need to ask yourself what it means to match the first observations within subjectid, the second observations within subjectid, and so on. Would it make equal sense to match the first with the third, the second with the fourth, or any other haphazard matching? If so, then there is no real ordering, so there is no real meaning to merging. You are about to obtain a haphazard result; you need to rethink your merge.

Examples

▷ Example 1

We have two datasets, one of which has information about the size of old automobiles, and the other of which has information about their expense:

```
. use http://www.stata-press.com/data/r12/autosize
(1978 Automobile Data)
. list
```

	make	weight	length
1.	Toyota Celica	2,410	174
2.	BMW 320i	2,650	177
3.	Cad. Seville	4,290	204
4.	Pont. Grand Prix	3,210	201
5.	Datsun 210	2,020	165
6.	Plym. Arrow	3,260	170

```
. use http://www.stata-press.com/data/r12/autoexpense
(1978 Automobile Data)
. list
```

	make	price	mpg
1.	Toyota Celica	5,899	18
2.	BMW 320i	9,735	25
3.	Cad. Seville	15,906	21
4.	Pont. Grand Prix	5,222	19
5.	Datsun 210	4,589	35

We can see that these datasets contain different information about nearly the same cars—the `autosize` file has one more car. We would like to get all the information about all the cars into one dataset.

Because we are adding new variables to old variables, this is a job for the `merge` command. We need only to decide what type of match merge we need.

Looking carefully at the datasets, we see that the `make` variable, which identifies the cars in each of the two datasets, also identifies individual observations within the datasets. What this means is that if you tell me the make of car, I can tell you the one observation that corresponds to that car. Because this is true for both datasets, we should use a `1:1` merge.

We will start with a clean slate to show the full process:

```
. use http://www.stata-press.com/data/r12/autosize
(1978 Automobile Data)

. merge 1:1 make using http://www.stata-press.com/data/r12/autoexpense

    Result                           # of obs.

    not matched                             1
        from master                         1  (_merge==1)
        from using                          0  (_merge==2)

    matched                                 5  (_merge==3)

. list
```

	make	weight	length	price	mpg	_merge
1.	BMW 320i	2,650	177	9,735	25	matched (3)
2.	Cad. Seville	4,290	204	15,906	21	matched (3)
3.	Datsun 210	2,020	165	4,589	35	matched (3)
4.	Plym. Arrow	3,260	170	.	.	master only (1)
5.	Pont. Grand Prix	3,210	201	5,222	19	matched (3)
6.	Toyota Celica	2,410	174	5,899	18	matched (3)

The merge is successful—all the data are present in the combined dataset, even that from the one car that has only size information. If we wanted only those makes for which all information is present, it would be up to us to `drop` the observations for which `_merge < 3`.

◁

▷ Example 2

Suppose we had the same setup as in the previous example, but we erroneously think that we have all the information on all the cars. We could tell `merge` that we expect only matches by using the `assert` option.

```
. use http://www.stata-press.com/data/r12/autosize, clear
(1978 Automobile Data)

. merge 1:1 make using http://www.stata-press.com/data/r12/autoexpense, assert(match)
merge:  after merge, not all observations matched
        (merged result left in memory)
r(9);
```

`merge` tells us that there is a problem with our assumption. To see how many mismatches there were, we can tabulate `_merge`:

```
. tabulate _merge
```

_merge	Freq.	Percent	Cum.
master only (1)	1	16.67	16.67
matched (3)	5	83.33	100.00
Total	6	100.00	

If we would like to list the problem observation, we can type

```
. list if _merge < 3
```

	make	weight	length	price	mpg	_merge
4.	Plym. Arrow	3,260	170	.	.	master only (1)

If we were convinced that all data should be complete in the two datasets, we would have to rectify the mismatch in the original datasets.

◁

▷ Example 3

Once again, suppose that we had the same datasets as before, but this time we want the final dataset to have only those observations for which there is a match. We do not care if there are mismatches—all that is important are the complete observations. By using the keep(match) option, we will guarantee that this happens. Because we are keeping only those observations for which the key variable matches, there is no need to generate the _merge variable. We could do the following:

```
. use http://www.stata-press.com/data/r12/autosize, clear
(1978 Automobile Data)
. merge 1:1 make using http://www.stata-press.com/data/r12/autoexpense, keep(match)
> nogenerate
```

Result	# of obs.
not matched	0
matched	5

```
. list
```

	make	weight	length	price	mpg
1.	BMW 320i	2,650	177	9,735	25
2.	Cad. Seville	4,290	204	15,906	21
3.	Datsun 210	2,020	165	4,589	35
4.	Pont. Grand Prix	3,210	201	5,222	19
5.	Toyota Celica	2,410	174	5,899	18

◁

▷ Example 4

We have two datasets: one has salespeople in regions and the other has regional data about sales. We would like to put all the information into one dataset. Here are the datasets:

```
. use http://www.stata-press.com/data/r12/sforce, clear
(Sales Force)
. list
```

	region	name
1.	N Cntrl	Krantz
2.	N Cntrl	Phipps
3.	N Cntrl	Willis
4.	NE	Ecklund
5.	NE	Franks
6.	South	Anderson
7.	South	Dubnoff
8.	South	Lee
9.	South	McNeil
10.	West	Charles
11.	West	Cobb
12.	West	Grant

```
. use http://www.stata-press.com/data/r12/dollars
(Regional Sales & Costs)
. list
```

	region	sales	cost
1.	N Cntrl	419,472	227,677
2.	NE	360,523	138,097
3.	South	532,399	330,499
4.	West	310,565	165,348

We can see that the region would be used to match observations in the two datasets, and this time we see that region identifies individual observations in the dollars dataset but not in the sforce dataset. This means we will have to use either an m:1 or a 1:m merge. Here we will open the sforce dataset and then merge the dollars dataset. This will be an m:1 merge, because region does not identify individual observations in the dataset in memory but does identify them in the using dataset. Here is the command and its result:

```
. use http://www.stata-press.com/data/r12/sforce
(Sales Force)
. merge m:1 region using http://www.stata-press.com/data/r12/dollars
(label region already defined)
```

Result	# of obs.	
not matched	0	
matched	12	(_merge==3)

. list

	region	name	sales	cost	_merge
1.	N Cntrl	Krantz	419,472	227,677	matched (3)
2.	N Cntrl	Phipps	419,472	227,677	matched (3)
3.	N Cntrl	Willis	419,472	227,677	matched (3)
4.	NE	Ecklund	360,523	138,097	matched (3)
5.	NE	Franks	360,523	138,097	matched (3)
6.	South	Anderson	532,399	330,499	matched (3)
7.	South	Dubnoff	532,399	330,499	matched (3)
8.	South	Lee	532,399	330,499	matched (3)
9.	South	McNeil	532,399	330,499	matched (3)
10.	West	Charles	310,565	165,348	matched (3)
11.	West	Cobb	310,565	165,348	matched (3)
12.	West	Grant	310,565	165,348	matched (3)

We can see from the result that all the values of region were matched in both datasets. This is a rare occurrence in practice!

Had we had the dollars dataset in memory and merged in the sforce dataset, we would have done a 1:m merge.

◁

We would now like to use a series of examples that shows how merge treats nonkey variables, which have the same names in the two datasets. We will call these "overlapping" variables.

▷ Example 5

Here are two datasets whose only purpose is for this illustration:

. use http://www.stata-press.com/data/r12/overlap1, clear
. list, sepby(id)

	id	seq	x1	x2
1.	1	1	1	1
2.	1	2	1	.
3.	1	3	1	2
4.	1	4	.	2
5.	2	1	.	1
6.	2	2	.	2
7.	2	3	1	1
8.	2	4	1	2
9.	2	5	.a	1
10.	2	6	.a	2
11.	3	1	.	.a
12.	3	2	.	1
13.	3	3	.	.
14.	3	4	.a	.a
15.	10	1	5	8

. use http://www.stata-press.com/data/r12/overlap2

```
. list
```

	id	bar	x1	x2
1.	1	11	1	1
2.	2	12	.	1
3.	3	14	.	.a
4.	20	18	1	1

We can see that id can be used as the key variable for putting the two datasets together. We can also see that there are two overlapping variables: x1 and x2.

We will start with a simple m:1 merge:

```
. use http://www.stata-press.com/data/r12/overlap1
. merge m:1 id using http://www.stata-press.com/data/r12/overlap2
```

Result	# of obs.	
not matched	2	
from master	1	(_merge==1)
from using	1	(_merge==2)
matched	14	(_merge==3)

```
. list, sepby(id)
```

	id	seq	x1	x2	bar	_merge
1.	1	1	1	1	11	matched (3)
2.	1	2	1	.	11	matched (3)
3.	1	3	1	2	11	matched (3)
4.	1	4	.	2	11	matched (3)
5.	2	1	.	1	12	matched (3)
6.	2	2	.	2	12	matched (3)
7.	2	3	1	1	12	matched (3)
8.	2	4	1	2	12	matched (3)
9.	2	5	.a	1	12	matched (3)
10.	2	6	.a	2	12	matched (3)
11.	3	1	.	.a	14	matched (3)
12.	3	2	.	1	14	matched (3)
13.	3	3	.	.	14	matched (3)
14.	3	4	.a	.a	14	matched (3)
15.	10	1	5	8	.	master only (1)
16.	20	.	1	1	18	using only (2)

Careful inspection shows that for the matched id, the values of x1 and x2 are still the values that were originally in the overlap1 dataset. This is the default behavior of merge—the data in the master dataset is the authority and is kept intact.

◁

▷ Example 6

Now we would like to investigate the update option. Used by itself, it will replace missing values in the master dataset with values from the using dataset:

```
. use http://www.stata-press.com/data/r12/overlap1, clear
. merge m:1 id using http://www.stata-press.com/data/r12/overlap2, update
```

Result	# of obs.	
not matched	2	
from master	1	(_merge==1)
from using	1	(_merge==2)
matched	14	
not updated	5	(_merge==3)
missing updated	4	(_merge==4)
nonmissing conflict	5	(_merge==5)

```
. list, sepby(id)
```

	id	seq	x1	x2	bar	_merge
1.	1	1	1	1	11	matched (3)
2.	1	2	1	1	11	missing updated (4)
3.	1	3	1	2	11	nonmissing conflict (5)
4.	1	4	1	2	11	nonmissing conflict (5)
5.	2	1	.	1	12	matched (3)
6.	2	2	.	2	12	nonmissing conflict (5)
7.	2	3	1	1	12	matched (3)
8.	2	4	1	2	12	nonmissing conflict (5)
9.	2	5	.	1	12	missing updated (4)
10.	2	6	.	2	12	nonmissing conflict (5)
11.	3	1	.	.a	14	matched (3)
12.	3	2	.	1	14	matched (3)
13.	3	3	.	.a	14	missing updated (4)
14.	3	4	.	.a	14	missing updated (4)
15.	10	1	5	8	.	master only (1)
16.	20	.	1	1	18	using only (2)

Looking through the resulting dataset observation by observation, we can see both what the update option updated as well as how the _merge variable gets its values.

The following is a listing that shows what is happening, where x1_m and x2_m come from the master dataset (overlap1), x1_u and x2_u come from the using dataset (overlap2), and x1 and x2 are the values that appear when using merge with the update option.

	id	x1_m	x1_u	x1	x2_m	x2_u	x2	_merge
1.	1	1	1	1	1	1	1	matched (3)
2.	1	1	1	1	.	1	1	missing updated (4)
3.	1	1	1	1	2	1	2	nonmissing conflict (5)
4.	1	.	1	1	2	1	2	nonmissing conflict (5)
5.	2	.	.	.	1	1	1	matched (3)
6.	2	.	.	.	2	1	2	nonmissing conflict (5)
7.	2	1	.	1	1	1	1	matched (3)
8.	2	1	.	1	2	1	2	nonmissing conflict (5)
9.	2	.a	.	.	1	1	1	missing updated (4)
10.	2	.a	.	.	2	1	2	nonmissing conflict (5)
11.	3	.	.	.	.a	.a	.a	matched (3)
12.	3	.	.	.	1	.a	1	matched (3)
13.	3	.	.	.	.	.a	.a	missing updated (4)
14.	3	.a	.	.	.a	.a	.a	missing updated (4)
15.	10	5	.	5	8	.	8	master only (1)
16.	20	.	1	1	.	1	1	using only (2)

From this, we can see two important facts: if there are both a conflict and an updated value, the value of _merge will reflect that there was a conflict, and missing values in the master dataset are updated by missing values in the using dataset.

◁

> ## Example 7

We would like to see what happens if the update and replace options are specified. The replace option extends the action of update to use nonmissing values of the using dataset to replace values in the master dataset. The values of _merge are unaffected by using both update and replace.

```
. use http://www.stata-press.com/data/r12/overlap1, clear
. merge m:1 id using http://www.stata-press.com/data/r12/overlap2, update replace
```

Result	# of obs.	
not matched	2	
from master	1	(_merge==1)
from using	1	(_merge==2)
matched	14	
not updated	5	(_merge==3)
missing updated	4	(_merge==4)
nonmissing conflict	5	(_merge==5)

. list, sepby(id)

	id	seq	x1	x2	bar	_merge
1.	1	1	1	1	11	matched (3)
2.	1	2	1	1	11	missing updated (4)
3.	1	3	1	1	11	nonmissing conflict (5)
4.	1	4	1	1	11	nonmissing conflict (5)
5.	2	1	.	1	12	matched (3)
6.	2	2	.	1	12	nonmissing conflict (5)
7.	2	3	1	1	12	matched (3)
8.	2	4	1	1	12	nonmissing conflict (5)
9.	2	5	.	1	12	missing updated (4)
10.	2	6	.	1	12	nonmissing conflict (5)
11.	3	1	.	.a	14	matched (3)
12.	3	2	.	1	14	matched (3)
13.	3	3	.	.a	14	missing updated (4)
14.	3	4	.	.a	14	missing updated (4)
15.	10	1	5	8	.	master only (1)
16.	20	.	1	1	18	using only (2)

◁

> ## Example 8

Suppose we would like to use the update option, as we did above, but we would like to keep only those observations for which the value of the key variable, id, was found in both datasets. This will be more complicated than in our earlier example, because the update option splits the matches into matches, match_updates, and match_conflicts. We must either use all these code words in the keep option or use their numerical equivalents, 3, 4, and 5. Here the latter is simpler.

. use http://www.stata-press.com/data/r12/overlap1, clear

. merge m:1 id using http://www.stata-press.com/data/r12/overlap2, update keep(3 4 5)

Result	# of obs.	
not matched	0	
matched	14	
not updated	5	(_merge==3)
missing updated	4	(_merge==4)
nonmissing conflict	5	(_merge==5)

```
. list, sepby(id)
```

	id	seq	x1	x2	bar	_merge
1.	1	1	1	1	11	matched (3)
2.	1	2	1	1	11	missing updated (4)
3.	1	3	1	2	11	nonmissing conflict (5)
4.	1	4	1	2	11	nonmissing conflict (5)
5.	2	1	.	1	12	matched (3)
6.	2	2	.	2	12	nonmissing conflict (5)
7.	2	3	1	1	12	matched (3)
8.	2	4	1	2	12	nonmissing conflict (5)
9.	2	5	.	1	12	missing updated (4)
10.	2	6	.	2	12	nonmissing conflict (5)
11.	3	1	.	.a	14	matched (3)
12.	3	2	.	1	14	matched (3)
13.	3	3	.	.a	14	missing updated (4)
14.	3	4	.	.a	14	missing updated (4)

◁

▷ Example 9

As a final example, we would like show one example of a 1:m merge. There is nothing conceptually different here; what is interesting is the order of the observations in the final dataset:

```
. use http://www.stata-press.com/data/r12/overlap2, clear
. merge 1:m id using http://www.stata-press.com/data/r12/overlap1
```

Result	# of obs.	
not matched	2	
from master	1	(_merge==1)
from using	1	(_merge==2)
matched	14	(_merge==3)

```
. list, sepby(id)
```

	id	bar	x1	x2	seq	_merge
1.	1	11	1	1	1	matched (3)
2.	2	12	.	1	1	matched (3)
3.	3	14	.	.a	1	matched (3)
4.	20	18	1	1	.	master only (1)
5.	1	11	1	1	2	matched (3)
6.	1	11	1	1	3	matched (3)
7.	1	11	1	1	4	matched (3)
8.	2	12	.	1	2	matched (3)
9.	2	12	.	1	3	matched (3)
10.	2	12	.	1	4	matched (3)
11.	2	12	.	1	5	matched (3)
12.	2	12	.	1	6	matched (3)
13.	3	14	.	.a	2	matched (3)
14.	3	14	.	.a	3	matched (3)
15.	3	14	.	.a	4	matched (3)
16.	10	.	5	8	1	using only (2)

We can see here that the first four observations come from the master dataset, and all additional observations, whether matched or unmatched, come below these observations. This illustrates that the master dataset is always in the upper-left corner of the merged dataset.

◁

Methods and formulas

merge is implemented as an ado-file.

References

Golbe, D. L. 2010. Stata tip 83: Merging multilingual datasets. *Stata Journal* 10: 152–156.

Gould, W. W. 2011a. Merging data, part 1: Merges gone bad. The Stata Blog: Not Elsewhere Classified. http://blog.stata.com/2011/04/18/merging-data-part-1-merges-gone-bad/

——. 2011b. Merging data, part 2: Multiple-key merges. The Stata Blog: Not Elsewhere Classified. http://blog.stata.com/2011/05/27/merging-data-part-2-multiple-key-merges/

Nash, J. D. 1994. dm19: Merging raw data and dictionary files. *Stata Technical Bulletin* 20: 3–5. Reprinted in *Stata Technical Bulletin Reprints*, vol. 4, pp. 22–25. College Station, TX: Stata Press.

Weesie, J. 2000. dm75: Safe and easy matched merging. *Stata Technical Bulletin* 53: 6–17. Reprinted in *Stata Technical Bulletin Reprints*, vol. 9, pp. 62–77. College Station, TX: Stata Press.

Also see

[D] **save** — Save Stata dataset

[D] **sort** — Sort data

[D] **append** — Append datasets

[D] **cross** — Form every pairwise combination of two datasets

[D] **joinby** — Form all pairwise combinations within groups

[U] **22 Combining datasets**

Title

> **missing values** — Quick reference for missing values

Description

This entry provides a quick reference for Stata's missing values.

Remarks

Stata has 27 numeric missing values:

., the default, which is called the *system missing value* or `sysmiss`

and

.a, .b, .c, ..., .z, which are called the *extended missing values.*

Numeric missing values are represented by large positive values. The ordering is

$$\text{all nonmissing numbers} < \, . \, < \, .a < \, .b < \cdots < \, .z$$

Thus the expression

$$\text{age} > 60$$

is true if variable `age` is greater than 60 or missing.

To exclude missing values, ask whether the value is less than '.'.

```
. list if age > 60 & age < .
```

To specify missing values, ask whether the value is greater than or equal to '.'. For instance,

```
. list if age >=.
```

Stata has one string missing value, which is denoted by `""` (blank).

Reference

Cox, N. J. 2010. Stata tip 84: Summing missings. *Stata Journal* 10: 157–159.

Also see

[U] **12.2.1 Missing values**

455

Title

---

mkdir — Create directory

Syntax

mkdir *directoryname* $\left[\, , \, \underline{pub}lic \right]$

Double quotes may be used to enclose *directoryname*, and the quotes must be used if *directoryname* contains embedded spaces.

Description

mkdir creates a new directory (folder).

Option

public specifies that *directoryname* be readable by everyone; otherwise, the directory will be created according to the default permissions of your operating system.

Remarks

Examples:

Windows

. mkdir myproj
. mkdir c:\projects\myproj
. mkdir "c:\My Projects\Project 1"

Mac and Unix

. mkdir myproj
. mkdir ~/projects/myproj

Also see

[D] **cd** — Change directory

[D] **copy** — Copy file from disk or URL

[D] **dir** — Display filenames

[D] **erase** — Erase a disk file

[D] **rmdir** — Remove directory

[D] **shell** — Temporarily invoke operating system

[D] **type** — Display contents of a file

[U] **11.6 Filenaming conventions**

Title

> **mvencode** — Change missing values to numeric values and vice versa

Syntax

Change missing values to numeric values

mvencode *varlist* [*if*] [*in*] , mv(# | *mvc* = # [\ *mvc* = #...] [\ else = #]) [override]

Change numeric values to missing values

mvdecode *varlist* [*if*] [*in*] , mv(*numlist* | *numlist* = *mvc* [\ *numlist* = *mvc* ...])

where *mvc* is one of . | .a | .b | ... | .z.

Menu

mvencode

Data > Create or change data > Other variable-transformation commands > Change missing values to numeric

mvdecode

Data > Create or change data > Other variable-transformation commands > Change numeric values to missing

Description

mvencode changes missing values in the specified *varlist* to numeric values.

mvdecode changes occurrences of a numlist in the specified *varlist* to a missing-value code.

Missing-value codes may be sysmiss (.) and the extended missing-value codes .a, .b, ..., .z.

String variables in *varlist* are ignored.

Options

####### Main

mv(# | *mvc* = # [\ *mvc* = #...] [\ else = #]) is required and specifies the numeric values to which the missing values are to be changed.

mv(#) specifies that all types of missing values be changed to #.

mv(*mvc*=#) specifies that occurrences of missing-value code *mvc* be changed to #. Multiple transformation rules may be specified, separated by a backward slash (\). The list may be terminated by the special rule else=#, specifying that all types of missing values not yet transformed be set to #.

Examples: mv(9), mv(.=99\.a=98\.b=97), mv(.=99\ else=98)

mv(*numlist* | *numlist*=*mvc* [\ *numlist* = *mvc* ...]) is required and specifies the numeric values that are to be changed to missing values.

457

mv(*numlist=mvc*) specifies that the values in *numlist* be changed to missing-value code *mvc*. Multiple transformation rules may be specified, separated by a backward slash (\). See [P] **numlist** for the syntax of a numlist.

Examples: mv(9), mv(99=.\98=.a\97=.b), mv(99=.\ 100/999=.a)

override specifies that the protection provided by mvencode be overridden. Without this option, mvencode refuses to make the requested change if any of the numeric values are already used in the data.

Remarks

You may occasionally read data in which missing (for example, a respondent failed to answer a survey question or the data were not collected) is coded with a special numeric value. Popular codings are 9, 99, −9, −99, and the like. If missing were encoded as −99, then

```
. mvdecode _all, mv(-99)
```

would translate the special code to the Stata missing value ".". Use this command cautiously because, even if −99 were not a special code, all −99s in the data would be changed to missing.

Sometimes different codes are used to represent different reasons for missing values. For instance, 98 may be used for "refused to answer" and 99 for "not applicable". Extended missing values (.a, .b, and so on) may be used to code these differences.

```
. mvdecode _all, mv(98=.a\ 99=.b)
```

Conversely, you might need to export data to software that does not understand that "." indicates a missing value, so you might code missing with a special numeric value. To change all missings to −99, you could type

```
. mvencode _all, mv(-99)
```

To change extended missing values back to numeric values, type

```
. mvencode _all, mv(.a=98\ .b=99)
```

This would leave sysmiss and all other extended missing values unchanged. To encode in addition sysmiss . to 999 and all other extended missing values to 97, you might type

```
. mvencode _all, mv(.=999\ .a=98\ .b=99\ else=97)
```

mvencode will automatically recast variables upward, if necessary, so even if a variable is stored as a byte, its missing values can be recoded to, say, 999. Also mvencode refuses to make the change if # (−99 here) is already used in the data, so you can be certain that your coding is unique. You can override this feature by including the override option.

Be aware of another potential problem with encoding and decoding missing values: value labels are not automatically adapted to the changed codings. You have to do this yourself. For example, the value label divlabor maps the value 99 to the string "not applicable". You used mvdecode to recode 99 to .a for all variables that are associated with this label. To fix the value label, clear the mapping for 99 and define it again for .a.

```
. label define divlabor 99 "", modify
. label define divlabor .a "not applicable", add
```

▷ Example 1

Our automobile dataset contains 74 observations and 12 variables. Let's first attempt to translate the missing values in the data to 1:

```
. use http://www.stata-press.com/data/r12/auto
(1978 Automobile Data)
. mvencode _all, mv(1)
        make: string variable ignored
       rep78: already 1 in     2 observations
     foreign: already 1 in    22 observations
no action taken
r(9);
```

Our attempt failed. `mvencode` first informed us that `make` is a string variable—this is not a problem but is reported merely for our information. String variables are ignored by `mvencode`. It next informed us that `rep78` was already coded 1 in 2 observations and that `foreign` was already coded 1 in 22 observations. Thus 1 would be a poor choice for encoding missing values because, after encoding, we could not tell a real 1 from a coded missing value 1.

We could force `mvencode` to encode the data with 1, anyway, by typing `mvencode _all, mv(1) override`. That would be appropriate if the 1s in our data already represented missing data. They do not, however, so we code missing as 999:

```
. mvencode _all, mv(999)
        make: string variable ignored
       rep78: 5 missing values
```

This worked, and we are informed that the only changes necessary were to 5 observations of `rep78`.

◁

▷ Example 2

Let's now pretend that we just read in the automobile data from some raw dataset in which all the missing values were coded 999. We can convert the 999s to real missings by typing

```
. mvdecode _all, mv(999)
        make: string variable ignored
       rep78: 5 missing values
```

We are informed that `make` is a string variable, so it was ignored, and that `rep78` contained 5 observations with 999. Those observations have now been changed to contain missing.

◁

Methods and formulas

`mvencode` and `mvdecode` are implemented as ado-files.

Acknowledgment

These versions of `mvencode` and `mvdecode` were written by Jeroen Weesie, Department of Sociology, Utrecht University, The Netherlands.

Also see

[D] **generate** — Create or change contents of variable

[D] **recode** — Recode categorical variables

Title

> **notes** — Place notes in data

Syntax

Attach notes to dataset

> <u>note</u>s [*evarname*] : *text*

List all notes

> <u>note</u>s

List specific notes

> <u>note</u>s [<u>l</u>ist] *evarlist* [in #[/#]]

Search for a text string across all notes in all variables and _dta

> <u>note</u>s search [*sometext*]

Replace a note

> <u>note</u>s replace *evarname* in # : *text*

Drop notes

> <u>note</u>s drop *evarlist* [in #[/#]]

Renumber notes

> <u>note</u>s renumber *evarname*

where *evarname* is _dta or a varname, *evarlist* is a varlist that may contain the _dta, and # is a number or the letter l.

If *text* includes the letters TS surrounded by blanks, the TS is removed, and a time stamp is substituted in its place.

Menu

notes (add)

Data > Variables Manager

notes list and notes search

Data > Data utilities > Notes utilities > List or search notes

notes replace

Data > Variables Manager

notes drop

Data > Variables Manager

notes renumber

Data > Data utilities > Notes utilities > Renumber notes

Description

notes attaches notes to the dataset in memory. These notes become a part of the dataset and are saved when the dataset is saved and retrieved when the dataset is used; see [D] **save** and [D] **use**. notes can be attached generically to the dataset or specifically to a variable within the dataset.

Remarks

Remarks are presented under the following headings:

> *How notes are numbered*
> *Attaching and listing notes*
> *Selectively listing notes*
> *Searching and replacing notes*
> *Deleting notes*
> *Warnings*

How notes are numbered

Notes are numbered sequentially, with the first note being 1. Say the myvar variable has four notes numbered 1, 2, 3, and 4. If you type notes drop myvar in 3, the remaining notes will be numbered 1, 2, and 4. If you now add another note, it will be numbered 5. That is, notes are not renumbered and new notes are added immediately after the highest numbered note. Thus, if you now dropped notes 4 and 5, the next note added would be 3.

You can renumber notes by using notes renumber. Going back to when myvar had notes numbered 1, 2, and 4 after dropping note 3, if you typed notes renumber myvar, the notes would be renumbered 1, 2, and 3. If you added a new note after that, it would be numbered 4.

Attaching and listing notes

A note is nothing formal; it is merely a string of text reminding you to do something, cautioning you against something, or saying anything else you might feel like jotting down. People who work with real data invariably end up with paper notes plastered around their terminal saying things like, "Send the new sales data to Bob", "Check the income variable in salary95; I don't believe it", or "The gender dummy was significant!" It would be better if these notes were attached to the dataset.

Adding a note to your dataset requires typing note or notes (they are synonyms), a colon (:), and whatever you want to remember. The note is added to the dataset currently in memory.

```
. note:  Send copy to Bob once verified.
```

You can replay your notes by typing `notes` (or `note`) by itself.

```
. notes
_dta:
  1.  Send copy to Bob once verified.
```

Once you resave your data, you can replay the note in the future, too. You add more notes just as you did the first:

```
. note: Mary wants a copy, too.
. notes
_dta:
  1.  Send copy to Bob once verified.
  2.  Mary wants a copy, too.
```

You can place time stamps on your notes by placing the word TS (in capitals) in the text of your note:

```
. note: TS merged updates from JJ&F
. notes
_dta:
  1.  Send copy to Bob once verified.
  2.  Mary wants a copy, too.
  3.  19 Apr 2011 15:38 merged updates from JJ&F
```

Notes may contain SMCL directives:

```
. use http://www.stata-press.com/data/r12/auto
(1978 Automobile Data)
. note: check reason for missing values in {cmd:rep78}
. notes
_dta:
  1.  from Consumer Reports with permission
  2.  check reason for missing values in rep78
```

The notes we have added so far are attached to the dataset generically, which is why Stata prefixes them with _dta when it lists them. You can attach notes to variables:

```
. note mpg: is the 44 a mistake?  Ask Bob.
. note mpg: what about the two missing values?
. notes
_dta:
  1.  Send copy to Bob once verified.
  2.  Mary wants a copy, too.
  3.  19 Apr 2011 15:38 merged updates from JJ&F
mpg:
  1.  is the 44 a mistake? Ask Bob.
  2.  what about the two missing values?
```

Up to 9,999 generic notes can be attached to _dta, and another 9,999 notes can be attached to each variable.

Selectively listing notes

Typing notes by itself lists all the notes. In full syntax, notes is equivalent to typing notes _all in 1/l. Here are some variations:

notes _dta	list all generic notes
notes mpg	list all notes for variable mpg
notes _dta mpg	list all generic notes and mpg notes
notes _dta in 3	list generic note 3
notes _dta in 3/5	list generic notes 3–5
notes mpg in 3/5	list mpg notes 3–5
notes _dta in 3/l	list generic notes 3 through last

Searching and replacing notes

You had a bad day yesterday, and you want to recheck the notes that you added to your dataset. Fortunately, you always put a time stamp on your notes.

```
. notes search "29 Jan"
_dta:
  2.  29 Jan 2011 13:40 check reason for missing values in foreign
```

Good thing you checked. It is rep78 that has missing values.

```
. notes replace _dta in 2: TS check reason for missing values in rep78
(note 2 for _dta replaced)
. notes
_dta:
  1.  from Consumer Reports with permission
  2.  30 Jan 2011 12:32 check reason for missing values in rep78
```

Deleting notes

notes drop works much like listing notes, except that typing notes drop by itself does not delete all notes; you must type notes drop _all. Here are some variations:

notes drop _dta	delete all generic notes
notes drop _dta in 3	delete generic note 3
notes drop _dta in 3/5	delete generic notes 3–5
notes drop _dta in 3/l	delete generic notes 3 through last
notes drop mpg in 4	delete mpg note 4

Warnings

- Notes are stored with the data, and as with other updates you make to the data, the additions and deletions are not permanent until you save the data; see [D] **save**.

- The maximum length of one note is 67,784 characters for Stata/MP, Stata/SE, and Stata/IC; it is 8,681 characters for Small Stata.

Methods and formulas

`notes` is implemented as an ado-file.

References

Gleason, J. R. 1998. dm57: A notes editor for Windows and Macintosh. *Stata Technical Bulletin* 43: 6–9. Reprinted in *Stata Technical Bulletin Reprints*, vol. 8, pp. 10–13. College Station, TX: Stata Press.

Long, J. S. 2009. *The Workflow of Data Analysis Using Stata*. College Station, TX: Stata Press.

Also see

[D] **describe** — Describe data in memory or in file

[D] **ds** — List variables matching name patterns or other characteristics

[D] **save** — Save Stata dataset

[D] **codebook** — Describe data contents

[D] **varmanage** — Manage variable labels, formats, and other properties

[U] **12.8 Characteristics**

Title

> **obs** — Increase the number of observations in a dataset

Syntax

```
set obs #
```

Description

set obs changes the number of observations in the current dataset. # must be at least as large as the current number of observations. If there are variables in memory, the values of all new observations are set to missing.

Remarks

> Example 1

set obs can be useful for creating artificial datasets. For instance, if we wanted to graph the function $y = x^2$ over the range 1–100, we could type

```
. drop _all
. set obs 100
obs was 0, now 100
. generate x = _n
. generate y = x^2
. scatter y x
```
(*graph not shown*)

◁

> Example 2

If we want to add an extra data point in a program, we could type

```
. local np1 = _N + 1
. set obs 'np1'
```

or

```
. set obs '=_N + 1'
```

◁

Also see

[D] **describe** — Describe data in memory or in file

466

Title

odbc — Load, write, or view data from ODBC sources

Syntax

List ODBC sources to which Stata can connect

> odbc <u>list</u>

Retrieve available names from specified data source

> odbc <u>query</u> ["*DataSourceName*", <u>verb</u>ose <u>sche</u>ma *connect_options*]

List column names and types associated with specified table

> odbc <u>des</u>cribe ["*TableName*", *connect_options*]

Import data from an ODBC data source

> odbc <u>load</u> [*extvarlist*] [*if*] [*in*], { <u>t</u>able("*TableName*") | <u>exec</u>("*SqlStmt*") }
> [*load_options connect_options*]

Export data to an ODBC data source

> odbc <u>insert</u> [*varlist*], <u>t</u>able("*TableName*")
> {dsn("*DataSourceName*") | <u>con</u>nectionstring("*ConnectionStr*")}
> [*insert_options connect_options*]

Allow SQL statements to be issued directly to ODBC data source

> odbc <u>exec</u>("*SqlStmt*") ,
> {dsn("*DataSourceName*") | <u>con</u>nectionstring("*ConnectionStr*")}
> [*connect_options*]

Batch job alternative to odbc exec

> odbc <u>sql</u>file("*filename*") ,
> {dsn("*DataSourceName*") | <u>con</u>nectionstring("*ConnectionStr*")}
> [loud *connect_options*]

Specify ODBC driver manager (Unix only)

> set <u>odbc</u>mgr { iodbc | unixodbc } [, <u>perma</u>nently]

467

where

> *DataSourceName* is the name of the ODBC source (database, spreadsheet, etc.)

> *ConnectionStr* is a valid ODBC connection string

> *TableName* is the name of a table within the ODBC data source

> *SqlStmt* is an SQL SELECT statement

> *filename* is pure SQL commands separated by semicolons

and where *extvarlist* contains
> *sqlvarname*
> *varname* = *sqlvarname*

connect_options	Description
<u>user</u>(*UserID*)	user ID of user establishing connection
<u>password</u>(*Password*)	password of user establishing connection
<u>dial</u>og(noprompt)	do not display ODBC connection-information dialog, and do not prompt user for connection information
<u>dial</u>og(prompt)	display ODBC connection-information dialog
<u>dial</u>og(complete)	display ODBC connection-information dialog only if there is not enough information
<u>dial</u>og(required)	display ODBC connection-information dialog only if there is not enough mandatory information provided
* <u>dsn</u>("*DataSourceName*")	name of data source
* <u>connec</u>tionstring("*ConnectionStr*")	ODBC connection string

* <u>dsn</u>("*DataSourceName*") is not allowed with odbc query. You may not specify both *DataSourceName* and connectionstring() with odbc query. Either dsn() or connectionstring() is required with odbc insert, odbc exec, and odbc sqlfile.

load_options	Description
* <u>table</u>("*TableName*")	name of table stored in data source
* <u>exec</u>("*SqlStmt*")	SQL SELECT statement to generate a table to be read into Stata
clear	load dataset even if there is one in memory
<u>noq</u>uote	alter Stata's internal use of SQL commands; seldom used
<u>lower</u>case	read variable names as lowercase
sqlshow	show all SQL commands issued
<u>alls</u>tring	read all variables as strings
<u>dates</u>tring	read date-formatted variables as strings

*Either table("*TableName*") or exec("*SqlStmt*") must be specified with odbc load.

insert_options	Description
* table("*TableName*")	name of table stored in data source
create	create a simple ODBC table
overwrite	clear data in ODBC table before data in memory is written to the table
insert	default mode of operation for the odbc insert command
quoted	quote all values with single quotes as they are inserted in ODBC table
sqlshow	show all SQL commands issued
as("*varlist*")	ODBC variables on the data source that correspond to the variables in Stata's memory

*table("*TableName*") is required for odbc insert.

Menu

odbc load

File > Import > ODBC data source

odbc insert

File > Export > ODBC data source

Description

odbc allows you to load, write, and view data from Open DataBase Connectivity (ODBC) sources into Stata. ODBC is a standardized set of function calls for accessing data stored in both relational and nonrelational database-management systems. By default on Unix platforms, iODBC is the ODBC driver manager Stata uses, but you can use unixODBC by using the command set odbcmgr unixodbc.

ODBC's architecture consists of four major components (or layers): the client interface, the ODBC driver manager, the ODBC drivers, and the data sources. Stata provides odbc as the client interface. The system is illustrated as follows:

odbc list produces a list of ODBC data source names to which Stata can connect.

odbc query retrieves a list of table names available from a specified data source's system catalog.

odbc describe lists column names and types associated with a specified table.

odbc load reads an ODBC table into memory. You can load an ODBC table specified in the table() option or load an ODBC table generated by an SQL SELECT statement specified in the exec() option. In both cases, you can choose which columns and rows of the ODBC table to read by specifying *extvarlist* and if and in conditions. *extvarlist* specifies the columns to be read and allows you to rename variables. For example,

```
. odbc load id=EmployeeID LastName, table(Employees) dsn(Northwind)
```

reads two columns, EmployeeID and LastName, from the Employees table of the Northwind data source. It will also rename variable EmployeeID to id.

odbc insert writes data from memory to an ODBC table. The data can be appended to an existing table, replace an existing table, or be placed in a newly created ODBC table.

odbc exec allows for most SQL statements to be issued directly to any ODBC data source. Statements that produce output, such as SELECT, have their output neatly displayed. By using Stata's ado language, you can also generate SQL commands on the fly to do positional updates or whatever the situation requires.

odbc sqlfile provides a "batch job" alternative to the odbc exec command. A file is specified that contains any number of any length SQL commands. Every SQL command in this file should be delimited by a semicolon and must be constructed as pure SQL. Stata macros and ado-language syntax are not permitted. The advantage in using this command, as opposed to odbc exec, is that only one connection is established for multiple SQL statements. A similar sequence of SQL commands used via odbc exec would require constructing an ado-file that issued a command and, thus, a connection for every SQL command. Another slight difference is that any output that might be generated from an SQL command is suppressed by default. A loud option is provided to toggle output back on.

set odbcmgr iodbc specifies that the ODBC driver manager is iODBC (the default). set odbcmgr unixodbc specifies that the ODBC driver manager is unixODBC.

Options

user(*UserID*) specifies the user ID of the user attempting to establish the connection to the data source. By default, Stata assumes that the user ID is the same as the one specified in the previous odbc command or is empty if user() has never been specified in the current session of Stata.

password(*Password*) specifies the password of the user attempting to establish the connection to the data source. By default, Stata assumes that the password is the same as the one previously specified or is empty if the password has not been used during the current session of Stata. Typically, the password() option will not be specified apart from the user() option.

dialog(noprompt | prompt | complete | required) specifies the mode the ODBC Driver Manager uses to display the ODBC connection-information dialog to prompt for more connection information.

noprompt is the default value. The ODBC connection-information dialog is not displayed, and you are not prompted for connection information. If there is not enough information to establish a connection to the specified data source, an error is returned.

prompt causes the ODBC connection-information dialog to be displayed.

complete causes the ODBC connection-information dialog to be displayed only if there is not enough information, even if the information is not mandatory.

required causes the ODBC connection-information dialog to be displayed only if there is not enough mandatory information provided to establish a connection to the specified data source. You are prompted only for mandatory information; controls for information that is not required to connect to the specified data source are disabled.

dsn("*DataSourceName*") specifies the name of a data source, as listed by the odbc list command. If a name contains spaces, it must be enclosed in double quotes. By default, Stata assumes that the data source name is the same as the one specified in the previous odbc command. This option is not allowed with odbc query. Either the dsn() option or the connectionstring() option may be specified with odbc describe and odbc load, and one of these options must be specified with odbc insert, odbc exec, and odbc sqlfile.

connectionstring("*ConnectionStr*") specifies a connection string rather than the name of a data source. Stata does not assume that the connection string is the same as the one specified in the previous odbc command. Either *DataSourceName* or the connectionstring() option may be specified with odbc query; either the dsn() option or the connectionstring() option can be specified with odbc describe and odbc load, and one of these options must be specified with odbc insert, odbc exec, and odbc sqlfile.

table("*TableName*") specifies the name of an ODBC table stored in a specified data source's system catalog, as listed by the odbc query command. If a table name contains spaces, it must be enclosed in double quotes. Either the table() option or the exec() option—but not both—is required with the odbc load command.

exec("*SqlStmt*") allows you to issue an SQL SELECT statement to generate a table to be read into Stata. An error message is returned if the SELECT statement is an invalid SQL statement. The statement must be enclosed in double quotes. Either the table() option or the exec() option—but not both—is required with the odbc load command.

clear permits the data to be loaded, even if there is a dataset already in memory, and even if that dataset has changed since the data were last saved.

noquote alters Stata's internal use of SQL commands, specifically those relating to quoted table names, to better accommodate various drivers. This option has been particularly helpful for DB2 drivers.

lowercase causes all the variable names to be read as lowercase.

sqlshow is a useful option for showing all SQL commands issued to the ODBC data source from the odbc insert or odbc load command. This can help you debug any issues related to inserting or loading.

allstring causes all variables to be read as string data types.

datestring causes all date- and time-formatted variables to be read as string data types.

create specifies that a simple ODBC table be created on the specified data source and populated with the data in memory. Column data types are approximated based on the existing format in Stata's memory.

overwrite allows data to be cleared from an ODBC table before the data in memory are written to the table. All data from the ODBC table are erased, not just the data from the variable columns that will be replaced.

insert appends data to an existing ODBC table and is the default mode of operation for the odbc insert command.

quoted is useful for ODBC data sources that require all inserted values to be quoted. This option specifies that all values be quoted with single quotes as they are inserted into an ODBC table.

as("*varlist*") allows you to specify the ODBC variables on the data source that correspond to the variables in Stata's memory. If this option is specified, the number of variables must equal the number of variables being inserted, even if some names are identical.

loud specifies that output be displayed for SQL commands.

verbose specifies that odbc query list any data source alias, nickname, typed table, typed view, and view along with tables so that you can load data from these table types.

schema specifies that odbc query return schema names with the table names from a data source. Note: The schema names returned from odbc query will also be used with the odbc describe and odbc load commands. When using odbc load with a schema name, you might also need to specify the noquote option because some drivers do not accept quotes around table or schema names.

permanently (set odbcmgr only) specifies that, in addition to making the change right now, the setting be remembered and become the default setting when you invoke Stata.

Remarks

When possible, the examples in this manual entry are developed using the Northwind sample database that is automatically installed with Microsoft Access. If you do not have Access, you can still use odbc, but you will need to consult the documentation for your other ODBC sources to determine how to set them up.

Remarks are presented under the following headings:

> *Setting up the data sources*
> *Listing ODBC data source names*
> *Listing available table names from a specified data source's system catalog*
> *Describing a specified table*
> *Loading data from ODBC sources*

Setting up the data sources

Before using Stata's ODBC commands, you must register your ODBC database with the *ODBC Data Source Administrator*. This process varies depending on platform, but the following example shows the steps necessary for Windows.

Using Windows 7, XP, or Vista, follow these steps to create an ODBC User Data Source for the Northwind sample database:

1. From the *Start Menu*, select the *Control Panel*.

2. In the *Control Panel* window, click on *Administrative Tools*.

3. In the *Administrative Tools* window, double-click on *Data Sources (ODBC)*. Vista users will have to click on *Classic View* on the left side of the Control Panel window before *Administrative Tools* is visible.

4. In the *Data Sources (ODBC)* dialog box,

   a. click on the *User DSN* tab;

   b. click on **Add...**;

   c. choose *Microsoft Access Driver (*.mdb)* on the *Create New Data Source* dialog box; and

   d. click on **Finish**.

5. In the *ODBC Microsoft Access Setup* dialog box, type Northwind in the *Data Source Name* field and click on **Select...**. Locate the Northwind.mdb database and click on **OK** to finish creating the data source.

❑ Technical note

In earlier versions of Windows, the exact location of the *Data Source (ODBC)* dialog varies, but it is always somewhere within the *Control Panel*.

❑

Listing ODBC data source names

odbc list is used to produce a list of data source names to which Stata can connect. For a specific data source name to be shown in the list, the data source has to be registered with the *ODBC Data Source Administrator*. See *Setting up the data sources* for information on how to do this.

▷ Example 1

```
. odbc list
Data Source Name                        Driver
Visual FoxPro Database                  Microsoft Visual FoxPro Driver
Visual FoxPro Tables                    Microsoft Visual FoxPro Driver
dBase Files - Word                      Microsoft dBase VFP Driver (*.dbf)
FoxPro Files - Word                     Microsoft FoxPro VFP Driver (*.dbf)
MS Access Database                      Microsoft Access Driver (*.mdb)
Northwind                               Microsoft Access Driver (*.mdb)
dBASE Files                             Microsoft dBase Driver (*.dbf)
DeluxeCD                                Microsoft Access Driver (*.mdb)
Excel Files                             Microsoft Excel Driver (*.xls)
ECDCMusic                               Microsoft Access Driver (*.mdb)
```

In the above list, Northwind is one of the sample Microsoft Access databases that Access installs by default.

◁

Listing available table names from a specified data source's system catalog

odbc query is used to list table names available from a specified data source.

▷ Example 2

```
. odbc query "Northwind"
DataSource: Northwind
Path      : C:\Program Files\Microsoft Office\Office\Samples\Northwind
Categories
Customers
Employees
Order Details
Orders
Products
Shippers
Suppliers
```

◁

❑ Technical note

To query a *Microsoft Excel* data source, you must define a database as a named range within Excel. Multiple name ranges can exist within an Excel file, and each one is treated as a separate table.

To define a named range within Excel, highlight the entire range, including all columns of interest; from the Excel menu, select *Insert*, select *Name*, click on *Define*, enter the desired name, and save the file.

You can also describe a worksheet associated with an Excel file without defining a named range for the worksheet. To do so, you must specify the name of the worksheet in the odbc describe command followed by a dollar sign ($).

❑

Describing a specified table

odbc describe is used to list column (variable) names and their SQL data types that are associated with a specified table.

➢ Example 3

Here we specify that we want to list all variables in the Employees table of the Northwind data source.

```
. odbc describe "Employees", dsn("Northwind")
DataSource: Northwind (query)
Table:      Employees (load)
```

Variable Name	Variable Type
EmployeeID	COUNTER
LastName	VARCHAR
FirstName	VARCHAR
Title	VARCHAR
TitleOfCourtesy	VARCHAR
BirthDate	DATETIME
HireDate	DATETIME
Address	VARCHAR
City	VARCHAR
Region	VARCHAR
PostalCode	VARCHAR
Country	VARCHAR
HomePhone	VARCHAR
Extension	VARCHAR
Photo	LONGBINARY
Notes	LONGCHAR
ReportsTo	INTEGER

◁

Loading data from ODBC sources

odbc load is used to load an ODBC table into memory.

To load an ODBC table listed in the odbc query output, specify the table name in the table() option and the data source name in the dsn() option.

▷ Example 4

We want to load the Employees table from the Northwind data source.

```
. clear

. odbc load, table("Employees") dsn("Northwind")
note: Photo is of a type not supported in Stata; skipped

. describe
Contains data
  obs:              9
  vars:            16
  size:         3,222
```

variable name	storage type	display format	value label	variable label
EmployeeID	long	%12.0g		
LastName	str20	%20s		
FirstName	str10	%10s		
Title	str30	%30s		
TitleOfCourtesy	str25	%25s		
BirthDate	double	%td		
HireDate	double	%td		
Address	str60	%60s		
City	str15	%15s		
Region	str15	%15s		
PostalCode	str10	%10s		
Country	str15	%15s		
HomePhone	str24	%24s		
Extension	str4	%9s		
Notes	str80	%80s		
ReportsTo	long	%12.0g		

```
Sorted by:
     Note:  dataset has changed since last saved
```

❑

❑ Technical note

When Stata loads the ODBC table, data are converted from SQL data types to Stata data types. Stata does not support all SQL data types. If the column cannot be read because of incompatible data types, Stata will issue a note and skip a column. The following table lists the supported SQL data types and their corresponding Stata data types:

SQL data type	Stata data type
SQL_BIT SQL_TINYINT	byte
SQL_SMALLINT	int
SQL_INTEGER SQL_BIGINT	long
SQL_DECIMAL SQL_NUMERIC	double
SQL_REAL	float
SQL_FLOAT SQL_DOUBLE	double
SQL_CHAR SQL_VARCHAR SQL_LONGVARCHAR SQL_WCHAR SQL_WVARCHAR SQL_WLONGVARCHAR	string
SQL_TIME SQL_DATE SQL_TIMESTAMP SQL_TYPE_TIME SQL_TYPE_DATE SQL_TYPE_TIMESTAMP	double
SQL_BINARY SQL_VARBINARY SQL_LONGVARBINARY	*not supported* *not supported* *not supported*

❑

You can also load an ODBC table generated by an SQL SELECT statement specified in the exec() option.

▷ Example 5

Suppose that, from the Northwind data source, we want a list of all the customers who have placed orders. We might use the SQL SELECT statement

```
SELECT DISTINCT c.CustomerID, c.CompanyName
FROM Customers c
INNER JOIN Orders o
      ON c.CustomerID = o.CustomerID
```

To load the table into Stata, we use odbc load with the exec() option.

```
. odbc load, exec("SELECT DISTINCT c.CustomerID, c.CompanyName FROM Customers c
> INNER JOIN Orders o ON c.CustomerID = o.CustomerID") dsn("Northwind") clear
. describe
Contains data
  obs:            89
  vars:            2
  size:        4,361
```

variable name	storage type	display format	value label	variable label
CustomerID	str5	%9s		
CompanyName	str40	%40s		

```
Sorted by:
     Note:  dataset has changed since last saved
```
◁

The *extvarlist* is optional. It allows you to choose which columns (variables) are to be read and to rename variables when they are read.

▷ Example 6

Suppose that we want to load the EmployeeID column and the LastName column from the Employees table of the Northwind data source. Moreover, we want to rename EmployeeID as id and LastName as name.

```
. odbc load id=EmployeeID name=LastName, table("Employees") dsn("Northwind") clear
. describe
Contains data
  obs:             9
  vars:            2
  size:          252
```

variable name	storage type	display format	value label	variable label
id	long	%12.0g		EmployeeID
name	str20	%20s		LastName

```
Sorted by:
     Note:  dataset has changed since last saved
```
◁

The if and in qualifiers allow you to choose which rows are to be read. You can also use a WHERE clause in the SQL SELECT statement to select the rows to be read.

▷ Example 7

Suppose that we want the information from the Order Details table, where Quantity is greater than 50. We can specify the if and in qualifiers,

```
. odbc load if Quantity>50, table("Order Details") dsn("Northwind") clear
. summarize Quantity
```

Variable	Obs	Mean	Std. Dev.	Min	Max
Quantity	159	72.56604	18.38255	52	130

or we can issue the SQL SELECT statement directly:

```
. odbc load, exec("SELECT * FROM [Order Details] WHERE Quantity>50")
> dsn("Northwind") clear
. summarize Quantity
```

Variable	Obs	Mean	Std. Dev.	Min	Max
Quantity	159	72.56604	18.38255	52	130

◁

Also see

[D] **export** — Overview of exporting data from Stata

[D] **import** — Overview of importing data into Stata

Title

order — Reorder variables in dataset

Syntax

order *varlist* [, *options*]

options	Description
first	move *varlist* to beginning of dataset; the default
last	move *varlist* to end of dataset
before(*varname*)	move *varlist* before *varname*
after(*varname*)	move *varlist* after *varname*
alphabetic	alphabetize *varlist* and move it to beginning of dataset
sequential	alphabetize *varlist* keeping numbers sequential and move it to beginning of dataset

Menu

Data > Data utilities > Change order of variables

Description

order relocates *varlist* to a position depending on which option you specify. If no option is specified, order relocates *varlist* to the beginning of the dataset in the order in which the variables are specified.

Options

first shifts *varlist* to the beginning of the dataset. This is the default.

last shifts *varlist* to the end of the dataset.

before(*varname*) shifts *varlist* before *varname*.

after(*varname*) shifts *varlist* after *varname*.

alphabetic alphabetizes *varlist* and moves it to the beginning of the dataset. For example, here is a varlist in alphabetic order: a x7 x70 x8 x80 z. If combined with another option, alphabetic just alphabetizes *varlist*, and the movement of *varlist* is controlled by the other option.

sequential alphabetizes *varlist*, keeping variables with the same ordered letters but with differing appended numbers in sequential order. *varlist* is moved to the beginning of the dataset. For example, here is a varlist in sequential order: a x7 x8 x70 x80 z.

479

Remarks

▷ Example 1

When using `order`, you must specify a *varlist*, but you do not need to specify all the variables in the dataset. For example, we want to move the `make` and `mpg` variables to the front of the `auto` dataset.

```
. use http://www.stata-press.com/data/r12/auto4
(1978 Automobile Data)

. describe

Contains data from http://www.stata-press.com/data/r12/auto4.dta
  obs:            74                          1978 Automobile Data
 vars:             6                          6 Apr 2011 00:20
 size:         2,072
```

variable name	storage type	display format	value label	variable label
price	int	%8.0gc		Price
weight	int	%8.0gc		Weight (lbs.)
mpg	int	%8.0g		Mileage (mpg)
make	str18	%-18s		Make and Model
length	int	%8.0g		Length (in.)
rep78	int	%8.0g		Repair Record 1978

```
Sorted by:

. order make mpg

. describe

Contains data from http://www.stata-press.com/data/r12/auto4.dta
  obs:            74                          1978 Automobile Data
 vars:             6                          6 Apr 2011 00:20
 size:         2,072
```

variable name	storage type	display format	value label	variable label
make	str18	%-18s		Make and Model
mpg	int	%8.0g		Mileage (mpg)
price	int	%8.0gc		Price
weight	int	%8.0gc		Weight (lbs.)
length	int	%8.0g		Length (in.)
rep78	int	%8.0g		Repair Record 1978

```
Sorted by:
```

We now want `length` to be the last variable in our dataset, so we could type `order make mpg price weight rep78 length`, but it would be easier to use the `last` option:

```
. order length, last

. describe
```

Contains data from http://www.stata-press.com/data/r12/auto4.dta
```
  obs:            74                          1978 Automobile Data
  vars:            6                          6 Apr 2011 00:20
  size:         2,072
```

variable name	storage type	display format	value label	variable label
make	str18	%-18s		Make and Model
mpg	int	%8.0g		Mileage (mpg)
price	int	%8.0gc		Price
weight	int	%8.0gc		Weight (lbs.)
rep78	int	%8.0g		Repair Record 1978
length	int	%8.0g		Length (in.)

Sorted by:

We now change our mind and decide that we prefer that the variables be alphabetized.

```
. order _all, alphabetic

. describe
```

Contains data from http://www.stata-press.com/data/r12/auto4.dta
```
  obs:            74                          1978 Automobile Data
  vars:            6                          6 Apr 2011 00:20
  size:         2,072
```

variable name	storage type	display format	value label	variable label
length	int	%8.0g		Length (in.)
make	str18	%-18s		Make and Model
mpg	int	%8.0g		Mileage (mpg)
price	int	%8.0gc		Price
rep78	int	%8.0g		Repair Record 1978
weight	int	%8.0gc		Weight (lbs.)

Sorted by:

◁

❑ Technical note

If your data contain variables named year1, year2, ..., year19, year20, specify the sequential option to obtain this ordering. If you specify the alphabetic option, year10 will appear between year1 and year11.

❑

Methods and formulas

order is implemented as an ado-file.

References

Gleason, J. R. 1997. dm51: Defining and recording variable orderings. *Stata Technical Bulletin* 40: 10–12. Reprinted in *Stata Technical Bulletin Reprints*, vol. 7, pp. 49–52. College Station, TX: Stata Press.

Weesie, J. 1999. dm74: Changing the order of variables in a dataset. *Stata Technical Bulletin* 52: 8–9. Reprinted in *Stata Technical Bulletin Reprints*, vol. 9, pp. 61–62. College Station, TX: Stata Press.

Also see

[D] **describe** — Describe data in memory or in file

[D] **ds** — List variables matching name patterns or other characteristics

[D] **edit** — Browse or edit data with Data Editor

[D] **rename** — Rename variable

Title

> **outfile** — Export dataset in text format

Syntax

<u>ou</u>tfile [*varlist*] using *filename* [*if*] [*in*] [, *options*]

options	Description
Main	
<u>dictionary</u>	write the file in Stata's dictionary format
<u>nol</u>abel	output numeric values (not labels) of labeled variables; the default is to write labels in double quotes
<u>noq</u>uote	do not enclose strings in double quotes
<u>c</u>omma	write file in comma-separated (instead of space-separated) format
<u>w</u>ide	force 1 observation per line (no matter how wide)
Advanced	
rjs	right-justify string variables; the default is to left-justify
fjs	left-justify if format width < 0; right-justify if format width > 0
runtogether	all on one line, no quotes, no space between, and ignore formats
<u>m</u>issing	retain missing values; use only with comma
replace	overwrite the existing file

replace does not appear in the dialog box.

Menu

File > Export > Text data (fixed- or free-format)

Description

outfile writes data to a disk file in plain-text format, which can be read by other programs. The new file is *not* in Stata format; see [D] **save** for instructions on saving data for later use in Stata.

The data saved by outfile can be read back by infile; see [D] **import**. If *filename* is specified without an extension, .raw is assumed unless the dictionary option is specified, in which case .dct is assumed. If your *filename* contains embedded spaces, remember to enclose it in double quotes.

Options

> ⌐ Main ⌐

dictionary writes the file in Stata's data dictionary format. See [D] **infile (fixed format)** for a description of dictionaries. comma, missing, and wide are not allowed with dictionary.

nolabel causes Stata to write the numeric values of labeled variables. The default is to write the labels enclosed in double quotes.

483

noquote prevents Stata from placing double quotes around the contents of strings, meaning string variables and value labels.

comma causes Stata to write the file in comma-separated–value format. In this format, values are separated by commas rather than by blanks. Missing values are written as two consecutive commas unless missing is specified.

wide causes Stata to write the data with 1 observation per line. The default is to split observations into lines of 80 characters or fewer, but strings longer than 80 characters are never split across lines.

⌐ Advanced ⌐

rjs and fjs affect how strings are justified; you probably do not want to specify either of these options. By default, outfile outputs strings left-justified in their field.

If rjs is specified, strings are output right-justified. rjs stands for "right-justified strings".

If fjs is specified, strings are output left- or right-justified according to the variable's format: left-justified if the format width is negative and right-justified if the format width is positive. fjs stands for "format-justified strings".

runtogether is a programmer's option that is valid only when all variables of the specified *varlist* are of type string. runtogether specifies that the variables be output in the order specified, without quotes, with no spaces between, and ignoring the display format attached to each variable. Each observation ends with a new line character.

missing, valid only with comma, specifies that missing values be retained. When comma is specified without missing, missing values are changed to null strings ("").

The following option is available with outfile but is not shown in the dialog box:

replace permits outfile to overwrite an existing dataset.

Remarks

outfile enables data to be sent to a disk file for processing by a non-Stata program. Each observation is written as one or more records that will not exceed 80 characters unless you specify the wide option. Each column other than the first is prefixed by two blanks.

outfile is careful to put the data in columns in case you want to read the data by using formatted input. String variables and value labels are output in left-justified fields by default. You can change this behavior by using the rjs or fjs options.

Numeric variables are output right-justified in the field width specified by their display format. A numeric variable with a display format of %9.0g will be right-justified in a nine-character field. Commas are not written in numeric variables, even if a comma format is used.

If you specify the dictionary option, the data are written in the same way, but preceding the data, outfile writes a data dictionary describing the contents of the file.

▷ Example 1

We have entered into Stata some data on seven employees in our firm. The data contain employee name, employee identification number, salary, and sex:

```
. list
```

	name	empno	salary	sex
1.	Carl Marks	57213	24,000	male
2.	Irene Adler	47229	27,000	female
3.	Adam Smith	57323	24,000	male
4.	David Wallis	57401	24,500	male
5.	Mary Rogers	57802	27,000	female
6.	Carolyn Frank	57805	24,000	female
7.	Robert Lawson	57824	22,500	male

The last variable in our data, sex, is really a numeric variable, but it has an associated value label.

If we now wish to use a program other than Stata with these data, we must somehow get the data over to that other program. The standard Stata-format dataset created by save will not do the job—it is written in a special format that only Stata understands. Most programs, however, understand plain-text datasets, such as those produced by a text editor. We can tell Stata to produce such a dataset by using outfile. Typing outfile using employee creates a dataset called employee.raw that contains all the data. We can use the Stata type command to review the resulting file:

```
. outfile using employee

. type employee.raw
"Carl Marks"         57213        24000   "male"
"Irene Adler"        47229        27000   "female"
"Adam Smith"         57323        24000   "male"
"David Wallis"       57401        24500   "male"
"Mary Rogers"        57802        27000   "female"
"Carolyn Frank"      57805        24000   "female"
"Robert Lawson"      57824        22500   "male"
```

We see that the file contains the four variables and that Stata has surrounded the string variables with double quotes.

◁

❑ Technical note

The nolabel option prevents Stata from substituting value-label strings for the underlying numeric values; see [U] **12.6.3 Value labels**. The last variable in our data is really a numeric variable:

```
. outfile using employ2, nolabel

. type employ2.raw
"Carl Marks"         57213        24000           0
"Irene Adler"        47229        27000           1
"Adam Smith"         57323        24000           0
"David Wallis"       57401        24500           0
"Mary Rogers"        57802        27000           1
"Carolyn Frank"      57805        24000           1
"Robert Lawson"      57824        22500           0
```

❑

❏ Technical note

If you do not want Stata to place double quotes around the contents of string variables, you can specify the `noquote` option:

```
. outfile using employ3, noquote
. type employ3.raw
Carl Marks         57213      24000  male
Irene Adler        47229      27000  female
Adam Smith         57323      24000  male
David Wallis       57401      24500  male
Mary Rogers        57802      27000  female
Carolyn Frank      57805      24000  female
Robert Lawson      57824      22500  male
```

❏

▷ Example 2

Stata never writes over an existing file unless explicitly told to do so. For instance, if the file `employee.raw` already exists and we attempt to overwrite it by typing `outfile using employee`, here is what would happen:

```
. outfile using employee
file employee.raw already exists
r(602);
```

We can tell Stata that it is okay to overwrite a file by specifying the `replace` option:

```
. outfile using employee, replace
```

◁

❏ Technical note

Some programs prefer data to be separated by commas rather than by blanks. Stata produces such a dataset if you specify the `comma` option:

```
. outfile using employee, comma replace
. type employee.raw
"Carl Marks",57213,24000,"male"
"Irene Adler",47229,27000,"female"
"Adam Smith",57323,24000,"male"
"David Wallis",57401,24500,"male"
"Mary Rogers",57802,27000,"female"
"Carolyn Frank",57805,24000,"female"
"Robert Lawson",57824,22500,"male"
```

❏

▷ Example 3

Finally, `outfile` can create data dictionaries that `infile` can read. Dictionaries are perhaps the best way to organize your raw data. A dictionary describes your data so that you do not have to remember the order of the variables, the number of variables, the variable names, or anything else. The file in which you store your data becomes self-documenting so that you can understand the data in the future. See [D] **infile (fixed format)** for a full description of data dictionaries.

When you specify the `dictionary` option, Stata writes a `.dct` file:

```
. outfile using employee, dict replace
. type employee.dct
dictionary {
        str15  name                   '"Employee name"'
        float  empno                  '"Employee number"'
        float  salary                 '"Annual salary"'
        float  sex        :sexlbl      '"Sex"'
}
"Carl Marks"              57213          24000   "male"
"Irene Adler"             47229          27000   "female"
"Adam Smith"              57323          24000   "male"
"David Wallis"            57401          24500   "male"
"Mary Rogers"            57802          27000   "female"
"Carolyn Frank"           57805          24000   "female"
"Robert Lawson"          57824          22500   "male"
```

◁

▷ Example 4

We have historical data on the S&P 500 for the month of January 2001.

```
. use http://www.stata-press.com/data/r12/outfilexmpl
(S&P 500)
. describe
Contains data from http://www.stata-press.com/data/r12/outfilexmpl.dta
  obs:            21                          S&P 500
  vars:            6                          6 Apr 2011 16:02
  size:          420                          (_dta has notes)
```

variable name	storage type	display format	value label	variable label
date	int	%td		Date
open	float	%9.0g		Opening price
high	float	%9.0g		High price
low	float	%9.0g		Low price
close	float	%9.0g		Closing price
volume	int	%12.0gc		Volume (thousands)

```
Sorted by:  date
```

The `date` variable has a display format of `%td` so that it is displayed as *ddmmmyyyy*.

```
. list
```

	date	open	high	low	close	volume
1.	02jan2001	1320.28	1320.28	1276.05	1283.27	11,294
2.	03jan2001	1283.27	1347.76	1274.62	1347.56	18,807
3.	04jan2001	1347.56	1350.24	1329.14	1333.34	21,310
4.	05jan2001	1333.34	1334.77	1294.95	1298.35	14,308
5.	08jan2001	1298.35	1298.35	1276.29	1295.86	11,155
6.	09jan2001	1295.86	1311.72	1295.14	1300.8	11,913
7.	10jan2001	1300.8	1313.76	1287.28	1313.27	12,965
8.	11jan2001	1313.27	1332.19	1309.72	1326.82	14,112
9.	12jan2001	1326.82	1333.21	1311.59	1318.55	12,760
10.	16jan2001	1318.32	1327.81	1313.33	1326.65	12,057
11.	17jan2001	1326.65	1346.92	1325.41	1329.47	13,491
12.	18jan2001	1329.89	1352.71	1327.41	1347.97	14,450
13.	19jan2001	1347.97	1354.55	1336.74	1342.54	14,078
14.	22jan2001	1342.54	1353.62	1333.84	1342.9	11,640
15.	23jan2001	1342.9	1362.9	1339.63	1360.4	12,326
16.	24jan2001	1360.4	1369.75	1357.28	1364.3	13,090
17.	25jan2001	1364.3	1367.35	1354.63	1357.51	12,580
18.	26jan2001	1357.51	1357.51	1342.75	1354.95	10,980
19.	29jan2001	1354.92	1365.54	1350.36	1364.17	10,531
20.	30jan2001	1364.17	1375.68	1356.2	1373.73	11,498
21.	31jan2001	1373.73	1383.37	1364.66	1366.01	12,953

We outfile our data and use the type command to view the result.

```
. outfile using sp
. type sp.raw
"02jan2001"    1320.28    1320.28    1276.05    1283.27         11294
"03jan2001"    1283.27    1347.76    1274.62    1347.56         18807
"04jan2001"    1347.56    1350.24    1329.14    1333.34         21310
"05jan2001"    1333.34    1334.77    1294.95    1298.35         14308
"08jan2001"    1298.35    1298.35    1276.29    1295.86         11155
"09jan2001"    1295.86    1311.72    1295.14     1300.8         11913
"10jan2001"     1300.8    1313.76    1287.28    1313.27         12965
"11jan2001"    1313.27    1332.19    1309.72    1326.82         14112
"12jan2001"    1326.82    1333.21    1311.59    1318.55         12760
"16jan2001"    1318.32    1327.81    1313.33    1326.65         12057
"17jan2001"    1326.65    1346.92    1325.41    1329.47         13491
"18jan2001"    1329.89    1352.71    1327.41    1347.97         14450
"19jan2001"    1347.97    1354.55    1336.74    1342.54         14078
"22jan2001"    1342.54    1353.62    1333.84     1342.9         11640
"23jan2001"     1342.9     1362.9    1339.63     1360.4         12326
"24jan2001"     1360.4    1369.75    1357.28     1364.3         13090
"25jan2001"     1364.3    1367.35    1354.63    1357.51         12580
"26jan2001"    1357.51    1357.51    1342.75    1354.95         10980
"29jan2001"    1354.92    1365.54    1350.36    1364.17         10531
"30jan2001"    1364.17    1375.68     1356.2    1373.73         11498
"31jan2001"    1373.73    1383.37    1364.66    1366.01         12953
```

The date variable, originally stored as an int, was outfiled as a string variable. Whenever Stata outfiles a variable with a date format, Stata outfiles the variable as a string.

◁

Also see

Title

> **outsheet** — Write spreadsheet-style dataset

Syntax

$$\underline{\text{out}}\text{sheet} \; \big[\textit{varlist}\big] \; \text{using} \; \textit{filename} \; \big[\textit{if}\big] \; \big[\textit{in}\big] \; \big[\text{, } \textit{options}\big]$$

options	Description
Main	
<u>comma</u>	output in comma-separated (instead of tab-separated) format
<u>delim</u>iter("*char*")	use *char* as delimiter
<u>non</u>ames	do not write variable names on the first line
<u>nol</u>abel	output numeric values (not labels) of labeled variables
<u>noq</u>uote	do not enclose strings in double quotes
<u>replace</u>	overwrite existing *filename*

If *filename* is specified without a suffix, .out is assumed.

If your *filename* contains embedded spaces, remember to enclose it in double quotes.

replace does not appear in the dialog box.

Menu

File > Export > Comma- or tab-separated data

Description

outsheet, by default, writes data into a file in tab-separated format. outsheet also allows users to specify comma-separated format or any separation character that they prefer. export excel may be a better option if you are exporting data to a program that can read Excel files; see [D] **import excel**.

Options

Main

comma specifies comma-separated format rather than the default tab-separated format.

delimiter("*char*") allows you to specify other separation characters. For instance, if you want the values in the file to be separated by a semicolon, specify delimiter(";").

nonames specifies that variable names not be written in the first line of the file; the file is to contain data values only.

nolabel specifies that the numeric values of labeled variables be written into the file rather than the label associated with each value.

noquote specifies that string variables not be enclosed in double quotes.

The following option is available with outsheet but is not shown in the dialog box:

replace specifies that *filename* be replaced if it already exists.

Remarks

If you wish to move your data into another program, you can do any of the following:

- Cut and paste from Stata's Data Editor; see [GS] **6 Using the Data Editor** (GSM, GSU, or GSW).
- Use outsheet.
- Use another Stata export method; see [D] **export**.
- Use an external data-transfer program; see [U] **21.4 Transfer programs**.

outsheet is typically preferred to outfile for moving the data to a spreadsheet, and outfile is probably better for moving data to another statistical program.

If your goal is to send data to another Stata user, you could use outsheet or outfile, but it is easiest to send the .dta dataset. This will work even if you use Stata for Windows and your colleague uses Stata for Mac. All Statas can read each others' .dta files.

▷ Example 1

outsheet copies the data currently loaded in memory into the specified file. It is easy to use.

```
. use http://www.stata-press.com/data/r12/auto
(1978 Automobile Data)
. keep make price mpg rep78 foreign
. keep in 1/10
(64 observations deleted)
```

Let's write our shortened version of the auto dataset in tab-separated text format to the file myauto.out:

```
. outsheet using myauto
. type myauto.out
make      price    mpg      rep78    foreign
"AMC Concord"   4099    22       3        "Domestic"
"AMC Pacer"     4749    17       3        "Domestic"
"AMC Spirit"    3799    22                "Domestic"
"Buick Century" 4816    20       3        "Domestic"
"Buick Electra" 7827    15       4        "Domestic"
"Buick LeSabre" 5788    18       3        "Domestic"
"Buick Opel"    4453    26                "Domestic"
"Buick Regal"   5189    20       3        "Domestic"
"Buick Riviera" 10372   16       3        "Domestic"
"Buick Skylark" 4082    19       3        "Domestic"
```

We remember that we are not copying our data to a spreadsheet, so we want to suppress the dataset names from the first line of the file.

```
. outsheet using myauto, nonames
file myauto.out already exists
r(602);
```

We can erase myauto.out (see [D] **erase**), specify the replace option, or use a different filename.

```
. outsheet using myauto, nonames replace
. type myauto.out
"AMC Concord"   4099    22      3       "Domestic"
"AMC Pacer"     4749    17      3       "Domestic"
"AMC Spirit"    3799    22              "Domestic"
"Buick Century" 4816    20      3       "Domestic"
"Buick Electra" 7827    15      4       "Domestic"
"Buick LeSabre" 5788    18      3       "Domestic"
"Buick Opel"    4453    26              "Domestic"
"Buick Regal"   5189    20      3       "Domestic"
"Buick Riviera" 10372   16      3       "Domestic"
"Buick Skylark" 4082    19      3       "Domestic"
```

◁

Also see

[D] **export** — Overview of exporting data from Stata

[D] **import** — Overview of importing data into Stata

[U] **21 Inputting and importing data**

Title

> **pctile** — Create variable containing percentiles

Syntax

Create variable containing percentiles

> pctile [*type*] *newvar* = *exp* [*if*] [*in*] [*weight*] [, *pctile_options*]

Create variable containing quantile categories

> xtile *newvar* = *exp* [*if*] [*in*] [*weight*] [, *xtile_options*]

Compute percentiles and store them in r()

> _pctile *varname* [*if*] [*in*] [*weight*] [, *_pctile_options*]

pctile_options	Description
Main	
nquantiles(*#*)	number of quantiles; default is nquantiles(2)
genp(*newvar_p*)	generate *newvar_p* variable containing percentages
altdef	use alternative formula for calculating percentiles

xtile_options	Description
Main	
nquantiles(*#*)	number of quantiles; default is nquantiles(2)
cutpoints(*varname*)	use values of *varname* as cutpoints
altdef	use alternative formula for calculating percentiles

_pctile_options	Description
nquantiles(*#*)	number of quantiles; default is nquantiles(2)
percentiles(*numlist*)	calculate percentiles corresponding to the specified percentages
altdef	use alternative formula for calculating percentiles

aweights, fweights, and pweights are allowed (see [U] **11.1.6 weight**), except when the altdef option is specified, in which case no weights are allowed.

Menu

pctile

Statistics > Summaries, tables, and tests > Summary and descriptive statistics > Create variable of percentiles

xtile

Statistics > Summaries, tables, and tests > Summary and descriptive statistics > Create variable of quantiles

Description

pctile creates a new variable containing the percentiles of *exp*, where the expression *exp* is typically just another variable.

xtile creates a new variable that categorizes *exp* by its quantiles. If the cutpoints(*varname*) option is specified, it categorizes *exp* using the values of *varname* as category cutpoints. For example, *varname* might contain percentiles of another variable, generated by pctile.

_pctile is a programmer's command that computes up to 1,000 percentiles and places the results in r(); see [U] **18.8 Accessing results calculated by other programs**. summarize, detail computes some percentiles (1, 5, 10, 25, 50, 75, 90, 95, and 99th); see [R] **summarize**.

Options

⌐ Main ⌐

nquantiles(*#*) specifies the number of quantiles. It computes percentiles corresponding to percentages $100\,k/m$ for $k = 1, 2, \ldots, m - 1$, where $m = \#$. For example, nquantiles(10) requests that the 10th, 20th, …, 90th percentiles be computed. The default is nquantiles(2); that is, the median is computed.

genp(*newvar$_p$*) (pctile only) specifies a new variable to be generated containing the percentages corresponding to the percentiles.

altdef uses an alternative formula for calculating percentiles. The default method is to invert the empirical distribution function by using averages, $(x_i + x_{i+1})/2$, where the function is flat (the default is the same method used by summarize; see [R] **summarize**). The alternative formula uses an interpolation method. See *Methods and formulas* at the end of this entry. Weights cannot be used when altdef is specified.

cutpoints(*varname*) (xtile only) requests that xtile use the values of *varname*, rather than quantiles, as cutpoints for the categories. All values of *varname* are used, regardless of any if or in restriction; see the technical note in the xtile section below.

percentiles(*numlist*) (_pctile only) requests percentiles corresponding to the specified percentages. Percentiles are placed in r(r1), r(r2), …, etc. For example, percentiles(10(20)90) requests that the 10th, 30th, 50th, 70th, and 90th percentiles be computed and placed into r(r1), r(r2), r(r3), r(r4), and r(r5). Up to 1,000 (inclusive) percentiles can be requested. See [P] **numlist** for the syntax of a numlist.

Remarks

Remarks are presented under the following headings:

> pctile
> xtile
> _pctile

pctile

pctile creates a new variable containing percentiles. You specify the number of quantiles that you want, and pctile computes the corresponding percentiles. Here we use Stata's auto dataset and compute the deciles of mpg:

```
. use http://www.stata-press.com/data/r12/auto
(1978 Automobile Data)
. pctile pct = mpg, nq(10)
. list pct in 1/10
```

	pct
1.	14
2.	17
3.	18
4.	19
5.	20
6.	22
7.	24
8.	25
9.	29
10.	.

If we use the `genp()` option to generate another variable with the corresponding percentages, it is easier to distinguish between the percentiles.

```
. drop pct
. pctile pct = mpg, nq(10) genp(percent)
. list percent pct in 1/10
```

	percent	pct
1.	10	14
2.	20	17
3.	30	18
4.	40	19
5.	50	20
6.	60	22
7.	70	24
8.	80	25
9.	90	29
10.	.	.

`summarize, detail` calculates standard percentiles.

```
. summarize mpg, detail
```

Mileage (mpg)

	Percentiles	Smallest		
1%	12	12		
5%	14	12		
10%	14	14	Obs	74
25%	18	14	Sum of Wgt.	74
50%	20		Mean	21.2973
		Largest	Std. Dev.	5.785503
75%	25	34		
90%	29	35	Variance	33.47205
95%	34	35	Skewness	.9487176
99%	41	41	Kurtosis	3.975005

`summarize, detail` can calculate only these particular percentiles. The `pctile` and `_pctile` commands allow you to compute any percentile.

Weights can be used with pctile, xtile, and _pctile:

```
. drop pct percent
. pctile pct = mpg [w=weight], nq(10) genp(percent)
(analytic weights assumed)
. list percent pct in 1/10
```

	percent	pct
1.	10	14
2.	20	16
3.	30	17
4.	40	18
5.	50	19
6.	60	20
7.	70	22
8.	80	24
9.	90	28
10.	.	.

The result is the same, no matter which weight type you specify—aweight, fweight, or pweight.

xtile

xtile creates a categorical variable that contains categories corresponding to quantiles. We illustrate this with a simple example. Suppose that we have a variable, bp, containing blood pressure measurements:

```
. use http://www.stata-press.com/data/r12/bp1, clear
. list bp, sep(4)
```

	bp
1.	98
2.	100
3.	104
4.	110
5.	120
6.	120
7.	120
8.	120
9.	125
10.	130
11.	132

xtile can be used to create a variable, quart, that indicates the quartiles of bp.

```
. xtile quart = bp, nq(4)
. list bp quart, sepby(quart)
```

	bp	quart
1.	98	1
2.	100	1
3.	104	1
4.	110	2
5.	120	2
6.	120	2
7.	120	2
8.	120	2
9.	125	3
10.	130	4
11.	132	4

The categories created are

$$(-\infty, x_{[25]}], \quad (x_{[25]}, x_{[50]}], \quad (x_{[50]}, x_{[75]}], \quad (x_{[75]}, +\infty)$$

where $x_{[25]}$, $x_{[50]}$, and $x_{[75]}$ are, respectively, the 25th, 50th (median), and 75th percentiles of bp. We could use the pctile command to generate these percentiles:

```
. pctile pct = bp, nq(4) genp(percent)
. list bp quart percent pct, sepby(quart)
```

	bp	quart	percent	pct
1.	98	1	25	104
2.	100	1	50	120
3.	104	1	75	125
4.	110	2	.	.
5.	120	2	.	.
6.	120	2	.	.
7.	120	2	.	.
8.	120	2	.	.
9.	125	3	.	.
10.	130	4	.	.
11.	132	4	.	.

xtile can categorize a variable on the basis of any set of cutpoints, not just percentiles. Suppose that we wish to create the following categories for blood pressure:

$$(-\infty, 100], \quad (100, 110], \quad (110, 120], \quad (120, 130], \quad (130, +\infty)$$

To do this, we simply create a variable containing the cutpoints,

```
. input class
          class
  1. 100
  2. 110
  3. 120
  4. 130
  5. end
```

and then use xtile with the cutpoints() option:

```
. xtile category = bp, cutpoints(class)
. list bp class category, sepby(category)
```

	bp	class	category
1.	98	100	1
2.	100	110	1
3.	104	120	2
4.	110	130	2
5.	120	.	3
6.	120	.	3
7.	120	.	3
8.	120	.	3
9.	125	.	4
10.	130	.	4
11.	132	.	5

The cutpoints can, of course, come from anywhere. They can be the quantiles of another variable or the quantiles of a subgroup of the variable. Suppose that we had a variable, case, that indicated whether an observation represented a case (case = 1) or control (case = 0).

```
. use http://www.stata-press.com/data/r12/bp2, clear
. list in 1/11, sep(4)
```

	bp	case
1.	98	1
2.	100	1
3.	104	1
4.	110	1
5.	120	1
6.	120	1
7.	120	1
8.	120	1
9.	125	1
10.	130	1
11.	132	1

We can categorize the cases on the basis of the quantiles of the controls. To do this, we first generate a variable, pct, containing the percentiles of the controls' blood pressure data:

```
. pctile pct = bp if case==0, nq(4)
. list pct in 1/4
```

	pct
1.	104
2.	117
3.	124
4.	.

Then we use these percentiles as cutpoints to classify bp: for all subjects.

```
. xtile category = bp, cutpoints(pct)
. gsort -case bp
. list bp case category in 1/11, sepby(category)
```

	bp	case	category
1.	98	1	1
2.	100	1	1
3.	104	1	1
4.	110	1	2
5.	120	1	3
6.	120	1	3
7.	120	1	3
8.	120	1	3
9.	125	1	4
10.	130	1	4
11.	132	1	4

❏ Technical note

In the last example, if we wanted to categorize only cases, we could have issued the command

```
. xtile category = bp if case==1, cutpoints(pct)
```

Most Stata commands follow the logic that using an *if exp* is equivalent to dropping observations that do not satisfy the expression and running the command. This is not true of xtile when the cutpoints() option is used. (When the cutpoints() option is not used, the standard logic is true.) xtile uses all nonmissing values of the cutpoints() variable whether or not these values belong to observations that satisfy the if expression.

If you do not want to use all the values in the cutpoints() variable as cutpoints, simply set the ones that you do not need to missing. xtile does not care about the order of the values or whether they are separated by missing values.

❏

❏ Technical note

Quantiles are not always unique. If we categorize our blood pressure data by quintiles rather than quartiles, we get

```
. use http://www.stata-press.com/data/r12/bp1, clear
. xtile quint = bp, nq(5)
. pctile pct = bp, nq(5) genp(percent)
```

```
. list bp quint pct percent, sepby(quint)
```

	bp	quint	pct	percent
1.	98	1	104	20
2.	100	1	120	40
3.	104	1	120	60
4.	110	2	125	80
5.	120	2	.	.
6.	120	2	.	.
7.	120	2	.	.
8.	120	2	.	.
9.	125	4	.	.
10.	130	5	.	.
11.	132	5	.	.

The 40th and 60th percentile are the same; they are both 120. When two (or more) percentiles are the same, they are given the lower category number.

❏

_pctile

_pctile is a programmer's command. It computes percentiles and stores them in r(); see [U] **18.8 Accessing results calculated by other programs**.

You can use _pctile to compute quantiles, just as you can with pctile:

```
. use http://www.stata-press.com/data/r12/auto, clear
(1978 Automobile Data)
. _pctile weight, nq(10)
. return list
scalars:
          r(r1)    =    2020
          r(r2)    =    2160
          r(r3)    =    2520
          r(r4)    =    2730
          r(r5)    =    3190
          r(r6)    =    3310
          r(r7)    =    3420
          r(r8)    =    3700
          r(r9)    =    4060
```

The percentiles() option (abbreviation p()) can be used to compute any percentile you wish:

```
. _pctile weight, p(10, 33.333, 45, 50, 55, 66.667, 90)
. return list
scalars:
          r(r1)    =    2020
          r(r2)    =    2640
          r(r3)    =    2830
          r(r4)    =    3190
          r(r5)    =    3250
          r(r6)    =    3400
          r(r7)    =    4060
```

_pctile, pctile, and xtile each have an option that uses an alternative definition of percentiles, based on an interpolation scheme; see *Methods and formulas* below.

```
. _pctile weight, p(10, 33.333, 45, 50, 55, 66.667, 90) altdef
. return list
scalars:
         r(r1)       =   2005
         r(r2)       =   2639.985
         r(r3)       =   2830
         r(r4)       =   3190
         r(r5)       =   3252.5
         r(r6)       =   3400.005
         r(r7)       =   4060
```

The default formula inverts the empirical distribution function. The default formula is more commonly used, although some consider the "alternative" formula to be the standard definition. One drawback of the alternative formula is that it does not have an obvious generalization to noninteger weights.

❑ Technical note

summarize, detail computes the 1st, 5th, 10th, 25th, 50th (median), 75th, 90th, 95th, and 99th percentiles. There is no real advantage in using _pctile to compute these percentiles. Both summarize, detail and _pctile use the same internal code. _pctile is slightly faster because summarize, detail computes a few extra things. The value of _pctile is its ability to compute percentiles other than these standard ones.

❑

Saved results

pctile and _pctile save the following in r():

Scalars
    r(r#)   value of #-requested percentile

Methods and formulas

pctile and xtile are implemented as ado-files.

The default formula for percentiles is as follows: Let $x_{(j)}$ refer to the x in ascending order for $j = 1, 2, \ldots, n$. Let $w_{(j)}$ refer to the corresponding weights of $x_{(j)}$; if there are no weights, $w_{(j)} = 1$. Let $N = \sum_{j=1}^{n} w_{(j)}$.

To obtain the pth percentile, which we will denote as $x_{[p]}$, let $P = Np/100$, and let

$$W_{(i)} = \sum_{j=1}^{i} w_{(j)}$$

Find the first index, i, such that $W_{(i)} > P$. The pth percentile is then

$$x_{[p]} = \begin{cases} \dfrac{x_{(i-1)} + x_{(i)}}{2} & \text{if } W_{(i-1)} = P \\ x_{(i)} & \text{otherwise} \end{cases}$$

When the `altdef` option is specified, the following alternative definition is used. Here weights are not allowed.

Let i be the integer floor of $(n+1)p/100$; that is, i is the largest integer $i \leq (n+1)p/100$. Let h be the remainder $h = (n+1)p/100 - i$. The pth percentile is then

$$x_{[p]} = (1-h)x_{(i)} + hx_{(i+1)}$$

where $x_{(0)}$ is taken to be $x_{(1)}$ and $x_{(n+1)}$ is taken to be $x_{(n)}$.

`xtile` produces the categories

$$(-\infty, x_{[p_1]}], \ (x_{[p_1]}, x_{[p_2]}], \ \dots \ , \ (x_{[p_{m-2}]}, x_{[p_{m-1}]}], \ (x_{[p_{m-1}]}, +\infty)$$

numbered, respectively, $1, 2, \dots, m$, based on the m quantiles given by the p_kth percentiles, where $p_k = 100\,k/m$ for $k = 1, 2, \dots, m-1$.

If $x_{[p_{k-1}]} = x_{[p_k]}$, the kth category is empty. All elements $x = x_{[p_{k-1}]} = x_{[p_k]}$ are put in the $(k-1)$th category: $(x_{[p_{k-2}]}, x_{[p_{k-1}]}]$.

If `xtile` is used with the `cutpoints(varname)` option, the categories are

$$(-\infty, y_{(1)}], \ (y_{(1)}, y_{(2)}], \ \dots \ , \ (y_{(m-1)}, y_{(m)}], \ (y_{(m)}, +\infty)$$

and they are numbered, respectively, $1, 2, \dots, m+1$, based on the m nonmissing values of *varname*: $y_{(1)}, y_{(2)}, \dots, y_{(m)}$.

Acknowledgment

`xtile` is based on a command originally posted on Statalist (see [U] **3.4 The Stata listserver**) by Philip Ryan, University of Adelaide, Australia.

Also see

[R] **centile** — Report centile and confidence interval

[R] **summarize** — Summary statistics

[U] **18.8 Accessing results calculated by other programs**

Title

> **putmata** — Put Stata variables into Mata and vice versa

Syntax

> putmata *putlist* [*if*] [*in*] [, *putmata_options*]

> getmata [*getlist*] [, *getmata_options*]

putmata_options	Description
omitmissing	omit observations with missing values
view	create vectors and matrices as views, not as copies
replace	replace existing Mata vectors and matrices

A *putlist* can be as simple as a list of Stata variable names. See below for details.

getmata_options	Description
double	create Stata variables as doubles
update	update existing Stata variables
replace	replace existing Stata variables
id(*name*)	match observations with rows based on equal values of variable *name* and matrix *name*. id(*varname=vecname*) is also allowed.
force	allow nonconformable matrices; usually, id() is preferable

A *getlist* can be as simple as a list of Mata vector names. See below for details.

Definition of *putlist* for use with putmata:

A *putlist* is one or more of any of the following:

> *
> *varname*
> *varlist*
> *vecname=varname*
> *matname=*(*varlist*)
> *matname=*([*varlist*] # [*varlist*] [...])

Example: putmata *
Creates a vector in Mata for each of the Stata variables in memory. Vectors contain the same data as Stata variables. Vectors have the same names as the corresponding variables.

Example: putmata mpg weight displ
Creates a vector in Mata for each variable specified. Vectors have the same names as the corresponding variables. In this example, displ is an abbreviation for the variable displacement; thus the vector will also be named displacement.

Example: putmata mileage=mpg pounds=weight
Creates a vector for each variable specified. Vector names differ from the corresponding variable names. In this example, vectors will be named mileage and pounds.

Example: `putmata y=mpg X=(weight displ)`
  Creates $N \times 1$ Mata vector **y** equal to Stata variable `mpg`, and creates $N \times 2$ Mata matrix **X** containing the values of Stata variables `weight` and `displacement`.

Example: `putmata y=mpg X=(weight displ 1)`
  Creates $N \times 1$ Mata vector **y** containing `mpg`, and creates $N \times 3$ Mata matrix **X** containing `weight`, `displacement`, and a column of 1s. After typing this example, you could enter Mata and type `invsym(X'X)*X'y` to obtain the regression coefficients.

Syntactical elements may be combined. It is valid to type

    . putmata mpg foreign X=(weight displ) Z=(foreign 1)

No matter how you specify the *putlist*, you will need to specify the `replace` option if some or all vectors already exist in Mata:

    . putmata mpg foreign X=(weight displ) Z=(foreign 1), replace

Definition of *getlist* for use with `getmata`:

A *getlist* is one or more of any of the following:

> *vecname*
> *varname=vecname*
> *(varname varname ... varname)=matname*
> *(varname*)=matname*

Example: `getmata x1 x2`
  Creates a Stata variable for each Mata vector specified. Variables will have the same names as the corresponding vectors. Names may not be abbreviated.

Example: `getmata myvar1=x1 myvar2=x2`
  Creates a Stata variable for each Mata vector specified. Variable names will differ from the corresponding vector names.

Example: `getmata (firstvar secondvar)=X`
  Creates one Stata variable corresponding to each column of the Mata matrix specified. In this case, the matrix has two columns, and corresponding variables will be named `firstvar` and `secondvar`. If the matrix had three columns, then three variable names would need to be specified.

Example: `getmata (myvar*)=X`
  Creates one Stata variable corresponding to each column of the Mata matrix specified. Variables will be named `myvar1`, `myvar2`, etc. The matrix may have any number of columns, even zero!

Syntactical elements may be combined. It is valid to type

    . getmata r1 r2 final=r3 (rplus*=X)

No matter how you specify the *getlist*, you will need to specify the `replace` or `update` option if some or all variables already exist in Stata:

    . getmata r1 r2 final=r3 (rplus*=X), replace

Description

`putmata` exports the contents of Stata variables to Mata vectors and matrices.

`getmata` imports the contents of Mata vectors and matrices to Stata variables.

putmata and getmata are useful for creating solutions to problems more easily solved in Mata. The commands are also useful in teaching.

Options for putmata

omitmissing specifies that observations containing a missing value in any of the numeric variables specified be omitted from the vectors and matrices created in Mata. In

```
. putmata y=mpg X=(weight displ 1), omitmissing
```

rows would be omitted from y and X in which the corresponding observation contained missing in any of mpg, weight, or displ. In this case, specifying omitmissing would be equivalent to typing

```
. putmata y=mpg X=(weight displ 1) if !missing(mpg) & !missing(weight) ///
      & !missing(displ)
```

All vectors and matrices created by a single putmata command will have the same number of rows (observations). That is true whether you specify if, in, or the omitmissing option.

view specifies that putmata create views rather than copies of the Stata data in the Mata vectors and matrices. Views require less memory than copies and offer the advantage (and disadvantage) that changes in the Stata data are immediately reflected in the Mata vectors and matrices, and vice versa.

If you specify numeric constants using the *matname*=(...) syntax, *matname* is created as a copy even if the view option is specified. Other vectors and matrices created by the command, however, would be views.

Use of the view option with putmata often obviates the need to use getmata to import results back into Stata.

Warning 1: Mata records views as "this vector is a view onto variable 3, observations 2 through 5 and 7". If you change the order of the variables, the order of the observations, or drop variables once the views are created, then the contents of the views will change.

Warning 2: When assigning values in Mata to view vectors, code

```
v[] = ...
```

not v = ...

To have changes reflected in the underlying Stata data, you must update the elements of the view v, not redefine it. To update all the elements of v, you literally code v[.]. In the matrix case, you code X[.,.].

replace specifies that existing Mata vectors or matrices be replaced should that be necessary.

Options for getmata

double specifies that Stata numeric variables be created as doubles. The default is that they be created as floats. Actually, variables start out as floats or doubles, but then they are compressed (see [D] **compress**).

update and replace are alternatives. They have the same meaning unless the id() or force option is specified.

When `id()` or `force` is not specified, both `replace` and `update` specify that it is okay to replace the values in existing Stata variables. By default, vectors can be posted to new Stata variables only.

When `id()` or `force` is specified, `replace` and `update` allow posting of values of existing variables, just as usual. The options differ in how the posting is performed when the `id()` or `force` option causes only a subset of the observations of the variables to be updated. `update` specifies that the remaining values be left as they are. `replace` specifies that the remaining values be set to missing, just as if the existing variable(s) were being created for the first time.

id(*name*) and id(*varname*=*vecname*) specify how the rows in the Mata vectors and matrices match the observations in the Stata data. Observation i matches row j if variable *name*[i] equals vector *name*[j], or in the second syntax, if *varname*[i] = *vecname*[j]. The ID variable (vector) must contain values that uniquely identify the observations (rows). Only in observations that contain matching values will the variable be modified. Values in observations that have no match will not be modified or will be set to missing, as appropriate; values in the ID vector that have no match will be ignored.

Example: You wish to run a regression of `y` on `x1` and `x2` on the males in the data and use that result to obtain the fitted values for the males. Stata already has commands that will do this, namely, `regress y x1 x2 if male` followed by `predict yhat if male`. For instructional purposes, let's say you wish to do this in Mata. You type

```
. putmata myid y  X=(x1 x2 1) if male
. mata
: b = invsym(X'X)*X'y
: yhat = X*b
: end
. getmata yhat, id(myid)
```

The new Stata variable `yhat` will contain the predicted values for males and missing values for the females. If the `yhat` variable already existed, you would type

```
. getmata yhat, id(myid) replace
```

or

```
. getmata yhat, id(myid) update
```

The `replace` option would set the female observations to missing. The `update` option would leave the female observations unchanged.

If you do not have an identification variable, create one first by typing `generate myid = _n`.

force specifies that it is okay to post vectors and matrices with fewer or with more rows than the number of observations in the data. The `force` option is an alternative to `id()`, and usually, `id()` is the appropriate choice.

If you specify `force` and if there are fewer rows in the vectors and matrices than observations in the data, new variables will be padded with missing values. If there are more rows than observations, observations will be added to the data and previously existing variables will be padded with missing values.

Remarks

Remarks are presented under the following headings:

      Use of putmata
      Use of putmata and getmata
      Using putmata and getmata on subsets of observations
      Using views
      Constructing do-files

Use of putmata

In this example, we will use Mata to make a calculation and report the result, but we will not post results back to Stata. We will use `putmata` but not `getmata`.

Consider solving for **b** the set of linear equations

$$\mathbf{y} = \mathbf{Xb} \tag{1}$$

where **y**: $N \times 1$, **X**: $N \times k$, and **b**: $k \times 1$. If $N = k$, then $\mathbf{y} = \mathbf{Xb}$ amounts to solving k equations for k unknowns, and the solution is

$$\mathbf{b} = \mathbf{X}^{-1}\mathbf{y} \tag{2}$$

That solution is obtained by premultiplying both sides of (1) by $\mathbf{X}^{-1}$.

When $N > k$, (2) can be used to obtain least-square results if matrix inversion is appropriately defined. Assume that you wish to demonstrate this when matrix inversion is defined as the Moore–Penrose generalized inverse for nonsquare matrices. The demonstration can be obtained by typing

```
. sysuse auto, clear
. regress mpg weight displacement
. putmata y=mpg X=(weight displacement 1)
. mata
: pinv(X)*y
: end

. _
```

The Mata expression `pinv(X)*y` will display a 3×1 column vector. The elements of the vector will equal the coefficients reported by `regress mpg weight displacement`.

For your information, the Moore–Penrose inverse of rectangular matrix **X**: $N \times k$ is a $k \times N$ rectangular matrix. Among other properties, `pinv(X)*X = I`, where **I** is the $k \times k$ identity matrix. You can demonstrate that using Mata, too:

```
. mata: pinv(X)*X
```

Use of putmata and getmata

In this example, we will use Mata to calculate a result that we wish to post back to Mata. We will use both `putmata` and `getmata`.

Some problems are more easily solved in Mata than in Stata. For instance, say that you need to create new Stata variable D from existing variable C, defined as

$$\mathtt{D}[\,i\,] = \mathrm{sum}(\mathtt{C}[\,j\,] - \mathtt{C}[\,i\,]) \text{ for all } \mathtt{C}[\,j\,] > \mathtt{C}[\,i\,]$$

where i and j index observations.

This problem can be solved in Stata, but the solution is elusive to most people. The solution is more natural in Mata because the Mata solution corresponds almost letter for letter with the mathematical statement of the problem. If C and D were Mata vectors rather than Stata variables, the solution would be

```
D = J(rows(C), 1, 0)
for (i=1; i<=rows(C); i++) {
        for (j=1; j<=rows(C); j++) {
                if (C[j]>C[i]) D[i] = D[i] + (C[j] - C[i])
        }
}
```

The most difficult part of this solution to understand is the first line, D = J(rows(C), 1, 0), and that is because you may not be familiar with Mata's J() function. D = J(rows(C), 1, 0) creates a rows(C) × 1 column vector of 0s. The arguments of J() are in just that order.

C and D are not vectors in Mata, or at least they are not yet. Using getmata, we can create vector C from variable C and run our Mata solution. Then using putmata, we can post Mata vector D back to new Stata variable D. The solution includes these three steps, also shown in the do-file below:

(1) In Stata, use putmata to create vector C in Mata equal to variable C in Stata: putmata C.

(2) Use Mata to solve the problem, creating new Mata vector D.

(3) In Stata again, use getmata to create new variable D equal to Mata vector D.

Because of the typing involved in the solution, we would package the code in a do-file:

-- begin myfile.do ----------

```
use mydata, clear
putmata C                                                          (1)

mata:                                                              (2)
D = J(rows(C), 1, 0)
for (i=1; i<=rows(C); i++) {
        for (j=1; j<=rows(C); j++) {
                if (C[j]>C[i]) D[i] = D[i] + (C[j] - C[i])
        }
}
end

getmata D                                                          (3)
save mydata, replace
```

-- end myfile.do ----------

With myfile.do now in place, in Stata we would type

```
. do myfile
```

Notes:

(1) Our program might be better if we changed putmata C to read putmata C, replace and if we changed getmata D to read getmata D, replace. As things are right now, typing do myfile works, but if we were then to run it a second time, it would not work. Stata would encounter the putmata command and issue an error that matrix C already exists. Even if Stata got through that, it would encounter the getmata command and issue an error that variable D already exists. Perhaps that is an advantage. You cannot run myfile.do again without dropping matrix C and variable D. If you consider that a disadvantage, however, include the replace option.

(2) In our solution, we entered Mata by typing `mata:`, which is to say, `mata` with a colon. Interactively, we usually enter Mata by just typing `mata`. The colon affects how Mata treats errors. When working interactively, we want Mata to note errors but then to continue running so we can correct ourselves. In do-files, we want Mata to note the error and stop. That is the difference between `mata` without the colon and `mata` with the colon. Remember to use `mata:` when writing do-files.

(3) Rather than specify the `replace` option, you could modify the do-file to drop any preexisting Mata vector C and any preexisting variable D. To drop vector C, in Mata you can type `mata drop C`, or in Stata, you can type `mata: mata drop C`. To drop variable D, in Stata you can type `drop D`. You must worry that the variables do not exist, so in your do-file, you would code

```
capture mata: mata drop C
capture drop D
```

Rather than dropping vector C, you might prefer just to clear Mata:

```
clear mata
```

Using putmata and getmata on subsets of observations

`putmata` can be used to create Mata vectors that contain a subset of the observations in the Stata data, and `getmata` can be used to fetch such vectors back into Stata. Thus you can work with only the males or only outcomes in which failures are observed, and so on. Below we work with only the observations in which C does not contain missing values.

In the create-variable-D-from-C example above, we assumed that there were no missing values in C, or at least we did not consider the issue. It turns out that our code produces several missing values in the presence of just one missing value in C. Perhaps, if there are missing values, we want to exclude them from our calculation. We could complicate our Mata code to handle that. We could modify our Mata code to read

```
use mydata, clear
putmata C

D = J(rows(C), 1, 0)
for (i=1; i<=rows(C); i++) {
        if (C[i]>=.) D[i] = .                    // new
        else for (j=1; j<=rows(C); j++) {
                if (C[j]<.) {                    // new
                        if (C[j]>C[i]) D[i] = D[i] + (C[j] - C[i])
                }
        }
}
end

getmata D
save mydata, replace
```

Easier, however, is simply to restrict Mata vector C to the nonmissing elements of Stata variable C, which we could do by replacing `putmata C` with

```
putmata C if !missing(C)
```

or, equivalently,

```
putmata C, omitmissing
```

Whichever way we coded it, if the data contained 100 observations and variable C contained 82 nonmissing values, new Mata vector C would contain 82 rows rather than 100. The observations corresponding to missing(C) would be omitted from the vector, and that means we could run our original Mata solution without modification.

There is, however, an issue. At the end of our code when we post the Mata solution vector D to Stata variable D—getmata D—we will need to specify which of the 100 observations are to receive the 82 results stored in the vector. getmata has an option to handle this situation—id(*varname*), where *varname* is the name of an identification variable.

An identification variable is a variable that takes on different values for each observation in the data. The values could be 1, 2, ..., 100; or they could be 1.25, −2, ..., 16.5; or they could be Nick, Bill, ..., Mary. The values can be numeric or string, and they need not be in order. All that is important is that the variable contain a unique (different) value in each observation. Possibly, the data already contain such a variable. If not, you can create one by typing

```
generate fid = _n
```

When we use putmata to create vector C, we will need simultaneously to create vector fid containing the selected values of variable fid, which we can do by adding fid to the *putlist*:

```
putmata fid C if !missing(C)
```

The above command creates two vectors in Mata: fid and C. When we post the resulting vector D back to Stata, we will specify the id(fid) option to indicate into which observations getmata is to post the results:

```
getmata D, id(fid)
```

The id(fid) option is taken to mean that there exists a variable named fid and a vector named fid. It is by comparing the values in each that getmata determines how the rows of the vectors correspond to the observations of the data.

The entire solution is

```
─────────────────────────────────────── begin myfile.do ──────────

use mydata, clear
putmata fid C if !missing(C)          // new: we put fid & add if !missing(C)

mata:
D = J(rows(C), 1, 0)
for (i=1; i<=rows(C); i++) {
        for (j=1; j<=rows(C); j++) {
                if (C[j]>C[i]) D[i] = D[i] + (C[j] - C[i])
        }
}
end

getmata D, id(fid)            // new: we add option id(fid)
save mydata, replace
─────────────────────────────────────────── end myfile.do ────────
```

The above code will run on data with or without missing values. New variable D will be missing in observations where C is missing, but D will otherwise contain nonmissing values.

Using views

When you type or code `putmata C`, vector C is created as a copy of the Stata data. The variable and the vector are separate things. An alternative is to make the Mata vector a view onto the Stata variable. By that, we mean that both the variable and the vector share the same recording of the values. Views save memory but are slightly less efficient in terms of execution time. Views have other advantages and disadvantages, too.

For instance, if you type `putmata mpg` and then, in Mata, type `mpg[1]=20`, you will change not only the Mata vector but also the Stata data! Or if, after typing `putmata mpg`, you typed `replace mpg = 20 in 1`, that would modify both the data and the Mata vector! This is an advantage if you are fixing real errors and a disadvantage if you intend to do something else.

If in the middle of your Mata session where you are working with views you take a break and return to Stata, it is important that you do not modify the Stata data in certain ways. Rather than recording copies of the data, views record notes about the mapping. A view might record that this Mata vector corresponds to variable 3, observations 2 through 20 and 39. If you change the sort order of the data, the view will still be working with observations 2 through 20 and 39 even though those physical observations now contain different data. If you drop the first or second variable, the view will still be working with the third variable even though that will now be a different variable!

The memory savings offered by views are considerable, at least when working with large datasets. Say that you have a dataset containing 200 variables and 1,000,000 observations. Your data might be 1 GB in size. Even so, typing `putmata *, view`, and thus creating 200 vectors each with 1,000,000 rows, would consume only a few dozen kilobytes of memory.

All the examples shown above work equally well with copies or views. We have been working with copies, but in the previous example, where we coded

        putmata fid C if !missing(C)

we could switch to working with views by coding

        putmata fid C if !missing(C), view

With that one change, our code would still work and it would use less memory.

With that one change, we would still not be working with views everywhere we could, however. Vector D—the vector we create in Mata and then post back to Stata—would still be a regular vector. We can save additional memory by making D a view, too. Before we do that, let us warn you that we do not recommend doing this unless the memory savings is vitally important. The result, when complete, will be elegant and memory efficient, but the extra memory savings is seldom worth the debugging effort.

No extra changes are required to your code when the vectors you make into views contain values that are not modified in the code. Vector C is such a vector. We use the values stored in C, but we do not change them. Vector D, on the other hand, is a vector in which we change values. It is usually easier if you do not convert such vectors into views.

With that proviso, we are going to make D into a view, too, and in the process, we will drop the use of fid altogether:

```
──────────────────────────────────────── begin myfile.do ─────────
    use mydata, clear
    generate D = .                      // new
    putmata C D if !missing(C), view    // changed

    mata:
    D[.] = J(rows(C), 1, 0)             // changed
    for (i=1; i<=rows(C); i++) {
            for (j=1; j<=rows(C); j++) {
                    if (C[j]>C[i]) D[i] = D[i] + (C[j] - C[i])
            }
    }
    end
                                        // we drop the getmata
    save mydata, replace
──────────────────────────────────────── end myfile.do ─────────
```

In this solution, we create new Stata variable D at the outset, and then we modify the putmata command to create view vectors for both C and D. Our code, which stores results in vector D, now simultaneously posts to variable D when we store results in vector D, so we can omit the getmata D at the end because results are already posted! Moreover, we no longer have to concern ourselves with matching observations to rows via fid. Rows of D now automatically align themselves with the selected observations in variable D by the mere fact of D being a view.

The beginning of our Mata code has an important change, however. We change

    D = J(rows(C), 1, 0)

to

    D[.] = J(rows(C), 1, 0)

That change is very important. What we coded previously created vector D. What we now code changes the values stored in existing vector D. If we left what we coded previously, Mata would discard the view currently stored in D and create a new D—a regular Mata vector unconnected to Stata—containing 0s.

Constructing do-files

putmata and getmata can be used interactively, but if you have much Mata code between the put and the get, you will be better off using a do-file because do-files can be easily edited when they have a mistake in them. We recommend the following outline for such do-files:

```
———————————————————————————————— begin outline.do ————————

version 12                             (1)
mata clear                             (2)
// Stata code for setup goes here      (3)
putmata ...                            (4)
mata:
// Mata code goes here                 (5)
end
getmata                                (6)
mata clear                             (7)

—————————————————————————————————— end outline.do ————————
```

Notes on do-file steps:

(1) A do-file should always start with a version statement; it ensures that the do-file continues to work in the years to come as new versions of Stata are released. See [P] **version**.

(2) The do-file should not depend on Mata having certain vectors, matrices, or programs already loaded and set up because if you attempt to run the do-file again later, what you assumed may not be true. A do-file should be self-contained. To ensure that is true the first time we write and run the do-file and to ensure on subsequent runs that nothing lying around in Mata gets in our way, we clear Mata.

(3) You may need to sort your data, create extra variables that your do-file will use, or drop variables that you are assuming do not already exist. In the last iteration of myfile.do, we needed to generate D = ., and it would not have been a bad idea to capture drop D before we did that. Our example did not depend on the sort order of the data, but if it had, we would have included the sort even if we were certain that the data would already be in the right order.

(4) Put the putmata command here. If putmata includes the omitmissing option, then put everything you need to put in a single putmata command. Otherwise, you can use multiple putmata commands if you find that more convenient. If you use multiple putmata commands, be sure to include the same if *expression* and in *range* qualifiers on each one.

(5) The Mata code goes here. Note that we type mata: (mata with a colon) to enter Mata. mata: ensures that errors stop Mata and thus our do-file.

(6) The getmata command goes here if you need it. Be sure to include getmata's id(*name*) or id(*vecname=varname*) option if, on the putmata command in step 4, you included the if *expression* qualifier or the in *range* qualifier or the omitmissing option. If you include id(), be sure you included the ID variable in the putmata command in step 4.

(7) We conclude by clearing Mata again to avoid leaving memory allocated needlessly and to avoid causing problems for poorly written do-files that we might subsequently run.

putmata and getmata are designed to work interactively and in do-files. The commands are not designed to work with ado-files. An ado-file is something like a do-file, but it defines a program that implements a new command of Stata, and well-written ado-files do not use globals such as the global vectors and matrices that putmata creates. Ado-files use local variables. Ado-file programmers should use the Mata functions st_data() and st_view() (see [M-5] **st_data( )** and [M-5] **st_view( )**) to create vectors and matrices, and if necessary, use st_store() (see [M-5] **st_store( )**) to post the contents of those vectors and matrices back to Stata.

Saved results

putmata saves the following in r():

Scalars
r(N)	number of rows in created vectors and matrices
r(K_views)	number of vectors and matrices created as views
r(K_copies)	number of vectors and matrices created as copies

The total number of vectors and matrices created is r(K_views) + r(K_copies).

r(N)=. if r(K_views) + r(K_copies) = 0. r(N)=0 means that zero-observation vectors and matrices were created, which is to say, vectors and matrices dimensioned 0×1 and $0 \times k$.

getmata saves the following in r():

Scalars
r(K_new)	number of new variables created
r(K_existing)	number of existing variables modified

The total number of variables modified is r(K_new) + r(K_existing).

Methods and formulas

putmata and getmata are implemented as ado-files.

Reference

Gould, W. W. 2010. Mata Matters: Stata in Mata. *Stata Journal* 10: 125–142.

Also see

[M-4] **stata** — Stata interface functions

[M-5] **st_data( )** — Load copy of current Stata dataset

[M-5] **st_view( )** — Make matrix that is a view onto current Stata dataset

[M-5] **st_store( )** — Modify values stored in current Stata dataset

Title

range — Generate numerical range

Syntax

range *varname* $\#_{\text{first}}$ $\#_{\text{last}}$ $\left[\,\#_{\text{obs}}\,\right]$

Menu

Data > Create or change data > Other variable-creation commands > Generate numerical range

Description

range generates a numerical range, which is useful for evaluating and graphing functions.

Remarks

range constructs the variable *varname*, taking on values $\#_{\text{first}}$ to $\#_{\text{last}}$, inclusive, over $\#_{\text{obs}}$. If $\#_{\text{obs}}$ is not specified, the number of observations in the current dataset is used.

range can be used to produce increasing sequences, such as

    . range x 0 12.56 100

or it can be used to produce decreasing sequences:

    . range z 100 1

▷ Example 1

To graph $y = e^{-x/6}\sin(x)$ over the interval $[0, 12.56]$, we can type

    . range x 0 12.56 100
    obs was 0, now 100
    . generate y = exp(-x/6)*sin(x)
    . scatter y x, yline(0) ytitle(y = exp(-x/6) sin(x))

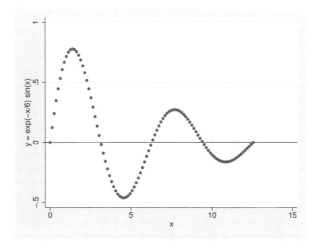

◁

▷ Example 2

Stata is not limited solely to graphing functions—it can draw parameterized curves as well. For instance, consider the curve given by the polar coordinate relation $r = 2\sin(2\theta)$. The conversion of polar coordinates to parameterized form is $(y, x) = (r\sin\theta, r\cos\theta)$, so we can type

```
. clear
. range theta 0 2*_pi 400
(obs was 100, now 400)
. generate r = 2*sin(2*theta)
. generate y = r*sin(theta)
. generate x = r*cos(theta)
. line y x, yline(0) xline(0) aspectratio(1)
```

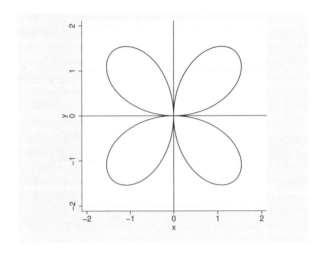

◁

Methods and formulas

range is implemented as an ado-file.

Also see

[D] **egen** — Extensions to generate

[D] **obs** — Increase the number of observations in a dataset

Title

> **recast** — Change storage type of variable

Syntax

> recast *type varlist* [, force]

where *type* is byte, int, long, float, double, or str1, str2, ..., str244.

Description

> recast changes the storage type of the variables identified in *varlist* to *type*.

Option

> force makes recast unsafe by causing the variables to be given the new storage type even if that will cause a loss of precision, introduction of missing values, or, for string variables, the truncation of strings.
>
> force should be used with caution. force is for those instances where you have a variable saved as a double but would now be satisfied to have the variable stored as a float, even though that would lead to a slight rounding of its values.

Remarks

> See [U] **12 Data** for a description of storage types. Also see [D] **compress** and [D] **destring** for alternatives to recast.

▷ Example 1

> recast refuses to change a variable's type if that change is inappropriate for the values actually stored, so it is always safe to try:
>
> ```
> . use http://www.stata-press.com/data/r12/auto
> (1978 Automobile Data)
> . describe headroom
> ```
>
> | variable name | storage type | display format | value label | variable label |
> |---|---|---|---|---|
> | headroom | float | %6.1f | | Headroom (in.) |
>
> ```
> . recast int headroom
> headroom:  37 values would be changed; not changed
> ```

Our attempt to change headroom from a float to an int was ignored—if the change had been made, 37 values would have changed. Here is an example where the type can be changed:

> ```
> . describe mpg
> ```
>
> | variable name | storage type | display format | value label | variable label |
> |---|---|---|---|---|
> | mpg | int | %8.0g | | Mileage (mpg) |

```
. recast byte mpg

. describe mpg
```

	storage	display	value	
variable name	type	format	label	variable label
mpg	byte	%8.0g		Mileage (mpg)

recast works with string variables as well as numeric variables, and it provides all the same protections:

```
. describe make
```

	storage	display	value	
variable name	type	format	label	variable label
make	str18	%-18s		Make and Model

```
. recast str16 make
make:  2 values would be changed; not changed
```

recast can be used both to promote and to demote variables:

```
. recast str20 make

. describe make
```

	storage	display	value	
variable name	type	format	label	variable label
make	str20	%-20s		Make and Model

◁

Methods and formulas

recast is implemented as an ado-file.

Also see

[D] **compress** — Compress data in memory

[D] **destring** — Convert string variables to numeric variables and vice versa

[U] **12.2.2 Numeric storage types**

[U] **12.4.4 String storage types**

Title

> **recode** — Recode categorical variables

Syntax

Basic syntax

> recode *varlist* (*rule*) $\big[$(*rule*) ...$\big]$ $\big[$, g̲enerate(*newvar*)$\big]$

Full syntax

> recode *varlist* (*erule*) $\big[$(*erule*) ...$\big]$ $\big[$*if*$\big]$ $\big[$*in*$\big]$ $\big[$, *options*$\big]$

where the most common forms for *rule* are

rule	Example	Meaning
# = #	3 = 1	3 recoded to 1
# # = #	2 . = 9	2 and . recoded to 9
#/# = #	1/5 = 4	1 through 5 recoded to 4
n̲onmissing = #	nonmiss = 8	all other nonmissing to 8
m̲issing = #	miss = 9	all other missings to 9

where *erule* has the form

> *element* $\big[$*element* ...$\big]$ = *el* $\big[$"*label*"$\big]$
>
> n̲onmissing = *el* $\big[$"*label*"$\big]$
>
> m̲issing = *el* $\big[$"*label*"$\big]$
>
> else | * = *el* $\big[$"*label*"$\big]$

element has the form

> *el* | *el*/*el*

and *el* is

> # | min | max

The keyword rules missing, nonmissing, and else must be the last rules specified. else may not be combined with missing or nonmissing.

options	Description
Options	
g̲enerate(*newvar*)	generate *newvar* containing transformed variables; default is to replace existing variables
prefix(*str*)	generate new variables with *str* prefix
l̲abel(*name*)	specify a name for the value label defined by the transformation rules
copyrest	copy out-of-sample values from original variables
test	test that rules are invoked and do not overlap

Menu

Data > Create or change data > Other variable-transformation commands > Recode categorical variable

Description

recode changes the values of numeric variables according to the rules specified. Values that do not meet any of the conditions of the rules are left unchanged, unless an *otherwise* rule is specified.

A range *#1/#2* refers to *all* (real and integer) values between *#1* and *#2*, including the boundaries *#1* and *#2*. This interpretation of *#1/#2* differs from that in numlists.

min and max provide a convenient way to refer to the minimum and maximum for each variable in *varlist* and may be used in both the from-value and the to-value parts of the specification. Combined with if and in, the minimum and maximum are determined over the restricted dataset.

The keyword rules specify transformations for values not changed by the previous rules:

<u>non</u>missing	all nonmissing values not changed by the rules
<u>miss</u>ing	all missing values (., .a, .b, ..., .z) not changed by the rules
else	all nonmissing and missing values not changed by the rules
*	synonym for else

recode provides a convenient way to define value labels for the generated variables during the definition of the transformation, reducing the risk of inconsistencies between the definition and value labeling of variables. Value labels may be defined for integer values and for the extended missing values (.a, .b, ..., .z), but not for noninteger values or for sysmiss (.).

Although this is not shown in the syntax diagram, the parentheses around the *rule*s and keyword clauses are optional if you transform only one variable and if you do not define value labels.

Options

┌─ Options ┐

generate(*newvar*) specifies the names of the variables that will contain the transformed variables. into() is a synonym for generate(). Values outside the range implied by if or in are set to missing (.), unless the copyrest option is specified.

If generate() is not specified, the input variables are overwritten; values outside the if or in range are not modified. Overwriting variables is dangerous (you cannot undo changes, value labels may be wrong, etc.), so we strongly recommend specifying generate().

prefix(*str*) specifies that the recoded variables be returned in new variables formed by prefixing the names of the original variables with *str*.

label(*name*) specifies a name for the value label defined from the transformation rules. label() may be defined only with generate() (or its synonym, into()) and prefix(). If a variable is recoded, the label name defaults to *newvar* unless a label with that name already exists.

copyrest specifies that out-of-sample values be copied from the original variables. In line with other data-management commands, recode defaults to setting *newvar* to missing (.) outside the observations selected by if *exp* and in *range*.

test specifies that Stata test whether rules are ever invoked or that rules overlap; for example, (1/5=1) (3=2).

Remarks

Remarks are presented under the following headings:

> *Simple examples*
> *Setting up value labels with recode*
> *Referring to the minimum and maximum in rules*
> *Recoding missing values*
> *Recoding subsets of the data*
> *Otherwise rules*
> *Test for overlapping rules*

Simple examples

Many users experienced with other statistical software use the `recode` command often, but easier and faster solutions in Stata are available. On the other hand, `recode` often provides simple ways to manipulate variables that are not easily accomplished otherwise. Therefore, we show other ways to perform a series of tasks with and without `recode`.

We want to change 1 to 2, leave all other values unchanged, and store the results in the new variable nx.

```
. recode x (1 = 2), gen(nx)
```

or

```
. gen nx = x
. replace nx = 2 if nx==1
```

or

```
. gen nx = cond(x==1,2,x)
```

We want to swap 1 and 2, saving them in nx.

```
. recode x (1 = 2) (2 = 1), gen(nx)
```

or

```
. gen nx = cond(x==1,2,cond(x==2,1,x))
```

We want to recode item by collapsing 1 and 2 into 1, 3 into 2, and 4 to 7 (boundaries included) into 3.

```
. recode item (1 2 = 1) (3 = 2) (4/7 = 3), gen(Ritem)
```

or

```
. gen Ritem = item
. replace Ritem = 1 if inlist(item,1,2)
. replace Ritem = 2 if item==3
. replace Ritem = 3 if inrange(item,4,7)
```

We want to change the "direction" of the $1, \ldots, 5$ valued variables x1, x2, x3, storing the transformed variables in nx1, nx2, and nx3 (that is, we form new variable names by prefixing old variable names with an "n").

```
. recode x1 x2 x3 (1=5) (2=4) (3=3) (4=2) (5=1), pre(n) test
```

or

```
. gen nx1 = 6-x1
. gen nx2 = 6-x2
```

```
. gen nx3 = 6-x3
. forvalues i = 1/3 {
        generate nx‘i’ = 6-x‘i’
  }
```

In the categorical variable `religion`, we want to change 1, 3, and the real and integer numbers 3 through 5 into 6; we want to set 2, 8, and 10 to 3 and leave all other values unchanged.

```
. recode religion 1 3/5 = 6 2 8 10 = 3
```

or

```
. replace religion = 6 if religion==1 | inrange(religion,3,5)
. replace religion = 3 if inlist(religion,2,8,10)
```

This example illustrates two features of `recode` that were included for backward compatibility with previous versions of `recode` but that we do not recommend. First, we omitted the parentheses around the rules. This is allowed if you recode one variable and you do not plan to define value labels with `recode` (see below for an explanation of this feature). Personally, we find the syntax without parentheses hard to read, although we admit that we could have used blanks more sensibly. Because difficulties in reading may cause us to overlook errors, we recommend always including parentheses. Second, because we did not specify a `generate()` option, we overwrite the `religion` variable. This is often dangerous, especially for "original" variables in a dataset. We recommend that you always specify `generate()` unless you want to overwrite your data.

Setting up value labels with recode

The `recode` command is most often used to transform categorical variables, which are many times value labeled. When a value-labeled variable is overwritten by `recode`, it may well be that the value label is no longer appropriate. Consequently, output that is labeled using these value labels may be misleading or wrong.

When `recode` creates one or more new variables with a new classification, you may want to put value labels on these new variables. It is possible to do this in three steps:

1. Create the new variables (`recode` ... , `gen()`).

2. Define the value label (`label define` ...).

3. Link the value label to the variables (`label value` ...).

Inconsistencies may emerge from mistakes between steps 1 and 2. Especially when you make a change to the recode 1, it is easy to forget to make a similar adjustment to the value label 2. Therefore, `recode` can perform steps 2 and 3 itself.

Consider recoding a series of items with values

> 1 = strongly agree
> 2 = agree
> 3 = neutral
> 4 = disagree
> 5 = strongly disagree

into three items:

> 1 = positive (= "strongly agree" or "agree")
> 2 = neutral
> 3 = negative (= "strongly disagree" or "disagree")

This is accomplished by typing

```
. recode item* (1 2 = 1 positive) (3 = 2 neutral) (4 5 = 3 negative), pre(R)
> label(Item3)
```

which is much simpler and safer than

```
. recode item1-item7 (1 2 = 1) (3 = 2) (4 5 = 3), pre(R)
. label define Item3  1 positive 2 neutral 3 negative
. forvalues i = 1/7 {
        label value Ritem`i' Item3
   }
```

▷ Example 1

As another example, let's recode vote (voting intentions) for 12 political parties in the Dutch parliament into left, center, and right parties. We then tabulate the original and new variables so that we can check that everything came out correctly.

```
. use http://www.stata-press.com/data/r12/recodexmpl
. label list pparty
pparty:
           1 pvda
           2 cda
           3 d66
           4 vvd
           5 groenlinks
           6 sgp
           7 rpf
           8 gpv
           9 aov
          10 unie55
          11 sp
          12 cd
. recode polpref (1 5 11 = 1 left) (2 3 = 2 center) (4 6/10 12 = 3 right),
> gen(polpref3)
(2020 differences between polpref and polpref3)
. tabulate polpref polpref3
```

pol party choice if elections	RECODE of polpref (pol party choice if elections)			Total
	left	center	right	
pvda	622	0	0	622
cda	0	525	0	525
d66	0	634	0	634
vvd	0	0	930	930
groenlinks	199	0	0	199
sgp	0	0	54	54
rpf	0	0	63	63
gpv	0	0	30	30
aov	0	0	17	17
unie55	0	0	23	23
sp	45	0	0	45
cd	0	0	25	25
Total	866	1,159	1,142	3,167

◁

Referring to the minimum and maximum in rules

recode allows you to refer to the minimum and maximum of a variable in the transformation rules. The keywords min and max may be included as a from-value, as well as a to-value.

For example, we might divide age into age categories, storing in iage.

```
. recode age (0/9=1) (10/19=2) (20/29=3) (30/39=4) (40/49=5) (50/max=6),
> gen(iage)
```

or

```
. gen iage = 1 + irecode(age,9,19,29,39,49)
```

or

```
. gen iage = min(6, 1+int(age/10))
```

As another example, we could set all incomes less than 10,000 to 10,000 and those more than 200,000 to 200,000, storing the data in ninc.

```
. recode inc (min/10000 = 10000) (200000/max = 200000), gen(ninc)
```

or

```
. gen ninc = inc
. replace ninc = 10000 if ninc<10000
. replace ninc = 200000 if ninc>200000 & !missing(ninc)
```

or

```
. gen ninc = max(min(inc,200000),10000)
```

or

```
. gen ninc = clip(inc,10000,200000)
```

Recoding missing values

You can also set up rules in terms of missing values, either as from-values or as to-values. Here recode mimics the functionality of mvdecode and mvencode (see [D] **mvencode**), although these specialized commands execute much faster.

Say that we want to change missing (.) to 9, storing the data in X:

```
. recode x (.=9), gen(X)
```

or

```
. gen X = cond(x==., 9, x)
```

or

```
. mvencode x, mv(.=9) gen(X)
```

We want to change 9 to .a and 8 to ., storing the data in z.

```
. recode x (9=.a) (8=.), gen(z)
```

or

```
. gen z = cond(x==9, .a, cond(x==8, ., x))
```

or

```
. mvdecode x, mv(9=.a, 8=.) gen(z)
```

Recoding subsets of the data

We want to swap in x the values 1 and 2 only for those observations for which age>40, leaving all other values unchanged. We issue the command

```
. recode x (1=2) (2=1) if age>40, gen(y)
```

or

```
. gen y = cond(x==1,2,cond(x==2,1,x)) if age>40
```

We are in for a surprise. y is missing for observations that do not satisfy the if condition. This outcome is in accordance with how Stata's data-manipulation commands usually work. However, it may not be what you intend. The copyrest option specifies that x be copied into y for all nonselected observations:

```
. recode x (1=2) (2=1) if age>40, gen(y) copy
```

or

```
. gen y = x
. recode y (1=2) (2=1) if age>40
```

or

```
. gen y = cond(age>40,cond(x==1,2,cond(x==2,1,x)),x)
```

Otherwise rules

In all our examples so far, recode had an implicit rule that specified that values that did not meet the conditions of any of the rules were to be left unchanged. recode also allows you to use an "otherwise rule" to specify how untransformed values are to be transformed. recode supports three kinds of otherwise conditions:

nonmissing	all nonmissing not yet transformed
missing	all missing values not yet transformed
else	all values, missing or nonmissing, not yet transformed

The otherwise rules are to be specified *after* the standard transformation rules. nonmissing and missing may be combined with each other, but not with else.

Consider a recode that swaps the values 1 and 2, transforms all other nonmissing values to 3, and transforms all missing values (that is, sysmiss and the extended missing values) to . (sysmiss). We could type

```
. recode x (1=2) (2=1) (nonmissing=3) (missing=.), gen(z)
```

or

```
. gen z = cond(x==1,2,cond(x==2,1,cond(!missing(x),3),.))
```

As a variation, if we had decided to recode all extended missing values to .a but to keep sysmiss . distinct at ., we could have typed

```
. recode x (1=2) (2=1) (.=.) (nonmissing=3) (missing=.a), gen(z)
```

Test for overlapping rules

recode evaluates the rules from left to right. Once a value has been transformed, it will not be transformed again. Thus if rules "overlap", the first matching rule is applied, and further matches are ignored. A common form of overlapping is illustrated in the following example:

>   ... (1/5 = 1) (5/10 = 2)

Here 5 occurs in the condition parts of both rules. Because rules are matched left to right, 5 matches the first rule, and the second rule will not be tested for 5, unless recode is instructed to test for rule overlap with the test option.

Other instances of overlapping rules usually arise because you mistyped the rules. For instance, you are recoding voting intentions for parties in elections into three groups of parties (left, center, right), and you type

>   ... (1/5 = 1) ... (3 = 2)

Party 3 matches the conditions 1/5 and 3. Because recode applies the first matching rule, party 3 will be mapped into party category 1. The second matching rule is ignored. It is not clear what was wrong in this example. You may have included party 3 in the range 1/5 or mistyped 3 in the second rule. Either way, recode did not notice the problem and your data analysis is in jeopardy. The test option specifies that recode display a warning message if values are matched by more than one rule. With the test option specified, recode also tests whether all rules were applied at least once and displays a warning message otherwise. Rules that never matched any data may indicate that you mistyped a rule, although some conditions may not have applied to (a selection of) your data.

Methods and formulas

recode is implemented as an ado-file.

Acknowledgment

This version of recode was written by Jeroen Weesie, Department of Sociology, Utrecht University, The Netherlands.

Also see

[D] **generate** — Create or change contents of variable

[D] **mvencode** — Change missing values to numeric values and vice versa

Title

Syntax

<u>ren</u>ame *old_varname new_varname*

Menu

Data > Data utilities > Rename variables

Description

rename changes the name of existing variable *old_varname* to *new_varname*; the contents of the variable are unchanged. Also see [D] **rename group** for renaming groups of variables.

Remarks

▷ Example 1

rename allows you to change variable names. Say that we have labor market data for siblings.

```
. use http://www.stata-press.com/data/r12/renamexmpl
. describe
Contains data from http://www.stata-press.com/data/r12/renamexmpl.dta
  obs:           277
  vars:            6                          9 Jan 2011 11:57
  size:        6,648
```

variable name	storage type	display format	value label	variable label
famid	float	%9.0g		
edu	float	%9.0g		
exp	float	%9.0g		
promo	float	%9.0g		
sex	float	%9.0g	sex	
inc	float	%9.0g		

```
Sorted by:  famid
```

We decide to rename the exp and inc variables.

```
. rename exp experience
. rename inc income
. describe
Contains data from http://www.stata-press.com/data/r12/renamexmpl.dta
  obs:           277
 vars:             6                          9 Jan 2011 11:57
 size:         6,648
```

variable name	storage type	display format	value label	variable label
famid	float	%9.0g		
edu	float	%9.0g		
experience	float	%9.0g		
promo	float	%9.0g		
sex	float	%9.0g	sex	
income	float	%9.0g		

```
Sorted by:  famid
     Note:  dataset has changed since last saved
```

The exp variable is now called experience, and the inc variable is now called income.

◁

References

Cox, N. J., and J. Weesie. 2001. dm88: Renaming variables, multiply and systematically. *Stata Technical Bulletin* 60: 4–6. Reprinted in *Stata Technical Bulletin Reprints*, vol. 10, pp. 41–44. College Station, TX: Stata Press.

Jenkins, S. P., and N. J. Cox. 2001. dm83: Renaming variables: Changing suffixes. *Stata Technical Bulletin* 59: 5–6. Reprinted in *Stata Technical Bulletin Reprints*, vol. 10, pp. 34–35. College Station, TX: Stata Press.

Also see

[D] **rename group** — Rename groups of variables

[D] **generate** — Create or change contents of variable

[D] **varmanage** — Manage variable labels, formats, and other properties

Title

---

rename group — Rename groups of variables

Syntax

Rename a single variable

    <u>ren</u>ame *old new* $\bigl[$, *options*$_1$ $\bigr]$

Rename groups of variables

    <u>ren</u>ame (*old*$_1$ *old*$_2$ ...) (*new*$_1$ *new*$_2$ ...) $\bigl[$, *options*$_1$ $\bigr]$

Change the case of groups of variable names

    <u>ren</u>ame *old*$_1$ *old*$_2$ ..., { <u>u</u>pper | <u>l</u>ower | <u>p</u>roper } $\bigl[$ *options*$_2$ $\bigr]$

where *old* and *new* specify the existing and the new variable names. The rules for specifying them
are

1. rename stat status: Renames stat to status.

   Rule 1: This is the same rename command documented in [D] **rename**, with which
   you are familiar.

2. rename (stat inc) (status income): Renames stat to status and inc to income.

   Rule 2: Use parentheses to specify multiple variables for *old* and *new*.

3. rename (v1 v2) (v2 v1): Swaps v1 and v2.

   Rule 3: Variable names may be interchanged.

4. rename (a b c) (b c a): Swaps names. Renames a to b, b to c, and c to a.

   Rule 4: There is no limit to how many names may be interchanged.

5. rename (a b c) (c b a): Renames a to c and c to a, but leaves b as is.

   Rule 5: Renaming variables to themselves is allowed.

6. rename jan* *1: Renames all variables starting with jan to instead end with 1, for example,
   janstat to stat1, janinc to inc1, etc.

   Rule 6.1: * in *old* selects the variables to be renamed. * means that zero or more characters
   go here.

   Rule 6.2: * in *new* corresponds with * in *old* and stands for the text that * in *old* matched.

   * in *new* or *old* is called a wildcard character, or just a wildcard.

   rename jan* *: Removes prefix jan.

   rename *jan *: Removes suffix jan.

7. rename jan? ?1: Renames all variables starting with jan and ending in one character by
   removing jan and adding 1 to the end; for example, jans is renamed to s1, but janstat
   remains unchanged. ? means that exactly one character goes here, just as * means that zero
   or more characters go here.

Rule 7: ? means exactly one character, ?? means exactly two characters, etc.

8. rename *jan* **: Removes prefix, midfix, and suffix jan, for example, janstat to stat, injanstat to instat, and subjan to sub.

> Rule 8: You may specify more than one wildcard in *old* and in *new*. They correspond in the order given.

rename jan*s* *s*1: Renames all variables that start with jan and contain s to instead end in 1, dropping the jan, for example, janstat to stat1 and janest to est1, but not janinc to inc1.

9. rename *jan* *: Removes jan and whatever follows from variable names, thereby renaming statjan to stat, incjan71 to inc, ...

> Rule 9: You may specify more wildcards in *old* than in *new*.

10. rename *jan* .*: Removes jan and whatever precedes it from variable names, thereby renaming midjaninc to inc, ...

> Rule 10: Wildcard . (dot) in *new* skips over the corresponding wildcard in *old*.

11. rename *pop jan=: Adds prefix jan to all variables ending in pop, for example, age1pop to janage1pop, ...

rename (status bp time) admit=: Renames status to admitstatus, bp to admitbp, and time to admittime.

rename *whatever* pre=: Adds prefix pre to all variables selected by *whatever*, however *whatever* is specified.

> Rule 11: Wildcard = in *new* specifies the original variable name.

rename *whatever* =jan: Adds suffix jan to all variables selected by *whatever*.

rename *whatever* pre=fix: Adds prefix pre and suffix fix to all variables selected by *whatever*.

12. rename v# stat#: Renames v1 to stat1, v2 to stat2, ..., v10 to stat10, ...

> Rule 12.1: # is like * but for digits. # in *old* selects one or more digits.

> Rule 12.2: # in *new* copies the digits just as they appear in the corresponding *old*.

13. rename v(#) stat(#): Renames v1 to stat1, v2 to stat2, ..., but does not rename v10, ...

> Rule 13.1: (#) in *old* selects exactly one digit. Similarly, (##) selects exactly two digits, and so on, up to ten # symbols.

> Rule 13.2: (#) in *new* means reformat to one or more digits. Similarly, (##) reformats to two or more digits, and so on, up to ten # symbols.

rename v(##) stat(##): Renames v01 to stat01, v02 to stat02, ..., v10 to stat10, ..., but does not rename v0, v1, v2, ..., v9, v100, ...

14. rename v# v(##): Renames v1 to v01, v2 to v02, ..., v10 to v10, v11 to v11, ..., v100 to v100, v101 to v101, ...

> Rule 14: You may combine #, (#), (##), ... in *old* with any of #, (#), (##), ... in *new*.

rename v(##) v(#): Renames v01 to v1, v02 to v2, ..., v10 to v10, ..., but does not rename v001, etc.

rename stat(##) stat_20(##): Renames stat10 to stat_2010, stat11 to stat_2011, ..., but does not rename stat1, stat2, ...

rename stat(#) to stat_200(#): Renames stat1 to stat_2001, stat2 to stat_2002, ..., but does not rename stat10 or stat_2010.

15. rename v# (a b c): Renames v1 to a, v10 to b, and v2 to c if variables v1, v10, v2 appear in that order in the data. Because three variables were specified in *new*, v# in *old* must select three variables or rename will issue an error.

   Rule 15.1: You may mix syntaxes. Note that the explicit and implied numbers of variables must agree.

   rename v# (a b c), sort: Renames (for instance) v1 to a, v2 to b, and v10 to c.

   Rule 15.2: The sort option places the variables selected by *old* in order and does so smartly. In the case where #, (#), (##), ... appear in *old*, sort places the variables in numeric order.

   rename v* (a b c), sort: Renames (for instance) valpha to a, vbeta to b, and vgamma to c regardless of the order of the variables in the data.

   Rule 15.3: In the case where * or ? appears in *old*, sort places the variables in alphabetical order.

16. rename v# v#, renumber: Renames (for instance) v9 to v1, v10 to v2, v8 to v3, ..., assuming that variables v9, v10, v8, ... appear in that order in the data.

   Rule 16.1: The renumber option resequences the numbers.

   rename v# v#, renumber sort: Renames (for instance) v8 to v1, v9 to v2, v10 to v3, ... Concerning option sort, see rule 15.2 above.

   rename v# v#, renumber(10) sort: Renames (for instance) v8 to v10, v9 to v11, v10 to v12, ...

   Rule 16.2: The renumber(#) option allows you to specify the starting value.

17. rename v* v#, renumber: Renames (for instance) valpha to v1, vgamma to v2, vbeta to v3, ..., assuming variables valpha, vgamma, vbeta, ... appear in that order in the data.

   Rule 17: # in *new* may correspond to *, ?, #, (#), (##), ... in *old*.

   rename v* v#, renumber sort: Renames (for instance) valpha to v1, vbeta to v2, vgamma to v3, ... Also see rule 15.3 above concerning the sort option.

   rename *stat stat#, renumber: Renames, for instance, janstat to stat1, febstat to stat2, ... Note that # in *new* corresponds to * in *old*, just as in the previous example.

   rename *stat stat(##), renumber: Renames, for instance, janstat to stat01, febstat to stat02, ...

   rename *stat stat#, renumber(0): Renames, for instance, janstat to stat0, febstat to stat1, ...

   rename *stat stat#, renumber sort: Renames, for instance, aprstat to stat1, augstat to stat2, ...

18. rename (a b c) v#, addnumber: Renames a to v1, b to v2, and c to v3.

   Rule 18: The addnumber option allows you to add numbering. More formally, if you specify addnumber, you may specify one more wildcard in *new* than is specified in *old*, and that extra wildcard must be #, (#), (##), ...

19. rename a(#)(#) a(#)[2](#)[1]: Renames a12 to a21, a13 to a31, a14 to a41, ...,
    a21 to a12, ...

> Rule 19.1: You may specify explicit subscripts with wildcards in *new* to make explicit its
> matching wildcard in *old*. Subscripts are specified in square brackets after a
> wildcard in *new*. The number refers to the number of the wildcard in *old*.

rename *stat* *[2]stat*[1]: Swaps prefixes and suffixes; it renames bpstata to
astatbp, rstater to erstatr, etc.

rename *stat* *[2]stat*: Does the same as above; it swaps prefixes and suffixes.

> Rule 19.2: After specifying a subscripted wildcard, subsequent unsubscripted wildcards
> correspond to the same wildcards in *old* as they would if you had removed the
> subscripted wildcards altogether.

rename v#a# v#_#[1]_a#[2]: Renames v1a1 to v1_1_a1, v1a2 to v1_1_a2, ..., v2a1
to v2_2_a1, ...

> Rule 19.3: Using subscripts, you may refer to the same wildcard in *old* more than once.

Subscripts are commonly used to interchange suffixes at the ends of variable names. For
instance, you have districts and schools within them, and many of the variable names in your
data match *_#_#. The first number records district and the second records school within
district. To reverse the ordering, you type rename *_#_# *_#[3]_#[2]. When specifying
subscripts, you refer to them by the position number in the original name. For example, our
original name was *_#_# so [1] refers to *, [2] refers to the first #, and [3] refers to the
last #.

In summary, the pattern specifiers are

Specifier	Meaning in *old*
*	0 or more characters
?	1 character exactly
#	1 or more digits
(#)	1 digit exactly
(##)	2 digits exactly
(###)	3 digits exactly
...	
(##########)	10 digits exactly

Specifier	May correspond in *old* with	Meaning in *new*
*	*, ?, #, (#), ...	copies matched text
?	?	copies a character
#	#, (#), ...	copies a number as is
(#)	#, (#), ...	reformats to 1 or more digits
(##)	#, (#), ...	reformats to 2 or more digits
...		
(##########)	#, (#), ...	reformats to 10 digits
.	*, ?, #, (#), ...	skip
=	*nothing*	copies entire variable name

Specifier # in any of its guises may also correspond with * or ? if the `renumber` option is specified.

The options are as follows:

options₁	Description
<u>add</u>number	add sequential numbering to end
<u>add</u>number(#)	addnumber, starting at #
<u>renum</u>ber	renumber sequentially
<u>renum</u>ber(#)	renumber, starting at #
<u>s</u>ort	sort before numbering
<u>dry</u>run	do not rename, but instead produce a report
r	save variable names in r() for programming use

These options correspond to the first and second syntaxes.

options₂	Description
<u>u</u>pper	uppercase variable names (UPPERCASE)
<u>l</u>ower	lowercase variable names (lowercase)
<u>p</u>roper	propercase variable names (Propercase)
<u>dry</u>run	do not rename, but instead produce a report
r	save variable names in r() for programming use

These options correspond to the third syntax. One of `upper`, `lower`, or `proper` must be specified.

Menu

Data > Data utilities > Rename variables

Description

`rename` changes the names of existing variables to the new names specified. See [D] **rename** for the base `rename` syntax. Documented here is the advanced syntax for renaming groups of variables.

Options for renaming variables

addnumber and addnumber(#) specify to add a sequence number to the variable names. See item 18 of *Syntax*. If # is not specified, the sequence number begins with 1.

renumber and renumber(#) specify to replace existing numbers or text in a set of variable names with a sequence number. See items 16 and 17 of *Syntax*. If # is not specified, the sequence number begins with 1.

sort specifies that the existing names be placed in order before the renaming is performed. See item 15 of *Syntax* for details. This ordering matters only when addnumber or renumber is also specified or when specifying a list of variable names for *old* or *new*.

dryrun specifies that the requested renaming not be performed but instead that a table be displayed showing the old and new variable names. It is often a good idea to specify this option before actually renaming the variables.

r is a programmer's option that requests that old and new variable names be saved in r(). This option may be specified with or without dryrun.

Options for changing the case of groups of variable names

upper, lower, and proper specify how the variables are to be renamed. upper specifies that variable names be changed to uppercase; lower, to lowercase; and proper, to having the first letter capitalized and the remaining letters in lowercase. One of these three options must be specified.

dryrun and r are the same options as documented directly above.

Remarks

Remarks are presented under the following headings:

> *Advice*
> *Explanation*
> ** matches 0 or more characters; use ?* to match 1 or more*
> ** is greedy*
> *# is greedier*

Advice

1. Read [D] **rename** before reading this entry.

2. Read items 1–19 (the Rules) under *Syntax* above before reading the rest of these remarks.

3. Specify the dryrun option when using complicated patterns. dryrun presents a table of the old and new variable names rather than actually renaming the variables, so you can check that the patterns you have specified produce the desired result.

Explanation

The rename command has three syntaxes; see *Syntax*. See [D] **rename** for details on the first syntax, renaming a single variable. The remaining two syntaxes are for renaming groups of variables and for changing the case of groups of variables. These two syntaxes are the ones we will focus on for the remainder of this manual entry. Here they are again:

rename $(old_1\ old_2\ \ldots)$ $(new_1\ new_2\ \ldots)$

rename $old_1\ old_2\ \ldots$, { upper | lower | proper }

The second syntax shown above merely changes the case of variables, such as MPG or mpg or Mpg. For instance, to rename all variables to be lowercase, type

    rename *, lower

The first syntax shown above is more daunting and more powerful. The first syntax has two styles, with and without parentheses:

    rename (bp_0 bp_1) (bp_1 bp_0)

    rename pop*80 pop_*_1980

You can combine the two styles whenever it is convenient.

    rename v* (mpg weight displacement)

    rename (mpg weight displacement) v#, addnumber

    rename (bp_0 bp_1 pop*80) (bp_1 bp_0 pop_*_1980)

We summarize all of this by simply writing the syntax as

    rename *old new*, ...

and referring to *old* and *new*.

Wildcards play different but related roles in *old* and *new*. When you type

    rename pop*80 pop_*_1980

the wildcard (* in this case) in *old* specifies which variables are to be renamed, and in *new* the wildcard stands for the text that appears in the variables to be renamed. In this case, there is just one wildcard, but sometimes there are more.

In *old*, * means zero or more characters go here. Specifying pop*80 means find all variables that begin with pop and end in 80. Say that doing so results in three variables being found: poplt2080, pop204080, and pop41plus80. To understand how * is interpreted in *new*, it is useful to write the three found variables like this:

pop*80	=	pop	+	*	+	80
poplt2080	=	pop	+	lt20	+	80
pop204080	=	pop	+	2040	+	80
pop41plus80	=	pop	+	41plus	+	80

* in *new* refers to what was found by * in *old*. So the new pattern pop_*_1980 will assemble the following new variable names for each of the old names:

old variable	* is	$\rightarrow$	pop_*_1980 is
poplt2080	lt20	$\rightarrow$	pop_lt20_1980
pop204080	2040	$\rightarrow$	pop_2040_1980
pop41plus80	41plus	$\rightarrow$	pop_41plus_1980

Thus typing `rename pop*80 pop_*_1980` is equivalent to typing

```
rename poplt2080    pop_lt20_1980
rename pop204080    pop_2040_1980
rename pop41plus80 pop_41plus_1980
```

There are three basic wildcard characters for specification in *old*, and they filter the variables to be renamed:

      *   0 or more characters go here
      ?   exactly 1 character goes here
      #   number goes here (this one comes in 11 flavors!)

The generic # listed above collects all the digits. The other 10 flavors are (#), which means exactly 1 digit goes here; (##), which means exactly 2 digits go here; and so on, up to exactly 10 digits go here.

All the above, the $3 + 10 = 13$ wildcard characters, can appear in *new*, where each has a different but related meaning:

      *   copy corresponding text from *old* as is
      ?   copy corresponding character from *old*
      #   copy corresponding number from *old* as is
     (#)   reformat corresponding number from *old* to 1 or more digits
    (##)   reformat corresponding number from *old* to 2 or more digits
      . . .

In addition, *new* allows two special wildcard characters of its own:

     =   copy the entire original variable name
     .   skip the corresponding text in *old*

With the above information and the definitions of the options, you can derive on your own the first eighteen rules given in *Syntax*. The nineteenth rule concerns subscripting. In *new*, you can specify explicitly to which wildcard in *old* you are referring. You can type

```
rename pop*80 pop_*_1980
```

or you can type

```
rename pop*80 pop_*[1]_1980
```

thus making it explicit that the * in *new* is referring to the text matched by the first wildcard in *old*. That * corresponds to * is hardly surprising, especially when there is only one * in *old*, so let's complicate the example:

```
rename v*_* outcome_*_*
```

You can type that command, or you can type

```
rename v*_* outcome_*[1]_*[2]
```

More importantly, you can specify the subscripts in whatever order you wish, so you could type

```
rename v*_* outcome_*[2]_*[1]
```

That command would interchange the text in *old* matched by the two wildcards.

* matches 0 or more characters; use ?* to match 1 or more

l*a in *old* matches louisiana and it matches la because * means zero or more characters. What if you want to match louisiana and lymphoma but not la?

For instance, say you have from–to variables named from*to* and from variables named from*. The problem is that variable fromtoledo would match from*to*. To avoid that, rather than describing the from–to pattern from*to*, you use from?*to?*. Thus you could type

       rename from?*to?* from_?*_to_?*

?* is not a secret wildcard we have yet to tell you about—it is merely the two wildcards ? and * in sequence. ? means exactly one character goes here, and * means zero or more characters go here, so ?* means one or more characters go here. In the same way, ??* means two or more characters go here, and so on.

* is greedy

Consider the existing variable assessment and pattern *s* in *old*. Clearly, *s* matches assessment, but how? That is, among these possibilities,

$$
\begin{array}{ccccc}
\texttt{assessment} & = & * & \texttt{s} & * \\
\hline
 & & \texttt{a} & + \texttt{s} + & \texttt{sessment} \\
 & & \texttt{as} & + \texttt{s} + & \texttt{essment} \\
 & & \texttt{asse} & + \texttt{s} + & \texttt{sment} \\
 & & \texttt{asses} & + \texttt{s} + & \texttt{ment}
\end{array}
$$

which one is true? We need to know the answer to know what each of the corresponding wildcards in *new* will mean. The answer is that * is greedy, and the pattern is matched from left to right. As we move through the variable name from left to right, at each step * takes the most characters possible, subject to the pattern working out.

$$
\begin{array}{ccccc}
 & & * & \texttt{s} & * \\
\hline
\texttt{assessment} & = & \texttt{asses} & + \texttt{s} + & \texttt{ment}
\end{array}
$$

Thus the first * in *new* would stand for asses and the second would stand for ment.

The "subject to the pattern working out" part is important. Variable sunglasses would be broken out by *s* as

$$
\begin{array}{ccccc}
 & & * & \texttt{s} & * \\
\hline
\texttt{sunglasses} & = & \texttt{sunglasse} & + \texttt{s} + & \textit{nothing}
\end{array}
$$

But by *s?*, the breakout would be

$$
\begin{array}{cccccc}
 & & * & \texttt{s} & ? & * \\
\hline
\texttt{sunglasses} & = & \texttt{sunglas} & + \texttt{s} + & \texttt{e} + & \texttt{s}
\end{array}
$$

is greedier

Wildcard # in *old* is greedier than *, which means that when * and # are up against each other, # wins.

Consider the pattern *# and the variable name v1234. Given that * is greedy and that the # specifies one or more digits, the possible solutions are

$$v1234 \;=\; \frac{*}{\begin{array}{ll} v123 + & 4 \\ v12 \;+ & 34 \\ v1 \;\;+ & 234 \\ v \;\;\;\;+ & 1234 \end{array}}$$

The solution chosen by rename is the last one, v + 1234. Thus you can type

```
rename *# period_#[2]
```

without concern that some digits might be lost.

Saved results

rename saves nothing in r() by default. If the r option is specified, then rename saves the following in r():

Scalar
    r(n)   number of variables to be renamed
Macros
    r(oldnames)   original variable names
    r(newnames)   new variable names

Variables that are renamed to themselves are omitted from the recorded lists.

Methods and formulas

rename is implemented as an ado-file that uses Mata.

Also see

[D] **rename** — Rename variable

[D] **generate** — Create or change contents of variable

[D] **varmanage** — Manage variable labels, formats, and other properties

Title

reshape — Convert data from wide to long form and vice versa

Syntax

Overview

long		
i	*j*	*stub*
1	**1**	4.1
1	**2**	4.5
2	**1**	3.3
2	**2**	3.0

wide		
i	*stub*1	*stub*2
1	4.1	4.5
2	3.3	3.0

reshape

To go from long to wide:

j existing variable

reshape wide *stub*, i(*i*) j(*j*)

To go from wide to long:

reshape long *stub*, i(*i*) j(*j*)

j new variable

To go back to long after using reshape wide:

reshape long

To go back to wide after using reshape long:

reshape wide

Basic syntax

Convert data from wide form to long form

reshape long *stubnames* , i(*varlist*) [*options*]

Convert data from long form to wide form

reshape wide *stubnames* , i(*varlist*) [*options*]

Convert data back to long form after using reshape wide

reshape long

Convert data back to wide form after using reshape long

reshape wide

List problem observations when reshape fails

reshape error

options	Description
* i(*varlist*)	use *varlist* as the ID variables
j(*varname* [*values*])	long→wide: *varname*, existing variable
	wide→long: *varname*, new variable
	optionally specify values to subset *varname*
<u>string</u>	*varname* is a string variable (default is numeric)

* i(*varlist*) is required.

where *values* is $\#\left[-\#\right]\left[\#\ \dots\right]$ if *varname* is numeric (default)

"*string*" ["*string*" ...] if *varname* is string

and where *stubnames* are variable names (long→wide), or stubs of variable names (wide→long), and either way, may contain @, denoting where *j* appears or is to appear in the name.

Advanced syntax

   reshape i *varlist*

   reshape j *varname* [*values*] [, <u>string</u>]

   reshape xij *fvarnames* [, <u>atw</u>l(*chars*)]

   reshape xi [*varlist*]

   reshape [<u>query</u>]

   reshape clear

Menu

Data > Create or change data > Other variable-transformation commands > Convert data between wide and long

Description

reshape converts data from *wide* to *long* form and vice versa.

Options

i(*varlist*) specifies the variables whose unique values denote a logical observation. i() is required.

j(*varname* [*values*]) specifies the variable whose unique values denote a subobservation. *values* lists the unique values to be used from *varname*, which typically are not explicitly stated because reshape will determine them automatically from the data.

string specifies that j() may contain string values.

atwl(*chars*), available only with the advanced syntax and not shown in the dialog box, specifies that *chars* be substituted for the @ character when converting the data from wide to long form.

Remarks

Remarks are presented under the following headings:

> Description of basic syntax
> Wide and long data forms
> Avoiding and correcting mistakes
> reshape long and reshape wide without arguments
> Missing variables
> Advanced issues with basic syntax: i()
> Advanced issues with basic syntax: j()
> Advanced issues with basic syntax: xij
> Advanced issues with basic syntax: String identifiers for j()
> Advanced issues with basic syntax: Second-level nesting
> Description of advanced syntax

See Mitchell (2010, chap. 8) for information and examples using `reshape`.

Description of basic syntax

Before using `reshape`, you need to determine whether the data are in long or wide form. You also must determine the logical observation (i) and the subobservation (j) by which to organize the data. Suppose that you had the following data, which could be organized in wide or long form as follows:

i		$\ldots\ldots X_{ij} \ldots\ldots$			i	j		X_{ij}
id	sex	inc80	inc81	inc82	id	year	sex	inc
1	0	5000	5500	6000	1	80	0	5000
2	1	2000	2200	3300	1	81	0	5500
3	0	3000	2000	1000	1	82	0	6000
					2	80	1	2000
					2	81	1	2200
					2	82	1	3300
					3	80	0	3000
					3	81	0	2000
					3	82	0	1000

Given these data, you could use `reshape` to convert from one form to the other:

```
. reshape long inc, i(id) j(year)        /* goes from left form to right */
. reshape wide inc, i(id) j(year)        /* goes from right form to left */
```

Because we did not specify `sex` in the command, Stata assumes that it is constant within the logical observation, here `id`.

Wide and long data forms

Think of the data as a collection of observations X_{ij}, where i is the logical observation, or group identifier, and j is the subobservation, or within-group identifier.

Wide-form data are organized by logical observation, storing all the data on a particular observation in one row. Long-form data are organized by subobservation, storing the data in multiple rows.

> ## Example 1

For example, we might have data on a person's ID, gender, and annual income over the years 1980–1982. We have two X_{ij} variables with the data in wide form:

```
. use http://www.stata-press.com/data/r12/reshape1
. list
```

	id	sex	inc80	inc81	inc82	ue80	ue81	ue82
1.	1	0	5000	5500	6000	0	1	0
2.	2	1	2000	2200	3300	1	0	0
3.	3	0	3000	2000	1000	0	0	1

To convert these data to the long form, we type

```
. reshape long inc ue, i(id) j(year)
(note: j = 80 81 82)
```

Data	wide	->	long
Number of obs.	3	->	9
Number of variables	8	->	5
j variable (3 values)		->	year
xij variables:			
inc80 inc81 inc82		->	inc
ue80 ue81 ue82		->	ue

There is no variable named `year` in our original, wide-form dataset. `year` will be a new variable in our long dataset. After this conversion, we have

```
. list, sep(3)
```

	id	year	sex	inc	ue
1.	1	80	0	5000	0
2.	1	81	0	5500	1
3.	1	82	0	6000	0
4.	2	80	1	2000	1
5.	2	81	1	2200	0
6.	2	82	1	3300	0
7.	3	80	0	3000	0
8.	3	81	0	2000	0
9.	3	82	0	1000	1

We can return to our original, wide-form dataset by using `reshape wide`.

```
. reshape wide inc ue, i(id) j(year)
(note:  j = 80 81 82)
```

Data	long	->	wide
Number of obs.	9	->	3
Number of variables	5	->	8
j variable (3 values)	year	->	(dropped)
xij variables:			
inc	inc	->	inc80 inc81 inc82
ue	ue	->	ue80 ue81 ue82

. list

	id	inc80	ue80	inc81	ue81	inc82	ue82	sex
1.	1	5000	0	5500	1	6000	0	0
2.	2	2000	1	2200	0	3300	0	1
3.	3	3000	0	2000	0	1000	1	0

Converting from wide to long creates the j (year) variable. Converting back from long to wide drops the j (year) variable.

◁

❑ Technical note

If your data are in wide form and you do not have a group identifier variable (the i(*varlist*) required option), you can create one easily by using generate; see [D] **generate**. For instance, in the last example, if we did not have the id variable in our dataset, we could have created it by typing

. generate id = _n

❑

Avoiding and correcting mistakes

reshape often detects when the data are not suitable for reshaping; an error is issued, and the data remain unchanged.

▷ Example 2

The following wide data contain a mistake:

. use http://www.stata-press.com/data/r12/reshape2, clear
. list

	id	sex	inc80	inc81	inc82
1.	1	0	5000	5500	6000
2.	2	1	2000	2200	3300
3.	3	0	3000	2000	1000
4.	2	0	2400	2500	2400

. reshape long inc, i(id) j(year)
(note: j = 80 81 82)
i=id does not uniquely identify the observations;
there are multiple observations with the same value of id.
Type "reshape error" for a listing of the problem observations.
r(9);

The i variable must be unique when the data are in the wide form; we typed i(id), yet we have 2 observations for which id is 2. (Is person 2 a male or female?)

◁

▷ Example 3

It is not a mistake when the i variable is repeated when the data are in long form, but the following data have a similar mistake:

```
. use http://www.stata-press.com/data/r12/reshapexp1
. list
```

	id	year	sex	inc
1.	1	80	0	5000
2.	1	81	0	5500
3.	1	81	0	5400
4.	1	82	0	6000

```
. reshape wide inc, i(id) j(year)
(note:  j = 80 81 82)
year not unique within id;
there are multiple observations at the same year within id.
Type "reshape error" for a listing of the problem observations.
r(9);
```

In the long form, i(id) does not have to be unique, but j(year) must be unique within i; otherwise, what is the value of inc in 1981 for which id==1?

reshape told us to type reshape error to view the problem observations.

```
. reshape error
(note: j = 80 81 82)

i (id) indicates the top-level grouping such as subject id.
j (year) indicates the subgrouping such as time.
The data are in the long form;  j should be unique within i.

There are multiple observations on the same year within id.

The following 2 of 4 observations have repeated year values:
```

	id	year
2.	1	81
3.	1	81

```
(data now sorted by id year)
```

◁

▷ Example 4

Consider some long-form data that have no mistakes. We list the first 4 observations.

```
. list in 1/4
```

	id	year	sex	inc	ue
1.	1	80	0	5000	0
2.	1	81	0	5500	1
3.	1	82	0	6000	0
4.	2	80	1	2000	1

Say that when converting the data to wide form, however, we forget to mention the ue variable (which varies within person).

```
. reshape wide inc, i(id) j(year)
(note:  j = 80 81 82)
ue not constant within id
Type "reshape error" for a listing of the problem observations.
r(9);
```

Here `reshape` observed that ue was not constant within i and so could not restructure the data so that there were single observations on i. We should have typed

```
. reshape wide inc ue, i(id) j(year)
```
◁

In summary, there are three cases in which `reshape` will refuse to convert the data:

1. The data are in wide form and i is not unique.

2. The data are in long form and j is not unique within i.

3. The data are in long form and an unmentioned variable is not constant within i.

▷ Example 5

With some mistakes, `reshape` will probably convert the data and produce a surprising result. Suppose that we forget to mention that the ue variable varies within id in the following wide data:

```
. use http://www.stata-press.com/data/r12/reshape1
. list
```

	id	sex	inc80	inc81	inc82	ue80	ue81	ue82
1.	1	0	5000	5500	6000	0	1	0
2.	2	1	2000	2200	3300	1	0	0
3.	3	0	3000	2000	1000	0	0	1

```
. reshape long inc, i(id) j(year)
(note: j = 80 81 82)
```

Data		wide	->	long
Number of obs.		3	->	9
Number of variables		8	->	7
j variable (3 values)			->	year
xij variables:				
	inc80 inc81 inc82		->	inc

```
. list, sep(3)
```

	id	year	sex	inc	ue80	ue81	ue82
1.	1	80	0	5000	0	1	0
2.	1	81	0	5500	0	1	0
3.	1	82	0	6000	0	1	0
4.	2	80	1	2000	1	0	0
5.	2	81	1	2200	1	0	0
6.	2	82	1	3300	1	0	0
7.	3	80	0	3000	0	0	1
8.	3	81	0	2000	0	0	1
9.	3	82	0	1000	0	0	1

We did not state that ue varied within i, so the variables ue80, ue81, and ue82 were left as is. `reshape` did not complain. There is no real problem here because no information has been lost. In fact, this may actually be the result we wanted. Probably, however, we simply forgot to include ue among the X_{ij} variables.

If you obtain an unexpected result, here is how to undo it:

1. If you typed `reshape long` ... to produce the result, type `reshape wide` (without arguments) to undo it.

2. If you typed `reshape wide` ... to produce the result, type `reshape long` (without arguments) to undo it.

So, we can type

```
. reshape wide
```

to get back to our original, wide-form data and then type the `reshape long` command that we intended:

```
. reshape long inc ue, i(id) j(year)
```

◁

reshape long and reshape wide without arguments

Whenever you type a `reshape long` or `reshape wide` command with arguments, reshape remembers it. Thus you might type

```
. reshape long inc ue, i(id) j(year)
```

and work with the data like that. You could then type

```
. reshape wide
```

to convert the data back to the wide form. Then later you could type

```
. reshape long
```

to convert them back to the long form. If you save the data, you can even continue using `reshape wide` and `reshape long` without arguments during a future Stata session.

Be careful. If you create new X_{ij} variables, you must tell `reshape` about them by typing the full `reshape` command, although no real damage will be done if you forget. If you are converting from long to wide form, `reshape` will catch your error and refuse to make the conversion. If you are converting from wide to long, `reshape` will convert the data, but the result will be surprising: remember what happened when we forgot to mention the ue variable and ended up with ue80, ue81, and ue82 in our long data; see example 5. You can `reshape long` to undo the unwanted change and then try again.

Missing variables

When converting data from wide form to long form, `reshape` does not demand that all the variables exist. Missing variables are treated as variables with missing observations.

▷ Example 6

Let's drop ue81 from the wide form of the data:

```
. use http://www.stata-press.com/data/r12/reshape1, clear
. drop ue81
. list
```

	id	sex	inc80	inc81	inc82	ue80	ue82
1.	1	0	5000	5500	6000	0	0
2.	2	1	2000	2200	3300	1	0
3.	3	0	3000	2000	1000	0	1

```
. reshape long inc ue, i(id) j(year)
(note:  j = 80 81 82)
(note: ue81 not found)
```

Data	wide	->	long
Number of obs.	3	->	9
Number of variables	7	->	5
j variable (3 values)		->	year
xij variables:			
inc80 inc81 inc82		->	inc
ue80 ue81 ue82		->	ue

```
. list, sep(3)
```

	id	year	sex	inc	ue
1.	1	80	0	5000	0
2.	1	81	0	5500	.
3.	1	82	0	6000	0
4.	2	80	1	2000	1
5.	2	81	1	2200	.
6.	2	82	1	3300	0
7.	3	80	0	3000	0
8.	3	81	0	2000	.
9.	3	82	0	1000	1

reshape placed missing values where ue81 values were unavailable. If we reshaped these data back to wide form by typing

```
. reshape wide inc ue, i(id) j(year)
```

the ue81 variable would be created and would contain all missing values.

◁

Advanced issues with basic syntax: i()

The i() option can indicate one i variable (as our past examples have illustrated) or multiple variables. An example of multiple i variables would be hospital ID and patient ID within each hospital.

```
. reshape ... , i(hid pid)
```

Unique pairs of values for hid and pid in the data define the grouping variable for reshape.

Advanced issues with basic syntax: j()

The j() option takes a variable name (as our past examples have illustrated) or a variable name and a list of values. When the values are not provided, reshape deduces them from the data. Specifying the values with the j() option is rarely needed.

reshape never makes a mistake when the data are in long form and you type reshape wide. The values are easily obtained by tabulating the j variable.

reshape can make a mistake when the data are in wide form and you type reshape long if your variables are poorly named. Say that you have the inc80, inc81, and inc82 variables, recording income in each of the indicated years, and you have a variable named inc2, which is not income but indicates when the area was reincorporated. You type

        . reshape long inc, i(id) j(year)

reshape sees the inc2, inc80, inc81, and inc82 variables and decides that there are four groups in which j = 2, 80, 81, and 82.

The easiest way to solve the problem is to rename the inc2 variable to something other than "inc" followed by a number; see [D] **rename**.

You can also keep the name and specify the j values. To perform the reshape, you can type

        . reshape long inc, i(id) j(year 80-82)

or

        . reshape long inc, i(id) j(year 80 81 82)

You can mix the dash notation for value ranges with individual numbers. reshape would understand 80 82-87 89 91-95 as a valid values specification.

At the other extreme, you can omit the j() option altogether with reshape long. If you do, the j variable will be named _j.

Advanced issues with basic syntax: xij

When specifying variable names, you may include @ characters to indicate where the numbers go.

▷ Example 7

Let's reshape the following data from wide to long form:

        . use http://www.stata-press.com/data/r12/reshape3, clear
        . list

	id	sex	inc80r	inc81r	inc82r	ue80	ue81	ue82
1.	1	0	5000	5500	6000	0	1	0
2.	2	1	2000	2200	3300	1	0	0
3.	3	0	3000	2000	1000	0	0	1

```
. reshape long inc@r ue, i(id) j(year)
(note:   j = 80 81 82)
```

Data	wide	->	long
Number of obs.	3	->	9
Number of variables	8	->	5
j variable (3 values)		->	year
xij variables:			
	inc80r inc81r inc82r	->	incr
	ue80 ue81 ue82	->	ue

```
. list, sep(3)
```

	id	year	sex	incr	ue
1.	1	80	0	5000	0
2.	1	81	0	5500	1
3.	1	82	0	6000	0
4.	2	80	1	2000	1
5.	2	81	1	2200	0
6.	2	82	1	3300	0
7.	3	80	0	3000	0
8.	3	81	0	2000	0
9.	3	82	0	1000	1

At most one @ character may appear in each name. If no @ character appears, results are as if the @ character appeared at the end of the name. So, the equivalent `reshape` command to the one above is

```
. reshape long inc@r ue@, i(id) j(year)
```

`inc@r` specifies variables named `inc#r` in the wide form and `incr` in the long form. The @ notation may similarly be used for converting data from long to wide format:

```
. reshape wide inc@r ue, i(id) j(year)
```

◁

Advanced issues with basic syntax: String identifiers for j()

The `string` option allows j to take on string values.

▷ Example 8

Consider the following wide data on husbands and wives. In these data, `incm` is the income of the man and `incf` is the income of the woman.

```
. use http://www.stata-press.com/data/r12/reshape4, clear
. list
```

	id	kids	incm	incf
1.	1	0	5000	5500
2.	2	1	2000	2200
3.	3	2	3000	2000

These data can be reshaped into separate observations for males and females by typing

```
. reshape long inc, i(id) j(sex) string
(note:  j = f m)
```

Data	wide	->	long
Number of obs.	3	->	6
Number of variables	4	->	4
j variable (2 values)		->	sex
xij variables:			
	incf incm	->	inc

The `string` option specifies that j take on nonnumeric values. The result is

```
. list, sep(2)
```

	id	sex	kids	inc
1.	1	f	0	5500
2.	1	m	0	5000
3.	2	f	1	2200
4.	2	m	1	2000
5.	3	f	2	2000
6.	3	m	2	3000

`sex` will be a string variable. Similarly, these data can be converted from long to wide form by typing

```
. reshape wide inc, i(id) j(sex) string
```

◁

Strings are not limited to being single characters or even having the same length. You can specify the location of the string identifier in the variable name by using the @ notation.

▷ Example 9

Suppose that our variables are named id, kids, incmale, and incfem.

```
. use http://www.stata-press.com/data/r12/reshapexp2, clear
. list
```

	id	kids	incmale	incfem
1.	1	0	5000	5500
2.	2	1	2000	2200
3.	3	2	3000	2000

```
. reshape long inc, i(id) j(sex) string
(note:  j = fem male)
```

Data	wide	->	long
Number of obs.	3	->	6
Number of variables	4	->	4
j variable (2 values)		->	sex
xij variables:			
	incfem incmale	->	inc

```
. list, sep(2)
```

	id	sex	kids	inc
1.	1	fem	0	5500
2.	1	male	0	5000
3.	2	fem	1	2200
4.	2	male	1	2000
5.	3	fem	2	2000
6.	3	male	2	3000

If the wide data had variables named minc and finc, the appropriate reshape command would have been

```
. reshape long @inc, i(id) j(sex) string
```

The resulting variable in the long form would be named inc.

We can also place strings in the middle of the variable names. If the variables were named incMome and incFome, the reshape command would be

```
. reshape long inc@ome, i(id) j(sex) string
```

Be careful with string identifiers because it is easy to be surprised by the result. Say that we have wide data having variables named incm, incf, uem, uef, agem, and agef. To make the data long, we might type

```
. reshape long inc ue age, i(id) j(sex) string
```

Along with these variables, we also have the variable agenda. reshape will decide that the sexes are m, f, and nda. This would not happen without the string option if the variables were named inc0, inc1, ue0, ue1, age0, and age1, even with the agenda variable present in the data.

◁

Advanced issues with basic syntax: Second-level nesting

Sometimes the data may have more than one possible j variable for reshaping. Suppose that your data have both a year variable and a sex variable. One logical observation in the data might be represented in any of the following four forms:

```
. list in 1/4    // The long-long form
```

	hid	sex	year	inc
1.	1	f	90	3200
2.	1	f	91	4700
3.	1	m	90	4500
4.	1	m	91	4600

```
. list in 1/2     // The long-year wide-sex form
```

	hid	year	minc	finc
1.	1	90	4500	3200
2.	1	91	4600	4700

```
. list in 1/2     // The wide-year long-sex form
```

	hid	sex	inc90	inc91
1.	1	f	3200	4700
2.	1	m	4500	4600

```
. list in 1       // The wide-wide form
```

	hid	minc90	minc91	finc90	finc91
1.	1	4500	4600	3200	4700

reshape can convert any of these forms to any other. Converting data from the long–long form to the wide–wide form (or any of the other forms) takes two reshape commands. Here is how we would do it:

From		To		Command
year	sex	year	sex	
long	long	long	wide	reshape wide @inc, i(hid year) j(sex) string
long	wide	long	long	reshape long @inc, i(hid year) j(sex) string
long	long	wide	long	reshape wide inc, i(hid sex) j(year)
wide	long	long	long	reshape long inc, i(hid sex) j(year)
long	wide	wide	wide	reshape wide minc finc, i(hid) j(year)
wide	wide	long	wide	reshape long minc finc, i(hid) j(year)
wide	long	wide	wide	reshape wide @inc90 @inc91, i(hid) j(sex) string
wide	wide	wide	long	reshape long @inc90 @inc91, i(hid) j(sex) string

Description of advanced syntax

The advanced syntax is simply a different way of specifying the reshape command, and it has one seldom-used feature that provides extra control. Rather than typing one reshape command to describe the data and perform the conversion, such as

```
. reshape long inc, i(id) j(year)
```

you type a sequence of reshape commands. The initial commands describe the data, and the last command performs the conversion:

```
. reshape i id
. reshape j year
. reshape xij inc
. reshape long
```

reshape i corresponds to i() in the basic syntax.

reshape j corresponds to j() in the basic syntax.

`reshape xij` corresponds to the variables specified in the basic syntax. `reshape xij` also accepts the `atwl()` option for use when @ characters are specified in the *fvarnames*. `atwl` stands for at-when-long. When you specify names such as `inc@r` or `ue@`, in the long form the names become `incr` and `ue`, and the @ character is ignored. `atwl()` allows you to change @ into whatever you specify. For example, if you specify `atwl(X)`, the long-form names become `incXr` and `ueX`.

There is also one more specification, which has no counterpart in the basic syntax:

. `reshape xi` *varlist*

In the basic syntax, Stata assumes that all unspecified variables are constant within i. The advanced syntax works the same way, unless you specify the `reshape xi` command, which names the constant-within-i variables. If you specify `reshape xi`, any variables that you do not explicitly specify are dropped from the data during the conversion.

As a practical matter, you should explicitly drop the unwanted variables before conversion. For instance, suppose that the data have variables `inc80`, `inc81`, `inc82`, `sex`, `age`, and `age2` and that you no longer want the `age2` variable. You could specify

. `reshape xi sex age`

or

. `drop age2`

and leave `reshape xi` unspecified.

`reshape xi` does have one minor advantage. It saves `reshape` the work of determining which variables are unspecified. This saves a relatively small amount of computer time.

Another advanced-syntax feature is `reshape query`, which is equivalent to typing `reshape` by itself. `reshape query` reports which `reshape` parameters have been defined. `reshape i`, `reshape j`, `reshape xij`, and `reshape xi` specifications may be given in any order and may be repeated to change or correct what has been specified.

Finally, `reshape clear` clears the definitions. `reshape` definitions are stored with the dataset when you save it. `reshape clear` allows you to erase these definitions.

The basic syntax of `reshape` is implemented in terms of the advanced syntax, so you can mix basic and advanced syntaxes.

Saved results

`reshape` stores the following characteristics with the data (see [P] **char**):

_dta[ReS_i]	i variable names
_dta[ReS_j]	j variable name
_dta[ReS_jv]	j values, if specified
_dta[ReS_Xij]	X_{ij} variable names
_dta[ReS_Xi]	X_i variable names, if specified
_dta[ReS_atwl]	atwl() value, if specified
_dta[ReS_str]	1 if option string specified; 0 otherwise

Methods and formulas

`reshape` is implemented as an ado-file.

Acknowledgment

This version of `reshape` was based in part on the work of Jeroen Weesie (1997), Utrecht University, The Netherlands.

References

Baum, C. F., and N. J. Cox. 2007. Stata tip 45: Getting those data into shape. *Stata Journal* 7: 268–271.

Gould, W. W. 1997. stata48: Updated reshape. *Stata Technical Bulletin* 39: 4–16. Reprinted in *Stata Technical Bulletin Reprints*, vol. 7, pp. 5–20. College Station, TX: Stata Press.

Jeanty, P. W. 2010. Using the world development indicators database for statistical analysis in Stata. *Stata Journal* 10: 30–45.

Mitchell, M. N. 2010. *Data Management Using Stata: A Practical Handbook*. College Station, TX: Stata Press.

Weesie, J. 1997. dm48: An enhancement of reshape. *Stata Technical Bulletin* 38: 2–4. Reprinted in *Stata Technical Bulletin Reprints*, vol. 7, pp. 40–43. College Station, TX: Stata Press.

——. 1998. dm58: A package for the analysis of husband–wife data. *Stata Technical Bulletin* 43: 9–13. Reprinted in *Stata Technical Bulletin Reprints*, vol. 8, pp. 13–20. College Station, TX: Stata Press.

Also see

[D] **save** — Save Stata dataset

[P] **char** — Characteristics

[D] **stack** — Stack data

[D] **xpose** — Interchange observations and variables

Title

> **rmdir** — Remove directory

Syntax

rmdir *directory_name*

Double quotes may be used to enclose the directory name, and the quotes must be used if the directory name contains embedded blanks.

Description

rmdir removes an empty directory (folder).

Remarks

Examples:

Windows

. rmdir myproj
. rmdir c:\projects\myproj
. rmdir "c:\My Projects\Project 1"

Mac and Unix

. rmdir myproj
. rmdir ~/projects/myproj

Also see

[D] **cd** — Change directory

[D] **copy** — Copy file from disk or URL

[D] **dir** — Display filenames

[D] **erase** — Erase a disk file

[D] **shell** — Temporarily invoke operating system

[D] **type** — Display contents of a file

[D] **mkdir** — Create directory

[U] **11.6 Filenaming conventions**

Title

> **sample** — Draw random sample

Syntax

> sample # $\left[\,if\,\right]$ $\left[\,in\,\right]$ $\left[\,,\ \underline{c}ount\ by(groupvars)\,\right]$

by is allowed; see [D] **by**.

Menu

Statistics > Resampling > Draw random sample

Description

sample draws random samples of the data in memory. "Sampling" here is defined as drawing observations without replacement; see [R] **bsample** for sampling with replacement.

The size of the sample to be drawn can be specified as a percentage or as a count:

- sample without the count option draws a #% pseudorandom sample of the data in memory, thus discarding $(100 - \#)\%$ of the observations.

- sample with the count option draws a #-observation pseudorandom sample of the data in memory, thus discarding $\_N - \#$ observations. # can be larger than $\_N$, in which case all observations are kept.

In either case, observations not meeting the optional if and in criteria are kept (sampled at 100%).

If you are interested in reproducing results, you must first set the random-number seed; see [R] **set seed**.

Options

count specifies that # in sample # be interpreted as an observation count rather than as a percentage. Typing sample 5 without the count option means that a 5% sample be drawn; typing sample 5, count, however, would draw a sample of 5 observations.

Specifying # as greater than the number of observations in the dataset is not considered an error.

by(groupvars) specifies that a #% sample be drawn within each set of values of groupvars, thus maintaining the proportion of each group.

count may be combined with by(). For example, typing sample 50, count by(sex) would draw a sample of size 50 for men and 50 for women.

Specifying by varlist: sample # is equivalent to specifying sample #, by(varlist); use whichever syntax you prefer.

Remarks

▷ Example 1

We have NLSY data on young women aged 14–26 years in 1968 and wish to draw a 10% sample of the data in memory.

```
. use http://www.stata-press.com/data/r12/nlswork
(National Longitudinal Survey.  Young Women 14-26 years of age in 1968)
. describe, short
Contains data from http://www.stata-press.com/data/r12/nlswork.dta
  obs:        28,534                      National Longitudinal Survey.
                                          Young Women 14-26 years of age
                                          in 1968
  vars:           21                      7 Dec 2010 17:02
  size:      941,622
Sorted by:  idcode   year
. sample 10
(25681 observations deleted)
. describe, short
Contains data from http://www.stata-press.com/data/r12/nlswork.dta
  obs:         2,853                      National Longitudinal Survey.
                                          Young Women 14-26 years of age
                                          in 1968
  vars:           21                      7 Dec 2010 17:02
  size:       94,149
Sorted by:
     Note:  dataset has changed since last saved
```

Our original dataset had 28,534 observations. The sample-10 dataset has 2.853 observations, which is the nearest number to 0.10×28534.

◁

▷ Example 2

Among the variables in our data is race; race = 1 denotes whites, race = 2 denotes blacks, and race = 3 denotes other races. We want to keep 100% of the nonwhite women but only 10% of the white women.

```
. use http://www.stata-press.com/data/r12/nlswork, clear
(National Longitudinal Survey.  Young Women 14-26 years of age in 1968)
. tab race
```

1=white, 2=black, 3=other	Freq.	Percent	Cum.
1	20,180	70.72	70.72
2	8,051	28.22	98.94
3	303	1.06	100.00
Total	28,534	100.00	

```
. sample 10 if race == 1
(18162 observations deleted)
```

```
. describe, short
Contains data from http://www.stata-press.com/data/r12/nlswork.dta
  obs:        10,372                        National Longitudinal Survey.
                                            Young Women 14-26 years of age
                                            in 1968
  vars:           21                        7 Dec 2010 17:02
  size:      342,276
Sorted by:
     Note:  dataset has changed since last saved
. display .10*20180 + 8051 + 303
10372
```

◁

▷ Example 3

Now let's suppose that we want to keep 10% of each of the three categories of race.

```
. use http://www.stata-press.com/data/r12/nlswork, clear
(National Longitudinal Survey.  Young Women 14-26 years of age in 1968)
. sample 10, by(race)
(25681 observations deleted)
. tab race
```

1=white, 2=black, 3=other	Freq.	Percent	Cum.
1	2,018	70.73	70.73
2	805	28.22	98.95
3	30	1.05	100.00
Total	2,853	100.00	

This differs from simply typing sample 10 in that with by(), sample holds constant the percentages of white, black, and other women.

◁

❑ Technical note

We have a large dataset on disk containing 125,235 observations. We wish to draw a 10% sample of this dataset without loading the entire dataset (perhaps because the dataset will not fit in memory). sample will not solve this problem—the dataset must be loaded first—but it is rather easy to solve it ourselves. Say that bigdata.dct contains the dictionary for this dataset; see [D] **import**. One solution is to type

```
. infile using bigdata if runiform()<=.1
dictionary {
    etc.
}
(12,580 observations read)
```

The if modifier on the end of infile drew uniformly distributed random numbers over the interval 0 and 1 and kept each observation if the random number was less than or equal to 0.1. This, however, did not draw an exact 10% sample—the sample was expected to contain only 10% of the observations, and here we obtained just more than 10%. This is probably a reasonable solution.

If the sample must contain precisely 12,524 observations, however, after getting too many observations, we could type

```
. generate u=runiform()
. sort u
. keep in 1/12524
(56 observations deleted)
```

That is, we put the resulting sample in random order and keep the first 12,524 observations. Now our only problem is making sure that, at the first step, we have more than 12,524 observations. Here we were lucky, but half the time we will not be so lucky — after typing infile ... if runiform()<=.1, we will have less than a 10% sample. The solution, of course, is to draw more than a 10% sample initially and then cut it back to 10%.

How much more than 10% do we need? That depends on the number of records in the original dataset, which in our example is 125,235.

A little experimentation with bitesti (see [R] **bitest**) provides the answer:

```
. bitesti 125235 12524 .102
          N    Observed k    Expected k    Assumed p    Observed p

     125235         12524      12773.97      0.10200       0.10000
  Pr(k >= 12524)                        = 0.990466  (one-sided test)
  Pr(k <= 12524)                        = 0.009777  (one-sided test)
  Pr(k <= 12524 or k >= 13025) = 0.019584  (two-sided test)
```

Initially drawing a 10.2% sample will yield a sample larger than 10% 99 times of 100. If we draw a 10.4% sample, we are virtually assured of having enough observations (type bitesti 125235 12524 .104 for yourself).

❑

Methods and formulas

sample is implemented as an ado-file.

References

Cox, N. J. 2001. dm86: Sampling without replacement: Absolute sample sizes and keeping all observations. *Stata Technical Bulletin* 59: 8–9. Reprinted in *Stata Technical Bulletin Reprints*, vol. 10, pp. 38–39. College Station, TX: Stata Press.

Weesie, J. 1997. dm46: Enhancement to the sample command. *Stata Technical Bulletin* 37: 6–7. Reprinted in *Stata Technical Bulletin Reprints*, vol. 7, pp. 37–38. College Station, TX: Stata Press.

Also see

[R] **bsample** — Sampling with replacement

Title

> **save** — Save Stata dataset

Syntax

Save data in memory to file

> s̲ave [*filename*] [, *save_options*]

Save data in memory to file in Stata 9/Stata 10 format

> saveold *filename* [, *saveold_options*]

save_options	Description
n̲ol̲abel	omit value labels from the saved dataset
replace	overwrite existing dataset
all	save e(sample) with the dataset; programmer's option
o̲rphans	save all value labels
emptyok	save dataset even if zero observations and zero variables

saveold_options	Description
n̲ol̲abel	omit value labels from the saved dataset
replace	overwrite existing dataset
all	save e(sample) with the dataset; programmer's option

Menu

File > Save As...

Description

save stores the dataset currently in memory on disk under the name *filename*. If *filename* is not specified, the name under which the data were last known to Stata (c(filename)) is used. If *filename* is specified without an extension, .dta is used. If your *filename* contains embedded spaces, remember to enclose it in double quotes.

saveold saves the dataset currently in memory on disk under the name *filename* in Stata 9/Stata 10 format. Stata 11 has the same dataset format as Stata 10, but Stata 11 is smart enough to read Stata 12 datasets.

If you are using Stata 12 and want to save a file so that it may be read by someone using Stata 9 or Stata 10, simply use the saveold command.

Options for save

nolabel omits value labels from the saved dataset. The associations between variables and value-label names, however, are saved along with the dataset label and the variable labels.

replace permits save to overwrite an existing dataset.

all is for use by programmers. If specified, e(sample) will be saved with the dataset. You could run a regression; save mydata, all; drop _all; use mydata; and predict yhat if e(sample).

orphans saves all value labels, including those not attached to any variable.

emptyok is a programmer's option. It specifies that the dataset be saved, even if it contains zero observations and zero variables. If emptyok is not specified and the dataset is empty, save responds with the message "no variables defined".

Options for saveold

nolabel omits value labels from the saved dataset. The associations between variables and value-label names, however, are saved along with the dataset label and the variable labels.

replace permits saveold to overwrite an existing dataset.

all is for use by programmers. If specified, e(sample) will be saved with the dataset. You could run a regression; save mydata, all; drop _all; use mydata; and predict yhat if e(sample).

Remarks

Stata keeps the data on which you are currently working in your computer's memory. You put the data there in the first place; see [U] **21 Inputting and importing data**. Thereafter, you can save the dataset on disk so that you can use it easily in the future. Stata stores your data on disk in a compressed format that only Stata understands. This does not mean, however, that you are locked into using only Stata. Any time you wish, you can export the data to a format other software packages understand; see [D] **export**.

Stata goes to a lot of trouble to keep you from accidentally losing your data. When you attempt to leave Stata by typing exit, Stata checks that your data have been safely stored on disk. If not, Stata refuses to let you leave. (You can tell Stata that you want to leave anyway by typing exit, clear.) Similarly, when you save your data in a disk file, Stata ensures that the disk file does not already exist. If it does exist, Stata refuses to save it. You can use the replace option to tell Stata that it is okay to overwrite an existing file.

▷ Example 1

We have entered data into Stata for the first time. We have the following data:

```
. describe
Contains data
    obs:            39                          Minnesota Highway Data, 1973
   vars:             5
   size:           936
```

variable name	storage type	display format	value label	variable label
acc_rate	float	%9.0g		Accident rate
spdlimit	float	%9.0g		Speed limit
acc_pts	float	%9.0g		Access points per mile
rate	float	%9.0g	rcat	Accident rate per million vehicle miles
spdcat	float	%9.0g	scat	Speed limit category

```
Sorted by:
     Note:  dataset has changed since last saved
```

We have a dataset containing 39 observations on five variables, and, evidently, we have gone to a lot of trouble to prepare this dataset. We have used the label data command to label the data Minnesota Highway Data, the label variable command to label all the variables, and the label define and label values commands to attach value labels to the last two variables. (See [U] **12.6.3 Value labels** for information about doing this.)

At the end of the describe, Stata notes that the "dataset has changed since last saved". This is Stata's way of gently reminding us that these data need to be saved. Let's save our data:

```
. save hiway
file hiway.dta saved
```

We type save hiway, and Stata stores the data in a file named hiway.dta. (Stata automatically added the .dta suffix.) Now when we describe our data, we no longer get the warning that our dataset has not been saved; instead, we are told the name of the file in which the data are saved:

```
. describe
Contains data from hiway.dta
    obs:            39                          Minnesota Highway Data, 1973
   vars:             5                          18 Jan 2011 11:42
   size:           936
```

variable name	storage type	display format	value label	variable label
acc_rate	float	%9.0g		Accident rate
spdlimit	float	%9.0g		Speed limit
acc_pts	float	%9.0g		Access points per mile
rate	float	%9.0g	rcat	Accident rate per million vehicle miles
spdcat	float	%9.0g	scat	Speed limit category

```
Sorted by:
```

Just to prove to you that the data have really been saved, let's eliminate the copy of the data in memory by typing drop _all:

```
. drop _all
. describe
Contains data
   obs:            0
   vars:           0
   size:           0
Sorted by:
```

We now have no data in memory. Because we saved our dataset, we can retrieve it by typing use hiway:

```
. use hiway
(Minnesota Highway Data, 1973)

. describe
Contains data from hiway.dta
   obs:           39                          Minnesota Highway Data, 1973
   vars:           5                          18 Jan 2011 11:42
   size:         936
```

variable name	storage type	display format	value label	variable label
acc_rate	float	%9.0g		Accident rate
spdlimit	float	%9.0g		Speed limit
acc_pts	float	%9.0g		Access points per mile
rate	float	%9.0g	rcat	Accident rate per million vehicle miles
spdcat	float	%9.0g	scat	Speed limit category

```
Sorted by:
```

◁

▷ Example 2

Continuing with our previous example, we have saved our data in the file hiway.dta. We continue to work with our data and discover an error; we made a mistake when we typed one of the values for the spdlimit variable:

```
. list in 1/3
```

	acc_rate	spdlimit	acc_pts	rate	spdcat
1.	1.61	50	2.2	Below 4	Above 60
2.	1.81	60	6.8	Below 4	55 to 60
3.	1.84	55	14	Below 4	55 to 60

In the first observation, the spdlimit variable is 50, whereas the spdcat variable indicates that the speed limit is more than 60 miles per hour. We check our original copy of the data and discover that the spdlimit variable ought to be 70. We can fix it with the replace command:

```
. replace spdlimit=70 in 1
(1 real change made)
```

If we were to describe our data now, Stata would warn us that our data have changed since they were last saved:

```
. describe

Contains data from hiway.dta
  obs:            39                          Minnesota Highway Data, 1973
 vars:             5                          18 Jan 2011 11:42
 size:           936

              storage   display    value
variable name   type    format     label     variable label

acc_rate        float   %9.0g                 Accident rate
spdlimit        float   %9.0g                 Speed limit
acc_pts         float   %9.0g                 Access points per mile
rate            float   %9.0g      rcat       Accident rate per million
                                                 vehicle miles
spdcat          float   %9.0g      scat       Speed limit category

Sorted by:
     Note:  dataset has changed since last saved
```

We take our cue and attempt to save the data again:

```
. save hiway
file hiway.dta already exists
r(602);
```

Stata refuses to honor our request, telling us instead that "file hiway.dta already exists". Stata will not let us accidentally overwrite an existing dataset. To replace the data, we must do so explicitly by typing save hiway, replace. If we want to save the file under the same name as it was last known to Stata, we can omit the filename:

```
. save, replace
file hiway.dta saved
```

Now our data are saved.   ◁

Methods and formulas

saveold is implemented as an ado-file.

Also see

[D] **compress** — Compress data in memory

[D] **import** — Overview of importing data into Stata

[D] **export** — Overview of exporting data from Stata

[D] **use** — Load Stata dataset

[P] **file formats .dta** — Description of .dta file format

[U] **11.6 Filenaming conventions**

Title

> **separate** — Create separate variables

Syntax

> separate *varname* $[if]$ $[in]$, by(*groupvar* | *exp*) $[options]$

options	Description
Main	
* by(*groupvar*)	categorize observations into groups defined by *groupvar*
* by(*exp*)	categorize observations into two groups defined by *exp*
Options	
generate(*stubname*)	name new variables by suffixing values to *stubname*; default is to use *varname* as prefix
sequential	use as name suffix categories numbered sequentially from 1
missing	create variables for the missing values
shortlabel	create shorter variable labels

* Either by(*groupvar*) or by(*exp*) must be specified.

Menu

Data > Create or change data > Other variable-transformation commands > Create separate variables

Description

separate creates new variables containing values from *varname*.

Options

___ Main ___

by(*groupvar* | *exp*) specifies one variable defining the categories or a logical expression that categorizes the observations into two groups.

If by(*groupvar*) is specified, *groupvar* may be a numeric or string variable taking on any values.

If by(*exp*) is specified, the expression must evaluate to true (1), false (0), or missing.

by() is required.

___ Options ___

generate(*stubname*) specifies how the new variables are to be named. If generate() is not specified, separate uses the name of the original variable, shortening it if necessary. If generate() is specified, separate uses *stubname*. If any of the resulting names is too long when the values are suffixed, it is not shortened and an error message is issued.

565

sequential specifies that categories be numbered sequentially from 1. By default, separate uses
the actual values recorded in the original variable, if possible, and sequential numbers otherwise.
separate can use the original values if they are all nonnegative integers smaller than 10,000.

missing also creates a variable for the category *missing*, if missing occurs (*groupvar* takes on the
value missing or *exp* evaluates to missing). The resulting variable is named in the usual manner but
with an appended underscore, for example, bp_. By default, separate creates no such variable.
The contents of the other variables are unaffected by whether missing is specified.

shortlabel creates a variable label that is shorter than the default. By default, when separate
generates the new variable labels, it includes the name of the variable being separated. shortlabel
specifies that the variable name be omitted from the new variable labels.

Remarks

▷ Example 1

We have data on the miles per gallon (mpg) and country of manufacture of 74 automobiles. We want
to compare the distributions of mpg for domestic and foreign automobiles by plotting the quantiles
of the two distributions (see [R] **diagnostic plots**).

```
. use http://www.stata-press.com/data/r12/auto
(1978 Automobile Data)

. separate mpg, by(foreign)
```

variable name	storage type	display format	value label	variable label
mpg0	byte	%8.0g		mpg, foreign == Domestic
mpg1	byte	%8.0g		mpg, foreign == Foreign

```
. list mpg* foreign
```

	mpg	mpg0	mpg1	foreign
1.	22	22	.	Domestic
2.	17	17	.	Domestic
3.	22	22	.	Domestic
	(*output omitted*)			
22.	16	16	.	Domestic
23.	17	17	.	Domestic
24.	28	28	.	Domestic
	(*output omitted*)			
73.	25	.	25	Foreign
74.	17	.	17	Foreign

```
. qqplot mpg0 mpg1
```

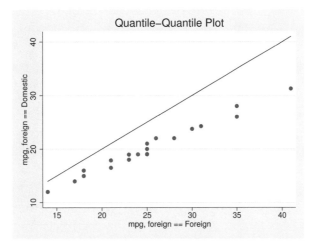

In our auto dataset, the foreign cars have better gas mileage.

◁

Saved results

separate saves the following in r():

Macros
    r(varlist)   names of the newly created variables

Methods and formulas

separate is implemented as an ado-file.

Acknowledgment

separate was originally written by Nicholas J. Cox, Durham University.

Reference

Baum, C. F. 2009. *An Introduction to Stata Programming*. College Station, TX: Stata Press.

Also see

[R] **tabulate oneway** — One-way tables of frequencies

[R] **tabulate twoway** — Two-way tables of frequencies

[R] **tabulate, summarize()** — One- and two-way tables of summary statistics

Title

shell — Temporarily invoke operating system

Syntax

$\{$ <u>sh</u>ell | ! $\}$ $[$ *operating_system_command* $]$

winexec *program_name* $[$ *program_args* $]$

$\{$ <u>xsh</u>ell | !! $\}$ $[$ *operating_system_command* $]$

Command availability:

| | Stata for ... | | | |
Command	Windows	Mac	Unix(GUI)	Unix(console)
shell	X	X	X	X
winexec	X	X	X	–
xshell	–	X	X	–

Description

shell (synonym: "!") allows you to send commands to your operating system or to enter your operating system for interactive use. Stata will wait for the shell to close or the *operating_system_command* to complete before continuing.

winexec allows you to start other programs (such as browsers) from Stata's command line. Stata will continue without waiting for the program to complete.

xshell (Stata for Mac and Unix(GUI) only) brings up an xterm in which the command is to be executed. On Mac OS X, xterm is available when *X11* is installed. *X11* may not be installed by default and can be installed by running the **Optional Installs** installer from your Mac OS X installation disc.

Remarks

Remarks are presented under the following headings:

> *Stata for Windows*
> *Stata for Mac*
> *Stata for Unix(GUI)*
> *Stata for Unix(console)*

Stata for Windows

shell, without arguments, preserves your session and invokes the operating system. The Command window will disappear, and a DOS window will appear, indicating that you may not continue in Stata until you exit the DOS shell. To reenter Stata, type exit at your operating system's prompt. Your Stata session is reestablished just as if you had never left.

568

Say that you are using Stata for Windows and you suddenly realize you need to do two things. You need to enter your operating system for a few minutes. Rather than exiting Stata, doing what you have to do, and then restarting Stata, you type `shell` in the Command window. A DOS window appears:

```
Microsoft Windows [Version 6.0.6000]
Copyright (c) 2006 Microsoft Corporation.  All rights reserved.
C:\data>
```

You can now do whatever you need to do in DOS, and Stata will wait until you exit the DOS window before continuing.

Experienced Stata users seldom type out the word `shell`. They type "!". Also you do not have to enter your operating system, issue a command, and then exit back to Stata. If you want to execute one command, you can type the command right after the word `shell` or the exclamation point:

```
. !rename try15.dta final.dta
```

If you do this, the DOS window will open and close as the command is executed.

Stata for Windows users can also use the `winexec` command, which allows you to launch any Windows application from within Stata. You can think of it as a shortcut for clicking on the Windows **Start** button, choosing **Run...**, and typing a command.

Assume that you are working in Stata and decide that you want to run a text editor:

```
. winexec notepad
```
> (*The Windows application Notepad will start and run at the same time as Stata*)

You could even pass a filename to your text editor:

```
. winexec notepad c:\docs\myfile.txt
```

You may need to specify a complete path to the executable that you wish to launch:

```
. winexec c:\windows\notepad c:\docs\myfile.txt
```

The important difference between `winexec` and `shell` is that Stata does not wait for whatever program `winexec` launches to complete before continuing. Stata will wait for the program `shell` launches to complete before performing any further commands.

Stata for Mac

`shell`, with arguments, invokes your operating system, executes one command, and redirects the output to the Results window. The command must complete before you can enter another command in the Command window.

Say that you are using Stata for Mac and suddenly realize that there are two things you have to do. You need to switch to the Finder or enter commands from a terminal for a few minutes. Rather than exiting Stata, doing what you have to do, and then switching back to Stata, you type `shell` and the command in the Command window to execute one command. You then repeat this step for each command that you want to execute from the shell.

Experienced Stata users seldom type out the word `shell`. They type "!".

```
. !mv try15.dta final.dta
```

Be careful not to execute commands, such as `vi`, that require interaction from you. Because all output is redirected to Stata's Results window, you will not be able to interact with the command from Stata. This will effectively lock up Stata because the command will never complete.

When you type `xshell vi myfile.do`, Stata invokes an `xterm` window (which in turn invokes a `shell`) and executes the command there. Typing `!!vi myfile.do` is equivalent to typing `xshell vi myfile.do`.

Stata for Mac users can also use the `winexec` command, which allows you to launch any native application from within Stata. You may, however, have to specify the absolute path to the application. If the application you wish to launch is a Mac OS X application bundle, you must specify an absolute path to the executable in the bundle.

Assume that you are working in Stata and decide that you want to run a text editor:

    . winexec /Applications/TextEdit.app/Contents/MacOS/TextEdit
            (*The OS X application TextEdit will start and run at the same time as Stata*)

You could even pass a filename to your text editor:

    . winexec /Applications/TextEdit.app/Contents/MacOS/TextEdit /Users/cnguyen/myfile.do

If you specify a file path as an argument to the program to be launched, you must specify an absolute path. Also using ~ in the path will not resolve to a home directory. `winexec` cannot launch PEF binaries such as those from Mac OS 9 and some Carbon applications. If an application cannot be launched from a terminal window, it cannot be launched by `winexec`.

The important difference between `winexec` and `shell` is that Stata does not wait for whatever program `winexec` launches to complete before continuing. Stata will wait for the program `shell` launches to complete before performing any further commands. `shell` is appropriate for executing shell commands; `winexec` is appropriate for launching applications.

Stata for Unix(GUI)

`shell`, without arguments, preserves your session and invokes the operating system. The Command window will disappear, and an `xterm` window will appear, indicating that you may not do anything in Stata until you exit the `xterm` window. To reenter Stata, type `exit` at the Unix prompt. Your Stata session is reestablished just as if you had never left.

Say that you are using Stata for Unix(GUI) and suddenly realize that you need to do two things. You need to enter your operating system for a few minutes. Rather than exiting Stata, doing what you have to do, and then restarting Stata, you type `shell` in the Command window. An `xterm` window will appear:

    mycomputer$ _

You can now do whatever you need to do, and Stata will wait until you exit the window before continuing.

Experienced Stata users seldom type out the word `shell`. They type "!". Also you do not have to enter your operating system, issue a command, and then exit back to Stata. If you want to execute one command, you can type the command right after the word `shell` or the exclamation point:

    . !mv try15.dta final.dta

Be careful because sometimes you will want to type

    . !!vi myfile.do

and in other cases,

    . winexec xedit myfile.do

`!!` is a synonym for `xshell`—a command different from, but related to, `shell`—and `winexec` is a different and related command, too.

Before we get into this, understand that if all you want is a shell from which you can issue Unix commands, type `shell` or `!`:

```
. !
mycomputer$ _
```

When you are through, type `exit` to the Unix prompt, and you will return to Stata:

```
mycomputer$ exit

.  _
```

If, on the other hand, you want to specify in Stata the Unix command that you want to execute, you need to decide whether you want to use `shell`, `xshell`, or `winexec`. The answer depends on whether the command you want to execute requires a terminal window or is an X application:

> ... does not need a terminal window: use `shell` ... (synonym: `!`...)
> ... needs a terminal window: use `xshell` ... (synonym: `!!`...)
> ... is an X application: use `winexec` ... (no synonym)

When you type `shell mv try15.dta final.dta`, Stata invokes your shell (`/bin/sh`, `/bin/csh`, etc.) and executes the specified command (`mv` here), routing the standard output and standard error back to Stata. Typing `!mv try15.dta final.dta` is the same as typing `shell mv try15.dta final.dta`.

When you type `xshell vi myfile.do`, Stata invokes an `xterm` window (which in turn invokes a shell) and executes the command there. Typing `!!vi myfile.do` is equivalent to typing `xshell vi myfile.do`.

When you type `winexec xedit myfile.do`, Stata directly invokes the command specified (`xedit` here). No `xterm` window is brought up nor is a shell invoked because, here, `xterm` does not need it. `xterm` is an X application that will create its own window in which to run. You could have typed `!!xedit myfile.do`. That would have brought up an unnecessary `xterm` window from which `xedit` would have been executed, and that would not matter. You could even have typed `!xedit myfile.do`. That would have invoked an unnecessary shell from which `xedit` would have been executed, and that would not matter, either. The important difference, however, is that `shell` and `xshell` wait until the process completes before allowing Stata to continue, and `winexec` does not.

❑ Technical note

You can set Stata global macros to control the behavior of `shell` and `xshell`. The macros are

> `$S_SHELL`   defines the shell to be used by `shell` when
> you type a command following `shell`.
> The default is something like "`/bin/sh -c`", although this can vary,
> depending on how your Unix environment variables are set.
>
> `$S_XSHELL`   defines shell to be used by `shell` and `xshell`
> when they are typed without arguments.
> The default is "`xterm`".
>
> `$S_XSHELL2`   defines shell to be used by `xshell` when it is
> typed with arguments.
> The default is "`xterm -e`".

For instance, if you type in Stata

```
. global S_XSHELL2 "/usr/X11R6/bin/xterm -e"
```

and then later type

```
. !!vi myfile.do
```

then Stata would issue the command /usr/X11R6/bin/xterm -e vi myfile.do to Unix.

If you do make changes, we recommend that you record the changes in your profile.do file.

❑

Stata for Unix(console)

shell, without arguments, preserves your session and then invokes your operating system. Your Stata session will be suspended until you exit the shell, at which point your Stata session is reestablished just as if you had never left.

Say that you are using Stata and you suddenly realize that you need to do two things. You need to enter your operating system for a few minutes. Rather than exiting Stata, doing what you have to do, and then restarting Stata, you type shell. A Unix prompt appears:

```
. shell
(Type exit to return to Stata)
$ _
```

You can now do whatever you need to do and type exit when you finish. You will return to Stata just as if you had never left.

Experienced Stata users seldom type out the word shell. They type '!'. Also you do not have to enter your operating system, issue a command, and then exit back to Stata. If you want to execute one command, you can type the command right after the word shell or the exclamation point. If you want to edit the file myfile.do, and if vi is the name of your favorite editor, you could type

```
. !vi myfile.do
```
              Stata opens your editor.
              When you exit your editor:
```
. _
```

Also see

[D] **cd** — Change directory

[D] **copy** — Copy file from disk or URL

[D] **dir** — Display filenames

[D] **erase** — Erase a disk file

[D] **mkdir** — Create directory

[D] **rmdir** — Remove directory

[D] **type** — Display contents of a file

Title

> **snapshot** — Save and restore data snapshots

Syntax

Save snapshot

    snapshot save [, label("label")]

Change snapshot label

    snapshot label snapshot# "label"

Restore snapshot

    snapshot restore snapshot#

List snapshots

    snapshot list [_all | numlist]

Erase snapshots

    snapshot erase _all | numlist

Menu

Data > Data Editor > Data Editor (Edit)

Description

snapshot saves to disk and restores from disk copies of the data in memory. snapshot's main purpose is to allow the Data Editor to save and restore data snapshots during an interactive editing session. A more popular alternative for programmers is preserve; see [P] **preserve**.

Snapshots are referred to by a *snapshot#*. If no snapshots currently exist, the next snapshot saved will receive a *snapshot#* of 1. If snapshots do exist, the next snapshot saved will receive a *snapshot#* one greater than the highest existing *snapshot#*.

snapshot save creates a temporary file containing a copy of the data currently in memory and attaches an optional label (up to 80 characters) to the saved snapshot. Up to 1,000 snapshots may be saved.

snapshot label changes the label on the specified snapshot.

snapshot restore replaces the data in memory with the data from the specified snapshot.

snapshot list lists specified snapshots.

snapshot erase erases specified snapshots.

Option

label(*label*) is for use with snapshot save and allows you to label a snapshot when saving it.

Remarks

snapshot was created to allow a user using the Data Editor to save and restore snapshots of their data while editing them interactively. It is similar to a checkpoint save in a video game, where after you have made a certain amount of progress, you wish to make sure you will be able to return to that point no matter what may happen in the future.

snapshot does not overwrite any copies of your data that you may have saved to disk. It saves a copy of the data currently in memory to a temporary file and allows you to later restore that copy to memory.

snapshot saves the date and time at which you create a snapshot. It is a good idea to also give a snapshot a label so that you will be better able to distinguish between multiple snapshots should you need to restore one.

❏ Technical note

Although we mention above the use of the Data Editor and we demonstrate below the use of snapshot, we recommend that data cleaning not be done interactively. Instead, we recommend that data editing and cleaning be done in a reproducible manner through the use of do-files; see [U] **16 Do-files**.

❏

▷ Example 1

You decide to make some changes to the auto dataset. You make a snapshot of the data before you begin making changes, and you make another snapshot after the changes:

```
. use http://www.stata-press.com/data/r12/auto
(1978 Automobile Data)
. snapshot save, label("before changes")
snapshot 1 (before changes) created at 19 Apr 2011 21:32
. generate gpm = 1/mpg
. label variable gpm "Gallons per mile"
. snapshot save, label("after changes")
snapshot 2 (after changes) created at 19 Apr 2011 21:34
```

You go on to do some analyses, but then, for some reason, you accidentally drop the variable you previously created:

```
. drop gpm
```

Luckily, you made some snapshots of your work:

```
. snapshot list
snapshot 1 (before changes) created at 19 Apr 2011 21:32
snapshot 2 (after changes) created at 19 Apr 2011 21:34
. snapshot restore 2
```

```
. describe gpm

                 storage   display    value
variable name    type      format     label      variable label

gpm              float     %9.0g                  Gallons per mile
```

◁

Saved results

snapshot save saves the following in r():

Scalars
    r(snapshot)   sequence number of snapshot saved

Also see

[D] **edit** — Browse or edit data with Data Editor

[P] **preserve** — Preserve and restore data

Title

sort — Sort data

Syntax

<u>so</u>rt *varlist* $\left[\,in\,\right]$ $\left[\,,\,\text{stable}\,\right]$

Menu

Data > Sort > Ascending sort

Description

sort arranges the observations of the current data into ascending order based on the values of the variables in *varlist*. There is no limit to the number of variables in the *varlist*. Missing numeric values are interpreted as being larger than any other number, so they are placed last with $. < .a < .b < \cdots < .z$. When you sort on a string variable, however, null strings are placed first. The dataset is marked as being sorted by *varlist* unless in *range* is specified. If in *range* is specified, only those observations are rearranged. The unspecified observations remain in the same place.

Option

stable specifies that observations with the same values of the variables in *varlist* keep the same relative order in the sorted data that they had previously. For instance, consider the following data:

```
x b
3 1
1 2
1 1
1 3
2 4
```

Typing sort x without the stable option produces one of the following six orderings:

x b	x b	x b	x b	x b	x b
1 2	1 2	1 1	1 1	1 3	1 3
1 1	1 3	1 3	1 2	1 1	1 2
1 3	1 1	1 2	1 3	1 2	1 1
2 4	2 4	2 4	2 4	2 4	2 4
3 1	3 1	3 1	3 1	3 1	3 1

Without the stable option, the ordering of observations with equal values of *varlist* is randomized. With sort x, stable, you will always get the first ordering and never the other five.

If your intent is to have the observations sorted first on x and then on b within tied values of x (the fourth ordering above), you should type sort x b rather than sort x, stable.

stable is seldom used and, when specified, causes sort to execute more slowly.

Remarks

Sorting data is one of the more common tasks involved in processing data. Sometimes, before Stata can perform some task, the data must be in a specific order. For example, if you want to use the by *varlist*: prefix, the data must be sorted in order of *varlist*. You use the `sort` command to fulfill this requirement.

▷ Example 1

Sorting data can also be informative. Suppose that we have data on automobiles, and each car's make and mileage rating (called `make` and `mpg`) are included among the variables in the data. We want to list the five cars with the lowest mileage rating in our data:

```
. use http://www.stata-press.com/data/r12/auto
(1978 Automobile Data)
. keep make mpg weight
. sort mpg, stable
. list make mpg in 1/5
```

	make	mpg
1.	Linc. Continental	12
2.	Linc. Mark V	12
3.	Cad. Deville	14
4.	Cad. Eldorado	14
5.	Linc. Versailles	14

◁

▷ Example 2: Tracking the sort order

Stata keeps track of the order of your data. For instance, we just sorted the above data on `mpg`. When we ask Stata to `describe` the data in memory, it tells us how the dataset is sorted:

```
. describe
Contains data from http://www.stata-press.com/data/r12/auto.dta
  obs:            74                          1978 Automobile Data
  vars:            3                          13 Apr 2011 17:45
  size:         1,628                         (_dta has notes)
```

	storage	display	value	
variable name	type	format	label	variable label
make	str18	%-18s		Make and Model
mpg	int	%8.0g		Mileage (mpg)
weight	int	%8.0gc		Weight (lbs.)

```
Sorted by:  mpg
     Note:  dataset has changed since last saved
```

Stata keeps track of changes in sort order. If we were to make a change to the `mpg` variable, Stata would know that the data are no longer sorted. Remember that the first observation in our data has `mpg` equal to 12, as does the second. Let's change the value of the first observation:

```
. replace mpg=13 in 1
(1 real change made)
```

```
. describe

Contains data from http://www.stata-press.com/data/r12/auto.dta
    obs:           74                          1978 Automobile Data
   vars:            3                          13 Apr 2011 17:45
   size:        1,628                          (_dta has notes)
```

variable name	storage type	display format	value label	variable label
make	str18	%-18s		Make and Model
mpg	int	%8.0g		Mileage (mpg)
weight	int	%8.0gc		Weight (lbs.)

```
Sorted by:
     Note:  dataset has changed since last saved
```

After making the change, Stata indicates that our dataset is "Sorted by:" nothing. Let's put the dataset back as it was:

```
. replace mpg=12 in 1
(1 real change made)
. sort mpg
```
◁

❏ Technical note

Stata does not track changes in the sort order and will sometimes decide that a dataset is not sorted when, in fact, it is. For instance, if we were to change the first observation of our auto dataset from 12 miles per gallon to 10, Stata would decide that the dataset is "Sorted by:" nothing, just as it did above when we changed mpg from 12 to 13. Our change in example 2 did change the order of the data, so Stata was correct. Changing mpg from 12 to 10, however, does not really affect the sort order.

As far as Stata is concerned, any change to the variables on which the data are sorted means that the data are no longer sorted, even if the change actually leaves the order unchanged. Stata may be dumb, but it is also fast. It sorts already-sorted datasets instantly, so Stata's ignorance costs us little.

❏

▷ Example 3: Sorting on multiple variables

Data can be sorted by more than one variable, and in such cases, the sort order is lexicographic. If we sort the data by two variables, for instance, the data are placed in ascending order of the first variable, and then observations that share the same value of the first variable are placed in ascending order of the second variable. Let's order our automobile data by mpg and within mpg by weight:

```
. sort mpg weight
. list in 1/8, sep(4)
```

	make	mpg	weight
1.	Linc. Mark V	12	4,720
2.	Linc. Continental	12	4,840
3.	Peugeot 604	14	3,420
4.	Linc. Versailles	14	3,830
5.	Cad. Eldorado	14	3,900
6.	Merc. Cougar	14	4,060
7.	Merc. XR-7	14	4,130
8.	Cad. Deville	14	4,330

The data are in ascending order of mpg, and, within each mpg category, the data are in ascending order of weight. The lightest car that achieves 14 miles per gallon in our data is the Peugeot 604.

◁

❑ Technical note

The sorting technique used by Stata is fast, but the order of variables not included in the *varlist* is not maintained. If you wish to maintain the order of additional variables, include them at the end of the *varlist*. There is no limit to the number of variables by which you may sort.

❑

▷ Example 4: Descending sorts

Sometimes you may want to order a dataset by descending sequence of something. Perhaps we wish to obtain a list of the five cars achieving the best mileage rating. The sort command orders the data only into ascending sequences. Another command, gsort, orders the data in ascending or descending sequences; see [D] **gsort**. You can also create the negative of a variable and achieve the desired result:

```
. generate negmpg = -mpg
. sort negmpg
. list in 1/5
```

	make	mpg	weight	negmpg
1.	VW Diesel	41	2,040	-41
2.	Subaru	35	2,050	-35
3.	Datsun 210	35	2,020	-35
4.	Plym. Champ	34	1,800	-34
5.	Toyota Corolla	31	2,200	-31

We find that the VW Diesel tops our list.

◁

▷ Example 5: Sorting on string variables

sort may also be used on string variables. The data are sorted alphabetically:

```
. sort make
. list in 1/5
```

	make	mpg	weight	negmpg
1.	AMC Concord	22	2,930	-22
2.	AMC Pacer	17	3,350	-17
3.	AMC Spirit	22	2,640	-22
4.	Audi 5000	17	2,830	-17
5.	Audi Fox	23	2,070	-23

◁

❑ Technical note

Bear in mind that Stata takes "alphabetically" to mean that all uppercase letters come before lowercase letters. As far as Stata is concerned, the following list is sorted alphabetically:

```
. list, sep(0)
```

	myvar
1.	ALPHA
2.	Alpha
3.	BETA
4.	Beta
5.	alpha
6.	beta

❑

References

Royston, P. 2001. Sort a list of items. *Stata Journal* 1: 105–106.

Schumm, L. P. 2006. Stata tip 28: Precise control of dataset sort order. *Stata Journal* 6: 144–146.

Also see

[D] **describe** — Describe data in memory or in file

[D] **gsort** — Ascending and descending sort

[U] **11 Language syntax**

Title

> **split** — Split string variables into parts

Syntax

split *strvar* $[if]$ $[in]$ $[, options]$

options	Description
Main	
generate(*stub*)	begin new variable names with *stub*; default is *strvar*
parse(*parse_strings*)	parse on specified strings; default is to parse on spaces
limit(*#*)	create a maximum of *#* new variables
notrim	do not trim leading or trailing spaces of original variable
Destring	
destring	apply destring to new string variables, replacing initial string variables with numeric variables where possible
ignore("*chars*")	remove specified nonnumeric characters
force	convert nonnumeric strings to missing values
float	generate numeric variables as type float
percent	convert percent variables to fractional form

Menu

Data > Create or change data > Other variable-transformation commands > Split string variables into parts

Description

split splits the contents of a string variable, *strvar*, into one or more parts, using one or more *parse_strings* (by default, blank spaces), so that new string variables are generated. Thus split is useful for separating "words" or other parts of a string variable. *strvar* itself is not modified.

Options

> Main

generate(*stub*) specifies the beginning characters of the new variable names so that new variables *stub*1, *stub*2, etc., are produced. *stub* defaults to *strvar*.

parse(*parse_strings*) specifies that, instead of using spaces, parsing use one or more *parse_strings*. Most commonly, one string that is one punctuation character will be specified. For example, if parse(,) is specified, "1,2,3" is split into "1", "2", and "3".

You can also specify 1) two or more strings that are alternative separators of "words" and 2) strings that consist of two or more characters. Alternative strings should be separated by spaces. Strings that include spaces should be bound by " ". Thus if parse(, " ") is specified, "1,2 3" is also split into "1", "2", and "3". Note particularly the difference between, say, parse(a b) and parse(ab): with the first, a and b are both acceptable as separators, whereas with the second, only the string ab is acceptable.

581

limit(*#*) specifies an upper limit to the number of new variables to be created. Thus limit(2) specifies that, at most, two new variables be created.

notrim specifies that the original string variable not be trimmed of leading and trailing spaces before being parsed. notrim is not compatible with parsing on spaces, because the latter implies that spaces in a string are to be discarded. You can either specify a parsing character or, by default, allow a trim.

⌐ Destring ⌐

destring applies destring to the new string variables, replacing the variables initially created as strings by numeric variables where possible. See [D] **destring**.

ignore(), force, float, percent; see [D] **destring**.

Remarks

split is used to split a string variable into two or more component parts, for example, "words". You might need to correct a mistake, or the string variable might be a genuine composite that you wish to subdivide before doing more analysis.

The basic steps applied by split are, given one or more separators, to find those separators within the string and then to generate one or more new string variables, each containing a part of the original. The separators could be, for example, spaces or other punctuation symbols, but they can in turn be strings containing several characters. The default separator is a space.

The key string functions for subdividing string variables and, indeed, strings in general, are strpos(), which finds the position of separators, and substr(), which extracts parts of the string. (See [D] **functions**.) split is based on the use of those functions.

If your problem is not defined by splitting on separators, you will probably want to use substr() directly. Suppose that you have a string variable, date, containing dates in the form "21011952" so that the last four characters define a year. This string contains no separators. To extract the year, you would use substr(date,-4,4). Again suppose that each woman's obstetric history over the last 12 months was recorded by a str12 variable containing values such as "npppppppbnn", where p, b, and n denote months of pregnancy, birth, and nonpregnancy. Once more, there are no separators, so you would use substr() to subdivide the string.

split discards the separators, because it presumes that they are irrelevant to further analysis or that you could restore them at will. If this is not what you want, you might use substr() (and possibly strpos()).

Finally, before we turn to examples, compare split with the egen function ends(), which produces the head, the tail, or the last part of a string. This function, like all egen functions, produces just one new variable as a result. In contrast, split typically produces several new variables as the result of one command. For more details and discussion, including comments on the special problem of recognizing personal names, see [D] **egen**.

split can be useful when input to Stata is somehow misread as one string variable. If you copy and paste into the Data Editor, say, under Windows by using the clipboard, but data are space-separated, what you regard as separate variables will be combined because the Data Editor expects comma- or tab-separated data. If some parts of your composite variable are numeric characters that should be put into numeric variables, you could use destring at the same time; see [D] **destring**.

```
. split var1, destring
```

Here no `generate()` option was specified, so the new variables will have names `var11`, `var12`, and so forth. You may now wish to use `rename` to produce more informative variable names. See [D] **rename**.

You can also use `split` to subdivide genuine composites. For example, email addresses such as tech-support@stata.com may be split at "@":

```
. split address, p(@)
```

This sequence yields two new variables: `address1`, containing the part of the email address before the "@", such as "tech-support", and `address2`, containing the part after the "@", such as "stata.com". The separator itself, "@", is discarded. Because `generate()` was not specified, the name `address` was used as a stub in naming the new variables. `split` displays the names of new variables created, so you will see quickly whether the number created matches your expectations.

If the details of individuals were of no interest and you wanted only machine names, either

```
. egen machinename = ends(address), tail p(@)
```

or

```
. generate machinename = substr(address, strpos(address,"@") + 1,.)
```

would be more direct.

Next suppose that a string variable holds names of legal cases that should be split into variables for plaintiff and defendant. The separators could be " V ", " V. ", " VS ", and " VS. ". (We assume that any inconsistency in the use of uppercase and lowercase has been dealt with by the string function `upper()`; see [D] **functions**.) Note particularly the leading and trailing spaces in our detailing of separators: the first separator is " V ", for example, not "V", which would incorrectly split "GOLIATH V DAVID" into "GOLIATH ", " DA", and "ID". The alternative separators are given as the argument to `parse()`:

```
. split case, p(" V " " V. " " VS " " VS. ")
```

Again with default naming of variables and recalling that separators are discarded, we expect new variables `case1` and `case2`, with no creation of `case3` or further new variables. Whenever none of the separators specified were found, `case2` would have empty values, so we can check:

```
. list case if case2 == ""
```

Suppose that a string variable contains fields separated by tabs. For example, `insheet` leaves tabs unchanged. Knowing that a tab is `char(9)`, we can type

```
. split data, p(`=char(9)') destring
```

`p(char(9))` would not work. The argument to `parse()` is taken literally, but evaluation of functions on the fly can be forced as part of macro substitution.

Finally, suppose that a string variable contains substrings bound in parentheses, such as (1 2 3) (4 5 6). Here we can split on the right parentheses and, if desired, those afterward. For example,

```
. split data, p(")")
. foreach v in `r(varlist)' {
        replace `v' = `v' + ")"
. }
```

Saved results

split saves the following in r():

Scalars
    r(nvars)   number of new variables created
    r(varlist)   names of the newly created variables

Methods and formulas

split is implemented as an ado-file.

Acknowledgments

split was written by Nicholas J. Cox, Durham University, who, in turn, thanks Michael Blasnik, M. Blasnik & Associates, for ideas contributed to an earlier jointly written program.

Also see

[D] **rename** — Rename variable

[D] **separate** — Create separate variables

[D] **destring** — Convert string variables to numeric variables and vice versa

[D] **egen** — Extensions to generate

[D] **functions** — Functions

Title

stack — Stack data

Syntax

stack *varlist* [*if*] [*in*], { <u>into</u>(*newvars*) | <u>g</u>roup(*#*) } [*options*]

options	Description
Main	
* <u>into</u>(*newvars*)	identify names of new variables to be created
* <u>g</u>roup(*#*)	stack *#* groups of variables in *varlist*
clear	clear dataset from memory
<u>wi</u>de	keep variables in *varlist* that are not specified in *newvars*

* Either into(*newvars*) or group(*#*) is required.

Menu

Data > Create or change data > Other variable-transformation commands > Stack data

Description

stack stacks the variables in *varlist* vertically, resulting in a dataset with variables *newvars* and $\_N \cdot (N_v/N_n)$ observations, where N_v is the number of variables in *varlist* and N_n is the number in *newvars*. stack creates the new variable _stack identifying the groups.

Options

⌐ Main ⌐

into(*newvars*) identifies the names of the new variables to be created. into() may be specified using variable ranges (for example, into(v1-v3)). Either into() or group(), but not both, must be specified.

group(*#*) specifies the number of groups of variables in *varlist* to be stacked. The created variables will be named according to the first group in *varlist*. Either group() or into(), but not both, must be specified.

clear indicates that it is okay to clear the dataset in memory. If you do not specify this option, you will be asked to confirm your intentions.

wide includes any of the original variables in *varlist* that are not specified in *newvars* in the resulting data.

585

Remarks

▷ Example 1

This command is best understood by examples. We begin with artificial but informative examples and end with useful examples.

```
. use http://www.stata-press.com/data/r12/stackxmpl
. list
```

	a	b	c	d
1.	1	2	3	4
2.	5	6	7	8

```
. stack  a b  c d, into(e f) clear
. list
```

	_stack	e	f
1.	1	1	2
2.	1	5	6
3.	2	3	4
4.	2	7	8

We formed the new variable e by stacking a and c, and we formed the new variable f by stacking b and d. _stack is automatically created and set equal to 1 for the first (a, b) group and equal to 2 for the second (c, d) group. (When _stack==1, the new data e and f contain the values from a and b. When _stack==2, e and f contain values from c and d.)

There are two groups because we specified four variables in the *varlist* and two variables in the into list, and $4/2 = 2$. If there were six variables in the *varlist*, there would be $6/2 = 3$ groups. If there were also three variables in the into list, there would be $6/3 = 2$ groups. Specifying six variables in the *varlist* and four variables in the into list would result in an error because $6/4$ is not an integer.

◁

▷ Example 2

Variables may be repeated in the *varlist*, and the *varlist* need not contain all the variables:

```
. use http://www.stata-press.com/data/r12/stackxmpl, clear
. list
```

	a	b	c	d
1.	1	2	3	4
2.	5	6	7	8

```
. stack  a b  a c, into(a bc) clear
```

```
. list
```

	_stack	a	bc
1.	1	1	2
2.	1	5	6
3.	2	1	3
4.	2	5	7

a was stacked on a and called a, whereas b was stacked on c and called bc.

If we had wanted the resulting variables to be called simply a and b, we could have used

```
. stack  a b  a c, group(2) clear
```

which is equivalent to

```
. stack  a b  a c, into(a b) clear
```

◁

▷ Example 3

In this artificial but informative example, the wide option includes the variables in the original dataset that were specified in *varlist* in the output dataset:

```
. use http://www.stata-press.com/data/r12/stackxmpl, clear
. list
```

	a	b	c	d
1.	1	2	3	4
2.	5	6	7	8

```
. stack  a b  c d, into(e f) clear wide
. list
```

	_stack	e	f	a	b	c	d
1.	1	1	2	1	2	.	.
2.	1	5	6	5	6	.	.
3.	2	3	4	.	.	3	4
4.	2	7	8	.	.	7	8

In addition to the stacked e and f variables, the original a, b, c, and d variables are included. They are set to missing where their values are not appropriate.

◁

▷ Example 4

This is the last artificial example. When you specify the wide option and repeat the same variable name in both the *varlist* and the into list, the variable will contain the stacked values:

```
. use http://www.stata-press.com/data/r12/stackxmpl, clear
. list
```

	a	b	c	d
1.	1	2	3	4
2.	5	6	7	8

```
. stack a b  a c, into(a bc) clear wide
. list
```

	_stack	a	bc	b	c
1.	1	1	2	2	.
2.	1	5	6	6	.
3.	2	1	3	.	3
4.	2	5	7	.	7

◁

▷ Example 5

We want one graph of y against x1 and y against x2. We might be tempted to type `scatter y x1 x2`, but that would graph y against x2 and x1 against x2. One solution is to type

```
. save mydata
. stack  y x1  y x2, into(yy x12) clear
. generate y1 = yy if _stack==1
. generate y2 = yy if _stack==2
. scatter y1 y2 x12
. use mydata, clear
```

The names yy and x12 are supposed to suggest the contents of the variables. yy contains (y,y), and x12 contains (x1,x2). We then make y1 defined at the x1 points but missing at the x2 points—graphing y1 against x12 is the same as graphing y against x1 in the original dataset. Similarly, y2 is defined at the x2 points but missing at x1—graphing y2 against x12 is the same as graphing y against x2 in the original dataset. Therefore, `scatter y1 y2 x12` produces the desired graph.

◁

▷ Example 6

We wish to graph y1 against x1 and y2 against x2 on the same graph. The logic is the same as above, but let's go through it. Perhaps we have constructed two cumulative distributions by using cumul (see [R] **cumul**):

```
. use http://www.stata-press.com/data/r12/citytemp
(City Temperature Data)
. cumul tempjan, gen(cjan)
. cumul tempjuly, gen(cjuly)
```

We want to graph both cumulatives in the same graph; that is, we want to graph cjan against tempjan and cjuly against tempjuly. Remember that we could graph the tempjan cumulative by typing

```
. scatter cjan tempjan, c(l) m(o) sort
  (output omitted )
```

We can graph the `tempjuly` cumulative similarly. To obtain both on the same graph, we must stack the data:

```
. stack  cjuly tempjuly    cjan tempjan, into(c temp) clear
. generate cjan  = c if _stack==1
(958 missing values generated)
. generate cjuly = c if _stack==2
(958 missing values generated)
. scatter cjan cjuly temp, c(l l) m(o o) sort
  (output omitted)
```

Alternatively, if we specify the `wide` option, we do not have to regenerate `cjan` and `cjuly` because they will be created automatically:

```
. use http://www.stata-press.com/data/r12/citytemp, clear
(City Temperature Data)
. cumul tempjan, gen(cjan)
. cumul tempjuly, gen(cjuly)
. stack  cjuly tempjuly    cjan tempjan, into(c temp) clear wide
. scatter cjan cjuly temp, c(l l) m(o o) sort
  (output omitted)
```

◁

❑ Technical note

There is a third way, not using the `wide` option, that is exceedingly tricky but is sometimes useful:

```
. use http://www.stata-press.com/data/r12/citytemp, clear
(City Temperature Data)
. cumul tempjan, gen(cjan)
. cumul tempjuly, gen(cjuly)
. stack cjuly tempjuly  cjan tempjan, into(c temp) clear
. sort _stack temp
. scatter c temp, c(L) m(o)
  (output omitted)
```

Note the use of `connect`'s capital L rather than lowercase l option. `c(L)` connects points only from left to right; because the data are sorted by `_stack temp`, `temp` increases within the first group (`cjuly` vs. `tempjuly`) and then starts again for the second (`cjan` vs. `tempjan`); see [G-4] *connectstyle*.

❑

Methods and formulas

`stack` is implemented as an ado-file.

Reference

Baum, C. F. 2009. *An Introduction to Stata Programming*. College Station, TX: Stata Press.

Also see

[D] **contract** — Make dataset of frequencies and percentages

[D] **reshape** — Convert data from wide to long form and vice versa

[D] **xpose** — Interchange observations and variables

Title

> **statsby** — Collect statistics for a command across a by list

Syntax

statsby $\begin{bmatrix} exp\_list \end{bmatrix}$ $\begin{bmatrix} , & options \end{bmatrix}$: *command*

options	Description
Main	
*by(*varlist* $\begin{bmatrix} , & \underline{\text{miss}}\text{ing} \end{bmatrix}$)	equivalent to interactive use of by *varlist*:
Options	
clear	replace data in memory with results
saving(*filename*, ...)	save results to *filename*; save statistics in double precision; save results to *filename* every # replications
total	include results for the entire dataset
subsets	include all combinations of subsets of groups
Reporting	
nodots	suppress replication dots
noisily	display any output from *command*
trace	trace *command*
nolegend	suppress table legend
verbose	display the full table legend
Advanced	
basepop(*exp*)	restrict initializing sample to *exp*; seldom used
force	do not check for svy commands; seldom used
forcedrop	retain only observations in by-groups when calling *command*; seldom used

* by(*varlist*) is required on the dialog box because statsby is useful to the interactive user only when using by(). All weight types supported by *command* are allowed except pweights; see [U] **11.1.6 weight**.

exp_list contains	(*name*: *elist*)
	elist
	eexp
elist contains	*newvarname* = (*exp*)
	(*exp*)
eexp is	*specname*
	[*eqno*]*specname*
specname is	_b
	_b[]
	_se
	_se[]
eqno is	# #
	name

exp is a standard Stata expression; see [U] **13 Functions and expressions**.

Distinguish between [], which are to be typed, and [], which indicate optional arguments.

Menu

Statistics > Other > Collect statistics for a command across a by list

Description

statsby collects statistics from *command* across a by list. Typing

. statsby *exp_list*, by(*varname*): *command*

executes *command* for each group identified by *varname*, building a dataset of the associated values from the expressions in *exp_list*. The resulting dataset replaces the current dataset, unless the saving() option is supplied. *varname* can refer to a numeric or a string variable.

command defines the statistical command to be executed. Most Stata commands and user-written programs can be used with statsby, as long as they follow standard Stata syntax and allow the if qualifier; see [U] **11 Language syntax**. The by prefix cannot be part of *command*.

exp_list specifies the statistics to be collected from the execution of *command*. If no expressions are given, *exp_list* assumes a default depending upon whether *command* changes results in e() and r(). If *command* changes results in e(), the default is _b. If *command* changes results in r() (but not e()), the default is all the scalars posted to r(). It is an error not to specify an expression in *exp_list* otherwise.

Options

⌐ Main ⌐

by(*varlist* [, missing]) specifies a list of existing variables that would normally appear in the by *varlist*: section of the command if you were to issue the command interactively. By default, statsby ignores groups in which one or more of the by() variables is missing. Alternatively, missing causes missing values to be treated like any other values in the by-groups, and results from the entire dataset are included with use of the subsets option. If by() is not specified, *command* will be run on the entire dataset. *varlist* can contain both numeric and string variables.

⌐ Options ⌐

clear specifies that it is okay to replace the data in memory, even though the current data have not been saved to disk.

saving(*filename*[, *suboptions*]) creates a Stata data file (.dta file) consisting of (for each statistic in *exp_list*) a variable containing the replicates.

  double specifies that the results for each replication be stored as doubles, meaning 8-byte reals. By default, they are stored as floats, meaning 4-byte reals.

  every(#) specifies that results be written to disk every #th replication. every() should be specified in conjunction with saving() only when *command* takes a long time for each replication. This will allow recovery of partial results should your computer crash. See [P] **postfile**.

total specifies that *command* be run on the entire dataset, in addition to the groups specified in the by() option.

subsets specifies that *command* be run for each group defined by any combination of the variables in the by() option.

nodots suppresses display of the replication dots. By default, one dot character is printed for each by-group. A red 'x' is printed if *command* returns with an error or if one of the values in *exp_list* is missing.

noisily causes the output of *command* to be displayed for each by-group. This option implies the nodots option.

trace causes a trace of the execution of *command* to be displayed. This option implies the noisily option.

nolegend suppresses the display of the table legend, which identifies the rows of the table with the expressions they represent.

verbose requests that the full table legend be displayed. By default, coefficients and standard errors are not displayed.

basepop(*exp*) specifies a base population that statsby uses to evaluate the *command* and to set up for collecting statistics. The default base population is the entire dataset, or the dataset specified by any if or in conditions specified on the *command*.

One situation where basepop() is useful is collecting statistics over the panels of a panel dataset by using an estimator that works for time series, but not panel data, for example,

```
. statsby, by(mypanels) basepop(mypanels==2): arima ...
```

force suppresses the restriction that *command* not be a svy command. statsby does not perform subpopulation estimation for survey data, so it should not be used with svy. statsby reports an error when it encounters svy in *command* if the force option is not specified. This option is seldom used, so use it only if you know what you are doing.

forcedrop forces statsby to drop all observations except those in each by-group before calling *command* for the group. This allows statsby to work with user-written commands that completely ignore if and in but do not return an error when either is specified. forcedrop is seldom used.

Remarks

Remarks are presented under the following headings:

> *Collecting coefficients and standard errors*
> *Collecting saved results*
> *All subsets*

Collecting coefficients and standard errors

▷ Example 1

We begin with an example using the auto.dta dataset. In this example, we want to collect the coefficients from a regression in which we model the price of a car on its weight, length, and mpg. We want to run this model for both domestic and foreign cars. We can do this easily by using statsby with the extended expression _b.

```
. use http://www.stata-press.com/data/r12/auto
(1978 Automobile Data)
. statsby _b, by(foreign) verbose nodots: regress price weight length mpg
     command:  regress price weight length mpg
   _b_weight:  _b[weight]
   _b_length:  _b[length]
      _b_mpg:  _b[mpg]
     _b_cons:  _b[_cons]
          by:  foreign
. list
```

	foreign	_b_wei~t	_b_length	_b_mpg	_b_cons
1.	Domestic	6.767233	-109.9518	142.7663	2359.475
2.	Foreign	4.784841	13.39052	-18.4072	-6497.49

If we were interested only in the coefficient of a particular variable, such as mpg, we would specify that particular coefficient; see [U] **13.5 Accessing coefficients and standard errors**.

```
. use http://www.stata-press.com/data/r12/auto, clear
(1978 Automobile Data)
. statsby mpg=_b[mpg], by(foreign) nodots: regress price weight length mpg
     command:  regress price weight length mpg
         mpg:  _b[mpg]
          by:  foreign
. list
```

	foreign	mpg
1.	Domestic	142.7663
2.	Foreign	-18.4072

The extended expression _se indicates that we want standard errors.

```
. use http://www.stata-press.com/data/r12/auto, clear
(1978 Automobile Data)
. statsby _se, by(foreign) verbose nodots: regress price weight length mpg
     command:  regress price weight length mpg
  _se_weight:  _se[weight]
  _se_length:  _se[length]
     _se_mpg:  _se[mpg]
    _se_cons:  _se[_cons]
          by:  foreign
```

. list

	foreign	_se_we~t	_se_le~h	_se_mpg	_se_cons
1.	Domestic	1.226326	39.48193	134.7221	7770.131
2.	Foreign	1.670006	50.70229	59.37442	6337.952

◁

▷ Example 2

For multiple-equation estimations, we can use [*eqno*]_b ([*eqno*]_se) to get the coefficients (standard errors) of a specific equation or use _b (_se) to get the coefficients (standard errors) of all the equations. To demonstrate, we use heckman and a slightly different dataset.

```
. use http://www.stata-press.com/data/r12/statsby, clear
. statsby _b, by(group) verbose nodots: heckman price mpg, sel(trunk)
        command:  heckman price mpg, sel(trunk)
    price_b_mpg:  [price]_b[mpg]
   price_b_cons:  [price]_b[_cons]
   select_b_tr~k: [select]_b[trunk]
   select_b_cons: [select]_b[_cons]
   athrho_b_cons: [athrho]_b[_cons]
  lnsigma_b_c~s:  [lnsigma]_b[_cons]
             by:  group
. list, compress noobs
```

group	price_b~g	price_~s	select_~k	select~s	athrho_~s	lnsigm~s
1	-253.9293	11836.33	-.0122223	1.248342	-.3107811	7.895351
2	-242.5759	11906.46	-.0488969	1.943078	-1.399222	8.000272
3	-172.6499	9813.357	-.0190373	1.452783	-.3282423	7.876059
4	-250.7318	10677.31	.0525965	.3502012	.6133645	7.96349

To collect the coefficients of the first equation only, we would specify [price]_b instead of _b.

```
. use http://www.stata-press.com/data/r12/statsby, clear
. statsby [price]_b, by(group) verbose nodots: heckman price mpg, sel(trunk)
        command:  heckman price mpg , sel(trunk)
    price_b_mpg:  [price]_b[mpg]
   price_b_cons:  [price]_b[_cons]
             by:  group
. list
```

	group	price_b~g	price_~s
1.	1	-253.9293	11836.33
2.	2	-242.5759	11906.46
3.	3	-172.6499	9813.357
4.	4	-250.7318	10677.31

◁

❑ Technical note

> If *command* fails on one or more groups, statsby will capture the error messages and ignore those groups.

❑

Collecting saved results

> Many Stata commands save results of calculations; see [U] **13.6 Accessing results from Stata commands**. statsby can collect the saved results and expressions involving these saved results, too. Expressions must be bound in parentheses.

▷ Example 3

> Suppose that we want to collect the mean and the median of price, as well as their ratios, and we want to collect them for both domestic and foreign cars. We might type

```
. use http://www.stata-press.com/data/r12/auto, clear
(1978 Automobile Data)
. statsby mean=r(mean) median=r(p50) ratio=(r(mean)/r(p50)), by(foreign) nodots:
> summarize price, detail
      command:  summarize price, detail
         mean:  r(mean)
       median:  r(p50)
        ratio:  r(mean)/r(p50)
           by:  foreign
. list
```

	foreign	mean	median	ratio
1.	Domestic	6072.423	4782.5	1.269717
2.	Foreign	6384.682	5759	1.108644

◁

❑ Technical note

> In *exp_list*, *newvarname* is not required. If no new variable name is specified, statsby names the new variables _stat_1, _stat_2, and so forth.

❑

All subsets

▷ Example 4

> When there are two or more variables in by (*varlist*), we can execute *command* for any combination, or subset, of the variables in the by() option by specifying the subsets option.

```
. use http://www.stata-press.com/data/r12/auto, clear
(1978 Automobile Data)

. statsby mean=r(mean) median=r(p50) n=r(N), by(foreign rep78) subsets nodots:
> summarize price, detail
        command:  summarize price, detail
           mean:  r(mean)
         median:  r(p50)
             n:   r(N)
             by:  foreign rep78

. list
```

	foreign	rep78	mean	median	n
1.	Domestic	1	4564.5	4564.5	2
2.	Domestic	2	5967.625	4638	8
3.	Domestic	3	6607.074	4749	27
4.	Domestic	4	5881.556	5705	9
5.	Domestic	5	4204.5	4204.5	2
6.	Domestic	.	6179.25	4853	48
7.	Foreign	3	4828.667	4296	3
8.	Foreign	4	6261.444	6229	9
9.	Foreign	5	6292.667	5719	9
10.	Foreign	.	6070.143	5719	21
11.	.	1	4564.5	4564.5	2
12.	.	2	5967.625	4638	8
13.	.	3	6429.233	4741	30
14.	.	4	6071.5	5751.5	18
15.	.	5	5913	5397	11
16.	.	.	6165.257	5006.5	74

In the above dataset, observation 6 is for domestic cars, regardless of the repair record; observation 10 is for foreign cars, regardless of the repair record; observation 11 is for both foreign cars and domestic cars given that the repair record is 1; and the last observation is for the entire dataset.

◁

❏ Technical note

To see the output from *command* for each group identified in the by() option, we can use the noisily option.

```
. use http://www.stata-press.com/data/r12/auto, clear
(1978 Automobile Data)

. statsby mean=r(mean) se=(r(sd)/sqrt(r(N))), by(foreign) noisily nodots:
> summarize price
statsby: First call to summarize with data as is:
```

```
. summarize price
    Variable |      Obs        Mean    Std. Dev.       Min        Max
-------------+--------------------------------------------------------
       price |       74    6165.257    2949.496       3291      15906
```

statsby legend:
```
       command:  summarize price
          mean:  r(mean)
            se:  r(sd)/sqrt(r(N))
            by:  foreign
```
Statsby groups
running (summarize price) on group 1

```
. summarize price
    Variable |      Obs        Mean    Std. Dev.       Min        Max
-------------+--------------------------------------------------------
       price |       52    6072.423    3097.104       3291      15906
```

running (summarize price) on group 2

```
. summarize price
    Variable |      Obs        Mean    Std. Dev.       Min        Max
-------------+--------------------------------------------------------
       price |       22    6384.682    2621.915       3748      12990
```

```
. list

     +-----------------------------------+
     |  foreign       mean          se   |
     |-----------------------------------|
  1. | Domestic    6072.423    429.4911  |
  2. |  Foreign    6384.682    558.9942  |
     +-----------------------------------+
```

Methods and formulas

statsby is implemented as an ado-file.

Acknowledgment

Speed improvements in statsby were based on code written by Michael Blasnik, M. Blasnik & Associates.

References

Cox, N. J. 2010. Speaking Stata: The statsby strategy. *Stata Journal* 10: 143–151.

Hardin, J. W. 1996. dm42: Accrue statistics for a command across a by list. *Stata Technical Bulletin* 32: 5–9. Reprinted in *Stata Technical Bulletin Reprints*, vol. 6, pp. 13–18. College Station, TX: Stata Press.

Newson, R. 1999a. dm65.1: Update to a program for saving a model fit as a dataset. *Stata Technical Bulletin* 58: 25. Reprinted in *Stata Technical Bulletin Reprints*, vol. 10, p. 7. College Station, TX: Stata Press.

——. 1999b. dm65: A program for saving a model fit as a dataset. *Stata Technical Bulletin* 49: 2–5. Reprinted in *Stata Technical Bulletin Reprints*, vol. 9, pp. 19–23. College Station, TX: Stata Press.

——. 2003. Confidence intervals and p-values for delivery to the end user. *Stata Journal* 3: 245–269.

Also see

Title

sysuse — Use shipped dataset

Syntax

Use example dataset installed with Stata

> sysuse $\left[\,"\,\right]$*filename*$\left[\,"\,\right]$ $\left[\,,\ \texttt{clear}\,\right]$

List example Stata datasets installed with Stata

> sysuse dir $\left[\,,\ \texttt{all}\,\right]$

Menu

File > Example Datasets...

Description

sysuse *filename* loads the specified Stata-format dataset that was shipped with Stata or that is stored along the ado-path. If *filename* is specified without a suffix, .dta is assumed.

sysuse dir lists the names of the datasets shipped with Stata plus any other datasets stored along the ado-path.

Options

clear specifies that it is okay to replace the data in memory, even though the current data have not been saved to disk.

all specifies that all datasets be listed, even those that include an underscore (_) in their name. By default, such datasets are not listed.

Remarks

Remarks are presented under the following headings:

> *Typical use*
> *A note concerning shipped datasets*
> *Using user-installed datasets*
> *How sysuse works*

Typical use

A few datasets are included with Stata and are stored in the system directories. These datasets are often used in the help files to demonstrate a certain feature.

Typing

```
. sysuse dir
```

600

lists the names of those datasets. One such dataset is `lifeexp.dta`. If you simply type `use lifeexp`, you will see

```
. use lifeexp
file lifeexp.dta not found
r(601);
```

Type `sysuse`, however, and the dataset is loaded:

```
. sysuse lifeexp
(Life expectancy, 1998)
```

The datasets shipped with Stata are stored in different folders (directories) so that they do not become confused with your datasets.

A note concerning shipped datasets

Not all the datasets used in the manuals are shipped with Stata. To obtain the other datasets, see [D] **webuse**.

The datasets used to demonstrate Stata are often fictional. If you want to know whether a dataset is real or fictional, and its history, load the dataset and type

```
. notes
```

A few datasets have no notes. This means that the datasets are believed to be real, but that they were created so long ago that information about their original source has been lost. Treat such datasets as if they were fictional.

Using user-installed datasets

Any datasets you have installed using `net` or `ssc` (see [R] **net** and [R] **ssc**) can be listed by typing `sysuse dir` and can be loaded using `sysuse` *filename*.

Any datasets you store in your personal ado folder (see [P] **sysdir**) are also listed by `sysuse dir` and can be loaded using `sysuse` *filename*.

How sysuse works

`sysuse` simply looks across the ado-path for `.dta` files; see [P] **sysdir**.

By default, `sysuse dir` does not list a dataset that contains an underscore (_) in its name. By convention, such datasets are used by ado-files to achieve their ends and probably are not of interest to you. If you type `sysuse dir, all` all datasets are listed.

Saved results

`sysuse dir` saves in the macro `r(files)` the list of dataset names.

`sysuse` *filename* saves in the macro `r(fn)` the *filename*, including the full path specification.

Methods and formulas

`sysuse` is implemented as an ado-file.

Also see

[D] **webuse** — Use dataset from Stata website

[D] **use** — Load Stata dataset

[P] **findfile** — Find file in path

[P] **sysdir** — Query and set system directories

[R] **net** — Install and manage user-written additions from the Internet

[R] **ssc** — Install and uninstall packages from SSC

Title

type — Display contents of a file

Syntax

<u>typ</u>e ["] *filename* ["] [, *options*]

Note: Double quotes must be used to enclose *filename* if the name contains blanks.

options	Description
<u>asis</u>	show file as is; default is to display files with suffixes .smcl or .sthlp as SMCL
smcl	display file as SMCL; default for files with suffixes .smcl or .sthlp
<u>showtabs</u>	display tabs as <T> rather than being expanded
<u>star</u>bang	list lines in the file that begin with "*!"

Description

type lists the contents of a file stored on disk. This command is similar to the Windows type command and the Unix more(1) or pg(1) commands.

In Stata for Mac and Stata for Unix, cat is a synonym for type.

Options

asis specifies that the file be shown exactly as it is. The default is to display files with suffixes .smcl or .sthlp as SMCL, meaning that the SMCL directives are interpreted and properly rendered. Thus type can be used to look at files created by the log using command.

smcl specifies that the file be displayed as SMCL, meaning that the SMCL directives are interpreted and properly rendered. This is the default for files with suffixes .smcl or .sthlp.

showtabs requests that any tabs be displayed as <T> rather than being expanded.

starbang lists only the lines in the specified file that begin with the characters "*!". Such comment lines are typically used to indicate the version number of ado-files, class files, etc. starbang may not be used with SMCL files.

Remarks

▷ Example 1

We have raw data containing the level of Lake Victoria Nyanza and the number of sunspots during the years 1902–1921 stored in a file called sunspots.raw. We want to read this dataset into Stata by using infile, but we cannot remember the order in which we entered the variables. We can find out by using the type command:

```
. type sunspots.raw
1902 -10   5     1903   13 24     1904   18 42
1905  15  63     1906   29 54     1907   21 62
1908  10  49     1909    8 44     1910    1 19
1911  -7   6     1912  -11  4     1913   -3  1
1914  -2  10     1915    4 47     1916   15 57
1917  35 104     1918   27 81     1919    8 64
1920   3  38     1921   -5 25
```

Looking at this output, we now remember that the variables are entered year, level, and number of sunspots. We can read this dataset by typing `infile year level spots using sunspots`.

If we wanted to see the tabs in `sunspots.raw`, we could type

```
. type sunspots.raw, showtabs
1902 -10   5<T>1903   13 24<T>1904   18 42
1905  15  63<T>1906   29 54<T>1907   21 62
1908  10  49<T>1909    8 44<T>1910    1 19
1911  -7   6<T>1912  -11  4<T>1913   -3  1
1914  -2  10<T>1915    4 47<T>1916   15 57
1917  35 104<T>1918   27 81<T>1919    8 64
1920   3  38<T>1921   -5 25
```

◁

▷ Example 2

In a previous Stata session, we typed `log using myres` and created `myres.smcl`, containing our results. We can use `type` to list the log:

```
. type myres.smcl
```

name:	<unnamed>
log:	/work/peb/dof/myres.smcl
log type:	smcl
opened on:	20 Jan 2011, 15:37:48

```
. use lbw
(Hosmer & Lemeshow data)

. logistic low age lwt i.race smoke ptl ht ui
```

Logistic regression

Number of obs	=	189
LR chi2(8)	=	33.22
Prob > chi2	=	0.0001
Pseudo R2	=	0.1416

Log likelihood =   -100.724

(*output omitted*)

```
. estat gof
```
Logistic model for low, goodness-of-fit test

(*output omitted*)

```
. log close
```

name:	<unnamed>
log:	/work/peb/dof/myres.smcl
log type:	smcl
closed on:	20 Jan 2011, 15:38:30

We could also use `view` to look at the log; see [R] **view**.

◁

Also see

[R] **translate** — Print and translate logs

[R] **view** — View files and logs

[P] **viewsource** — View source code

[D] **cd** — Change directory

[D] **copy** — Copy file from disk or URL

[D] **dir** — Display filenames

[D] **erase** — Erase a disk file

[D] **mkdir** — Create directory

[D] **rmdir** — Remove directory

[D] **shell** — Temporarily invoke operating system

[U] **11.6 Filenaming conventions**

Title

> **use** — Load Stata dataset

Syntax

Load Stata-format dataset

> u̲se *filename* [, clear no̲label]

Load subset of Stata-format dataset

> u̲se [*varlist*] [*if*] [*in*] using *filename* [, clear no̲label]

Menu

File > Open...

Description

use loads into memory a Stata-format dataset previously saved by save. If *filename* is specified without an extension, .dta is assumed. If your *filename* contains embedded spaces, remember to enclose it in double quotes.

In the second syntax for use, a subset of the data may be read.

Options

clear specifies that it is okay to replace the data in memory, even though the current data have not been saved to disk.

nolabel prevents value labels in the saved data from being loaded. It is unlikely that you will ever want to specify this option.

Remarks

▷ Example 1

We have no data in memory. In a previous session, we issued the command save hiway to save the Minnesota Highway Data that we had been analyzing. We retrieve it now:

```
. use hiway
(Minnesota Highway Data, 1973)
```

Stata loads the data into memory and shows us that the dataset is labeled "Minnesota Highway Data, 1973".

◁

> ## Example 2

We continue to work with our `hiway` data and find an error in our data that needs correcting:

```
. replace spdlimit=70 in 1
(1 real change made)
```

We remember that we need to forward some information from another dataset to a colleague. We use that other dataset:

```
. use accident
no; data in memory would be lost
r(4);
```

Stata refuses to load the data because we have not saved the `hiway` data since we changed it.

```
. save hiway, replace
file hiway.dta saved
. use accident
(Minnesota Accident Data)
```

After we save our `hiway` data, Stata lets us load our `accident` dataset. If we had not cared whether our changed `hiway` dataset were saved, we could have typed `use accident, clear` to tell Stata to load the accident data without saving the changed dataset in memory.

◁

❏ Technical note

In example 2, you saved a revised `hiway.dta` dataset, which you forward to your colleague. Your colleague issues the command

```
. use hiway
```

and gets the message

```
file hiway.dta not Stata format
r(610);
```

Your colleague is using a version of Stata older than Stata 11. If your colleague is using Stata 9 or 10, you can save the dataset in Stata 9 or 10 format by using the `saveold` command; see [D] **save**.

Newer versions of Stata can always read datasets created by older versions of Stata. Stata/MP and Stata/SE can read datasets created by Stata/IC. Stata/IC can read datasets created by Stata/MP and Stata/SE if those datasets conform to Stata/IC's limits; type `help limits`.

❏

▷ Example 3

If you are using a dataset that is too large for the amount of memory on your computer, you could load only some of the variables:

```
. use ln_wage grade age tenure race using nlswork
(National Longitudinal Survey.  Young Women 14-26 years of age in 1968)
. describe
Contains data from nlswork.dta
  obs:          28,534                        National Longitudinal Survey.
                                              Young Women 14-26 years of age
                                              in 1968
  vars:              5                        7 Dec 2010 17:02
  size:        313,874

              storage  display    value
variable name   type   format     label      variable label

age             byte   %8.0g                 age in current year
race            byte   %8.0g                 1=white, 2=black, 3=other
grade           byte   %8.0g                 current grade completed
tenure          float  %9.0g                 job tenure, in years
ln_wage         float  %9.0g                 ln(wage/GNP deflator)

Sorted by:
```

Stata successfully loaded the five variables.

◁

▷ Example 4

You are new to Stata and want to try working with a Stata dataset that was used in example 1 of [XT] **xtlogit**. You load the dataset:

```
. use http://www.stata-press.com/data/r12/union
(NLS Women 14-24 in 1968)
```

The dataset is successfully loaded, but it would have been shorter to type

```
. webuse union
(NLS Women 14-24 in 1968)
```

webuse is a synonym for use http://www.stata-press.com/data/r12/; see [D] **webuse**.

◁

Also see

[D] **compress** — Compress data in memory

[D] **datasignature** — Determine whether data have changed

[D] **import** — Overview of importing data into Stata

[D] **save** — Save Stata dataset

[D] **sysuse** — Use shipped dataset

[D] **webuse** — Use dataset from Stata website

[U] **11.6 Filenaming conventions**

[U] **21 Inputting and importing data**

Title

> **varmanage** — Manage variable labels, formats, and other properties

Syntax

varmanage

Menu

Data > Variables Manager

Description

varmanage opens the Variables Manager. The Variables Manager allows for the sorting and filtering of variables for the purpose of setting properties on one or more variables at a time. Variable properties include the name, label, storage type, format, value label, and notes. The Variables Manager also can be used to create *varlists* for the Command window.

Remarks

A tutorial discussion of varmanage can be found in [GS] **7 Using the Variables Manager** (GSM, GSU, or GSW).

Also see

[D] **drop** — Eliminate variables or observations

[D] **edit** — Browse or edit data with Data Editor

[D] **format** — Set variables' output format

[D] **label** — Manipulate labels

[D] **notes** — Place notes in data

[D] **rename** — Rename variable

Title

---
webuse — Use dataset from Stata website

Syntax

Load dataset over the web

    webuse ["]*filename*["] [, clear]

Report URL from which datasets will be obtained

    webuse query

Specify URL from which dataset will be obtained

    webuse set [http://]*url*[/]

Reset URL to default

    webuse set

Menu

File > Example Datasets...

Description

webuse *filename* loads the specified dataset, obtaining it over the web. By default, datasets are obtained from http://www.stata-press.com/data/r12/. If *filename* is specified without a suffix, .dta is assumed.

webuse query reports the URL from which datasets will be obtained.

webuse set allows you to specify the URL to be used as the source for datasets. webuse set without arguments resets the source to http://www.stata-press.com/data/r12/.

Option

clear specifies that it is okay to replace the data in memory, even though the current data have not been saved to disk.

Remarks

Remarks are presented under the following headings:

        Typical use
        A note concerning example datasets
        Redirecting the source

Typical use

In the examples in the Stata manuals, we see things such as

```
. use http://www.stata-press.com/data/r12/lifeexp
```

The above is used to load—in this instance—the dataset `lifeexp.dta`. You can type that, and it will work:

```
. use http://www.stata-press.com/data/r12/lifeexp
(Life expectancy, 1998)
```

Or you may simply type

```
. webuse lifeexp
(Life expectancy, 1998)
```

`webuse` is a synonym for use `http://www.stata-press.com/data/r12/`.

A note concerning example datasets

The datasets used to demonstrate Stata are often fictional. If you want to know whether a dataset is real or fictional, and its history, load the dataset and type

```
. notes
```

A few datasets have no notes. This means that the datasets are believed to be real but that they were created so long ago that information about their original source has been lost. Treat such datasets as if they were fictional.

Redirecting the source

By default, `webuse` obtains datasets from http://www.stata-press.com/data/r12/, but you can change that. Say that the site http://www.zzz.edu/users/sue/ has several datasets that you wish to explore. You can type

```
. webuse set http://www.zzz.edu/users/~sue
```

`webuse` will become a synonym for use `http://www.zzz.edu/users/~sue/` for the rest of the session or until you give another `webuse` command.

When you set the URL, you may omit the trailing slash (as we did above), or you may include it:

```
. webuse set http://www.zzz.edu/users/~sue/
```

You may also omit `http://`:

```
. webuse set www.zzz.edu/users/~sue
```

If you type `webuse set` without arguments, the URL will be reset to the default, http://www.stata-press.com/data/r12/:

```
. webuse set
```

Methods and formulas

`webuse` is implemented as an ado-file.

Also see

[D] **sysuse** — Use shipped dataset

[D] **use** — Load Stata dataset

Title

---

xmlsave — Export or import dataset in XML format

Syntax

Export dataset in memory to XML format

> <u>xmlsav</u>e *filename* [*if*] [*in*] [, *xmlsave_options*]

Export subset of dataset in memory to XML format

> <u>xmlsav</u>e *varlist* <u>us</u>ing *filename* [*if*] [*in*] [, *xmlsave_options*]

Import XML-format dataset

> xmluse *filename* [, *xmluse_options*]

xmlsave_options	Description
Main	
<u>doc</u>type(dta)	save XML file by using Stata's .dta format
<u>doc</u>type(excel)	save XML file by using Excel XML format
dtd	include Stata DTD in XML file
<u>leg</u>ible	format XML to be more legible
<u>rep</u>lace	overwrite existing *filename*

xmluse_options	Description
<u>doc</u>type(dta)	load XML file by using Stata's .dta format
<u>doc</u>type(excel)	load XML file by using Excel XML format
<u>shee</u>t("*sheetname*")	Excel worksheet to load
<u>cell</u>s(*upper-left*:*lower-right*)	Excel cell range to load
<u>dates</u>tring	import Excel dates as strings
<u>alls</u>tring	import all Excel data as strings
<u>firs</u>trow	treat first row of Excel data as variable names
<u>miss</u>ing	treat inconsistent Excel types as missing
nocompress	do not compress Excel data
clear	replace data in memory

Menu

xmlsave

File > Export > XML data

xmluse

File > Import > XML data

Description

xmlsave and xmluse allow datasets to be exported or imported in XML file formats for Stata's .dta and Microsoft Excel's SpreadsheetML format. XML files are advantageous because they are structured text files that are highly portable between applications that understand XML.

Stata can directly import files in Microsoft Excel .xls or .xlsx format. If you have files in that format or you wish to export files to that format, see [D] **import excel**.

xmlsave exports the data in memory in the dta XML format by default. To export the data, type

. xmlsave *filename*

although sometimes you will want to explicitly specify which document type definition (DTD) to use by typing

. xmlsave *filename*, doctype(dta)

xmluse can read either an Excel-format XML or a Stata-format XML file into Stata. You type

. xmluse *filename*

Stata will read into memory the XML file *filename*.xml, containing the data after determining whether the file is of document type dta or excel. As with the xmlsave command, the document type can also be explicitly specified with the doctype() option.

. xmluse *filename*, doctype(dta)

It never hurts to specify the document type; it is actually recommended because there is no guarantee that Stata will be able to determine the document type from the content of the XML file. Whenever the doctype() option is omitted, a note will be displayed that identifies the document type Stata used to load the dataset.

If *filename* is specified without an extension, .xml is assumed.

Options for xmlsave

⌐‾‾‾| Main |‾‾‾

doctype(dta | excel) specifies the DTD to use when exporting the dataset.

doctype(dta), the default, specifies that an XML file will be exported using Stata's .dta format (see [P] **file formats .dta**). This is analogous to Stata's binary dta format for datasets. All data that can normally be represented in a normal dta file will be represented by this document type.

doctype(excel) specifies that an XML file will be exported using Microsoft's SpreadsheetML DTD. SpreadsheetML is the term given by Microsoft to the Excel XML format. Specifying this document type produces a generic spreadsheet with variable names as the first row, followed by data. It can be imported by any version of Microsoft Excel that supports Microsoft's SpreadsheetML format.

dtd when combined with doctype(dta) embeds the necessary DTD into the XML file so that a validating parser of another application can verify the dta XML format. This option is rarely used, however, because it increases file size with information that is purely optional.

legible adds indents and other optional formatting to the XML file, making it more legible for a person to read. This extra formatting, however, is unnecessary and in larger datasets can significantly increase the file size.

replace permits xmlsave to overwrite existing *filename*.xml.

Options for xmluse

doctype(dta | excel) specifies the DTD to use when loading data from *filename*.xml. Although it is optional, use of doctype() is encouraged. If this option is omitted with xmluse, the document type of *filename*.xml will be determined automatically. When this occurs, a note will display the document type used to translate *filename*.xml. This automatic determination of document type is not guaranteed, and the use of this option is encouraged to prevent ambiguity between various XML formats. Specifying the document type explicitly also improves speed, as the data are only passed over once to load, instead of twice to determine the document type. In larger datasets, this advantage can be noticeable.

doctype(dta) specifies that an XML file will be loaded using Stata's dta format. This document type follows closely Stata's binary .dta format (see [P] **file formats .dta**).

doctype(excel) specifies that an XML file will be loaded using Microsoft's SpreadsheetML DTD. SpreadsheetML is the term given by Microsoft to the Excel XML format.

sheet("*sheetname*") imports the worksheet named *sheetname*. Excel files can contain multiple worksheets within one document, so using the sheet() option specifies which of these to load. The default is to import the first worksheet to occur within *filename*.xml.

cells(*upper-left*:*lower-right*) specifies a cell range within an Excel worksheet to load. The default range is the entire range of the worksheet, even if portions are empty. Often the use of cells() is necessary because data are offset within a spreadsheet, or only some of the data need to be loaded. Cell-range notation follows the letter-for-column and number-for-row convention that is popular within all spreadsheet applications. The following are valid examples:

> . xmluse *filename*, doctype(excel) cells(A1:D100)

> . xmluse *filename*, doctype(excel) cells(C23:AA100)

datestring forces all Excel SpreadsheetML date formats to be imported as strings to retain time information that would otherwise be lost if automatically converted to Stata's date format. With this option, time information can be parsed from the string after loading it.

allstring forces Stata to import all Excel SpreadsheetML data as string data. Although data type information is dictated by SpreadsheetML, there are no constraints to keep types consistent within columns. When such inconsistent use of data types occurs in SpreadsheetML, the only way to resolve inconsistencies is to import data as string data.

firstrow specifies that the first row of data in an Excel worksheet consist of variable names. The default behavior is to generate generic names. If any name is not a valid Stata variable name, a generic name will be substituted in its place.

missing forces any inconsistent data types within SpreadsheetML columns to be imported as missing data. This can be necessary for various reasons but often will occur when a formula for a particular cell results in an error, thus inserting a cell of type ERROR into a column that was predominantly of a NUMERIC type.

nocompress specifies that data not be compressed after loading from an Excel SpreadsheetML file. Because data type information in SpreadsheetML can be ambiguous, Stata initially imports with broad data types and, after all data are loaded, performs a compress (see [D] **compress**) to reduce data types to a more appropriate size. The following table shows the data type conversion used before compression and the data types that would result from using nocompress:

SpreadsheetML type	Initial Stata type
String	str244
Number	double
Boolean	double
DateTime	double
Error	str244

clear clears data in memory before loading from *filename*.xml.

Remarks

XML stands for Extensible Markup Language and is a highly adaptable text format derived from SGML. The World Wide Web Consortium is responsible for maintaining the XML language standards. See http://www.w3.org/XML/ for information regarding the XML language, as well as a thorough definition of its syntax.

The document type dta, used by both xmlsave and xmluse, represents Stata's own DTD for representing Stata .dta files in XML. Stata reserves the right to modify the specification for this DTD at any time, although this is unlikely to be a frequent event.

The document type excel, used by both xmlsave and xmluse, corresponds to the DTD developed by Microsoft for use in modern versions of Microsoft Excel spreadsheets. This product may incorporate intellectual property owned by Microsoft Corporation. The terms and conditions under which Microsoft is licensing such intellectual property may be found at

http://msdn.microsoft.com/library/en-us/odcXMLRef/html/odcXMLRefLegalNotice.asp

For more information about Microsoft Office and XML, see http://www.microsoft.com/office/xml/.

❏ Technical note

When you import data from Excel to Stata, a common hurdle is handling Excel's use of inconsistent data types within columns. Numbers, strings, and other types can be mixed freely within a column of Excel data. Stata, however, requires that all data in a variable be of one consistent type. This can cause problems when a column of data from Excel is imported into Stata and the data types vary across rows.

By default, xmluse attempts to import Excel data by using the data type information stored in the XML file. If an error due to data type inconsistencies is encountered, you can use the options firstrow, missing, and cells() to isolate the problem while retaining as much of the data-type information as possible.

However, identifying the problem and determining which option to apply can sometimes be difficult. Often you may not care in what format the data are imported into Stata, as long as you can import them. The quick solution for these situations is to use the allstring option to guarantee that all the data are imported as strings, assuming that the XML file itself was valid. Often converting the data back into numeric form after they are imported into Stata is easier, given Stata's vast data-management commands.

❏

▷ Example 1: Saving XML files

To export the current Stata dataset to a file, `auto.xml`, type

```
. xmlsave auto
```

To overwrite an existing XML dataset with a new file containing the variables make, mpg, and weight, type

```
. xmlsave make mpg weight using auto, replace
```

To export the dataset to an XML file for use with Microsoft Excel, type

```
. xmlsave auto, doctype(excel) replace
```

◁

▷ Example 2: Using XML files

Assuming that we have a file named `auto.xml` exported using the `doctype(dta)` option of xmlsave, we can read in this dataset with the command

```
. xmluse auto, doctype(dta) clear
```

If the file was exported from Microsoft Excel to a file called `auto.xml` that contained the worksheet Rollover Data, with the first row representing column headers (or variable names), we could import the worksheet by typing

```
. xmluse auto, doctype(excel) sheet("Rollover Data") firstrow clear
```

Continuing with the previous example: if we wanted just the first column of data in that worksheet, and we knew that there were only 75 rows, including one for the variable name, we could have typed

```
. xmluse auto, doc(excel) sheet("Rollover Data") cells(A1:A75) first clear
```

◁

Also see

[D] **compress** — Compress data in memory

[D] **export** — Overview of exporting data from Stata

[D] **import** — Overview of importing data into Stata

[P] **file formats .dta** — Description of .dta file format

Title

> **xpose** — Interchange observations and variables

Syntax

xpose , clear [*options*]

options	Description
*clear	reminder that untransposed data will be lost if not previously saved
format	use largest numeric display format from untransposed data
format(%*fmt*)	apply specified format to all variables in transposed data
varname	add variable _varname containing original variable names
promote	use the most compact data type that preserves numeric accuracy

* clear is required.

Menu

Data > Create or change data > Other variable-transformation commands > Interchange observations and variables

Description

xpose transposes the data, changing variables into observations and observations into variables. All new variables—that is, those created by the transposition—are made the default storage type. Thus any original variables that were strings will result in observations containing missing values. (If you transpose the data twice, you will lose the contents of string variables.)

Options

clear is required and is supposed to remind you that the untransposed data will be lost (unless you have saved the data previously).

format specifies that the largest numeric display format from your untransposed data be applied to the transposed data.

format(%*fmt*) specifies that the specified numeric display format be applied to all variables in the transposed data.

varname adds the new variable _varname to the transposed data containing the original variable names. Also, with or without the varname option, if the variable _varname exists in the dataset before transposition, those names will be used to name the variables after transposition. Thus transposing the data twice will (almost) yield the original dataset.

promote specifies that the transposed data use the most compact numeric data type that preserves the original data accuracy.

If your data contain any variables of type double, all variables in the transposed data will be of type double.

If variables of type float are present, but there are no variables of type double or long, the transposed variables will be of type float. If variables of type long are present, but there are no variables of type double or float, the transposed variables will be of type long.

618

Remarks

▷ Example 1

We have a dataset on something by county and year that contains

```
. use http://www.stata-press.com/data/r12/xposexmpl
. list
```

	county	year1	year2	year3
1.	1	57.2	11.3	19.5
2.	2	12.5	8.2	28.9
3.	3	18	14.2	33.2

Each observation reflects a county. To change this dataset so that each observation reflects a year, type

```
. xpose, clear varname
. list
```

	v1	v2	v3	_varname
1.	1	2	3	county
2.	57.2	12.5	18	year1
3.	11.3	8.2	14.2	year2
4.	19.5	28.9	33.2	year3

We would now have to drop the first observation (corresponding to the previous county variable) to make each observation correspond to one year. Had we not specified the varname option, the variable _varname would not have been created. The _varname variable is useful, however, if we want to transpose the dataset back to its original form.

```
. xpose, clear
. list
```

	county	year1	year2	year3
1.	1	57.2	11.3	19.5
2.	2	12.5	8.2	28.9
3.	3	18	14.2	33.2

◁

Methods and formulas

xpose is implemented as an ado-file.

See Hamilton (2009, chap. 2) for an introduction to Stata's data-management features.

References

Baum, C. F. 2009. *An Introduction to Stata Programming*. College Station, TX: Stata Press.

Hamilton, L. C. 2009. *Statistics with Stata (Updated for Version 10)*. Belmont, CA: Brooks/Cole.

Also see

[D] **reshape** — Convert data from wide to long form and vice versa

[D] **stack** — Stack data

Title

> **zipfile** — Compress and uncompress files and directories in zip archive format

Syntax

Add files or directories to a zip file

> `zipfile` *file* | *directory* [*file* | *directory*] ... `, saving(`*zipfilename*[`, replace`]`)`

Extract files or directories from a zip file

> `unzipfile` *zipfilename* [`, replace`]

Note: Double quotes must be used to enclose *file* and *directory* if the name or path contains blanks. *file* and *directory* may also contain the ? and * wildcard characters.

Description

    `zipfile` compresses files and directories into a zip file that is compatible with WinZip, PKZIP 2.04g, and other applications that use the zip archive format.

    `unzipfile` extracts files and directories from a file in zip archive format into the current directory. `unzipfile` can open zip files created by WinZip, PKZIP 2.04g, and other applications that use the zip archive format.

Option for zipfile

`saving(`*zipfilename*[`, replace`]`)` is required. It specifies the filename to be created or replaced. If *zipfilename* is specified without an extension, `.zip` will be assumed.

Option for unzipfile

`replace` overwrites any file or directory in the current directory with the files or directories in the zip file that have the same name.

Remarks

> ▷ Example 1: Creating a zip file

    Suppose that we would like to zip all the `.dta` files in the current directory into the file `myfiles.zip`. We would type

        `. zipfile *.dta, saving(myfiles)`

But we notice that we did not want the files in the current directory; instead, we wanted the files in the dta, abc, and eps subdirectories. We can easily zip all the .dta files from all three-character subdirectories of the current directory and overwrite the file myfiles.zip if it exists by typing

```
. zipfile ???/*.dta, saving(myfiles, replace)
```

◁

⊳ Example 2: Unzipping a zip file

Say, for example, we send myfiles.zip to a colleague, who now wants to unzip the file in the current directory, overwriting any files or directories that have the same name as the files or directories in the zip file. The colleague should type

```
. unzipfile myfiles, replace
```

◁

Subject and author index

This is the subject and author index for the *Data-Management Reference Manual*. Readers interested in topics other than data management should see the combined subject index (and the combined author index) in the *Quick Reference and Index*.

Semicolons set off the most important entries from the rest. Sometimes no entry will be set off with semicolons, meaning that all entries are equally important.

C

E

X

Y

Z